Ecological Studies, Vol. 166

Analysis and Synthesis

Ecological Studies

Volumes published since 1996 are listed at the end of this book.

Springer
New York
Berlin
Heidelberg
Hong Kong
London
Milan
Paris
Tokyo

Paul J. Hanson Stan D. Wullschleger
Editors

North American Temperate Deciduous Forest Responses to Changing Precipitation Regimes

Foreword by Jerry W. Elwood

With 129 Figures, 4 in Full Color

Springer

Paul J. Hanson
Environmental Sciences Division
Oak Ridge National Laboratory
Bethel Valley Road, Building 1059
P.O. Box 2008
Oak Ridge, TN 37831-6422
USA
hansonpj@ornl.gov

Stan D. Wullschleger
Environmental Sciences Division
Oak Ridge National Laboratory
Bethel Valley Road, Building 1059
P.O. Box 2008
Oak Ridge, TN 37831-6422
USA
wullschlegsd@ornl.gov

Cover illustration: Designed by LeJean Hardin of the Oak Ridge National Laboratory's Creative Media Services Division, Oak Ridge, Tennessee.

Library of Congress Cataloging-in-Publication Data

North America temperate deciduous forest responses to changing precipitation regimes / editors, Paul J. Hanson, Stan D. Wullschleger.
p. cm. — (Ecological studies ; 166)
Includes bibliographical references (p.).
ISBN 0-387-00309-6 (alk. paper)
1. Rain and rainfall—North America. 2. Forest ecology—North America. 3. Precipitation variability—North America. 4. Climatic changes—North America. I. Hanson, Paul J. (Paul James) II. Wullschleger, Stan D. III. Ecological studies ; v. 166.

SD390.7.R34N67 2003
577.3′097—dc21

2002044525

ISSN 0070-8356
ISBN 0-387-00309-6 Printed on acid-free paper.

Printed in the United States of America.

9 8 7 6 5 4 3 2 1 SPIN 10906231

www.springer-ny.com

Springer-Verlag New York Berlin Heidelberg
A member of BertelsmannSpringer Science+Business Media GmbH

Foreword

Ecologists have long been active in conducting observational studies and experimental research aimed at better understanding the response of terrestrial ecosystems to changes in their chemical and physical environment, including changes in climate. Early studies that focused on primary and secondary response of plants to air pollution gave way to studies that dealt with acidic precipitation, that in turn gave way to studies on pressing concerns of global change. Understanding the effects of increases in atmospheric CO_2 concentration, global warming, and regional changes in precipitation offers a unique set of challenges for the scientific community.

The interagency United States Global Change Research Program (USGCRP), established in 1989 and authorized by Congress in the Global Change Research Act of 1990, has been an advocate for research in these and other areas. A goal of the USGCRP is to understand and predict the consequences of global environmental change in terrestrial ecosystems. As a participating agency in the USGCRP, the Department of Energy (DOE) shares this goal through our Program for Ecosystem Research (PER) which seeks to improve the scientific basis for predicting and assessing effects of environmental changes on terrestrial ecosystems and their component organisms and processes. Specific goals of PER include understanding (1) how organisms within terrestrial ecosystems respond to environmental change, (2) how these responses are controlled, (3) how the responses and their underlying control mechanisms can be detected and measured, and (4) how responses of organisms can be integrated across organizational levels from the results of studies of specific research organisms to ecosystems.

Given these many objectives, the DOE Office of Biological and Environmental Research in the Office of Science began funding the Throughfall Displacement Experiment (TDE) in 1992. The results were expected to provide an improved mechanistic understanding of how upland mixed deciduous forests respond to altered precipitation amount. The 26 chapters of this book demonstrate that many of our expectations have been fulfilled. They represent an important source of information regarding the resiliency and/or vulnerability of our nation's deciduous forests to long-term (10 years to date) altered precipitation amount. Results from the TDE are important not only from the perspective of understanding the many mechanisms by which temperate deciduous forests respond to changes in the amount of precipitation, but also provide new information and insights needed for predicting and assessing whole forest responses to possible future changes. Such is the opportunity afforded by long-term studies like the TDE that are conducted at ecosystem spatial scales. It remains a challenge to incorporate the data and conclusions reported herein into future national assessments of climate change impacts, and make them available for discussion within the policy arena. Comprehensive and long-term data, from the TDE and similar studies, present the scientific communities with a unique opportunity to judge the efficacy of ecosystem models that are ultimately useful for forecasting effects of possible future environmental changes on the structure and functioning of ecosystems.

Finally, operation of the TDE since 1992 has required a long-term commitment of resources, and periodic peer-review has been required to support this sustained research. The goal has been to operate such studies, with continuing input and guidance from the scientific community, so that potentially important, but slow-to-develop, ecosystem responses have time to express themselves and be detected. As the TDE progresses towards its completion over the next several years, some conclusions will be strengthened but other insights may need to be modified or abandoned in favor of alternative explanations. Such is the scientific process and such is the return that can only come from dedicated investments in long-term studies.

Jerry W. Elwood
Director, Climate Change Research Division
Office of Biological and Environmental Research
Office of Science
U.S. Department of Energy

Preface

The current volume is in many respects the next issue in a series of books by Springer-Verlag that summarize fundamental research on functional processes operating in eastern deciduous forested ecosystems. *Analysis of Temperate Forest Ecosystems* (Reichle 1970) provided a broad description of temperate forest ecosystems and their nutrient and hydrologic cycles. That book resulted mainly from research on the Eastern Deciduous Forest Biome as a part of the International Biological Programme. Four books, *Biogeochemistry of a Forested Ecosystem* (Likens et al. 1977), *Pattern and Process in a Forested Ecosystem* (Borman and Likens 1979), *Forest Hydrology and Ecology at Coweeta* (Swank and Crossley 1988), and *Analysis of Biogeochemical Cycling Processes in Walker Branch Watershed* (Johnson and Van Hook 1989), include long-term studies of watershed-scale processes, including forest-management perspectives. *Atmospheric Deposition and Forest Nutrient Cycling* (Johnson and Lindberg 1992), motivated by the acid rain issue of the 1980s, includes thorough descriptions of nutrient cycling processes and ozone exposure in a variety of forested ecosystems including southern and northern hardwood sites. The present book was based on a need to better understand deciduous forested ecosystem responses to potential climatic conditions of the future. We hope that this work, when combined with its predecessor volumes, will provide important information on mechanisms of eastern deciduous forest plant and ecosystem responses to altered precipitation patterns, and that this will enhance assessments of potential impacts of environmental change on forested ecosystems.

Data from the Walker Branch Throughfall Displacement Experiment (TDE) detailed in this volume has been archived by the Carbon Dioxide Information and Analysis Center (CDIAC) and can be found at the following web address: http://cdiac.esd.ornl.gov/epubs/ndp/ndp078a/ndp078a.html. The archive includes data for site characterization, environment conditions, soil-water content, growth, seasonal phenology, plant physiology, soil respiration and decomposition, and nutrient cycling necessary for future development and testing of ecosystem models of upland oak deciduous forests.

Paul J. Hanson
Stan D. Wullschleger

References

Borman FH, Likens GE (1979) *Pattern and process in a forested ecosystem.* Springer-Verlag, New York.
Likens GE, Bormann FH, Pierce RS, Eaton JS, Johnson NM (1977) *Biogeochemistry of a Forested Ecosystem.* Springer-Verlag, New York.
Johnson DW, Lindberg SE Editors (1992) *Atmospheric deposition and forest nutrient cycling.* Springer-Verlag, New York.
Johnson DW, Van Hook RI Editors (1989) *Analysis of biogeochemical cycling processes in Walker Branch Watershed.* Springer-Verlag, New York.
Reichle DE Editor (1970) *Analysis of temperate forest ecosystems.* Springer-Verlag, New York.
Swank WT, Crossley DA Jr. Editors(1988) *Forest hydrology and ecology at Coweeta.* Springer-Verlag, New York.

Acknowledgments

All research on the Walker Branch Watershed Throughfall Displacement Experiment (TDE) was sponsored in full or in part by the Office of Science, Biological and Environmental Research (BER) Program, U.S. Department of Energy (DOE) as a part of the Program for Ecosystem Research. Work conducted by Oak Ridge National Laboratory personnel was conducted under contract No. DE-ACO5 00OR22725 with UT-Battelle, LLC. The TDE is on the Oak Ridge National Environmental Research Park.

Many people contributed to the completion of this book. Frederick M. O'Hara, Jr., did an excellent job as technical editor of the volume. Not only did he carefully edit each chapter, but in some cases he suggested major revisions of sections. The authors sincerely appreciate his knowledge of science, his humor, and his tenacity in performing the many tasks needed to bring this book to publication.

We thank the following students, faculty participants, and co-workers for participating in the long-term collection of data used in several chapters of this book: Philip Allen, Steve Armes, James E. Arnett III, Stephanie Bohlman, Bridgette Boudreaux, Mogan Castner, Jason Fults, Kristin Harter, Michael Hayes, David Holt, Joshua Johnson, Sabina Madsen, Howard Marotto, Tracy Misek, Mark Scannell, Kevin Smith, James B. Stringfellow, Todd Tabberer, and Karen Voiles.

The authors of Chapter 5 thank Donald E. Todd for the extensive effort involved in sample collection and Ramie V. Wilkerson for excellent technical support.

Work on Chapter 7 was supported by The University of Tennessee Agricultural Experiment Station and by the DOE's BER, National Institute for Global Environmental Change (NIGEC), Southeast Regional Center under Cooperative Agreement No. DE-FC03–90ER61010. Ann Stodola, Craig Green, Jenny Moore, Xiangrong Duan, Brendon Johnson, Dr. Arnold Saxton, Dr. Richard M. Evans, and Joshua Olinick were instrumental in accomplishing the work.

Support for G. Michael Gebre in the development of Chapter 8 was provided by an appointment to the Oak Ridge National Laboratory (ORNL) Postdoctoral Research Associates Program administered jointly by ORNL and the Oak Ridge Institute for Science and Education. The authors of Chapter 8 thank Donald E. Todd and Paul J. Hanson for the extensive effort involved in sample collection and Ramie V. Wilkerson for excellent technical support.

The authors of Chapter 9 thank Paul J. Hanson, Michael Huston, Donald E. Todd, and Scott Dolvin for assistance in obtaining samples from the TDE site and for assistance with sample processing. We thank the staff at the Biology Station of Chamela, Universidad Nacional Autónoma de México for assistance with our work at Chamela. Sandy McLaughlin provided assistance with the tree core sectioning.

The authors of Chapter 13 thank Patrice Cole, Caroline DeVan, Leigh Thomas, and Don Todd who helped collect, process, and plant seeds and assisted in the construction and maintenance of the seed-planting experiment. Paul Hanson encouraged our initiation of this project on the TDE, assisted in the development of our soil-water monitoring protocol, loaned us TDR probes and temperature dataloggers, and collated and summarized the temperature dataset. Travis Belote improved earlier drafts of the manuscript. Research on Chapter 13 was supported in part by the Department of Ecology and Evolutionary Biology, University of Tennessee, Knoxville.

The authors of Chapter 14 thank the following individuals for their facilitation of the collection of field data: D. Boldt, Danim Moe, Nilong R. Parikh, Jorge Quiles, Shawnette Warner, and Donald E. Todd.

The authors of Chapter 16 acknowledge the cooperation of the Public Power Institute of the Tennessee Valley Authority with Oak Ridge National Laboratory. They also thank Emily Childs, Brian Hale, Alan Mays, Jennifer Martin, Anita Rose, Larry Shelton, Brian Sunderland, Erich Webber, and numerous students for assistance in the collection and extraction of root samples and in the photographing and analyzing of minirhizotron data. Thanks are also due to E.G. O'Neill, Ron Hendricks, and Kurt Pregitzer for supplying minirhizotron equipment, assistance, and advice.

The authors of Chapter 18 thank Kenny Peavy, Larry Wilson, and Sue Carstensen for help in field sampling and/or laboratory analyses. George Cotsonis provided statistical advice and assistance in analyzing the data, and Chris Mowry assisted with the graphical presentation of the final results. Funding was provided by DOE's BER, NIGEC, Southeast Regional Center under Cooperative Agreement No. DE-FC03–90ER61010.

The authors of Chapter 19 thank Dr. James A Brenneman and James J. Wetteroff, Jr. for their work in the field and laboratory, and Dr. Theodor D. Leininger as well for valuable contributions to study planning and preparation of an earlier manuscript. Work described in Chapter 19 was partially supported by the USDA Forest Service, Missouri Department of Conservation, Missouri Ozark Forest Ecosystem Project, and the Missouri Agricultural Experiment Station (Project PSSL 0112).

The author of Chapter 20 thanks the Associated Colleges of the Midwest for making possible his stay at ORNL as part of the resident faculty for the Oak Ridge Science Semester. Dr. Michael Draney identified problematic specimens of linyphiids and Dr. William Ehmann provided valuable input on an early version of this chapter.

Chapter 23 was sponsored in part by the Nevada Agricultural Experiment Station, publication number 52031032.

Work on Chapter 24 was also sponsored by the US. Environmental Protection Agency's National Center for Ecological Research (NCER) STAR program grant number R825157–01 to the University of Tennessee.

Work associated with Chapter 26 was supported by the DOE's BER, NIGEC Southeast Regional Center. Dan Tinker, Jerry Winslow, and Sharon Stewart are thanked for assistance in developing the BGC++ model.

Paul J. Hanson
Stan D. Wullschleger

Contents

Contributors

Philip B. Allen	Department of Ecology and Evolutionary Biology, University of Tennessee, Knoxville, TN 37996-1610, USA. E-mail: pallen1@utk.edu
Robert M. Augé	Department of Plant Sciences and Landscape Systems, University of Tennessee, Knoxville, TN 37901-1071, USA. E-mail: auge@utk.edu
D. Gregory Barlar	Environmental Sciences Division, Oak Ridge National Laboratory, P.O. Box 2008, Oak Ridge, TN 37831, USA.
Johann N. Bruhn	Department of Plant Microbiology and Pathology, University of Missouri–Columbia, Columbia, MO 65211, USA. E-mail: BruhnJ@missouri.edu

M. Lala Chambers — Idaho National Engineering and Environmental Laboratory, Ecological and Cultural Resources Group, Idaho Falls, ID 83415-2213, USA. E-mail: CHAMML@inel.gov

Rebekah E. Chapman — Institute of Ecology, University of Georgia, Athens, GA 30602, USA. E-mail: rebekahc@arches.uga.edu

Lee W. Cooper — Department of Ecology and Evolutionary Biology, University of Tennessee, Knoxville, TN 37996, USA. E-mail: lcooper1@utk.edu

Kenneth L. Cramer — Department of Biology, Monmouth College, Monmouth, IL 61462, USA. E-mail: kenc@monm.edu

Nelson T. Edwards — Environmental Sciences Division, Oak Ridge National Laboratory, Oak Ridge, TN 37831, USA. E-mail: edwardsnt@ornl.gov

Jerry W. Elwood — Environmental Sciences Division, ER-74, U.S. Department of Energy, Germantown, MD 20874, USA. E-mail: Jerry.Elwood@science.doe.gov

Julia B. Gaudinski — Department of Integrative Biology, University of California Berkeley, Berkeley, CA 94720, USA. E-mail: gaudinsk@socrates.berkeley.edu

G. Michael Gebre — Environmetal Sciences Division, Oak Ridge National Laboratory, Oak Ridge, TN 37831, USA.

Carla A. Gunderson — Environmetal Sciences Division, Oak Ridge National Laboratory, Oak Ridge, TN 37831-6422, USA. E-mail: gundersonca@ornl.gov

Paul J. Hanson — Environmental Sciences Division, Oak Ridge National Laboratory, Oak Ridge, TN 37831-6422, USA. E-mail: hansonpj@ornl.gov

Milena Holmgren — Forest Ecology and Forest Management Group, Department of Environmental Sciences, Wageningen University, PO Box 342, 6700 AH Wageningen, The Netherlands. E-mail: Milena.Holmgren@btbo.bosb.wau.nl

E. Raymond Hunt, Jr. — USDA Agricultural Research Service Hydrology and Remote Sensing Laboratory, Building 007, Room 104, BARC-West, 10300 Baltimore Ave., Beltsville, MD 20705, USA. E-mail: erhunt@hydrolab.arsusda.gov

Michael A. Huston — Environmental Sciences Division, Oak Ridge National Laboratory, P.O. Box 2008, Oak Ridge, TN 37831, USA. E-mail: hustonma@ornl.gov.

Dale W. Johnson — Environmental and Resource Sciences, Fleischmann Ag Building, MS 370, University of Nevada, Reno; Reno, NV 89506, USA. E-mail: dwj@unr.edu

J. Devereux Joslin — Belowground Forest Research, 112 Newcrest Lane, Oak Ridge, TN 37830, USA, and, Public Power Institute, Tennessee Valley Authority, LAB 1C-N, Pine St., Norris, TN 37828-0920, USA. E-mail: jdjoslin@esper.com

Michelle B. Kreh — Environmental Sciences Division, Oak Ridge National Laboratory, P.O. Box 2008, Oak Ridge, TN 37831, USA.

Jeanne D. Mihail — Department of Plant Microbiology and Pathology, University of Missouri–Columbia, Columbia, MO 65211, USA.

Peter D. Mooreside	The Nature Conservancy, Fort Indiantown Gap Office, Environmental Unit, Building 11-19, Annville, PA 17003, USA. E-mail: pmooreside@tnc.org
Frederick M. O'Hara, Jr.	Consultant in Technical Communication, P.O. Box 4273, Oak Ridge, TN 37831, USA. E-mail: fohara@fohara.com
Elizabeth G. O'Neill	Environmental Sciences Division, Oak Ridge National Laboratory, Oak Ridge, TN 37831, USA.
Wilfred M. Post	Environmental Sciences Division, Oak Ridge National Laboratory, Oak Ridge, TN 37831, USA. E-mail: postwmiii@ornl.gov
Jeffery S. Riggs	Environmental Sciences Division, Oak Ridge National Laboratory, Oak Ridge, TN 37831, USA. E-mail: riggsjs@ornl.gov
Donald J. Shure	Biology Department, Emory University, Atlanta, GA 30322, USA. E-mail: dshure@biology.emory.edu
William A. Simpson, Jr.	Metals and Ceramics Division, Oak Ridge National Laboratory, Oak Ridge, TN 37831, USA. E-mail: simpsonwajr1@ornl.gov
Corina Solis	Instituo de Física, Universidad Nacional Autónoma de México, A.P. 20634, México, D.F. 01000. E-mail: corina@fenix.ifiscacu.unam.mx
M. Lynn Tharp	Oak Ridge National Laboratory, Oak Ridge, TN 37831, USA. E-mail: tharpml@ornl.gov

Donald E. Todd — Environmental Sciences Division, Oak Ridge National Laboratory, Oak Ridge, TN 37831, USA. E-mail: toddejr@ornl.gov

Susan E. Trumbore — Department of Earth System Science, University of California-Irvine, Irvine, CA 92697, USA. E-mail: setrumbo@uci.edu

Timothy J. Tschaplinski — Oak Ridge National Laboratory, Oak Ridge, TN 37831, USA. E-mail: tschaplinstj@ornl.gov

Jake F. Weltzin — Department of Ecology and Evolutionary Biology, University of Tennessee, Knoxville, TN 37996-1610, USA. E-mail: jweltzin@utk.edu

Darrell C. West — Environmental Sciences Division, Oak Ridge National Laboratory, Oak Ridge, TN 37831, USA.

Allan D. Wilson — Biology Department, Emory University, Atlanta, GA 30322, USA.

Kell B. Wilson — NOAA, Atmospheric Turbulence and Diffusion Division, Oak Ridge, TN 37831-2456, USA. E-mail: wilson@atdd.noaa.gov

Mark H. Wolfe — Tennessee Valley Authority, Norris, TN 37828, USA. E-mail: mhwolfe@tva.gov

Stan D. Wullschleger — Environmental Sciences Division, Oak Ridge National Laboratory, P.O. Box 2008, Oak Ridge, TN 37831-6422, USA. E-mail: wullschlegsd@ornl.gov.

1. Introduction

1. Introduction

Paul J. Hanson and Frederick M. O'Hara, Jr.

The Need for Large-Scale Climate-Change Studies

The Intergovernmental Panel on Climate Change (IPCC) concluded that climate changed during the past century, that human activities influenced the changes, and that climate is expected to continue to change in the future (Watson et al. 2002). The different scenarios studied indicate that continued increases in greenhouse gas concentrations will result in temperature increases from 1.4°C to 5.8°C (Houghton et al. 2001). This warming is expected to modify the global hydrologic cycle, resulting in increased winter precipitation at high latitudes, to result in more hot days and fewer cold days, and to change the frequency of droughts and floods, depending on location (Houghton et al. 2001). These predicted changes raise concerns about terrestrial ecosystem productivity, biogeochemical cycling, and the availability of water resources (Kirschbaum and Fischlin 1996; Melillo et al. 1990). As a result, the IPCC Working Group II Third Assessment Report (McCarthy et al. 2001) requested further research on the response of ecosystems to multiple stresses (e.g., increased temperature and drought).

The National Assessment Synthesis Team's report on climate-change impacts on the United States (NAST 2000) emphasizes the impacts that future precipitation regimes will likely have on ecosystem processes in most regions of the United States. In addition, concerns regarding vegetation impacts have been amplified because ecological change is expected to occur much faster than species can adapt to that change or disperse (migrate) to avoid its effects; climate-induced ecological change is also expected to outpace the rates at which successional processes occur in forests and thus long-term changes in species composition are likely (Davis 1989; Overpeck et al. 1991; Pastor and Post 1988; Solomon 1986).

Unfortunately, the direction and magnitude of expected changes in precipitation remain highly uncertain (Cooter et al. 1993; Schneider 1989; Houghton et al. 2001). Given this uncertainty, manipulative field experiments can play a powerful role in the identification of gradual and threshold ecosystem responses that might result from future precipitation changes. Many authors (Graham et al. 1990; Mooney 1991; Woodward 1992; Hanson 2000; Hanson and Weltzin 2000) have indicated the importance of large-scale and long-term manipulation experiments to adequately address the impacts of changing climates on ecosystems, and the North American deciduous forests are an important ecosystem in need of such large-scale experiments.

Deciduous Forests

North American deciduous forests are sensitive plant communities that are known to respond quickly to environmental perturbations (Smith 1987; Joyce and Birdsey 2000). Forests throughout the United States could be significantly effected by climatic change (Neilson et al. 1989), especially those forests in the southern states, where evapotranspiration is predicted to increase significantly with warming (Smith and Tirpak 1990). How such forests respond to decreased water availability or increased occurrence of drought is a key issue in developing predictive climate-change scenarios (Wigley et al. 1984). However, such predictions include many uncertainties, and the magnitude and even direction of response to climate change by temperate deciduous forests is highly dependent on the anticipated extent of that change (Bachelet et al. 2001): Moderate levels of change would likely lead to forest expansion, and severe climatic changes would likely lead to forest loss.

Concerns about the response of temperate deciduous forests to climatic change are not trivial. These forests cover 297 million hectares of North America and represent 41% of the total forest lands in the United States (Shriner and Street 1998). They appear across the landscape of the eastern United States in a variety of forms (Smith 1987):

Location (Eastern U.S.)	Dominant Species	Area (million hectares)
Northern	*Acer, Betula,* and *Fagus*	14
Central	*Quercus, Carya,* and *Liriodendron*	46.5
Southern	*Quercus/Pinus*	14

In addition to their extensive land cover, these forests are among the most highly productive ecosystems (Whittaker and Likens 1975), they store substantial amounts of carbon (Fan et al. 1998; Canadell et al. 2000), they are a key source of clean water (Satterlund 1972; Gleick and Adams 2000), they represent a major "closed forest" region of the globe (Singh et al. 2001), and they support a high state of biodiversity (Whittaker 1956; Iverson and Prasad 2001). The importance of temperate deciduous forests to society as a source of forest products, recreational opportunities, and clean water is increased by their proximity to many large population centers of eastern North America.

It is therefore important to know if these forests are sensitive to climatic changes such as warming and changes in precipitation, wind, and atmospheric composition. To address concerns about the sensitivity of deciduous forests to variations in precipitation and soil-water availability, a water-manipulation experiment appropriate to the large stature of eastern deciduous forests was initiated at the Walker Branch Watershed in east Tennessee. The goals of the experiment were to evaluate the impacts of above- and below-normal levels of precipitation on the forest trees and other plants, on the native and invasive pests and pathogens, and on the microbial processes that drive the forest's carbon, water, and nutrient cycles. This project, the Throughfall Displacement Experiment (TDE), became the focal point for a range of process-level and mechanistic studies of forest response to precipitation change. The results of the TDE are reported in this book and supplemented with related data from the scientific literature. The goal is to yield a comprehensive picture of deciduous-forest physiology, growth, and nutrient cycling responses to changing moisture regimes. Incorporating such response mechanisms into hierarchical models of ecosystems should provide (1) a better understanding of the consequences of changing precipitation patterns on eastern deciduous forests and (2) a framework within which to test hypotheses about future climate scenarios.

Overview of Contents

This book is organized into 8 sections, which include 26 chapters and a brief appendix. Section 1 introduces the TDE and the multiyear patterns of precipitation and variation in soil moisture that were observed during the study. Sections 2, 3, and 4 discuss short-term process responses to precipitation variability and resulting soil-moisture changes. Section 2 includes chapters on the carbon-cycle processes of photosynthesis, autotrophic respiration, and dormant-season carbohydrate storage. Section 3 includes chapters on water-cycle components, including whole-plant water use, leaf stomatal behavior, leaf water potential and osmotic adjustment, and isotopic signatures of ^{18}O and ^{13}C in leaf litter and wood cellulose as indicators of changing water use. Section 4 includes chapters on decomposition and soil–carbon turnover, including estimates of soil respiration, litter decomposition, soil–carbon turnover from background ^{14}C observations, and the rate of wood decomposition. Growth responses to precipitation change are covered in Section 5 and include discussions of seedling recruitment and growth, tree and sapling growth, fine-root growth, and stand-level canopy production. Section 6 summarizes the responses of other organisms to precipitation change, including chapters on herbivory, pathogenic root fungi, and detrital predator populations. The three chapters in Section 7 summarize stand-level and multiyear integrative responses to precipitation change. Key variables covered in Section 7 include net primary production and net ecosystem production, forest respiration, stand-level water-cycle responses to precipitation change (with a focus on transpiration), and elemental cycling of nitrogen and other potentially limiting elements (e.g., potassium, calcium, and magnesium). Section 8 completes the book with a series of chapters that place the specific results into a long-term perspective with respect to (1) past growth on Walker Branch Watershed, (2) forest succession, and (3) extrapolation of the findings across the southeastern United States.

References

Bachelet D, Neilson RP, Lenihan JM, Drapek RJ (2001) Climate change effects on vegetation distribution and carbon budget in the United States. Ecosystems 4:164–185.

Canadell JG, Mooney HA, Baldocchi DD, Berry JA, Ehleringer JR, Field CB, Gower ST, Hollinger DY, Hunt JE, Jackson RB, Running SW, Shaver GR, Steffen W, Trumbore SE, Valentini R, Bond BY (2000) Carbon metabolism of the terrestrial biosphere: A multitechnique approach for improved understanding. Ecosystems 3:115–130.

Cooter EJ, Eder BK, LeDuc SK, Truppi L (1993) *General circulation model output for forest climate change research and applications.* General Technical Report SE-85. U.S. Department of Agriculture, Forest Service, Southeastern Forest Experiment Station, Asheville, North Carolina.

Davis MB (1989) Lags in vegetation response to greenhouse warming. Clim Change 15:75–82.

Fan S, Gloor M, Mahlman J, Pacala S, Sarmiento J, Takahashi T, Tans P (1998) A large terrestrial carbon sink in North America implied by atmospheric and oceanic carbon dioxide data and models. Science 282:442–446.

Gleick PH, Adams DB (2000) *Water: Tthe potential consequences of climate variability and change for the water resources of the United States. Report of the Water Sector Assessment Team of the National Assessment of the Potential Consequences of Climate Variability and Change,* US Global Change Research Program, Washington, District of Columbia.

Graham RL, Turner MG, Dale VH (1990) How increasing CO_2 and climate change affect forests. Bioscience 40: 575–587.

Hanson PJ (2000) Large-scale water manipulations. In Sala OE, Jackson RB, Mooney HA, Howarth RW (Eds) *Methods in ecosystem science.* Springer-Verlag, New York, pp 341–352.

Hanson PJ, Weltzin JF (2000) Drought disturbance from climate change: Response of United States forests. Sci Total Environ 262:205–220.

Houghton JT, Ding Y, Griggs DJ, Noguer M, van der Linden PJ, Xiaosu D (Eds) (2001) *Climate Change 2001: The Scientific Basis: Contribution of Working Group I to the Third Assessment Report of the Intergovernmental Panel on Climate Change (IPCC).* Cambridge University Press, Cambridge, England.

Iverson LR, Prasad AM (2001) Potential changes in tree species richness and forest community types following climate change. Ecosystems 4:186–199.

Joyce LA, Birdsey R (Eds) (2000) *The impact of climate change on America's forests: A technical document supporting the 2000 USDA Forest Service RPA Assessment.* General Technical Report RMRS-GTR-59. USDA Forest Service, Rocky Mountain Research Station, Ft. Collins, Colorado.

Kirschbaum MUF, Fischlin A (1996) Climate change impacts on forests. In Watson RT, Zinyowera MC, Moss RH (eds) *Climate change 1995 impacts, adaptations and mitigation of climate change: Scientific-technical analysis,* Cambridge University Press, New York, pp 95–129.

McCarthy JJ, Canziani OF, Leary NA, Dokken DJ, White KS [Eds] (2001) *Climate Change 2001: Impacts, Adaptation, and Vulnerability: Contribution of Working Group II to the Third Assessment Report of the Intergovernmental Panel on Climate Change.* Cambridge University Press, Cambridge, England.

Melillo JM, Callaghan TV, Woodward FI, Salati E, and Sinha SK (1990) Effects on ecosystems, In Houghton JT, Jenkins GJ, Ephraums JJ (Eds) *Climate change the IPCC scientific assessment.* Cambridge University Press, New York, pp 282–310.

Mooney H (1991) Biological response to climate change: An agenda for research. Ecol Appl 1:112–117.

National Assessment Synthesis Team [NAST] (2000) *Climate change impacts on the United States: The potential consequences of climate variability and change,* US Global Change Research Program, Washington, District of Columbia.

Neilson RP, King GA, DeVelice RL, Lenihan J, Marks D, Dolph J, Campbell B, Glick G (1989) *Sensitivity of ecological landscapes and regions to global climatic change.* EPA-600-3-89-073. US Environmental Protection Agency, Washington, District of Columbia.

Overpeck JT, Bartlein PJ, Webb T III (1991) Potential magnitude of future vegetation change in eastern North America: Comparisons with the past. Science 254:692–695.

Pastor J, Post WM (1988) Response of northern forests to CO_2-induced climate change. Nature 334:55–58.

Satterlund DR (1972) *Wildland watershed management,* Ronald Press, New York.

Schneider SH (1989) The greenhouse effect: Science and policy. Science 243:771–781.

Shriner DS, Street RB (1998) North America. In Watson RT, Zinyowera MC, Moss RH, Dokken DJ (Eds) *The regional impacts of climate change: An assessment of vulnerability.* Cambridge University Press, New York, pp 253–330.

Singh A, Shi H, Zhu Z, Foresman T (2001) *An assessment of the status of the world's remaining closed forests.* UNEP/DEWA TR 01–2. United Nations Environment Programme, Division of Early Warning and Assessment, Nairobi, Kenya.

Smith JB, Tirpak DA (1990) *The potential effects of global climate change on the United States.* Hemisphere, New York.

Smith WH (1987) Future of the hardwood forest: Some problems with declines and air quality. In Hay RL, Woods FW, DeSelm H (Eds) *Proceedings of the central hardwood forest conference VI.* , February 24–26, 1997, Knoxville, Tennessee, pp 3–13.

Solomon AM (1986) Transient response of forests to CO_2-induced climate change: Simulation modeling experiments in eastern North America. Oecologia 68:567–579.

Watson RT, Dokken DJ, Noguer M, van der Linden P, Johnson C, Pan J, [Eds] (2002) *Climate Change 2001: Synthesis Report. Third assessment report of the Intergovernmental Panel on Climate Change (IPCC).* Cambridge University Press, Cambridge, England.

Whittaker RH (1956) The vegetation of the Great Smoky Mountains. Ecol Monogr 26:1–80.

Whittaker RH, Likens GE (1975) The biosphere and man. In Lieth H, Whittaker RH (Eds), *Primary productivity of the biosphere.* Springer-Verlag, Berlin, pp 305–328.

Wigley TML, Briffa KR, Jones PD (1984) Predicting plant productivity and water resources. Nature 312:102–103.

Woodward FI (1992) Tansley Review No. 41. Predicting plant responses to global environmental change. New Phytol 122:239–251.

2. Walker Branch Throughfall Displacement Experiment

Paul J. Hanson, Michael A. Huston, and Donald E. Todd

The Walker Branch Forest

The Walker Branch Watershed (~100 ha), located at 35°58′ N latitude and 84°17′ W longitude, is a part of the U.S. Department of Energy's (DOE's) National Environmental Research Park near Oak Ridge, Tennessee (Johnson and Van Hook 1989). Long-term (50-year) mean annual precipitation is 1352 mm, and mean annual temperature is 14.2°C. The soils are primarily Typic Paleudults derived from dolomitic bedrock. The soils are highly weathered and very deep (> 10 m) on ridge tops and therefore retain little evidence of their carbonate parent material. Plant-extractable water (water held between 0 and −2.5 MPa) for the upper meter of soil is approximately 183 mm. A large fraction of this water (44%) is held in the upper 0.35 m of the soil profile, where 74% of all fine roots in the upper 0.90 m of the profile are located (Joslin and Wolfe 1998; Chapter 16, this volume). Depth to bedrock at this location is approximately 30 m (McMaster 1967), and deep rooting may be a source of some water.

The forest on Walker Branch Watershed is a centrally located example of the eastern broadleaf forest province as defined by Bailey (1983) and historically has been characterized as a *Quercus/Carya* forest (Fig. 2.1). Insect outbreaks in the early 1980s, however, decimated the *Carya* populations (Dale et al. 1990), and the current forests are better termed "upland oak forests."

The Throughfall Displacement Experiment (TDE) was constructed across a 2-ha portion of an upper subcatchment of the Walker Branch Watershed in 1992

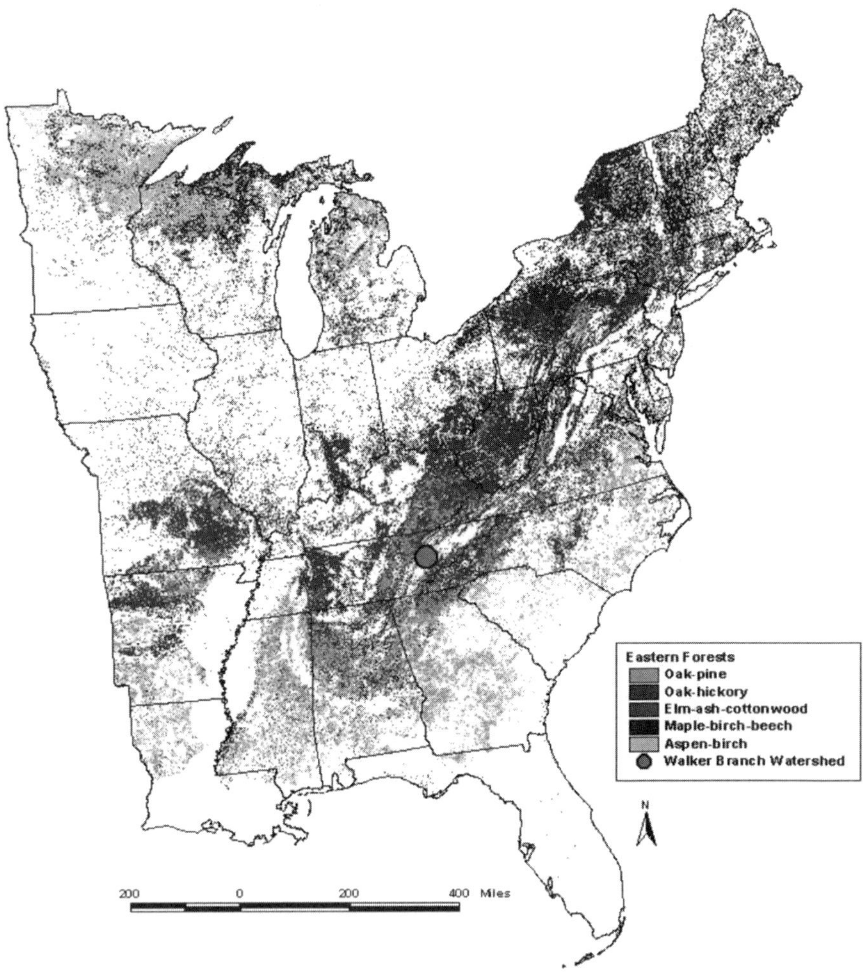

Figure 2.1. Current distribution of deciduous hardwood forests throughout the eastern United States and the location of the Walker Branch Throughfall Displacement Experiment (circle). Individual pixels represent 1 km^2. [Data are from the United States Department of Agriculture, Forest Service (http://www.srsfia.usfs.msstate.edu/rpa/rpa93.htm) and are based on Advanced Very High Resolution Radiometer data from the National Oceanic and Atmospheric Administration satellites. Map prepared by Holly K. Gibbs, Environmental Sciences Division, Oak Ridge National Laboratory.] (See also color plate.)

and 1993. In 1992, all trees greater than 0.1 m in diameter at 1.3 m [diameter at breast height (dbh)] were identified to species (729 individual trees), and heights and crown widths were measured directly or derived from allometric relationships from a subset of the measured data. *Quercus* spp and *Acer* spp were the major canopy dominants across all slope positions, *Liriodendron tulipifera* L. was a canopy dominant on the lower slope positions, and *Acer rubrum* L., *Nyssa syl-*

vatica Marsh. and *Oxydendrum arboreum* [L.] DC were the predominant species occupying midcanopy locations. In March of 1994, the stand basal area for trees greater than 0.1 m dbh averaged 21 m^2 ha^{-1}, with nearly identical basal areas on each of the treatment plots. By December 1999, the mean basal area across all plots had increased to 22.8 m^2 ha^{-1}(Hanson et al. 2001a). Saplings (i.e., trees $<$ 0.1 m dbh) contributed an additional 3.0 and 2.6 m^2 ha^{-1} to stand basal area in 1994 and 1999, respectively (Hanson et al. 2001a). A full listing of scientific and common names for the species discussed in this volume is included in the Appendix.

TDE Goal and General Hypotheses

The primary goal of the TDE was to develop a mechanistic understanding of the adjustments of forest ecosystems and organisms to changes in precipitation inputs resulting from a warming global climate. Such an understanding was considered essential for the hierarchical integration of physiological response mechanisms into assessments of the influence of climatic change on ecosystems. Figure 2.2 diagrams the general hypotheses that the TDE experiment was designed to address. Alterations in precipitation inputs were hypothesized to result in short-term direct effects on both soil microbial processes and plant physiology, resulting in (1) modified carbon assimilation, allocation and transpiration by individual plants and (2) altered rates of organic matter decomposition and nutrient turnover. Rapid biochemical feedbacks may allow some species to regulate their level of metabolic activity for more competitive use of available resources. Cumulative net effects of changes in the assimilatory and respiratory capacity of forest plants may lead to changes in the rate of production of leaves, wood, and fine roots and ultimately result in modified rates of detrital inputs to the soil. These anticipated alterations in physiology are determined by the genetic composition of the affected individuals. Genetic differences among individuals may ultimately lead to differential survival. Over longer periods of time (i.e., multiple years), changes in canopy leaf area production and canopy structure will result in modified radiation penetration and modified microclimates of the forest community. Altered understory microclimates suggest the potential for future changes in species composition over successional time scales (i.e., multiple decades). Long-term changes in soil carbon and nutrient availability of the forest soils may lead to altered nutrient uptake by the forest vegetation and feedback to changes in the assimilation potential of the forest canopy.

Experimental design

The experimental system was described in detail by Hanson et al. (1995, 1998, 2001a). The specific site on Walker Branch was chosen because of its uniform slope, consistent soils, and reasonably uniform distribution of vegetation. Early

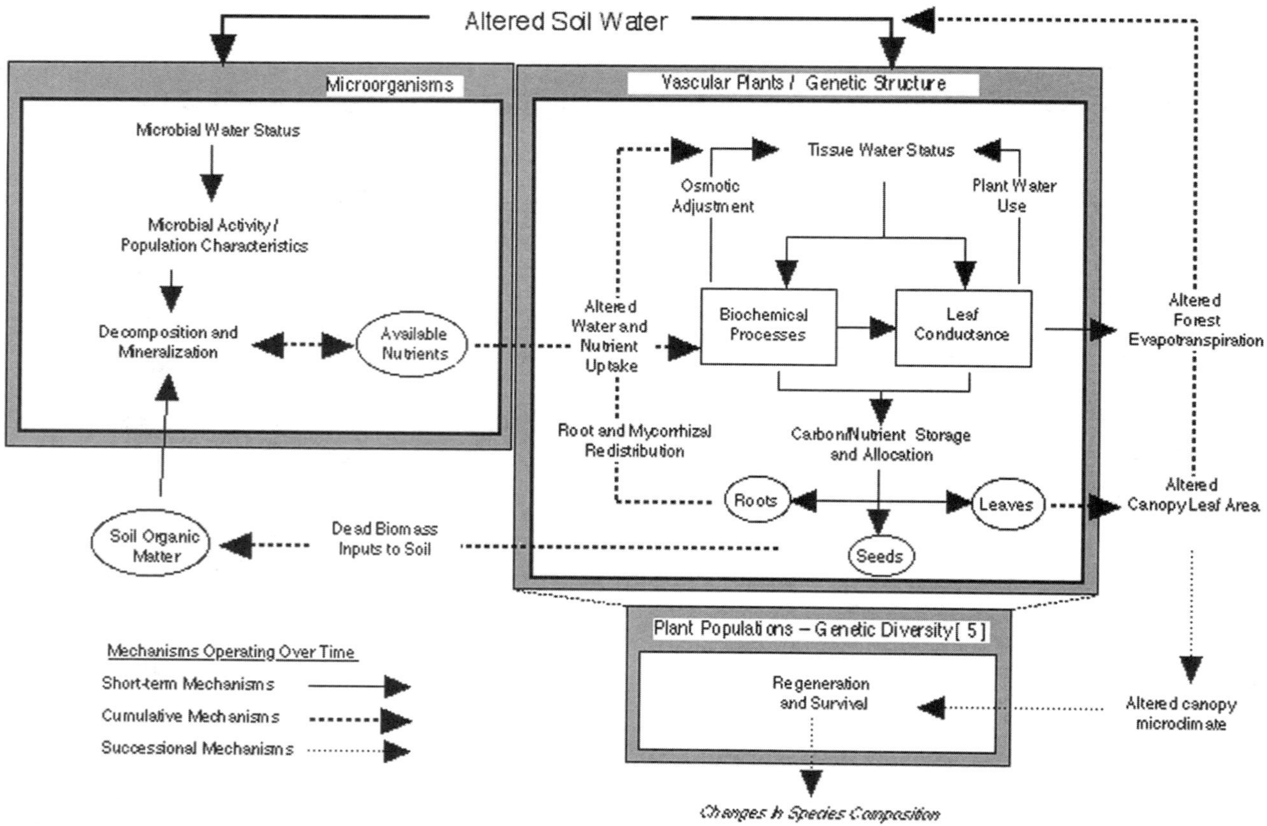

Figure 2.2. Hypothesized adjustments of forest organisms and ecosystem processes to altered precipitation. Processes and feedbacks within a box indicate mechanisms operating within an organism. Processes and feedbacks external to the boxes represent mechanisms operating at the ecosystem level. Type of arrow indicates the time frame for the mechanism or process indicated (see legend).

aerial photographs show that the TDE study area was forested in 1935, but several large dominant trees show open-growth characteristics, confirming that selective cutting along the ridge had been done before that time (Dale et al. 1990).

Manipulations of precipitation reaching the forest floor (hereafter termed "throughfall") were made by passively transferring a fraction of the throughfall from one experimental plot to another (Fig. 2.3). An advantage of this passive design was that all modifications of ambient throughfall were conducted in real time so that they were coordinated with appropriate atmospheric conditions (i.e., artificial rainfall was not added under high-light, high-humidity conditions). To ensure that mature trees of each plot had their canopies and root systems fully contained within a given soil-water condition, the experimental plots were made as large as possible. Three side-by-side experimental plots each measured 80 × 80 m; one wet, one dry, and one ambient. Each plot was divided into 100 subplots each 8 × 8 m that served as locations for repetitive, nondestructive measurements of soil and plant characteristics. An 8-m buffer zone around the edge of each plot (16 m between plots) was treated but not used for the observations of tree and sapling growth.

On the dry plot, throughfall precipitation was intercepted in ~1900 below-canopy troughs (0.3 × 5 m; Fig. 2.4) made of greenhouse-grade polyethylene that was suspended above the forest floor at an angle that followed the slope of the terrain (~33% of the ground area was covered). The lower and upper end of each trough were ~1.5 and 2.5 m off the ground, respectively. The polyethylene

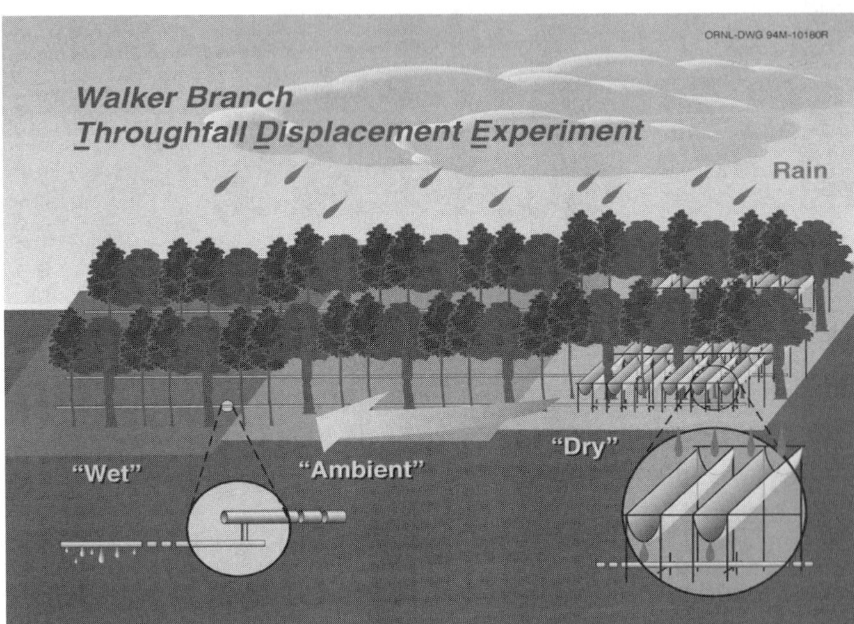

Figure 2.3. Schematic diagram of the TDE experimental design.

Figure 2.4. Understory throughfall collection troughs.

troughs were supported on a frame of ~20-mm-diameter galvanized electrical conduit and attached with large (~25.4 mm) binder clips. A series of standard aluminum gutters (~150 mm wide) positioned directly below the trough outlets collected the throughfall from the troughs. The troughs and gutters were constructed in 12.2-m modular sections so that sections damaged by wind and/or fallen large branches would not compromise throughfall interception and transfer of an entire collection line. Water collected in each modular trough/gutter assembly was drained via a reinforced flexible hose into a continuous 160-m pipeline made of ~100-mm schedule-40 PVC pipe. This pipeline served as the accumulation container and transfer pipeline for the collected throughfall.

The intercepted throughfall was transferred by gravity flow across the ambient plot and distributed onto the wet treatment plot through paired drip holes spaced ~1 m apart. The distribution pipeline laid out across the wet plot was constructed of ~50-mm schedule-40 PVC pipe. Dry plot troughs were arranged in 21 rows of ~80–90 troughs each.

The experimental area was located at the upper divide of the watershed so that lateral flow of water into the plots from upslope positions did not occur. The site had a southern aspect. Reductions in soil moisture anticipated from the experimental removal of 33% of the throughfall were designed to be comparable to the growing season having the lowest recorded rainfall during the dry 1980s decade (Cook et al. 1988), which resulted in sapling mortality and reduced growth of some vegetation (Jones et al. 1993).

The experimental design was not replicated because of cost and logistical limitations. This forced a reliance on "pseudoreplication" (Eberhardt and Thomas 1991). Addressing pseudoreplication in our sampling design is critical (Hurlbert 1984). Eberhardt and Thomas (1991) recommend that unreplicated experiments be supported by adequate sampling of site environmental parameters (including climatic conditions), comparable ambient areas, and pretreatment sampling of key variables. To provide information for those variables, the topography, soils, soil-water patterns, microclimate, and vegetation were extensively characterized before setting up the throughfall displacement system. Although not without limitations, unreplicated experimental designs have been used to provide invaluable understanding of forest ecosystem processes in manipulative studies at the watershed scale. Classic examples are the Hubbard Brook (Borman and Likens 1979) and Coweeta watershed studies (Swank and Crossley 1988).

TDE Environment and Performance from 1993 to 2000

The following section describes weather conditions, the record of soil-water content and soil-water potential by treatment, and indices of drought for the period of study. It also includes a discussion of TDE performance and potential confounding influences of the TDE infrastructure on stand microclimate (i.e., litter and soil temperatures). Supplemental information on atmospheric deposition and ozone exposure at the TDE site from 1993 through 2000 is provided as a reference to other environmental variables that can affect forest processes. A complete record of TDE weather data for hourly, daily, and annual time periods, and the complete spatial record of percent soil water data are described by Hanson et al. (2001b) and archived online at URL: http://cdiac.esd.ornl.gov/epubs/ndp/ndp078a/ndp078a.html.

Environmental Conditions

Weather

Weather conditions exhibited substantial interannual variation from 1993 through 2000 (Table 2.1). Lower-than-average annual precipitation was measured in 1993 (−16%), 1995 (−16%), 1998 (−9%), 1999 (−15%), and 2000 (−11%), and above-average precipitation was observed in 1994 (+24%), 1996 (+21%), and 1997 (+8%). Growing-season precipitation was near normal in 1994, 1999,

Table 2.1. Weather variables for each year of the study including (A) mean air temperature for the year and growing season, (B) cumulative radiation for the full year and the growing season, and (C) total annual precipitation

	Year of observations							
Variable	1993	1994	1995	1996	1997	1998	1999	2000
A. Air temperature (°C)								
Annual	14.3	14.4	14.4	13.6	14.1	15.8	15.0	14.5
GS	22.9	21.6	22.7	22.0	21.5	23.6	22.2	22.6
B. Cumulative radiation (MJ m^{-2})								
Annual	4724	4737	4960	4824	4849	4928	5600	5401
GS	2830	2643	2717	2777	2765	2928	3155	3010
C. Ambient precipitation (mm)								
Annual	1139	1675	1135	1638	1462	1225	1152	1207
GS	383	569	394	783	650	333	522	513

Note: The growing season (GS) is defined as May through September.

and 2000, but it was 26–38% less during the drought years of 1993, 1995, and 1998. Growing-season precipitation was 47% and 22% higher than normal in 1996 and 1997, respectively. Mean annual air temperature and annual incident solar radiation were not as variable as annual precipitation, but mean annual air temperatures in 1998 and 1999 were warmer than in the other years. Cumulative annual incident solar radiation at the site was similar across years ranging from 2643 to 3155 MJ m^{-2}.

Litter and Soil Temperatures

To judge the impact of the dry plot troughs on understory microclimates, supplemental transect surveys of instantaneous litter and soil temperatures were conducted periodically from 1994 through 1996. These observations were conducted on four transects (one upslope, two midslope, and one downslope) across the treatment plots (31 observations per transect). Transect observations of soil temperature at 0.15 m indicated no consistent significant trends with respect to TDE treatments. Transect observations showed that soil temperatures were occasionally warmer on the dry and ambient plots when compared to the wet treatment plot (Table 2.2), but such differences seldom exceeded a few tenths of a degree. If present throughout the year, this small temperature increase might increase soil and litter layer respiration by ~1%. However, because temperature differences occur only periodically, it is not likely that significant sustained impacts on litter/soil decomposition and root activity were occurring.

Litter layer temperatures obtained from periodic transect observations (Table 2.2) or continuous data loggers (data not shown) showed greater differences between treatment plots. Depending on the sampling event, hourly litter layer temperature differentials between plots could be as much as several degrees Celsius, but dry plot 24-h mean litter temperatures were seldom more than 1°C from ambient plot or wet plot litter temperatures. Observed differences did have a seasonal trend, however, with dry plot litter tending to be warmer than other plots before leaf production in the spring, cooler during the middle of the growing

Table 2.2. Mean litter surface and soil (0.15 m) temperatures (°C) from transect observations for each treatment area of the Walker Branch TDE on selected dates

	Dormant season			Growing season			
Variable/treatment	2/1/94	2/1/95	4/17/94	4/25/94	6/29/95	7/1/94	10/1/96
Litter surface							
Wet	4.2	15.6	—	—	25.3	24.4	17.3
Ambient	5.9	16.4	—	—	24.9	23.4	19.6
Dry	5.9	15.8	—	—	26.3	23.3	19.8
p-Value	<0.01	0.40	—	—	<0.01	<0.01	0.95
0.15-m soil							
Wet	6.5	4.3	9.2	14.6	19.5	18.9	16.4
Ambient	6.5	4.1	9.2	14.6	19.5	18.8	16.4
Dry	6.6	4.9	9.5	14.7	19.6	18.8	16.7
p-Value	0.15	<0.01	<0.01	0.12	0.06	0.67	<0.01

Note: Significant treatment differences are indicated by *p*-values < 0.05.

season, and warmer again toward the onset of winter. Little evidence of temperature differences were observed at night using either method (data not shown).

Evaluating TDE Treatments

The effectiveness of the TDE system for producing distinct soil moisture treatments was judged from 1993 to 2000 by comparing periodic measurements of an extensive network of nondestructive soil water measurements to pretreatment patterns of soil-water across the experimental area. Probes for continuous monitoring of soil-water content and matric potential were installed in 1998 to track seasonal dynamics.

Soil-Water Measurements

Soil-water content (%, v/v) was measured with a time-domain reflectometer (TDR) following the procedure of Topp and Davis (1985) as documented for soils with high coarse fraction content (Drungil et al. 1989). Hanson et al. (1998) conducted simultaneous sampling of soil-water content by destructive gravimetric and TDR approaches to verify that the TDR observations were consistent with past data based on gravimetric observations on Walker Branch (Peters et al. 1970). Three hundred ten sampling locations were installed at an 8×8-m spacing across the site, giving more than 100 soil-water monitoring locations per plot. At each location, two pairs of TDR waveguides were installed in a vertical orientation (0–0.35 m and 0–0.7 m). The surface (0–0.35 m) TDR measurements coincided with the zone of maximum root density in these soils (Joslin and Wolfe 1998). TDR measurements were normally conducted biweekly during the growing season and approximately monthly during the dormant season. Because the integrated measurement of the percent soil-water from 0 to 0.7 m (SW70) contained all of the soil water measured by the 0–0.35-m waveguides (SW35), the percent soil-water content between 0.35 and 0.7 m (SW3570) was estimated by subtraction from the following equation (Hanson et al. 1998):

$$\text{SW3570} = \frac{(\text{SW70} \times 0.7 \text{ m}^3) - (\text{SW35} \times 0.35 \text{ m}^3)}{0.35\text{m}^3}, \tag{1}$$

where 0.7 and 0.35 m^3 are the total and partial volume of a hypothetical soil sample 1 × 1 × (0.7 or 0.35) m.

In addition to the TDR soil-water measurements, automated soil-water sensors were added to each treatment plot in 1998 and 1999 for continuous hourly mean soil-water and soil-water potential measurements. Eight water content sensors (CS615 water content reflectometer; Campbell Scientific, Logan, Utah) were deployed at four depths in each of two soil pits per treatment plot. In two additional soil pits per treatment, eight heat-dissipation soil matric potential probes (CS229, Campbell Scientific, Logan, Utah) were also installed (four per pit). In each pit, the soil-water probes were placed according to soil horizons at the following approximate depths: 10, 30, 50, and 70–100 cm.

Statistical Analysis of Soil-Water Content Data

Pretreatment measurements (April 1992 through July 12, 1993) of soil-water content indicated the existence of gradients in water content across the TDE area. There was a decreasing gradient of soil water from the lower to the upper slope positions, and the mean soil-water content of the eastern third of the experimental area was lower than the rest. To judge the effectiveness of the throughfall transfer system, these spatial patterns had to be captured in a covariate matrix that could describe the influence of a specific slope position with respect to the overall site mean soil-water content. The pretreatment observations of the TDR grid network were used to produce this covariate matrix according to the methods of Hanson (2000). A contour plot of the covariate ranks for each of the measurement locations across the TDE are shown in Fig. 2.5. Positive and negative values represent locations that exhibit greater or lesser water contents than the grand mean, respectively. Note that the overall tendency is for the dry treatment area to already be drier than the ambient and wet plots prior to beginning the experiment. Slope position, soil texture, and coarse fraction may contribute to the observed pretreatment tendencies across the TDE area. Although we hoped that a single covariate rank based on an entire years worth of data would be robust enough to apply to all subsequent dates, two covariate ranks were required: one for the dormant season when soil water conditions are near saturation and one for summer periods. These separate ranks were not substantially different than the overall ranks shown in Fig. 2.5B.

Observed Patterns of Soil-Water Content

The seasonal patterns of mean TDR soil-water content by treatment in the 0–0.35-m and 0.35–0.7-m depth increments from 1993 through 2000 are shown in Fig. 2.6. Since the initiation of treatments in 1993, significant differences in

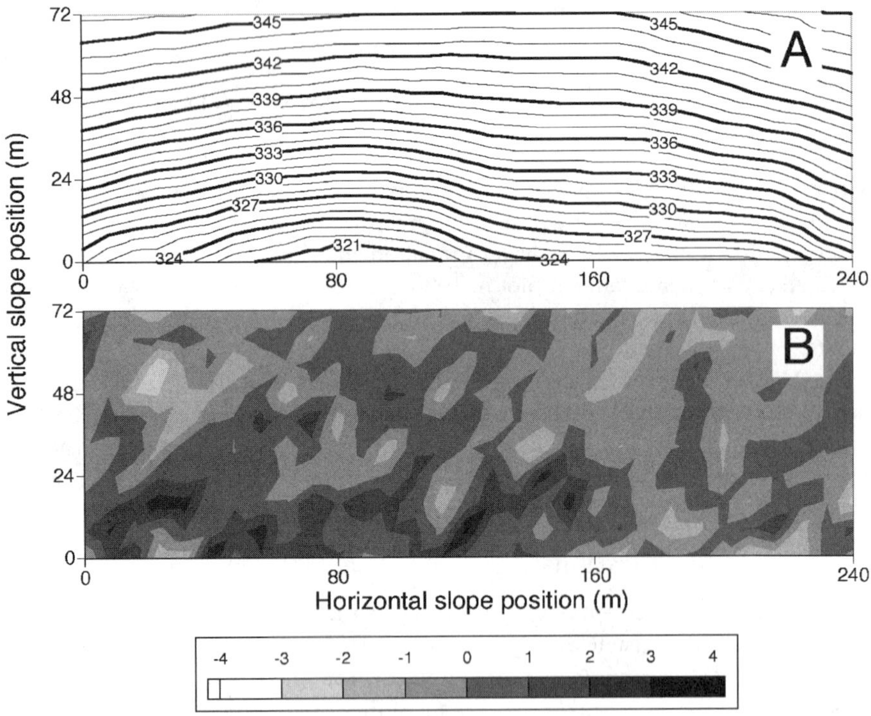

Figure 2.5. Elevation map of the experimental area in meters (A) and the pretreatment pattern of soil water across the experimental site expressed relative to the mean for the entire experimental area (B). A location with a rank of zero would be in agreement with the overall site mean.

the soil-water content in the top 0.35-m depth were present except during dormant seasons when all soils were at or near saturation, and at the depth of the drought in 1995. During a typical winter rain event on February 11, 1994, a subset of the 0.35-m waveguides were measured under saturated conditions, and we found the water content of the soils of the wet plot to be significantly elevated above those of the ambient and dry treatment areas (data not shown). Calculated soil-water contents for the 0.35–0.7-m depth show a similar separation of treatment means, but the differences are not as great nor are they as sustained in any one season as those in the top 0.35 m.

Contour plots of the 0–0.35-m soil-water contents for selected dates in 1994 and 1995 are provided in Figs. 2.7 and 2.8, respectively, to give a visual impression of the spatial heterogeneity of soil water content across the TDE in wet (1994) and dry (1995) years. From the contour plots and Fig. 2.7, it is clear that each growing season begins with saturated soils. Subsequently, with the initiation of evaporative demand and canopy transpiration in mid to late May, the dry plot water contents are drawn down faster than those in the ambient and wet plots. In years with significant dry periods (1993, 1995, 1998, and 1999), long periods

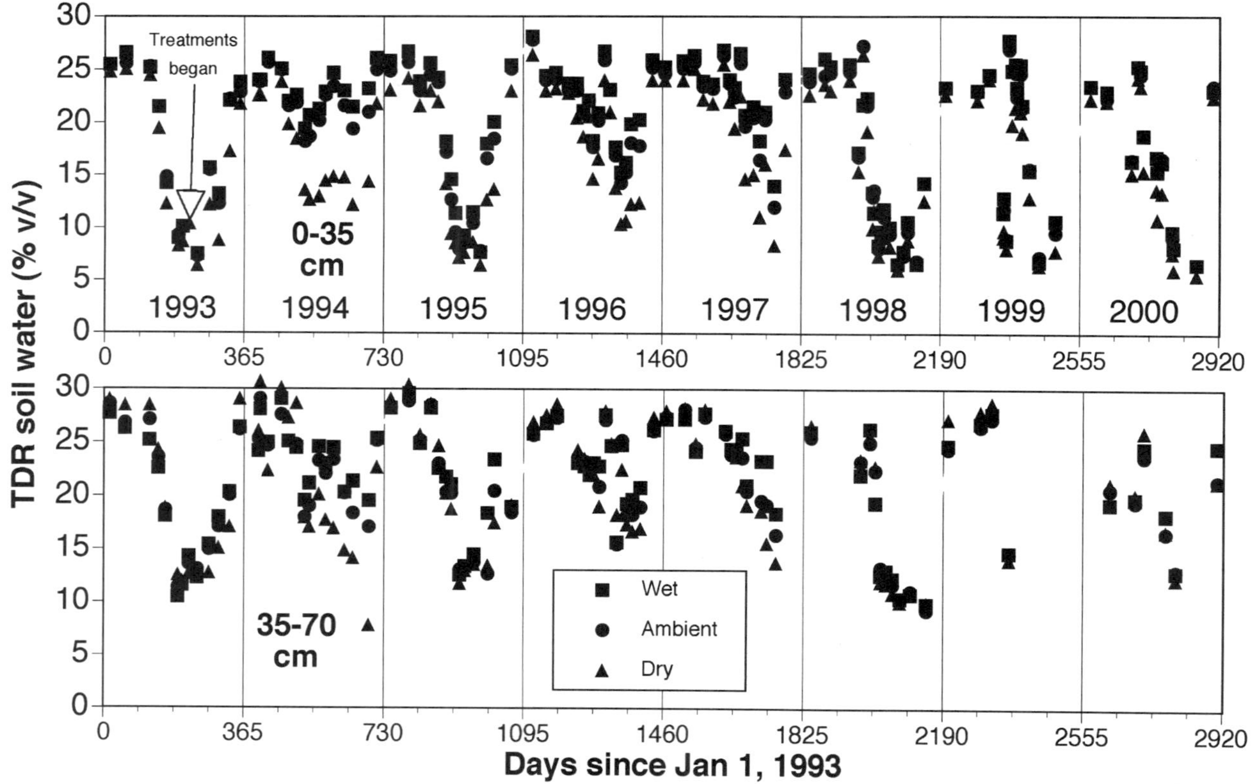

Figure 2.6. Soil-water content (% v/v) of the 0–0.35-m and 0.35–0.7-m soil depth from 1993 through 2000. Data are the mean values for the wet, ambient, and dry plots on the Walker Branch Throughfall Displacement Experiment. Treatments were initiated on July 14, 1993 (see arrow).

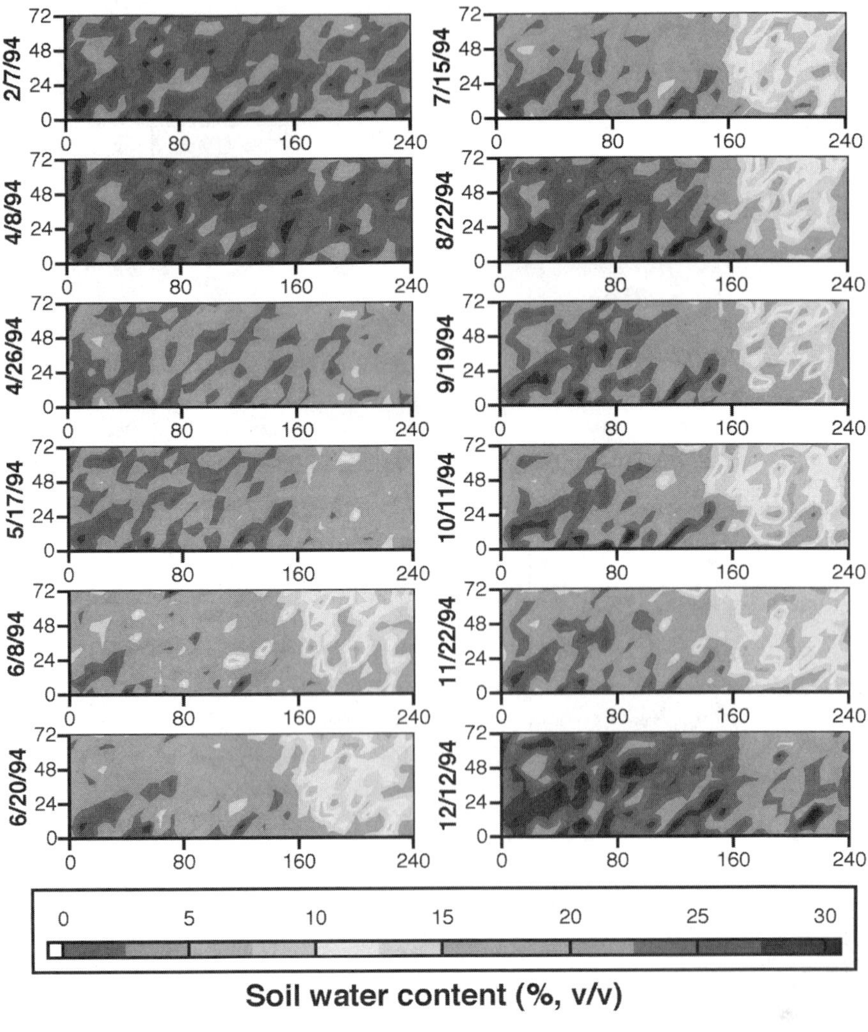

Figure 2.7. Contour plots of the mean soil water content from 0 to 0.35 m across the TDE area throughout 1994. March 4 and August 8 data were left off the figure. Numbers along the *x* and *y* axes of each plot are meters, as in Fig. 2.5. (See also color plate.)

without rainfall cause treatment differentials to disappear. However, following the depths of droughts, treatment differences redeveloped as the soils refilled at a faster and slower rate on the wet and dry plots, respectively. The nature of the dry treatment effects, including accelerated drying and slow recovery after drought, is nicely demonstrated by TDR data in Fig. 2.6 and by the continuous soil water data for 1999 in Fig. 2.9.

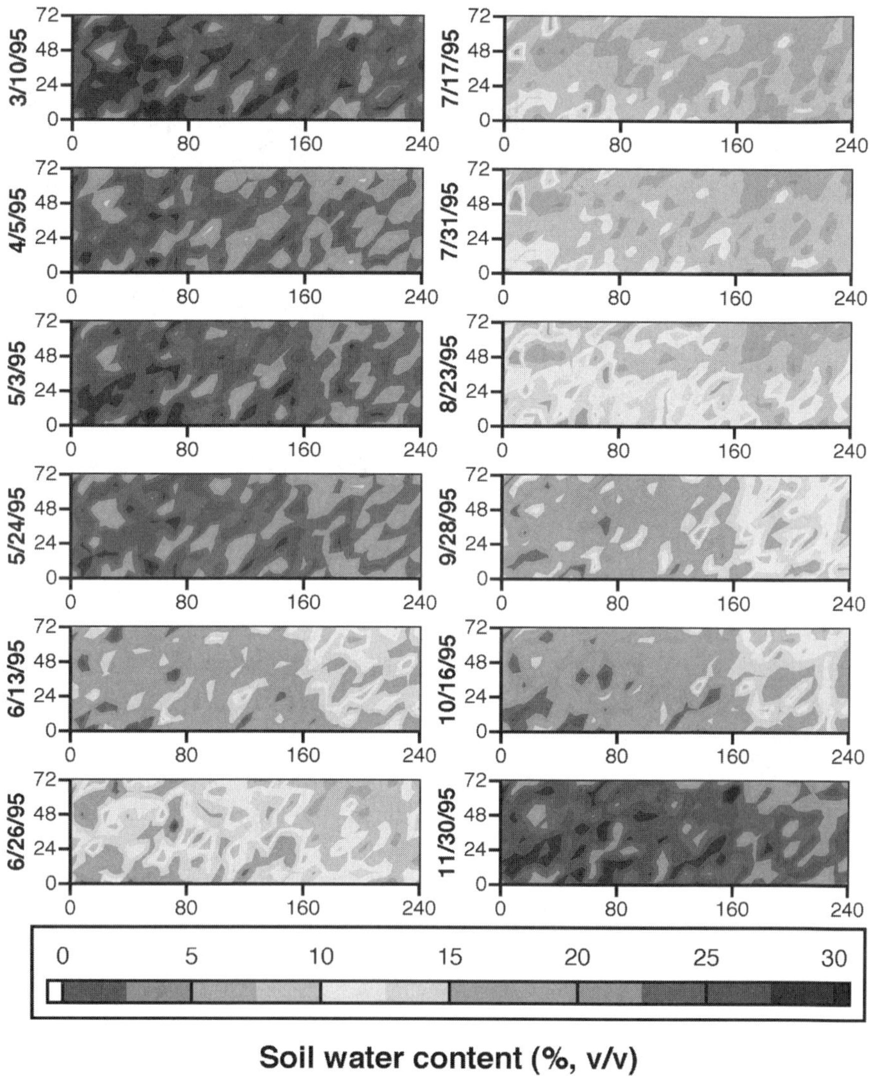

Figure 2.8. Contour plots of the mean soil-water content from 0 to 0.35 m across the TDE area throughout 1995. The January 17 data were left off the figure. Numbers along the x and y axes of each plot are meters, as in Fig. 2.5. (See also color plate.)

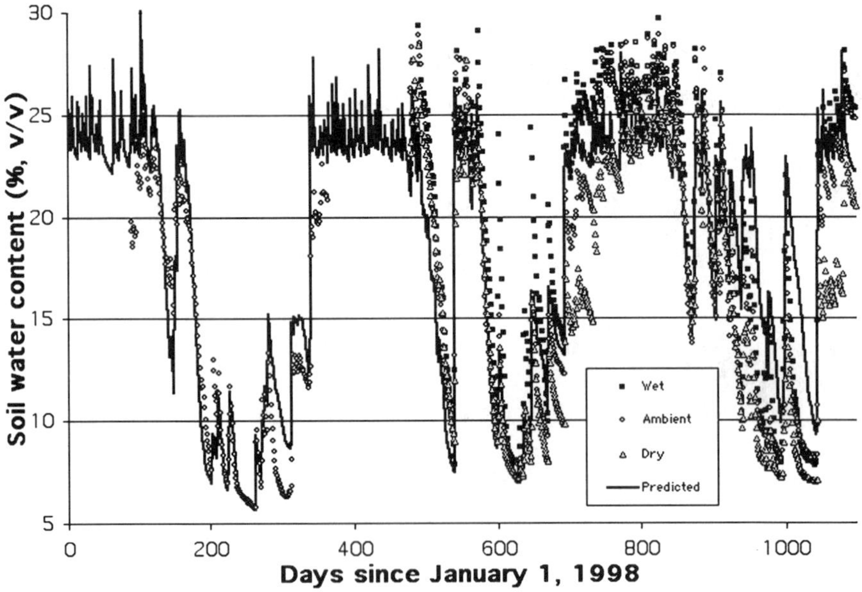

Figure 2.9. Multiyear seasonal pattern of wet, ambient, and dry plot measured soil-water content and predicted ambient-plot soil-water content for 1998 to 2000. Measured data and predictions are for the surface soil (0–35 cm) from hourly observations of *in situ* frequency domain reflectometer probes and represent the mean of three to four sensors per treatment plot.

Defining and Quantifying Drought

The research literature and the popular press are filled with examples of qualitative statements regarding the moisture status under which experiments are conducted. Common phrases used to describe the weather conditions include abnormally dry or wet, subambient precipitation, drought, severe drought, spring versus fall drought, and many others. Unfortunately, these phrases do little to allow comparisons across studies or between years within a given study. Further confounding of the characterization of drought develops from different applications of terminology by the public or different scientific disciplines. From the standpoint of climate and weather forecasting, drought is usually considered to be rainfall in quantities below the normal mean for an area. Whereas such a definition can be relevant to shallow-rooted plants, such as garden annuals and lawn grass, it often ignores the capacity of soils to store large amounts of water that can be available to mature forest for extended periods of time. For our studies and in this volume, we use soil-water potential (Kramer 1983) as our primary measurement of soil water status and the key variable for developing relationships between plant responses and variable precipitation inputs. Here, we describe indices of drought that we have applied to our studies for seasonal or year-to-year comparisons and contrast them with other available drought indices.

Measurements of Soil-Water Potential

Soil moisture retention curves for representative A-, AE-, and EB-horizon soils collected adjacent to the TDE site were obtained with psychrometric measurements of soil-water potential and gravimetric measurements of soil-water content (Hanson et al. 1998). These curves are provided in Fig. 2.10. The AE-soil retention curve was assumed to be applicable to the 0–0.35-m increment, and the EB-soil retention curve was assumed appropriate for the 0.35–0.7-m increment. These assumptions are subject to error because of variations in the thickness of soil horizons across the TDE.

To convert raw volumetric water content data (TDRr) collected on the TDE site to soil-water potential (in MPa), the TDRr must first be corrected to account for the mean coarse fraction of the soil in question according to the equation of Hanson (2000):

$$\text{TDRc} = \frac{\text{TDRr}}{100 - \text{Cf}} \times 100, \tag{2}$$

where Cf is the appropriate mean coarse fraction for the TDE location of interest. The mean Cf for the top 0.35-m and the 0.35–0.7-m increment was 14.2% and 17%, respectively (Hanson et al. 1998). Subsequent to the Cf correction, TDRc is used as an input to the equations in Fig. 2.10 to yield soil-water potentials for the depth increment.

Seasonal patterns of soil water potential (Fig. 2.11) provide a measure of the interannual variability of soil-water availability between years. Significant drought periods were evident in 1993, 1995, 1998, 1999, and 2000. Although the water content data of Fig. 2.6 showed substantial seasonal changes in the water content in both the 0–0.35-m and the 0.35–0.7-m depths of the soil profile, conversion of those data to soil-water potential (Fig. 2.11) showed that water in the 0.35–0.7-m depth remained largely available for plant use throughout the drought periods. Consistent evidence of reduced water availability at depth was only observed during the severe late-season drought of 1998.

Measurements of Soil-Water Stress

To facilitate comparisons of the severity of soil-water deficits between years, we report two indices of soil moisture deficits: (1) the minimum annual soil water potential (MPa) and (2) a water stress integral (units of MPa d; Myers 1988). The first index shows the greatest magnitude of drought in any year, and the second provides a measure of the duration and severity of droughts. Although Myers (1988) used integrated leaf water potentials, we sum daily mean soil-water potentials for the top 0.35-m depth associated with >70% of the forest roots, to obtain the water stress integral. The same approach could be used for other depth increments, but the integral values corresponding to various plant responses would differ. Because continuous monitoring of soil-water potentials in the top 0.35 m was not available prior to 1998 and because the spatial coverage of our

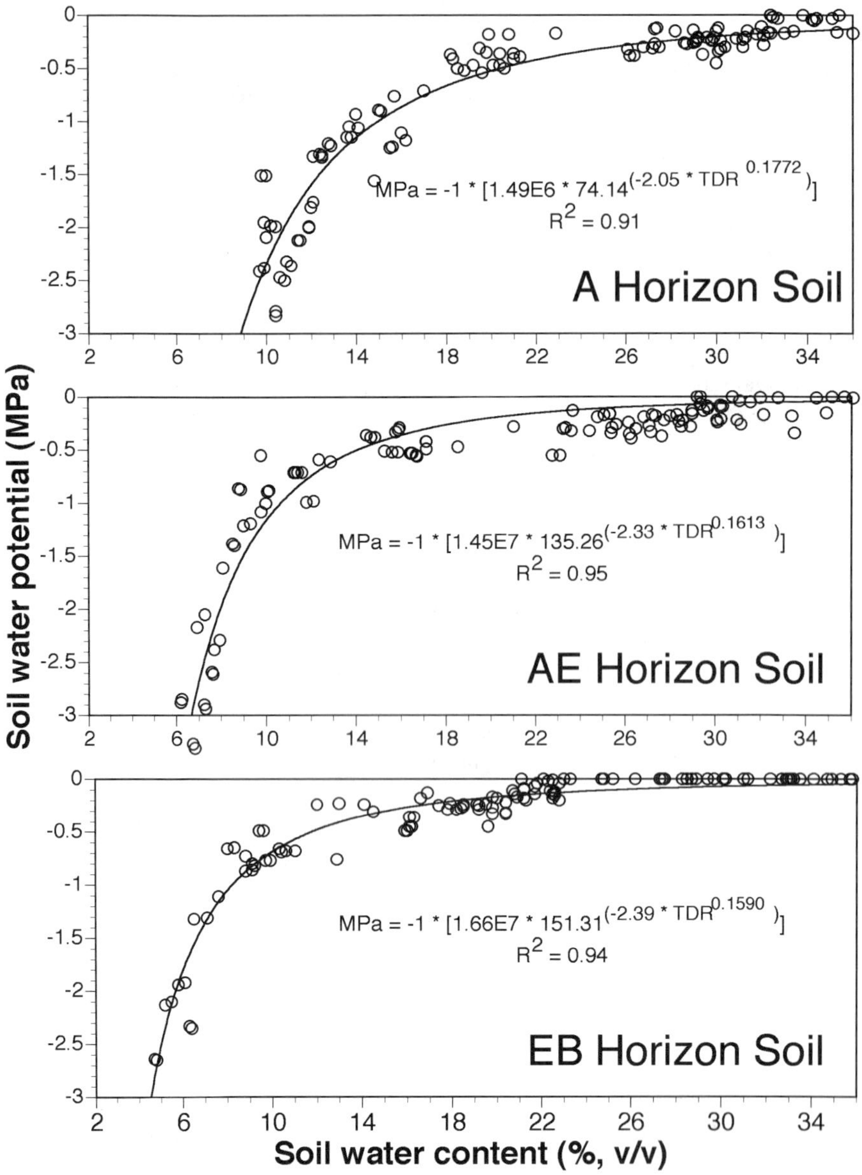

Figure 2.10. Soil moisture release curves for representative A-, AE-, and EB-horizon soils of the TDE. The fitted curves were generated with nonlinear least squares approaches.

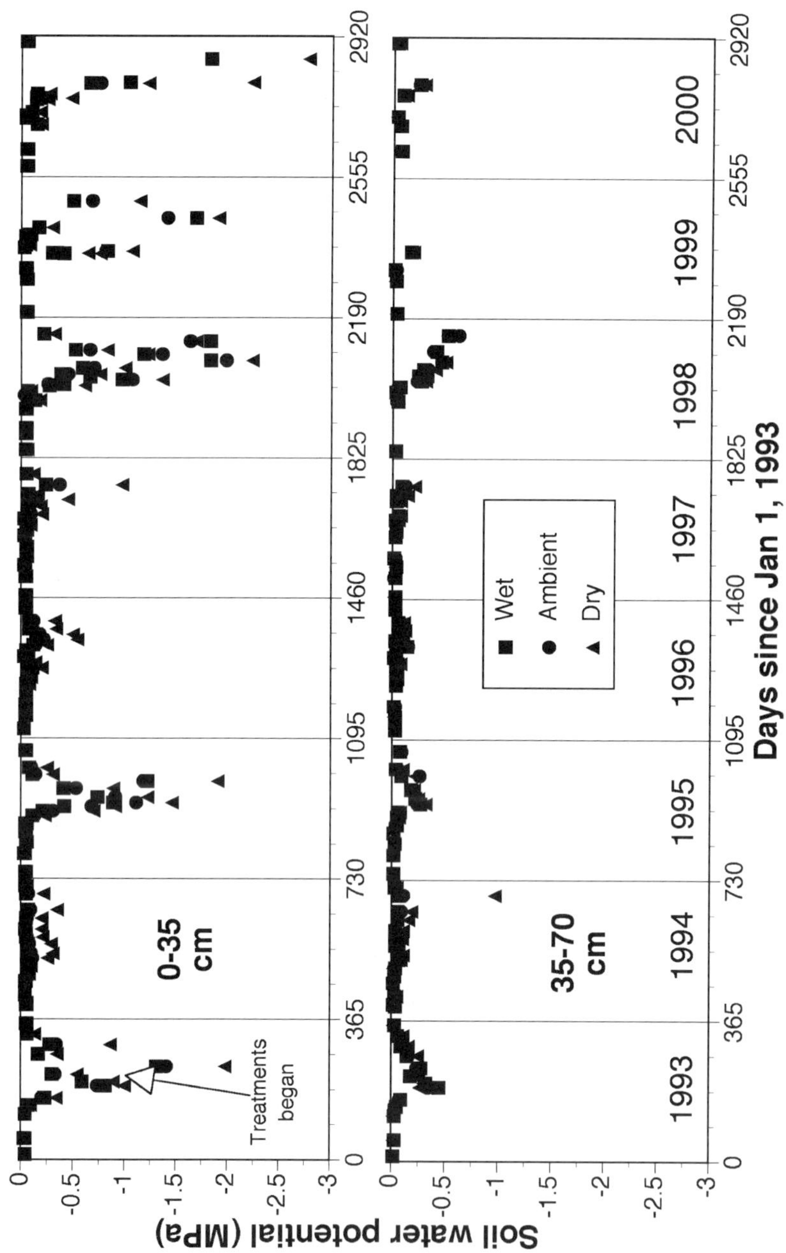

Figure 2.11. Soil-water potentials for the 0–0.35-m and 0.35–0.7-m soil depth from 1993 through 2000. Data are the mean value for the wet, ambient, and dry plots on the Walker Branch Throughfall Displacement Experiment. Treatments were initiated on July 14, 1993 (see arrow).

Table 2.3. Soil-water variables used to characterize drought severity for each year of the study including (A) minimum soil-water potential by treatment and (B) water stress integrals by treatment

Variable	Year of observations							
	1993	1994	1995	1996	1997	1998	1999	2000
A. Minimum daily soil water potential (MPa)[a]								
Wet	—	−0.24	−1.84	−0.40	−0.44	−2.44	−1.86	−1.21
Ambient	−2.18	−0.43	−2.22	−0.60	−0.86	−2.76	−1.91	−1.17
Dry	—	−1.74	−2.59	−1.08	−1.41	−2.91	−1.99	−1.97
B. Water stress integral (MPa d)[a]								
Wet plot								
Jan./July	—	−13	−46	−13	−11	−39	−17	−13
Aug./Dec.	—	−9	−44	−12	−14	−82	−65	−27
Annual	—	−22	−90	−25	−25	−121	−82	−40
Ambient plot								
Jan./July	−69	−15	−66	−14	−12	−48	−33	−15
Aug./Dec.	−11	−9	−67	−15	−17	−128	−85	−34
Annual	−81	−24	−133	−29	−29	−175	−118	−49
Dry plot								
Jan./July	—	−30	−91	−15	−13	−70	−47	−19
Aug./Dec.	—	−11	−99	−33	−33	−176	−132	−104
Annual	—	−41	−190	−48	−46	−246	−179	−123

[a] Values for the water stress integral and minimum daily soil water potential are estimates from a model of the water budget for these stands.

continuous sensors is limited, we used a combination of measured soil-water potential data and modeled interpolations for periods without data to provide a complete record of the 0–0.35-m water stress integral. Modeled interpolations were executed with a revised version of the TEHM/PROSPER model (Huff et al. 1977; Luxmoore 1983) coded with STELLA modeling software (High Performance Systems, Hanover, NH). Using data from 1998 to 2000, ambient predictions and the data in Fig. 2.9 demonstrate the efficacy of the modeling approach for interpolating soil-water conditions over time.

Minimum daily soil-water potentials showed that significant drought occurred in 1993, 1995, 1998, 1999, and 2000 (sustained values below −0.7 MPa in Fig. 2.11). However, the annual water stress integrals (Table 2.3), which account for the duration of drought, demonstrate that the 1993 and 2000 droughts (−81 and −49 MPa d, respectively) were not sustained as long as those in 1995, 1998, and 1999 (−133, −175, and −118 MPa d, respectively). In the wet growing seasons of 1994, 1996, and 1997, ambient soil-water potentials never fell below −0.9 MPa, but dry plot soil-water potentials reached substantially lower levels (Table 2.3). Calculated water stress integrals for January through July versus August through December further demonstrate year-to-year differences in the characteristics of droughts (Table 2.3). Whereas the droughts of 1998 and 1999 were most severe in the second half of the growing season (i.e., exhibited the most negative soil-water potentials), the drought of 1993 was most severe during the earlier half.

Atmospheric Deposition and Ozone Exposure

The National Acid Deposition Program/National Trends Network (NADP 2000) operates a site (TN00) for monitoring wet deposition chemistry on the Walker Branch Watershed within 1 km of the TDE site. Data from this site (Fig. 2.12) show that wet deposition of sulfur, nitrogen and base cations to the Walker Branch forest canopy were relatively stable from 1993 to 2000. Variability of wet deposition is driven in part by year-to-year differences in rainfall quantity and by local and regional changes in anthropogenic emissions to the atmosphere. Wet deposition inputs reflected in Fig. 2.12 represent approximately half of the total atmospheric deposition to Walker Branch Watershed (Johnson and Lindberg 1989, 1992) with dry particulate and vapor forms contributing the rest. Combining estimates of dry nitrogen deposition (Hanson et al. 1989; Lindberg et al. 1986; Kelly and Meagher 1986) with the 8-year mean wet nitrogen deposition numbers from the NADP site (5.5 kg N ha^{-1} y^{-1}), we estimate mean annual total nitrogen deposition to the Walker Branch forest to range from 10 to 15 kg N ha^{-1} y^{-1}. For an earlier period from 1981 to 1983, Johnson and Lindberg (1989) reported inputs of sulfate, nitrate, ammonium, calcium, and potassium of 1.6, 0.52, 0.18, 0.43, and 0.022 kmol ha^{-1} y^{-1}, respectively.

Hourly ozone data from the Freels Bend study area located 6 km from the TDE, which are collected as a part of Environmental Protection Agency's Aerometric Information Retrieval System (AIRS 2001), were used to estimate cu-

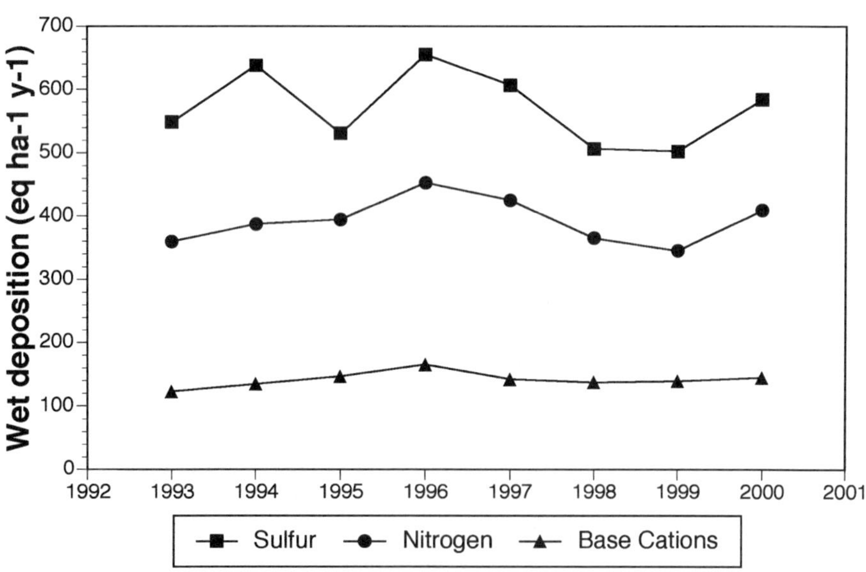

Figure 2.12. Annual sulfur, nitrogen, and base cation (K, Ca, and Mg) wet deposition to the forest canopy of Walker Branch Watershed from 1993 through 2000. Data are for site TN00 of the National Atmospheric Deposition Program/National Trends Network. (From NADP 2000.)

mulative growing-season ozone exposure (ppm h). Cumulative ozone exposure from April through September (i.e., the sum of all hourly ozone levels) was quite stable from year to year, ranging from 158 to 165 ppm h; most years had 10 or more hourly occurrences of ozone concentrations greater than 100 nL L^{-1}. The wet 1996 and 1997 growing seasons had less than 10 hours of ozone exceeding 100 nL L^{-1}. Seasonal levels of cumulative ozone exposure around 160 ppm h are sufficient to reduce mature tree (*Quercus rubra* L.) photosynthetic rates by as much as 50% over the growing season (Hanson et al. 1994).

Conclusions on TDE operations

The TDE manipulations were maintained continuously from July 1993 through December 2000, and soil-water measurements showed that the manipulations produced significantly different hydrologic budgets for the wet, ambient, and dry plots in years having either high or low growing-season precipitation (Hanson et al. 1995, 1998, 2001a). In addition, because of interannual variation in growing-season rainfall inputs, we obtained a much broader range of treatments than originally anticipated. The combination of treatment manipulations and variable annual and growing-season precipitation has allowed us not only to test the impact of a constant ±33% alteration of throughfall resulting from all incoming rainfall events but also to determine the influence of sustained droughts as well.

Related Studies

Readers of this volume are encouraged to contrast the results from the TDE with those from similar works that reflect different scenarios of precipitation change and vegetation types. Hanson (2000) summarized important characteristics of large-scale precipitation-manipulation studies and provided a brief summary of field-based precipitation studies including the TDE. Typical large-scale precipitation manipulations used either complete understory roofs (Gundersen et al. 1995) or partial coverage of the ground area with gutters or troughs (Holstener-Jørgensen and Holmsgaard 1994; Hanson et al. 1995, 1998) to manipulate precipitation levels. As of that review, the TDE represented the largest field manipulation of water attempted to date (1.3 ha), but other large-scale studies of similar magnitude have been conducted. A 2-ha rainfall exclusion study located in the Amazon forest (Nepstad et al. 2002) used an approach very similar to that of the TDE, but included only an ambient and reduced throughfall comparison. Nepstad et al. (2002) removed 72–75% of the throughfall from the dry plot throughout the rainy season (February to August). Although their scenario is a more severe change than the one used in the TDE study, soil-water deficits were not induced in their dry plot treatment because of an abundance of precipitation during the first year, and deep soil-water storage capacity.

Hultberg et al. (1993) and Moldan et al. (1995) used a 0.64-ha understory roof to study the impacts of rainfall chemistry on a *Picea* forest. Although their study did not address moisture stress, they studied changes in rainfall chemistries with 100% removal of ambient throughfall followed by the addition of artificial throughfall having variable chemistries. Conclusions based on differential chemical inputs are relevant to issues of nutrient cycling observations on the TDE experiment discussed in Chapter 23. Smaller-scale roofed catchment studies in other *Picea* forests (Rasmussen et al. 1995; Beier et al. 1995) also used 100% throughfall removal with variable duration. Where 100% throughfall removal was applied, attempts were commonly made to resaturate the soils periodically, especially during dormant periods. Although these studies did attempt to draw conclusions about the impact of precipitation change on growth processes, direct comparisons to the TDE results (Chapter 15, this volume) are not possible because adequate characterization of soil-water status was not a part of these studies.

Total control of water quantity and chemistry is easy to attain for small plants in greenhouses and growth chambers, but is much more difficult in the field. A notable attempt to achieve such control at the level of whole catchments is the CLIMEX project (Jenkins and Wright 1995), which used a 1200-m^2 greenhouse to attempt complete control over system precipitation, temperature, and atmospheric CO_2 concentrations. A recent mesic grassland study (Fay et al. 2000) combined passive rainfall collection and subsequent active redistribution to generate multiple precipitation treatments at the plot scale. They used a number of permanently installed plastic greenhouse roofs (9 $\times$ 14 m) as simultaneous shields from ambient rainfall and throughfall collectors. Together with large rainfall collection tanks, pumps, irrigation nozzles, and barriers to horizontal rooting and water movement (with plots measuring 7.6 $\times$ 7.6 m under each roof), their roofed plots represent a cost-effective approach for manipulating both the quantity and timing of rainfall events over a 6 $\times$ 6-m treatment area. Their approach allowed them to manipulate precipitation quantity and timing, including 30% reductions in precipitation and 30% reductions and/or extensions of dry periods. Initial results suggest that timing of rainfall inputs is critical to understanding ecosystem responses for grassland systems. A similar conclusion has been reached for mature tree responses to the TDE manipulations (Hanson et al. 2001a; Chapter 15, this volume).

Mechanistic biogeography and biogeochemistry models continue to be used to predict the future response of forest ecosystems and plant species to climate change scenarios (Bachelet et al. 2001; Iverson and Prasad 2001), but such predictions remain largely untested outside of retrospective analysis of historical vegetation patterns. Experimental manipulations as described in this book and the experimental studies in this chapter are needed to provide additional validation information for scenarios of climatic change that exceeds historical conditions and/or rates of change.

References

AIRS (2001) Aerometric Information Retrieval System, US Environmental Protection Agency, http://www.epa.gov/ttnairs1/index.html.

Bachelet D, Neilson RP, Lenihan JM, Drapek RJ (2001) Climate change effects on vegetation distribution and carbon budget in the United States. Ecosystems 4:164–185.

Bailey RG (1983) Delineation of ecosystem regions. Environ Manage 7: 365–373.

Beier C, Gundersen P, Hansen K, Rasmussen, L (1995) Experimental manipulations of water and nutrient input to a Norway spruce plantation at Klosterhede, Denmark, II. Effects on tree growth and nutrition. Plant Soil 168–169:613–622.

Borman FH, Likens GE (1979) *Pattern and process in a forested ecosystem.* Springer-Verlag, New York.

Cole DW, Rapp M (1981) Elemental cycling in forest ecosystems, In Reichle DE (Ed) *Dynamic principles of forest ecosystems,* Cambridge University Press, London, pp 341–409.

Cook ER, Kablack MA, Jacoby GC (1988) The 1986 drought in the Southeastern United States: How rare an event was it? J Geophys Res 93(D11):14257–14260.

Dale VH, Mann LK, Olson RJ, Johnson DW, Dearstone KC (1990) The long-term influence of past land use on the Walker Branch forest. Landscape Ecol 4:211–224.

Drungil CEC, Abt K, Gish TJ (1989) Soil moisture determination in gravelly soils with time domain reflectometry. Transactions ASAE 32(1):177–180.

Eberhardt LL, Thomas JM (1991) Designing environmental field studies. Ecol Monogr 61:53–73.

Fay PA, Carlisle JD, Knapp AK, Blair JM, Collins SL (2000) Altering rainfall timing and quantity in a mesic grassland ecosystem: design and performance of rainfall manipulation shelters. Ecosystems 3:308–319.

Gundersen P, Andersen BR, Beier C, Rasmussen L (1995) Experimental manipulations of water and nutrient input to a Norway spruce plantation at Klosterhede, Denmark, I. Unintended physical and chemical changes by roof experiments. Plant Soil 168–169:601–611.

Hanson PJ (2000) Large-scale water manipulations. In: Sala OE, Jackson RB, Mooney HA, Howarth RW (Eds) *Methods in ecosystem science,* Springer-Verlag, New York, pp 341–352.

Hanson PJ, Rott K, Taylor GE Jr., Lindberg SE, Gunderson CA, Ross-Todd BM (1989) NO_2 deposition to elements of a forest landscape. Atmos Environ 23:1783–1794.

Hanson PJ, Samuelson LJ, Wullschleger SD, Tabberer TA, Edwards GS (1994) Seasonal patterns of light-saturated photosynthesis and leaf conductance for mature and seedling *Quercus rubra* L. foliage: Differential sensitivity to ozone exposure. Tree Physiol 14:1351–1366.

Hanson PJ, Todd DE, Amthor JS (2001a) A six year study of sapling and large-tree growth and mortality responses to natural and induced variability in precipitation and throughfall. Tree Physiol 21:345–358.

Hanson PJ, Todd DE, Riggs JS, Wolfe ME, O'Neill EG (2001b) *Walker Branch Throughfall Displacement Experiment data report: Site characterization, system performance, weather, species composition, and growth.* ORNL/CDIAC–134; NDP–078A. Carbon Dioxide Information Analysis Center, Oak Ridge National Laboratory, Oak Ridge, Tennessee.

Hanson PJ, Todd DE, Edwards NT, Huston MA (1995) Field performance of the Walker Branch Throughfall Displacement Experiment, In Jenkins A, Ferrier RC, Kirby C (Eds) *Ecosystem manipulation experiments: Scientific approaches, experimental design and relevant results.* Ecosystem Research Report #20. Commission of the European Communities, Brussels, Belgium, pp 307–313.

Hanson PJ, Todd DE, Huston MA, Joslin JD, Croker J, Augé RM (1998) *Description and field performance of the Walker Branch Throughfall Displacement Experiment: 1993–1996.* ORNL/TM–13586. Oak Ridge National Laboratory, Oak Ridge, Tennessee.

Holstener-Jørgensen H, Holmsgaard E (1994) Effects of irrigation and drought on growth of young Norway spruce on east Danish moraine. For Landscape Res 1:127–142.

Huff DD, Luxmoore RJ, Mankin JB, Begovich CL (1977) *TEHM: A terrestrial ecosystem hydrology model.* EDFB/IBP-76/8; ORNL/NSF/EATC-27. International Biological Program, Oak Ridge National Laboratory, Oak Ridge, Tennessee.

Hultberg HA, Andersson BI, Moldan F (1993) The covered catchment—An experimental approach to reversal of acidification in a forest ecosystem. In Rasmussen L, Brydges T, Mathy P (Eds) *Experimental manipulations of biota and biogeochemical cycling in ecosystems.* Ecosystem Research Report No. 4. Commission of the European Communities, Brussels, Belgium, pp 46–54.

Hurlbert SH (1984) Pseudoreplication and the design of ecological field experiments. Ecol Monogr 54:187–211.

Iverson LR, Prasad AM (2001) Potential changes in tree species richness and forest community types following climate change. Ecosystems 4:186–199.

Jenkins A, Wright RF (1995) The CLIMEX project: performance of the experimental facility during the first year of treatment. In Jenkins A, Ferrier RC, Kirby C (Eds) *Ecosystem manipulation experiments: Scientific approaches, experimental design and relevant results,* Ecosystem Research Report #20. Commission of the European Communities, Brussels, Belgium, pp 323–327.

Johnson DW, Lindberg SE (1989) Acidic deposition on Walker Branch Watershed. In Adriano DC, Havas M (Eds) *Acidic precipitation,* Volume 1, Springer-Verlag, New York, pp 1–38.

Johnson DW, Lindberg SE (Eds) (1992) *Atmospheric deposition and forest nutrient cycling: A synthesis of the Integrated Forest Study.* Springer–Verlag, New York.

Johnson DW, VanHook RI (Eds) (1989) *Analysis of biogeochemical cycling processes in Walker Branch Watershed.* Springer-Verlag, New York.

Joslin JD, Wolfe MH (1998) Impacts of long-term water input manipulations on fine root production and mortality in mature hardwood forests. Plant Soil 204:165–174.

Jones EA, Reed DD, Mroz GD, Liechty HO, Cattelino PJ (1993) Climate stress as a precursor to forest decline: paper birch in northern Michigan, 1985–1990. Can J For Res 23:229–233.

Kelly JM, Meagher JF (1986) Nitrogen input/output relationships for three forested watersheds in eastern Tennessee, In Correl DL (Ed) *Watershed research perspectives.* Smithsonian Institution Press, Washington, District of Columbia, pp 360–391.

Kramer PJ (1983) *Water relations of plants.* Academic Press, New York.

Lindberg SE, Lovett GM, Richter DD, Johnson, DW (1986) Atmospheric deposition and canopy interactions of major ions in a forest. Science 231:141–145.

Luxmoore RJ (1983) Water budget of an eastern deciduous forest stand. Soil Sci Soc Am J 47:785–791.

McMaster WM (1967) *Hydrologic data for the Oak Ridge Area Tennessee.* Geologic Survey Water-Supply Paper 1839-N. US Government Printing Office, Washington, District of Columbia.

Moldan F, Hultberg H, Andersson A (1995) Gårdsjön covered catchment experiment: changes in runoff after three years of throughfall exclusion. In Jenkins A, Ferrier RC, Kirby C (Eds) *Ecosystem manipulation experiments: Scientific approaches, experimental design and relevant results,* Ecosystem Research Report #20. Commission of the European Communities, Brussels, Belgium, pp 113–118.

Myers BJ (1988) Water stress integral—A link between short-term stress and long-term growth. Tree Physiol 4: 315–323.

National Atmospheric Deposition Program [NADP] (2000) Annual report. http://nadp.sws.uiuc.edu/lib/program /apr2000.pdf

Nepstad DC, Moutinho P, Dias-Filho MB, Davidson E, Cardinot G, Markewitz D, Figueiredo R, Vianna N, Lefebvre P, Ray D, Chambers J, Sternberg L, Moreira M, Guerreiros JB, Barros L, Ishida FY, Belk E, Schwalbe K (2002) The effect of partial throughfall exclusion on canopy processes and biogeochemistry of an Amazon forest. J Geophys Res 107 (D20):8085–8085.

Peters LN, Grigal DF, Curlin JW, Selvidge WJ (1970) *Walker Branch Watershed Project: Chemical, physical and morphological properties of the soils of Walker Branch Watershed.* ORNL-TM-2968. Oak Ridge National Laboratory, Oak Ridge, Tennessee, pp 32–33.

Rasmussen L, Beier C, Andersen BR (1995) Effects of experimental manipulations with biogeochemical cycles in a Norway spruce plantation in Denmark: The Danish part of the EXMAN project. In Jenkins A, Ferrier RC, Kirby C (Eds) *Ecosystem manipulation experiments: Scientific approaches, experimental design and relevant results.* Ecosystem Research Report #20. Commission of the European Communities, Brussels, Belgium, pp 96–103.

Swank WT, Crossley DA Jr. (1988) *Forest hydrology and ecology at Coweeta.* Springer-Verlag, New York.

Topp GC, Davis JL (1985) Measurement of soil water content using time domain reflectometry (TDR): a field evaluation, Soil Sci Soc Amer J 49:19–24.

2. Carbon-Cycle Processes

3. Deciduous Hardwood Photosynthesis: Species Differences, Temporal Patterns, and Responses to Soil-Water Deficits

Kell B. Wilson and Paul J. Hanson

Changes in regional precipitation patterns will directly impact soil and foliage water status, resulting in physiological modifications in trees that can affect carbon assimilation rates (Briggs et al. 1986; Teskey et al. 1986; Ni and Pallardy 1992). Decreasing water potentials in the root and/or foliage can affect the carbon assimilation by adjusting stomatal conductance (Hinckley et al. 1978a; Epron and Dreyer 1993; Lowenstein and Pallardy 1998) or possibly by directly impacting the biochemical potential for carbon assimilation in the leaf (Lawlor 1995; Escalona et al. 1999). Tree species differ in their morphology and use of physiologic adaptations to avoid the negative impacts of drought on carbon assimilation (Hinckley et al. 1978b; Bahari et al. 1985; Briggs et al. 1986; Ni and Pallardy 1992; Abrams and Mostoller 1995; Lowenstein and Pallardy 1998; Tschaplinski et al. 1998). Species-specific differences in gas-exchange response to drought depend on characteristics such as rooting depth, stomatal sensitivity, osmotic adjustment, cavitation avoidance, and increased tolerance of desiccation (Abrams 1990; Lowenstein and Pallardy 1998; Tshaplinski et al. 1998). For example, because of their deep rooting habit, *Quercus* species are expected to outcompete the more mesic *Acer* and *Cornus* species in dry climates (Hinckley et al. 1979; Bahari et al. 1985; Abrams 1990). Species also may differ in their ability to recover predrought photosynthetic rates following a precipitation event (Ni and Pallardy 1992). Because of shallower rooting depths and limited water-storage capacity, photosynthesis of understory species and saplings is expected to be more sensitive to drying when compared to large overstory trees (Donovan and Ehleringer 1991; Flanagan et al. 1992).

In this chapter, we describe leaf gas-exchange measurements performed in the understory and overstory within the Throughfall Displacement Experiment (TDE) plots and adjacent forest. In addition to quantifying species differences in maximum assimilation rate, this chapter discusses species differences in response to drought in the following contexts: (1) response to TDE treatment, (2) year-to-year differences in response, and (3) differences in understory versus overstory trees. The reader is referred to articles by Larcher (1969), Ceulemans and Saugier (1991), Sullivan et al. (1996), Abrams (1998), and Wilson et al. (2000a, 2000b) for summaries of the photosynthetic characteristics of a wide range of tree species common to the North American temperate deciduous forest.

Methods

From 1994 through 1996, instantaneous light-saturated assimilation and foliar conductance were measured periodically on attached leaves of two species of understory saplings (*Acer rubrum* and *Cornus florida*), and detached mid- to upper-canopy leaves of three common overstory species (*Acer rubrum, Quercus alba,* and *Quercus prinus*). Measurements were conducted at seasonal intervals throughout 1994 and 1995 and periodically depending on soil-water status in 1996. Foliar gas-exchange measurements conducted from 1994 to 1996 were made with a portable gas-exchange system employing a 1-L chamber (Li-6200; LiCor Inc. Lincoln, Nebraska) that allows for instantaneous observations of CO_2 and H_2O exchange. For each of the sapling and tree species targeted, measurements on 6–10 individuals per treatment/slope position were attempted. Observations were made during the period from 0900 to 1500 to minimize diurnal impacts.

In 1997 and 1998, tower platforms were used to access upper-canopy foliage in each treatment of the TDE, offering a unique opportunity to directly evaluate the effects of soil-water content on foliar gas-exchange of mature deciduous trees in a natural environment. One tower was located on each of the three TDE treatments, allowing access to nine trees at heights exceeding 20 m. One additional *Q. alba* tree was measured from a nearby meteorological tower along the ridge less than 2 km from the TDE experiment. Although not all leaves were at the very top of the canopy, most leaves were in the sunlit or primarily sunlit portion of the canopy. On each tree, an attempt was made to sample on the same branch or branches at similar heights throughout the measurement period. Measured species include *Acer rubrum, Acer saccharum, Nyssa sylvatica, Quercus prinus,* and *Quercus alba.* These species comprise more than 75% of the overstory basal area in the TDE forest (Hanson et al. 2001). Understory saplings (1–5 m in height) were also sampled, within the three TDE plots and surrounding the meteorological tower. Nearly all of the important understory species were measured: *Acer rubrum, Carya* spp., *Cornus florida, Fagus grandifolia, Liriodendron tulipifera, Nyssa sylvatica, Sassafras albidum, Oxydendrum arboreum, Prunus serotina,* and *Quercus rubra.* The large majority of gas-exchange measurements in 1997 and 1998 were made with a steady-state system (LI6400 Li-Cor). Analyzers were regularly calibrated with CO_2 and H_2O free air and known standards. Measurements were performed during the 1997 and 1998 growing seasons on nearly 800 leaves on 65 trees encompassing 12 species.

Gas-exchange measurements were primarily done to evaluate light-saturated assimilation and photosynthetic capacity. Light-response curves indicated that when photosynthetically active radiation was at least 1100 μmol m^{-2} s^{-1}, photosynthesis was at least 95% of its maximum rate. Typically, each leaf was initially measured at near saturating light (>1100 μmol m^{-2} s^{-1}) and a reference "ambient" carbon dioxide concentration, which was always set to 360 μmol mol^{-1}. Following this light-saturated measurement at near-ambient CO_2, temperature, light, or carbon dioxide concentration in the leaf cuvette were sometimes manipulated. The response of assimilation to changes in internal-leaf CO_2 concentration (i.e., A/C_i response curves) were evaluated to examine the model parameters that indicate internal biological limitations on photosynthetic capacity (Farquhar et al. 1980), especially the maximum rate of carboxylation (V_{cmax}; Wilson et al. 2000a). Similar to other studies on *Acer* and *Quercus* trees (Briggs et al. 1986), temperature-response surfaces indicated a broad optimum. Temperatures at the time light-saturated assimilation rates were obtained was typically 25 ± 2.5°C and always between 20°C and 30°C, the broad range over which P_{sat} was stable. V_{cmax} estimates were normalized to 25°C (Wilson et al. 2000a). Foliar gas-exchange measurements were expressed per unit leaf area. Total leaf nitrogen concentration (%, g/g) of measured leaves was determined with a Carlo-Erba NA1500 Analyzer (Wilson et al. 2000a), but adjusted for leaf mass per unit area to yield leaf N per unit leaf area (N_{area}).

Results

Species Maximum Rates and Photosynthetic Capacity

Table 3.1 provides a summary of the light-saturated maximum net photosynthesis rate (P_{sat}), maximum carboxylation rate at 25°C (V_{cmax}), stomatal conductance

Table 3.1. For dominant trees, measured light-saturated photosynthetic rate P_{sat} (μmol m^{-2} s^{-1}), maximum rate of carboxylation V_{cmax} (μmol m^{-2} s^{-1}), light-saturated stomatal conductance g_{smax} (mol m^{-2} s^{-1}), and leaf N per unit leaf area (N_{area}, g m^{-2}) during the early summer (June 1 to July 15) for five species located on the TDE plots

Tree	P_{sat}	V_{cmax}	g_{smax}	No. of trees	N_{area}	n
Acer rubrum	9.6	37.6	0.136	3	1.50	34
Acer saccharum	10.4	42.8	0.150	1	1.50	9
Nyssa sylvatica	9.7	39.2	0.128	1	0.95	6
Quercus alba	15.1	60.0	0.284	2	1.99	45
Quercus prinus	16.0	63.1	0.284	2	1.92	24
Mean	12.2	48.5	0.196	1.8	1.57	23.6

Note: Data were collected at 25 ± 2.5°C, and V_{cmax} was normalized to 25°C (Wilson et al. 2000a). Also shown are the number of trees (# trees) and total number of leaves sampled (n) during the early summer period.

(g_{smax}), and leaf nitrogen per unit area (N_{area}) of the five dominant/subdominant species measured during the early summer before the effects of soil-water deficits or leaf aging were important (Wilson et al. 2000b). P_{sat}, V_{cmax}, g_{smax}, and N_{area} were greater for the two *Quercus* species ($P_{sat} \approx$ 15–16 μmol m^{-2} s^{-1}) than for the two *Acer* species ($P_{sat} \approx$ 10 μmol m^{-2} s^{-1}). For all species, the maximum carboxylation rate (V_{cmax}) at 25°C was approximately four times greater than the light-saturated net photosynthesis rate (P_{sat}), and there was a general positive correlation among N_{area}, V_{cmax}, P_{sat}, and g_{smax}.

Response to Drought

Gas-exchange responses to drought were evaluated in four ways: (1) differences among TDE treatments, (2) differences of all treatments between a nondrought and a drought year (1997 and 1998, especially), (3) species differences, and (4) differences between sapling and mature trees.

TDE and Seasonal Responses (1994–1996)

Overstory gas-exchange characteristics were not influenced by the TDE treatments in 1994 and 1996 when soil-water potentials of the upper soil (0–35 cm) were always greater than −0.6 MPa. However, at the depth of the 1995 drought, light-saturated photosynthesis of overstory tree foliage was 55% greater and 28% lower for the wet and dry treatment plots when compared to foliage on the control plot (data not shown). Foliar assimilation rates measured prior to 1997 were obtained from small branches shot from the canopy. Although every attempt was made to measure the leaves attached to these branches as soon as they reached the ground, a comparison of light-saturated assimilation values for the shotgun-sampled overstory leaves from 1994 to 1996 were lower than those described below measured for intact leaves. Therefore, pre-1997 canopy data are presented here in relative terms only.

Understory light-saturated photosynthetic rates of intact *A. rubrum* and *C. florida* saplings from 1994 to 1996 were consistently lower than rates for the overstory leaves (Fig. 3.1 versus Table 3.1). In September 1994, significant reductions in P_{sat} of *A. rubrum* and significant reductions in P_{sat} and g_{smax} of *C. florida* were observed for the dry treatment plot with respect to ambient and wet plot saplings (data not shown). In addition, predrought and drought data for understory sapling foliage in 1995 illustrate their responsiveness to available soil moisture (Fig. 3.1) Photosynthetic rates of wet plot saplings remained higher than rates for ambient or dry plot saplings, and photosynthetic rates were reduced by ~50% or more for both *A. rubrum* and *C. florida* species. Data for sapling leaf conductance showed a pattern similar to that in Fig. 3.1, suggesting that much of the drought-driven reduction in P_{sat} could be explained by stomatal closure (not shown). Rapid increases in sapling net photosynthesis with precipitation inputs following drought suggested that sapling foliar physiological function was very sensitive to changes in shallow soil moisture levels (1994–1996 data not shown, but see Fig. 3.4).

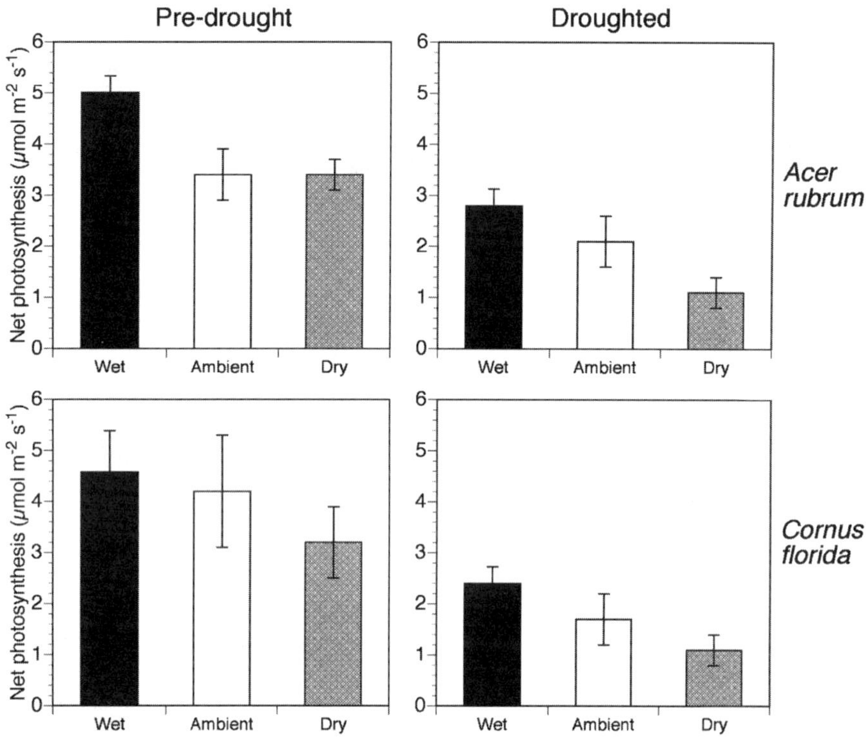

Figure 3.1. Light-saturated net photosynthesis of *A. rubrum* and *C. florida* saplings in 1995 for predrought and drought conditions.

TDE and Seasonal Responses (1997–1998)

Temporal patterns of the mean light-saturated assimilation rate for overstory foliage throughout 1997 and 1998 are shown in the first column of Fig. 3.2. A mean assimilation rate was not available for the "dry" treatment in 1997 because measurements were not performed on the dry plot *Q. prinus* in that year. A modest late summer decline in assimilation occurred in 1997, especially on the ambient treatment, although there were not significant soil-water deficits (Chapter 2, this volume, Fig. 2.11). Independent of treatment, light-saturated assimilation rates in 1998 were reduced during the period of most significant soil-water deficits (primarily between days 230 and 265, when soil-water potential was often less than about −1.2 MPa). Light-saturated assimilation rates during this period were only about one-half those during the similar period in 1997. Although the decline in P_{sat} during 1998 occurred earlier in the year for the dry compared to the ambient and wet plots, differences between the two years were more noticeable than differences between treatments. There was also little difference in the timing and magnitude of decline between the wet and ambient treatments. Assimilation rates recovered on all treatments following precipitation inputs sufficient to eliminate the severe soil-water deficits.

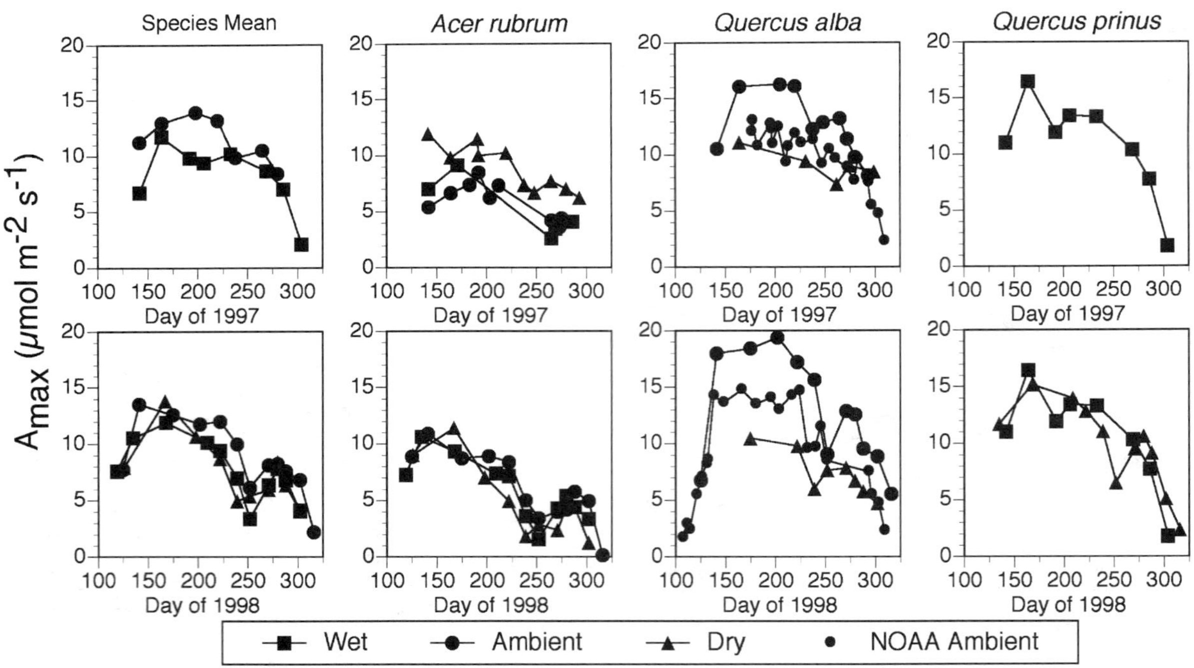

Figure 3.2. Mean light-saturated leaf assimilation rates for mean canopy, *A. rubrum, Q. alba,* and *Q. prinus* foliage during 1997 (top row) and 1998 (bottom row).

Statements concerning treatment effects from the 1997 and 1998 data are inconclusive because of limited sampling and the differences in species accessible from the canopy access towers. Limited access to intact leaves on multiple mature trees in the upper canopy was unavoidable, but the problem of differences in species accessibility can be addressed to some extent by comparing the mean assimilation rates when individual species were accessible across treatments (see further discussion below). We were also able to assess the impact of drought on canopy foliar processes from landscape-scale eddy covariance measures of net ecosystem CO_2 exchange (NEE) (Wilson et al. 2001). NEE is dominated by foliar processes of the forest canopy during mid-summer periods and can serve as another indication of foliar responses to soil moisture deficits. Figure 3.3 shows the relationship between NEE and photosynthetically active radiation based on eddy covariance data for a 2-week wet period in 1997 (days 231–245) and a 2-week drought period in 1998 (days 238–252). The observations demonstrate that the severe summer drought of 1998 reduced NEE by ~50% at high light levels. Because the 1998 growing season included the most severe drought for the 1993–

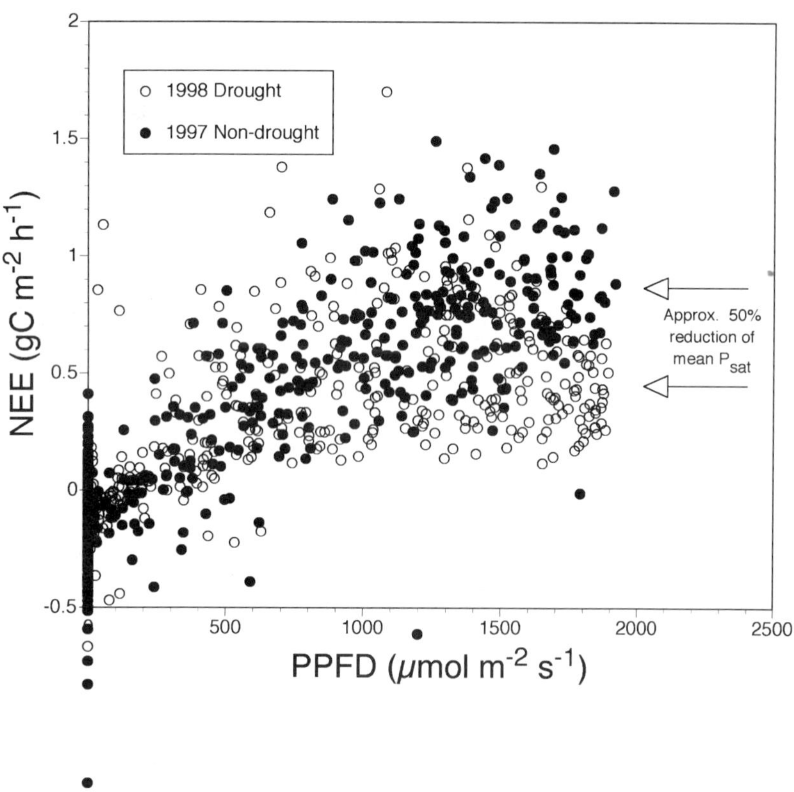

Figure 3.3. Mean hourly eddy covariance data for net ecosystem exchange of CO_2 (NEE) as a function of photosynthetic photon flux density (PPFD) for biweekly periods in 1998 (days 238–252) and 1997 (days 231–245).

2000 period (Chapter 2, this volume), a reduction of midday canopy photosynthesis of 50% may approximate the maximum drought response for upland oak systems growing on deep soils. It is important to note that the TDE forest canopy is made up of multiple species, some of which experience different responses than the canopy mean captured in the eddy covariance data (Fig. 3.3). Individual species responses are discussed further in the next subsection.

Species-Specific Seasonal Responses

Species-specific responses for *A. rubrum, Q. alba,* and *Q. prinus* are shown in Fig. 3.2. A single *A. rubrum* tree was accessible from each of the towers. P_{sat} of *A. rubrum* generally decreased in all treatments following day 200 (in the middle of July). Because water potential was high for most of 1997, the decrease during the summer of that year is presumably related to leaf aging (Wilson et al. 2000a, 2000b). Accelerated seasonal reductions in P_{sat} in 1998 were driven by drought conditions. Reductions of P_{sat} during the 1998 drought were reversible with subsequent recovery occurring prior to leaf senescence (days 250–300 in Fig. 3.2). An evaluation of the *A. rubrum* responses by treatment suggested that the decrease appeared sooner in the wet plot compared to the ambient and dry plots. Two mature *Q. prinus* trees were measured on the wet and dry treatments, but only the wet plot tree was measured during 1997 and 1998 (Fig. 3.2). During 1998, the assimilation rate of the *Q. prinus* trees showed a seasonal decline, and during the drought period (days 225–275), the decrease was greater for the tree exposed to the dry treatment. Three *Q. alba* trees were measured; one in the ambient TDE treatment, one in the dry TDE treatment and another accessible from a nearby meteorological tower (Fig. 3.2). P_{sat} for *Q. alba* decreased modestly in the late summer during both years, but the decline was greater during the most severe period of drought in 1998. Reductions in assimilation during the drought were not as large for the *Q. alba* trees compared to the *A. rubrum* and *Q. prinus* trees. Not shown were single *N. sylvatica* and *A. saccharum* trees located on the dry and the wet plots, respectively. P_{sat} was highly sensitive to drought for both species (Wilson et al. 2000b).

Sapling Versus Mature Tree Responses

Table 3.2 lists some gas exchange characteristics of understory species during the early growing season when assimilation rates were largest (generally mid-April through May). Light-saturated photosynthesis, the maximum rate of carboxylation, and stomatal conductance were lower in understory saplings relative to mature trees (Tables 3.1 and 3.2). P_{sat} of understory species was typically between 5 and 8 μmol m^{-2} s^{-1}, or about one-half that of the overstory. N_{area} and the model parameter V_{cmax} was also approximately one-half or less for sapling foliage relative to overstory foliage. Although limited replication precludes a comprehensive analysis of species differences in foliar gas-exchange characteristics, the data

Table 3.2. For understory species, measured light-saturated photosynthetic rate P_{sat} (μmol m^{-2} s^{-1}), maximum rate of carboxylation V_{cmax} (μmol m^{-2} s^{-1}), light-saturated stomatal conductance g_{smax} (mol m^{-2} s^{-1}), and leaf N per unit leaf area (N_{area}, g m^{-2})

Tree	P_{sat}	V_{cmax}	g_{smax}	No. of trees	N_{area}	*n*
Acer rubrum	5.3	18.1	0.096	6	0.69	10
Carya sp.	7.3	26.1	0.134	3	1.04	4
Cornus florida	5.7	20.3	0.144	4	NA	7
Fagus grandifolia	4.8	15.2	0.169	1	0.60	7
Liriodendron tulipifera	6.2	23.3	0.111	1	0.77	6
Oxydendrum arboreum	5.7	21.2	0.097	2	0.66	6
Prunus serotina	8.3	29.9	0.147	2	0.87	5
Quercus rubra	5.5	20.8	0.104	3	0.91	9
Quercus velutina	6.2	23.2	0.096	4	0.99	5
Sassafras albidum	6.1	22.3	0.114	3	0.61	6
Mean	6.2	22.2	0.119	2.8	0.79	6.4

Note: Maximum rates reported in this table often occurred earlier in the season (April and May) than did those for the overstory species (Table 3.1).

indicated that the difference in P_{sat} between *Acer* and *Quercus* was less in the understory compared to the overstory.

Figure 3.4 shows the temporal changes in sapling P_{sat} during the 1998 growing season normalized to the maximum species-specific values shown in Table 3.2. The highest P_{sat} values occurred very early in the season for some species (May to June), and although not all species were measured during this late-spring pe-

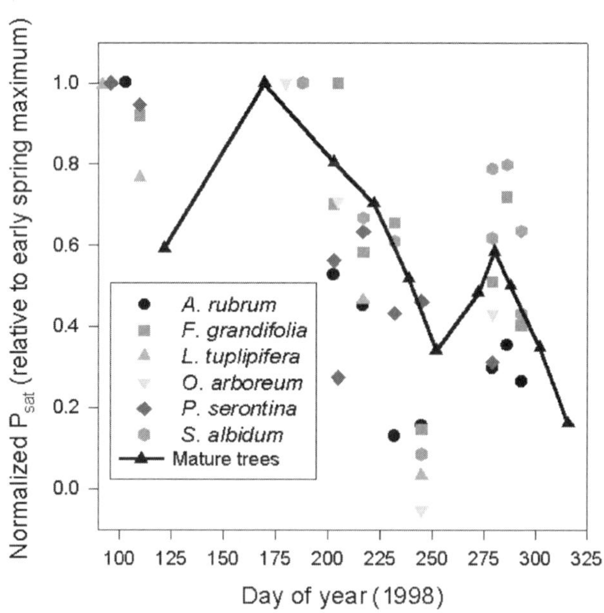

Figure 3.4. Seasonal pattern of the normalized light-saturated assimilation rate of understory species in 1998. Assimilation rates were normalized to the respective species maximum. The line represents the mean normalized rate for overstory species.

riod, we expect this to be the normal pattern for all species. For all species, the maximum rates shown in Table 3.2 were not maintained throughout the season and were often reduced to rates that were much less than one-half these maximum values during drought. P_{sat} for understory species generally declined sooner and to a greater extent during the drought than the seasonal decline for mature species (mean decline for mature species indicated by the line in Fig. 3.4). By day 200 in 1998, the relative decline in P_{sat} for the understory species was greater than that for the mature trees on the TDE plots. The greater sensitivity to drought in understory species continued throughout the remainder of the 1998 drought and was especially pronounced during the most severe portion of the soil drying, when assimilation rates were typically less than 20% of maximum and were often virtually zero (days 225–275; Fig. 3.4). In contrast, mature trees maintained a mean P_{sat} rate that was more than one-third the May–June maximum. Similar to the mature species, P_{sat} for saplings recovered following precipitation and replenished soil-water content (Fig. 3.4; days 275–300) prior to fall senescence.

Discussion

Maximum Light-Saturated Assimilation Rates

Gas-exchange measurements on the TDE plots, which indicate that *Quercus* species have a greater P_{sat} than co-occurring *Acer* trees, were consistent with most previous research (Hinckley et al. 1978b; Bahari et al. 1985; Abrams 1998). P_{sat} of *Quercus* species in the TDE study was ~55% greater than P_{sat} of *Acer,* within the range (50–70%) observed in other studies on mature deciduous trees in the eastern and midwestern United States (Reich et al. 1991; Bassow and Bazzaz 1998). Similar results were found for an Appalachian highland forest with some of the same *Quercus* and *Acer* species studied here (Sullivan et al. 1996). In contrast, P_{sat} for *Q. rubra* and *A. rubrum* were more similar in a Michigan forest (Briggs et al. 1986). However, the Michigan study was during a drought and *Q. rubra* is known to be more sensitive to drought that other *Quercus* species (Abrams 1990).

V_{cmax} was ~50% greater in *Quercus* than *Acer,* consistent with data for *Q. alba* and *A. saccharum* in Wullschleger (1993). The values of V_{cmax} for the TDE were about 70% greater than mean values reported in a comprehensive review (Wullschleger 1993), but the review included data for saplings and greenhouse plants which would have lowered the mean. The greater photosynthetic capacity and stomatal conductance of *Quercus* trees is correlated with the propensity of *Quercus* to have greater leaf nitrogen per unit area than *Acer,* both in this study and others (Reich 1991; Sullivan et al. 1996; Abrams 1998; Wilson et al. 2000a). When identical species are compared, photosynthetic rates from the TDE study were nearly twice that reported for a North Carolina hardwood forest (Sullivan et al. 1996), but the relative differences among species were similar. P_{sat} of *A. rubrum* reported for the TDE was slightly greater than was observed in a Massachusetts forest (Bassow and Bazzaz 1998), and the seasonal period of maximum P_{sat} was displaced 2–3 months (May or June in the TDE study versus August in Massachusetts).

Values of P_{sat} for understory species in natural closed canopies are also consistent with previous studies (Hinckley et al. 1978a; Jurik 1986; Naumburg and Ellsworth 2000), although lower maximum rates have been reported (Abrams 1998; Naumburg and Ellsworth 2000). Light-saturated assimilation rates in saplings growing without an overstory can be slightly greater than beneath closed canopies, but in open and closed canopies, the P_{sat} for *Acer* is generally less than for other species, such as *Carya, Quercus, Sassafras,* and *Prunus* (Hinckley et al. 1978a; Bahari et al. 1985; Abrams and Mostoller 1995; Kubiske and Abrams 1994; Abrams 1998). Hinckley et al. (1978a) and Abrams (1998) note that *Acer* can be more efficient than other species at low-light conditions common to the understory. Light-saturated assimilation is less meaningful for understory species because their leaves are frequently shaded, are rarely light saturated, and are often not in steady-state conditions when sunflecks occur (Naumburg and Ellsworth 2000). Because leaves are often shaded and the leaf area is low in the understory, the relative contribution of the understory to total canopy carbon assimilation is much less than that indicated by the relative differences in light-saturated photosynthetic rates provided in Tables 3.1 and 3.2.

Drought Response

As is the case with other biochemical and growth responses analyzed on the TDE, gas-exchange characteristics are influenced more by year-to-year differences in climate than by the drought treatment effects within years (Tschaplinski et al. 1998; Hanson et al. 2001; also Chapters 8 and 15, this volume). Assimilation rates generally decreased in all treatments during the drought in 1998. Smaller reductions during the late summer in 1997 are attributed primarily to normal leaf aging processes (Wilson et al. 2000a, 2000b). Limitation analyses (Wilson et al. 2000b) suggested that the decline from drought is primarily through stomatal control, but that smaller reductions resulting from nonstomatal limitations could not be ruled out.

A conclusive evaluation of how species respond differently to drought was not possible because of the limited replication of mature trees in this study. However, the observations in this study can be compared with studies in more controlled environments where sampling was less of a problem. It is commonly observed that *Quercus* species are effective at avoiding and/or tolerating drought (Dougherty and Hinckley 1980; Bahari et al. 1985; Hinckley et al. 1978b; Abrams 1990). *Quercus* trees have deep-penetrating root systems, ring-porous anatomy, and often have high osmotic-adjustment potential (Abrams 1990). Based on osmotic considerations, Tschaplinski et al. (1998) suggested that *Quercus* on the TDE plots were more drought tolerant than *A. rubrum* and *A. saccharum.* Consistent with this conclusion, gas-exchange characteristics of *Q. alba* were more resilient to drought than were those of *A. rubrum*. The two *Q. prinus* trees (one each on the wet and dry plots) demonstrated significant declines in assimilation during drought, but in earlier studies, *Q. prinus* has been shown to have features reducing drought effects (Abrams 1990). *Quercus prinus* also exhibited osmotic adjustment on the TDE manipulations (Tschaplinski et al. 1998), but osmotic adjustment is not always correlated with an overall drought response (Abrams 1990; Tschaplinski et al. 1998). Gas-exchange in *A. rubrum* has been shown to be more susceptible to drought than *Quercus* in other studies as well (Hinckley et al. 1978a; Bahari et al. 1985).

The decreased ability to extract water at greater soil depths and the lack of large water-storage capacity in trunks are likely causes of the more severe and earlier decline in assimilation of understory species. The earlier and larger decline in assimilation for saplings is also consistent with lower leaf water potential in understory trees at TDE during the drought of 1995 (Tschaplinski et al. 1998). A sap flow study of understory *A. rubrum* on the TDE during the 1995 drought also indicated that this species was highly susceptible to drought (Wullschleger et al. 1998; Chapter 6, this volume). In that study, transpiration was affected by the TDE treatment, a conclusion that is consistent with leaf conductance data observed both before and after the 1995 drought periods (data not shown). Throughout the 8-year TDE study period, soil-water potentials of the deep soil profile (35–70 cm) never fell below −1 MPa, and with the exception of the 1998 growing season, soil-water potentials in the lower soil profile never fell below −0.5 MPa (Fig. 2.11, Chapter 2, this volume). Clearly, established saplings and trees that support significant root systems deep within the soil profile have the potential to sustain photosynthetic capacity during dry periods. This places those plants and/or species at a distinct competitive advantage with others not exhibiting these characteristics.

References

Abrams MD (1990) Adaptations and responses to drought in *Quercus* species of North America. Tree Physiol 7: 227–238.

Abrams MD (1998) The red maple paradox. Bioscience 48:355–364.

Abrams MD, Mostoller SA (1995) Gas exchange, leaf structure and nitrogen in contrasting successional tree species growing in open and understory sites during a drought. Tree Physiol 15:361–370.

Bahari ZA, Pallardy SG, Parker WC (1985) Photosynthesis, water relations, and drought adaptation in six woody species of oak-hickory forests in central Missouri. For Sci 31:557–569.

Bassow SL, Bazzaz FA (1998) How environmental conditions affect canopy leaf-level photosynthesis in four deciduous tree species. Ecology 79:2660–2675.

Briggs TW, Jurik TW, Gates DM (1986) Non-stomatal limitation of CO_2 assimilation in three tree species during natural drought conditions. Physiol Plant 66:521–526.

Ceulemans RJ, Saugier B (1991) Photosynthesis. In Raghavendra AS (Ed) *Physiology of trees,* John Wiley & Sons, New York, pp 21–50.

Donovan LA, Ehleringer JR (1991) Ecophysiological differences among juvenile and reproductive plants of several woody species. Oecologia 86:594–597.

Dougherty PM, Hinckley TM (1980) The influence of a severe drought on net photosynthesis of white oak (*Quercus alba*). Can J Bot 59:335–341.

Epron D, Dreyer E (1993) Photosynthesis of oak leaves under water stress: Maintenance of high photochemical efficiency of photosystem II and occurrence of non-uniform CO_2 assimilation. Tree Physiol 13:107–117.

Escalona JM, Flexas J, Medrano H (1999) Stomatal and non-stomatal limitations of photosynthesis under water stress in field-grown grapevines. Aust J Plant Physiol 26:421–433.

Farquhar GD, von Caemmerer S, Berry JA (1980) A biochemical model of photosynthetic CO_2 assimilation in leaves of C_3 species. Planta 149:78–90.

Flanagan LB, Ehleringer JR, Marshall JD (1992) Differential uptake of summer precipitation among co-occurring trees and shrubs in a pinyon juniper woodland. Plant Cell Environ 15:831–836.

Hanson PJ , Todd DE, Amthor JS (2001) A six year study of sapling and large-tree growth and mortality responses to natural and induced variability in precipitation and throughfall. Tree Physiol 21:345–358.

Hinckley TM, Aslin RG, Aubuchon RR, Metcalf CL, Roberts JE (1978) Leaf conductance and photosynthesis in four species of the oak-hickory forest type. For Sci 24:73–84.

Hinckley TM, LassoieJP, Running SW (1978) Temporal and spatial variations in the water status of forest trees. For Sci 24:1–72.

Hinckley TM, Dougherty PM, Lassoie JP, Roberts JE, Teskey RO (1979) A severe drought: Impact on tree growth, phenology, net photosynthetic rate and water relations. Am Midl Nat 102:307–316.

Jurik TW (1986) Seasonal patterns of leaf photosynthetic capacity in successional northern hardwood tree species. Am J Bot 73:131–138.

Kubiske MD, Abrams MD (1994) Ecophysiological analysis of woody species in contrasting temperate communities during wet and dry years. Oecologia 98:303–312.

Larcher W (1969) The effect of environmental and physiological variables on the carbon dioxide gas exchange of trees. Photosynthetica 3:167–198.

Lawlor DW (1995) The effects of water deficit on photosynthesis. In Smirnoff N (Ed) *Environment and plant metabolism, Flexibility and acclimation.* Bios Scientific Publishers, Oxford, England, pp 129–160.

Loewenstein NJ, Pallardy SG (1998) Drought tolerance, xylem sap abscisic acid and stomatal conductance during soil drying: A comparison of canopy trees of three temperate deciduous angiosperms. Tree Physiol 18:431–439.

Naumberg E, Ellsworth DS (2000) Photosynthetic sunfleck utilization potential of understory saplings growing under elevated CO_2 in FACE. Oecologia 122:163–174.

Ni R, Pallardy SG (1992) Stomatal and nonstomatal limitations to net photosynthesis in seedlings of woody angiosperms. Plant Physiol 99:1502 1508.

Reich PB, Walters MB, Ellsworth DS (1991) Leaf age and season influence the relationships between leaf nitrogen, leaf mass per area and photosynthesis in maple and oak trees. Plant Cell Environ 14:251–259.

Sullivan NH, Bolstad PV, Vose JM (1996) Estimates of net photosynthetic parameters in mature forests of the Southern Appalachians. Tree Physiol. 16:397–406.

Teskey RO, Fites JA, Samuelson LJ, Bongarten BC (1986) Stomatal and nonstomatal limitations to net photosynthesis in *Pinus taeda* L. under different environmental conditions. Tree Physiol 2:131–142.

Tschaplinski TJ, Gebre GM, Shirshac TL (1998) Osmotic potential of several hardwood species as affected by manipulation of throughfall precipitation in an upland oak forest during a dry year. Tree Physiol 18:291–298.

Wilson KB, Baldocchi DD, Hanson PJ (2000a) Spatial and seasonal variability of photosynthetic parameters and their relationship to leaf nitrogen in a deciduous forest. Tree Physiol 20:565–578.

Wilson KB, Baldocchi DD, Hanson PJ (2000b) Quantifying stomatal and non-stomatal limitations to carbon assimilation resulting from leaf aging and drought in mature deciduous tree species. Tree Physiol 20:787–797.

Wilson KB, Baldocchi DD, Hanson PJ (2001) Leaf age affects the seasonal pattern of photosynthetic capacity and net ecosystem exchange of carbon in a deciduous forest. Plant Cell Environ 24:571–583.

Wullschleger S (1993) Biochemical limitations to carbon assimilation in C_3 plants—a retrospective analysis of the A/C_i curves from 109 species. J Exp Bot 44:907–920.

Wullschleger S, Hanson PJ, Tchaplinski TJ (1998) Whole-plant water flux in understory red maple exposed to altered precipitation regimes. Tree Physiol 18:71–79.

4. Aboveground Autotrophic Respiration

Nelson T. Edwards and Paul J. Hanson

Understanding the autotrophic respiratory activity of living tissues in a forest provides useful insights into the responses of the forest to its environment. Respiration measurements also contribute essential data for evaluating and constructing carbon budgets and models. Historically, studies have shown that the primary factors that control autotrophic respiration include temperature, phenology, and species. More recently, studies by Amthor (1989), Ryan (1995), Reich et al. (1998), Mitchell et al. (1999), Bolsted et al. (1999), and Maier (2001) have shown that nitrogen content plays a key role in autotrophic respiration rates. Aboveground autotrophic respiration can be conveniently divided into woody-tissue respiration and foliar respiration. Foliar respiration accounts for from about 30% of aboveground autotrophic respiration in some temperate deciduous forests (Edwards et al. 1981) to as much as 75% in some forests, such as in northern coniferous sites (Lavigne et al. 1997). Woody-tissue respiration is important because the bulk of the biomass in a mature forest stand is in woody tissue, and respiration in the living cells of this tissue occurs continuously, even when trees are "dormant." Woody-tissue respiration is also a good indicator of growth phenology and can be used to calculate growth rates if the respiration required for maintenance is taken into consideration. Woody-tissues account for from about 25% of the total aboveground autotrophic respiration in some forests, such as in northern coniferous sites (Lavigne et al. 1997), to more than half of the aboveground autotrophic respiration in some temperate deciduous forests (Edwards et al. 1981).

Maintenance Versus Growth Respiration

Respiration is generally partitioned into two functional components: growth respiration (R_g) and maintenance (R_m) respiration (McCree 1970). However, this is an oversimplification because growth and maintenance may share some biochemical reactions, and, in fact, some respiration may occur without benefit to the plant (i.e., heat and CO_2 are produced with no contribution to growth or maintenance) (Amthor 2000). Growth respiration involves O_2 consumption and CO_2 production resulting from energy production that contributes both directly (biosynthesis) and indirectly (e.g., translocation of substrates used in growth and ion uptake) to the production of new tissues. R_m involves gas exchange during energy-requiring processes within living cells, including protein turnover, regulation of ion and metabolite gradients, and physiological adaptation to a changing environment (Penning de Vries 1975). Partitioning these two components of respiration is important because R_g varies greatly through the year and rates vary with phenology and environmental factors that control growth. Growth rate may increase as temperatures increase, but R_g per unit of tissue produced is unaffected by temperature (Penning De Vries et al. 1974). By contrast, R_m varies primarily with changes in temperature, but R_m also increases with increases in relative growth rate (Amthor 1989; Lavigne and Ryan 1997). Although separation of the components of respiration is important, there is no highly accurate method for doing so. Amthor (2000) provides a comprehensive review of plant respiration with a detailed discussion of R_g and R_m.

Scaling to the Whole Plant

A major factor that affects the accuracy of scaling up to the whole plant is whether to express rates per unit volume or per unit surface area. Some researchers assume that respiration is primarily a function of cambial surface area; others contend that, because respiration occurs not only in the cambium but also in the living tissues of the sapwood, rates should be expressed per unit volume of sapwood. Sprugel (1990) discussed the pros and cons of both methods of scaling up and cites numerous examples of studies using either method. Sprugel (1990) and Edwards and Hanson (1996) reported good correlation between sapwood volume and R_m, but no significant correlation was shown between R_g and wood production or cambial area. Stockfors (2000) suggested that "the advantage of surface area compared with volume as a base for scaling up is, to a large extent, an artifact attributable to the combination of mean temperatures and temperature amplitudes increasing with height, and the volume of living cells per unit surface area decreasing with height. The two factors compensate each other." Carey et al. (1997) reported increases in R_m per unit volume as the stem diameter decreased. Maier et al. (1998) found in young loblolly pine trees that R_m was much higher in branches than in stems when expressed per unit volume but lower when expressed per unit surface area and suggested that scaling chamber-based stem-R_m measurements to the whole plant based on tissue volume may lead to serious errors. It is clear that if one is to scale up using respiration data expressed per unit volume of sapwood, one also needs to include the changing proportions of living cells in sapwood tissues in different parts of the tree as per Ryan (1990) or at least take measurements in as many different positions in each tree as feasible. If multiple measurements are not feasible, expressing the data per unit surface area may result in smaller errors when scaling up. In practice, however, measuring the surface area of a tree is much more time-consuming than measuring its sapwood volume. This is especially true in deciduous trees, which generally have more complex branching patterns than coniferous trees. In fact, the wood surface areas of few, if any, deciduous forest ecosystems have been measured. We express respiration data in this chapter as per unit sapwood volume because we have estimates for whole-plant sapwood volume but not for surface area.

Water Stress and Stem Respiration

Few studies have been performed with trees in natural stands that address questions relating water stress to growth and associated respiration rates. However, it is generally accepted that water stress in many trees promotes leaf senescence and decreased growth rates (Waisel 1991). Roberts (1964) reported a decrease in both the uptake of radiocarbon dioxide by leaves and the amount translocated out of the leaves with increasing water stress in *Liriodendron tulipifera* L. trees, and Edwards and McLaughlin (1978) found reduced stem respiration in *L. tulipifera* and *Quercus alba* L. to be related to lags in sucrose transport to the stem tissues during daily cycles. The primary objective of this chapter is to describe observed aboveground autotrophic woody-tissue respiration responses of the codominant mature trees and the predominant understory tree saplings to manipulative changes in soil moisture status and to intra-annual and interannual climatic conditions at the Walker Branch Throughfall Displacement Experiment (TDE). We also briefly discuss responses of foliar dark respiration based on measurements of a finite number of tree leaves.

Methods

Stem Respiration

Stem-respiration measurements were taken on 18 mature trees each of *Acer rubrum* L., *Quercus prinus* L., and *Quercus alba* L., at intervals during several months prior to establishment of the TDE in 1993 and continuing through 1996. Measurements were taken on 18 mature *L. tulipifera* L. trees at intervals in 1998 through 2000. Trees were selected to provide a wide range of diameter breast height (dbh) sizes within each treatment and within each species. All respiration measurements in 1993, except for one very late in the growing season and one during the dormant season, were taken prior to the beginning of water manipulation. Six overstory trees of each species were selected on each of the three TDE treatment plots. An effort was made to select trees in each treatment over a range of slope positions. This was not possible for *L. tulipifera* because this species primarily occupied lower slope positions (Chapter 2, this volume). Most of the data were collected during the growing season (April through August), but some measurements were made during the dormant season each year. Measurements were made with a closed infrared gas analysis system and modifications of previously described techniques (Sprugel and Benecke 1991; Edwards and Hanson 1996). Sapwood temperature was measured during each respiration measurement with a thermocouple inserted just deep enough to penetrate the cambium. Both respiration and temperature measurements were performed on the north-facing sides of the stems. In addition, some measurements were performed on the south-facing sides during 1993 and 1994. Stem temperatures were also monitored continuously on the north and south sides of a single *Q. alba* tree and a single *Q. prinus* tree through portions of the study.

Sapling Stem Respiration

Stem-respiration rates of *Cornus florida* L. and *A. rubrum* saplings were measured during 1994 and 1995. Ten saplings of each species in each treatment were measured at intervals throughout each year, with most of the measurements taken during the growing season. Measurement procedures followed those for stems, except that the chamber design was different. Chambers were longitudinally split 14-cm-long sections of 2.5-cm-diameter PVC pipes. The chambers were clamped around the stems of the saplings with band clamps. A nonporous foam was used to seal the chambers at each end and along the longitudinal edges.

Branch Respiration

Branch-respiration rates were measured in mature *A. rubrum, Q. alba,* and *Q. prinus* trees that were reachable from canopy-access towers in June and July 1997, June 1998, December 1998, and January 1999. Measurements were taken on three *Q. prinus,* three *A. rubrum,* and one *Q. alba,* with two branches measured on some of the trees. Only branches from the previous growing season were measured, and their diameters ranged from 1 to 2 cm. The December 1998 measurements were repeated measures of single trees of each species for the purpose of establishing dormant-season respiratory temperature coefficient (Q_{10}) values. Branch- and stem-respiration rates were measured simultaneously in June 1998 with two analyzers. Measurement techniques were the same as for saplings, except that branch-respiration chambers were shaded to prevent temperature increases within the chambers that might have been caused by direct sunlight in the upper canopy. Because of the limited number of trees that were accessible from the towers, no attempts were made to determine responses to treatments.

Branch and stem cross sections of *Q. alba, Q. prinus,* and *A. rubrum* adjacent to the study site were stained as a rough approximation of living tissues in branches versus stems. The staining technique was a modification of that reported by Ryan (1990). The cross sections were partially submerged in a 1% solution of triphenyl tetrazolium chloride for 2 h in dark, wide-mouth glass containers. A red, insoluble compound is formed when the tetrazolium solution reacts with dehydrogenase in the cytoplasm of living cells (Feist et al. 1971). The cross sections were blotted and then air-dried for 12 h before sanding with fine-grit sandpaper until smooth. The living tissue was visible with the naked eye as red-pigmented areas.

Stem-Respiration Calculations and Analysis

Respiration data for each sample date were normalized to the mean monthly temperature with Q_{10} values determined from repeated measures of individual trees of each species on each sample date. We express the respiration values per unit volume of sapwood or, in the case of branches and saplings, per unit volume of tissue. In saplings and branches, all of the wood enclosed by the chamber was assumed to contain living tissue. In stems, sapwood volume was calculated by multiplying sapwood thickness times the surface area defined by the length and width of the respiration chamber. Sapwood thickness was measured by taking a single-increment core adjacent to the respiration chamber on each tree.

For predictive purposes, we established two Q_{10} values and two base rates for each species. A dormant-season Q_{10} was determined from all the measurements taken in September through March, and a growing-season Q_{10} was determined for total respiration (R_t), where $R_t = R_g + R_m$, using all measurement taken in April through August. Temperatures were recorded during each individual measurement (air temperatures for saplings and branches; cambial zone temperatures for stems). All measurements for all trees of each species were regressed against temperatures to determine Q_{10} values and base rates. Base rates and Q_{10} values were calculated using

$$R_p = R_s Q_{10}{}^{(t_1 - t_2)/10}, \tag{1}$$

where R_p is the predicted respiration in μmol CO_2 m^{-3} sapwood s^{-1}, R_s is the measured reference respiration rate at reference temperature (t_2), Q_{10} is the respiratory temperature coefficient, and t_1 is the target temperature. For the dormant season, we calculated base rates at $t_1 = 0°C$, and for the growing season, $t_1 = 10°C$.

Annual total R_g for stems was calculated by three methods: (1) R_g = measured R_t minus predicted R_m; (2) R_g = predicted R_t minus predicted R_m; and (3) $R_g = 0.36 \times$ wood mass increment [assuming 36 g CO_2 efflux for each 100 g tissue constructed (Chapter 22, this volume)]. Wood-volume increments were calculated from dendrometer-band measurements assuming a circular stem cross section. Wood volume was converted to mass by multiplying wood densities for each species (US Forest Products Laboratory 1974) times wood volume. The densities used were 470, 605, 739, and 762 kg m^{-3} for *L. tulipifera, A. rubrum, Q. prinus,* and *Q. alba,* respectively.

Sapling and mature-tree stem-respiration data were analyzed with a one-way analysis of variance (ANOVA) to test treatment differences on each sample date. When the ANOVA indicated significant treatment differences in responses, the least significant difference method (LSD) was used to separate mean differences ($p \leq 0.05$).

Foliar Respiration

Foliar dark-respiration rates were measured in four overstory species and three species of saplings in mid to late summer after leaf growth was complete. On September 6 and 8, 1995, leaves were removed from 18 *A. rubrum* and 18 *Q. alba* trees that had been randomly selected across all three treatments. Branches at mid-canopy level were collected using shotgun sampling, and respiration was immediately measured in a darkened chamber with the leaves attached to the branches. On July 1 and August 21, 1997, foliar dark respiration was measured on 40 *A. rubrum* leaves, 5 *Q. alba* leaves, and 3 *Nyssa sylvatica* Marsh. leaves that were accessible from newly constructed canopy-access towers in each treatment plot (one tower per treatment). On June 17, 1999, foliar dark respiration was measured on 10 *Cornus* saplings, 7 *Carya glabra* (Mill.) Sweet saplings, and 10 *A. rubrum* saplings randomly scattered across all treatment plots. All tower-accessible and sapling leaves were measured *in situ* during the day in darkened chambers. Temperatures in the chambers were recorded each time a respiration measurement was taken. Respiration rates in 1995 were measured with a closed-system infrared gas analyzer. Rates after 1995 were measured with an open-system infrared gas analyzer.

Results

Species-Specific Responses

Annual patterns of stem-respiration rates by species and treatment are shown in Fig. 4.1. *Quercus prinus* (Fig. 4.1D) exhibited no significant treatment differences on most sample dates. Averaged across all sample dates, the dry treatment was 80% of the ambient treatment and the wet treatment was 90% of the ambient treatment. However, these differences were about the same as those observed during three sampling dates preceding water manipulations. The same trends were observed for *Q. alba* (Fig. 4.1C), with the highest rates observed in the ambient treatment both before and after initiation of treatments, with dry treatment rates averaging 86% of ambient before treatment and 95% after the start of treatments. In *A. rubrum* (Fig. 4.1A), the same trend was observed with both the dry and wet treatments, averaging only 70% of the ambient treatments across all sampling dates. Pretreatment average differences were 88% and 84% of ambient for the dry and wet treatments, respectively. *Liriodendron tulipifera* stem respiration (Fig. 4.1B) in the ambient and dry treatments averaged only 81% of the rates in the wet treatment, but differences were significant only during a single late-season

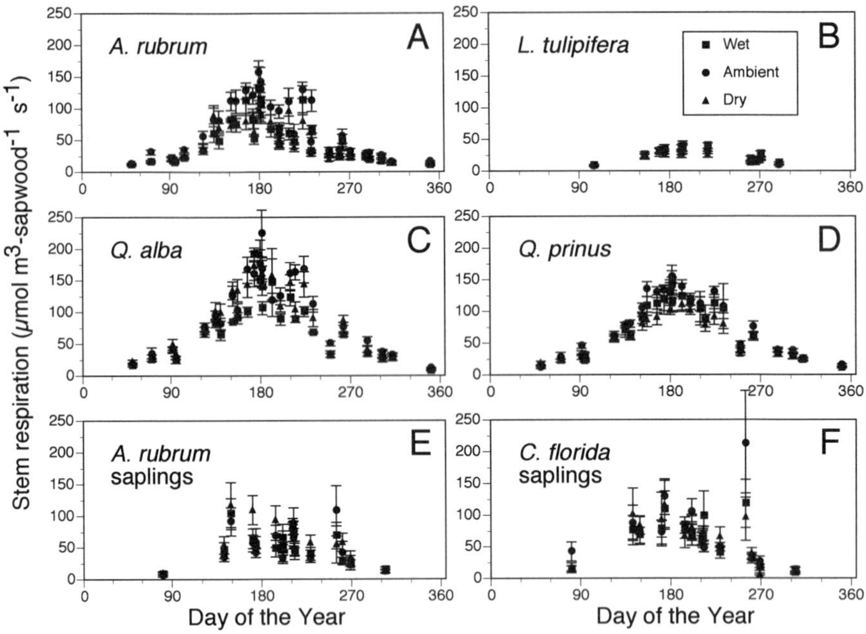

Figure 4.1. Respiration rates (means ± SE) by treatment plot for *A. rubrum* (A), *L. tulipifera* (B), *Q. prinus* (C), and *Q. alba* (D) trees, and *A. rubrum* (E) and *C. florida* (F) saplings. Rates are normalized to the mean monthly temperatures for the period during which measurements were taken with the use of within-single-day Q_{10} values.

drought in the year 2000. There were no treatment effects on sapling respiration, except on one sampling date in late September 1994 when *Cornus*-sapling rates in the dry treatment were significantly lower than rates in the ambient treatment (Fig. 4.1F). *Cornus* trended toward higher respiration rates in the ambient treatment than in the other treatments, but *A. rubrum* rates tended to be higher in the dry treatment plot (Fig. 4.1E). *Cornus*-sapling respiration averaged about 23% higher than *A. rubrum*-sapling respiration.

Temperature Responses

Respiration responses to temperature are provided for stems and branches of each of the canopy species measured and for stems of *A. rubrum* and *Cornus* saplings in Table 4.1. Dormant-season Q_{10} values ranged from 1.6 in *L. tulipifera* stems to 5.9 in *Q. prinus* branches. Dormant-season base rates at 0°C ranged from 1.9 μmol CO_2 m^{-3} s^{-1} in stems of *Cornus* saplings to 8.2 in *Q. prinus* and *A. rubrum* stems. Growing-season Q_{10} values ranged from 1.9 in *A. rubrum* stems in the dry treatment to 6.0 in *L. tulipifera* stems in the ambient treatment. Growing-season base rates at 10°C ranged from 3.0 μmol CO_2 m^{-3} s^{-1} in *L. tulipifera* stems in the ambient treatment to 158 in *Q. alba* branches. Growing-season Q_{10} values

Table 4.1. Dormant and growing season Q_{10} values and base respiration rates by species and treatment (A = ambient; D = dry; W = wet)

		Dormant Season (used to predict R_m)		Growing Season (used to predict R_t through growth period)	
Species	Structure	Q_{10}	Base rate at 0°C ($\mu mol\ CO_2\ m^{-3}\ s^{-1}$)	Q_{10}	Base rate at 10°C ($\mu mol\ CO_2\ m^{-3}\ s^{-1}$)
Q. prinus	Stems	2.6	8.2	2.5	39.6 (A)
				2.2	36.4 (D)
				2.6	33.6 (W)
	Branches	5.9	2.6	5.9	32.9
Q. alba	Stems	2.9	7.8	2.9	40.2 (A)
				2.5	44.6 (D)
				2.4	39.0 (W)
	Branches	2.3	5.3	2.3	158.0
L. tulipifera	Stems	1.6	3.3	6.0	3.0 (A)
				4.7	3.7 (D)
				3.9	5.9 (W)
A. rubrum	Stems	2.0	8.2	2.4	33.3 (A)
				1.9	30.0 (D)
				2.2	25.4 (W)
	Branches	3.3	6.6	3.3	34.3
	Saplings	1.8	3.6	2.1	16.1
C. florida	Saplings	2.3	1.9	2.6	14.5

Note: Q_{10} estimates were determined from all measurements taken from September through March, and a growing-season total respiration ($R_g + R_m$). Q_{10} was determined for each treatment from all measurements taken from April through August.

and base rates for canopy trees varied little by treatment except in *L. tulipifera.* In *L. tulipifera,* growing-season Q_{10} values declined from a very high value of 6 in the ambient treatment to 3.9 in the wet treatment, but the base rates showed the opposite trend and were very low.

Stem respiration during the growing season was strongly influenced by growth rates, as expected and as exemplified somewhat in Fig. 4.2, although the relationship between growth and respiration is more strongly demonstrated for some species than for others (e.g., *L. tulipifera*). Growth respiration varied somewhat by species, by year, and by the methods used for calculations (Figs. 4.3A–4.3D). In the *Quercus* species (Figs. 4.3A and 4.3B), Q_{10} predictions of respiration rates averaged slightly higher (by a factor of about 1.1) than measured rates in the drier years (1993 and 1995), but measured rates averaged 1.4 times higher than predicted rates in the wetter years (1994 and 1996). In *A. rubrum* (Fig. 4.3C), Q_{10} predicted rates were comparable to measured rates during dry years, but measured rates averaged 1.8 times the predicted rates during the wetter years. In *L. tulipifera* (Fig. 4.3D), Q_{10} predicted rates were slightly lower than measured rates in all 3 years that measurements were taken. In *Q. prinus,* measured rates averaged over 4 years were almost identical to rates calculated from diameter increments. For *Q. alba,* measured rates averaged 1.25 times the rates calculated from diameter increments. For both *L. tulipifera* and *A. rubrum,* measured rates averaged about

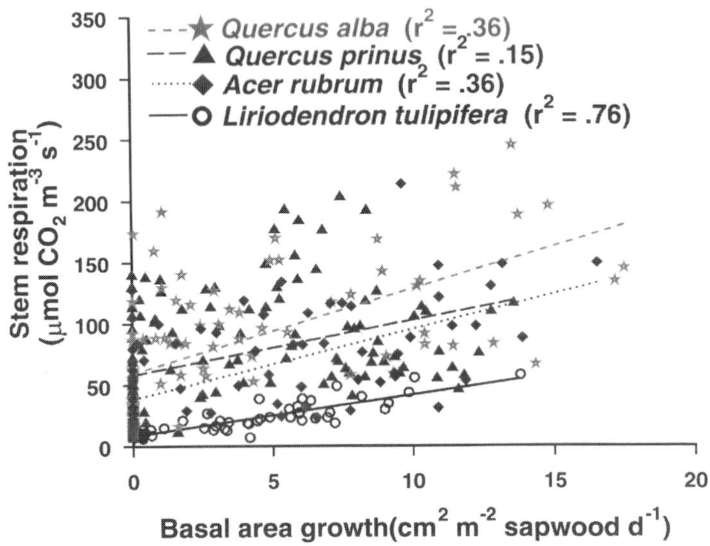

Figure 4.2. Stem respiration during the growing season of 1995 (normalized to monthly mean stem temperature) of each tree of each species across treatments is plotted against daily average basal-area growth for each tree. Basal-area growth was calculated from dendrometer-band measurements of diameter increments for the month during which respiration measurements were taken.

two times higher than rates calculated from diameter increments. Both methods demonstrated higher rates in the wetter years of 1994 and 1995 than in the drier years of 1993 and 1994. This pattern was not evident in Q_{10} predicted rates. In fact, the trend was the opposite, reflecting higher temperatures during the drier years. *L. tulipifera* rates ranged from 21% to 28% (depending on the method of calculation) of the average for the other three species. Comparisons of R_g calculated from diameter increments and respiration measurements during a single growing season are shown in Figs. 4.4A–4.4D. In all four species, the diameter increment underestimated R_g late in the growing season, whereas in spring and early summer, R_g was overestimated by diameter increments, especially in the *Quercus* species.

Branch-respiration rates in both *Quercus* species averaged half the rates of stem respiration during the dormant season; however, during the growing season, branch respiration exceeded stem respiration by a factor of 3 in *Q. prinus* and by a factor of 3.8 in *Q. alba* (Table 4.2). In *A. rubrum,* branch respiration was equal to stem respiration during the dormant season but was twice that of stems during the growing season. In trees with branches and stems measured simultaneously on June 8, 1998 (Fig. 4.5) branch respiration exceeded stem respiration on average by a factor of 5.4 (6.3, 5, and 2.8 in *Q. prinus, Q. alba,* and *A. rubrum,* respectively). Air temperature in the tree canopies averaged 4.1°C higher than stem temperatures below the canopy at the time measurements were taken.

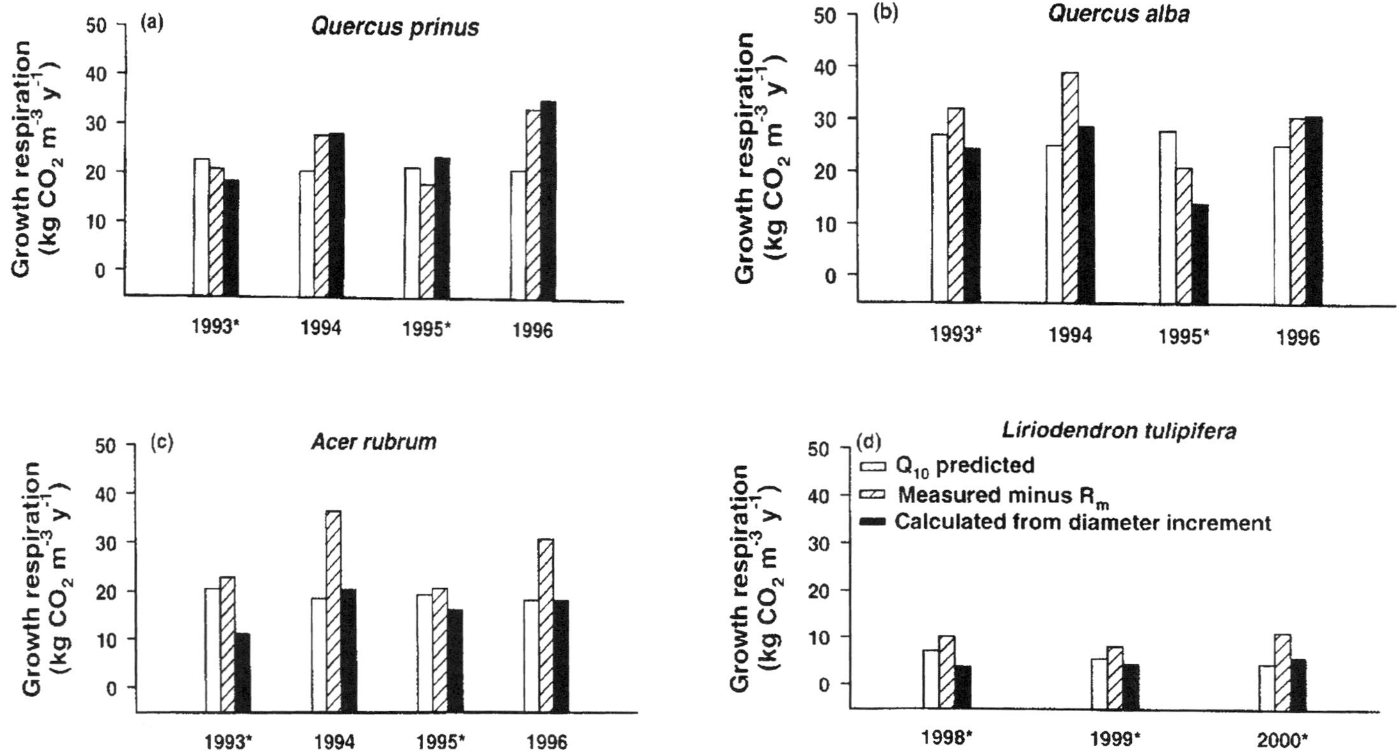

Figure 4.3. Annual total stem-growth respiration (R_g) is plotted for *Q. prinus* (a), *Q. alba* (b), and *A. rubrum* (c) in 1993 through 1996 and for *L. tulipifera* (d) in 1998 through 2000. An asterisk indicates a relatively dry year.

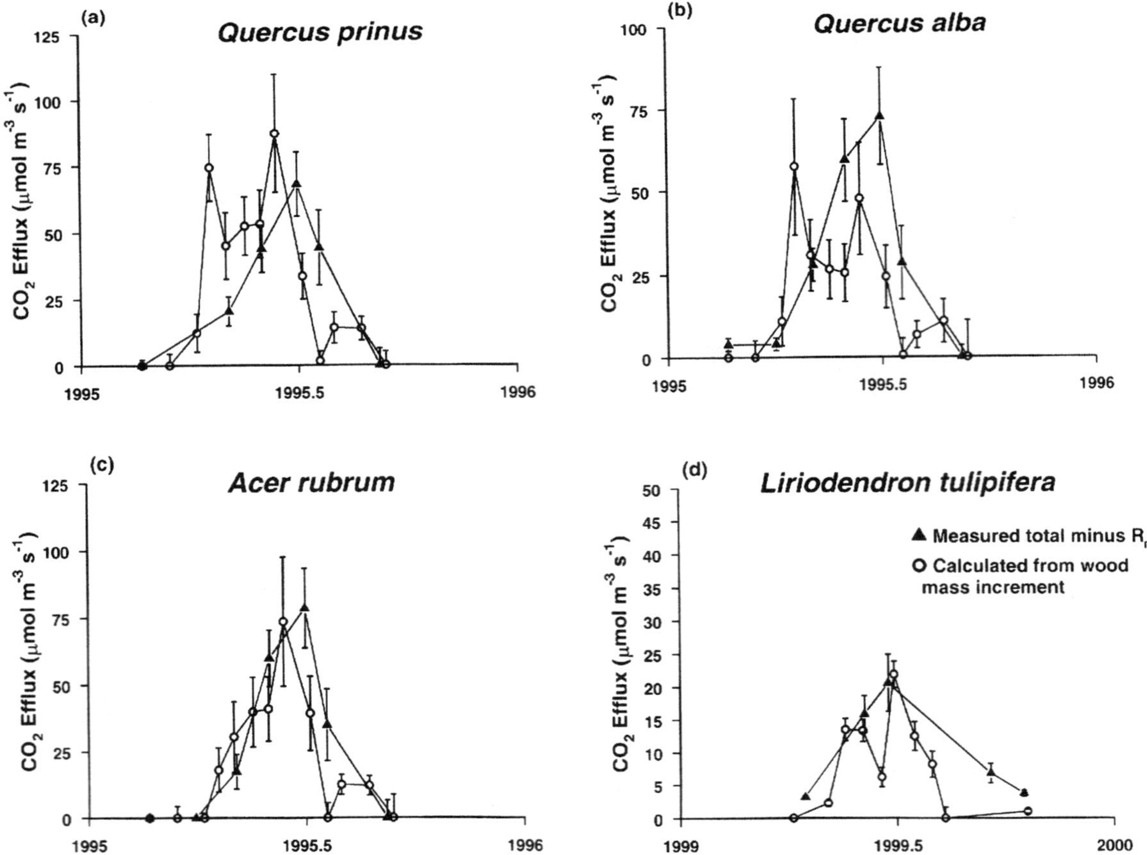

Figure 4.4. Seasonal trends in growth respiration rates in 1995 for (a) *Q. prinus,* (b) *Q. alba,* and (c) *A. rubrum,* and in 1999 for (d) *L. tulipifera.* Measured total respiration was normalized to monthly mean stem temperatures, and R_m was predicted from dormant-season base rates and Q_{10} values that used monthly mean stem temperatures.

Table 4.2. Stand-level annual total R_m, R_g, and total woody respiration (R_t)

	Respiration rate (kg C m^{-3} sapwood y^{-1})						Sapwood index (m^3 ha^{-1})		Stand-level R_t (g C m^{-2} y^{-1})		
	R_m		R_g		R_t						
Species and year	Stem	Branch	Stem	Branch	Stem	Branch	Stem	Branch	Stem	Branch	Total
Q. prinus											
1993	13.6	35.2	5.7	37.5	19.3	72.7	7.7	11.1	14.9	80.7	95.6
1994	12.9	26.5	7.8	37.1	20.7	63.6	8.1	11.7	16.8	74.4	91.2
1995	13.2	29.5	4.9	30.6	18.1	60.1	8.2	11.9	14.8	71.5	86.4
1996	12.1	24.6	9.1	40.1	21.2	64.7	8.9	12.8	18.9	82.8	101.7
Average	13.0	29.0	6.9	36.3	19.8	65.3	8.2	11.9	16.3	77.4	93.7
Q. alba											
1993	15.9	8.7	8.7	80.6	24.6	89.3	5.9	7.8	14.5	69.7	84.2
1994	14.8	8.0	10.6	110.0	25.4	118.0	6.4	8.4	16.3	99.1	115.4
1995	14.8	8.3	5.7	75.2	20.5	83.5	6.6	8.7	13.5	72.6	86.2
1996	14.0	7.6	8.3	84.7	22.3	92.3	6.6	8.8	14.7	81.2	95.9
Average	14.9	8.2	8.3	87.6	23.2	95.8	6.4	8.4	14.8	80.7	95.4
A. rubrum											
1993	8.8	23.4	6.2	11.4	15.0	34.8	14.9	5.3	22.4	18.4	40.8
1994	8.4	20.1	9.9	18.4	18.3	38.5	15.6	5.6	28.5	21.6	50.1
1995	8.6	21.2	5.6	10.2	14.2	31.4	16.1	5.8	22.9	18.2	41.1
1996	8.2	19.0	8.3	15.4	16.5	34.4	17.4	6.3	28.7	21.7	50.4
Average	8.5	20.9	7.5	13.9	16.0	34.8	16.0	5.8	25.6	20.0	45.6
L. tulipifera[a]											
1998	2.6	6.4	2.7	5.0	5.3	11.4	9.6	5.1	5.1	5.8	10.9
1999	2.5	6.2	2.2	4.1	4.7	10.3	9.7	5.2	4.6	5.4	9.9
2000	2.7	6.6	3.0	5.6	5.7	12.2	9.7	5.1	5.5	6.2	11.8
Average	2.6	6.4	2.6	4.9	5.2	11.3	9.7	5.1	5.1	5.8	10.9
A. rubrum saplings[b]											
1994	3.3	3.3	4.2	4.2	7.5	7.5	3.9	2.9		2.9	
1995	3.6	3.6	4.5	4.5	8.1	8.1	3.9	3.2		3.2	
Average	3.5	3.5	4.4	4.4	7.8	7.8	3.9	3.0		3.0	
C. florida saplings[b]											
1994	2.7	2.7	5.7	5.7	8.4	8.4	2.6	2.2		2.2	
1995	2.9	2.9	6.1	6.1	9.0	9.0	2.6	2.3		2.3	
Average	2.8	2.8	5.9	5.9	8.7	8.7	2.6	2.3		2.3	

Note: R_t and R_m data were normalized to monthly mean temperatures with the Q_{10} values in Table 4.1. $R_g = R_t - R_m$. Sapwood index data are from Chapter 22 (this volume).

[a] Branch respiration rates for *L. tulipifera* L. were not measured. Those rates were assumed to be higher than the stem respiration in *L. tulipifera* by the same proportions as in *A. rubrum*.

[b] Branch respiration rates in saplings were assumed to be the same as in the stems.

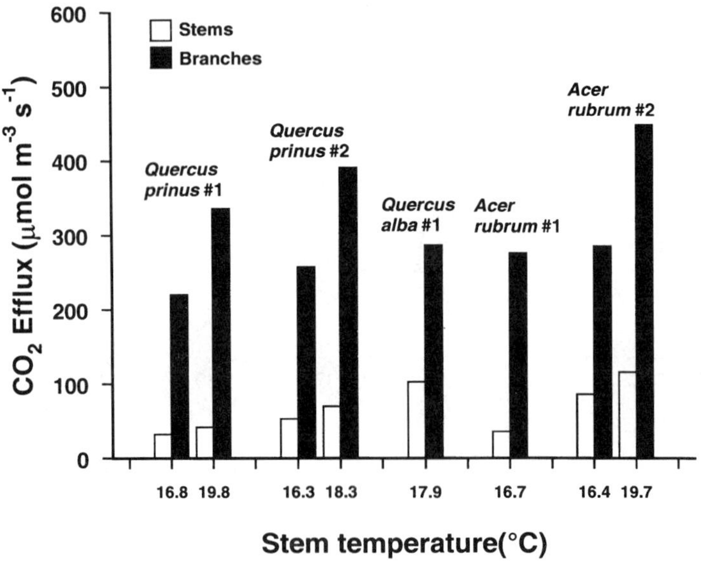

Figure 4.5. Respiration rates of stems and branches are compared for *Q. prinus, Q. alba, A. rubrum,* and *L. tulipifera.* Measurements of branches and stems of the same tree were recorded simultaneously with two analyzers. Temperatures of stems are at cambial depth, whereas branch temperatures were assumed to be equal to air temperature in the upper canopy.

Scaling Respiration to the Forest Stand

Annual total woody-tissue respiration at the stand level is provided in Table 4.2. Annual total respiration (R_t) averages were 93.7, 95.4, 45.6, 10.9, 3.0, and 2.3 g C m^{-2} soil surface y^{-1} for *Q. prinus, Q. alba, L. tulipifera, A. rubrum* saplings, and *C.* saplings, respectively. Branch respiration contributed 82%, 84%, 44%, and 53% of the total aboveground woody respiration in *Q. prinus, Q. alba, A. rubrum,* and *L. tulipifera,* respectively. In the *Quercus* species, most of the stand-level respiration resulted from R_g in branches (46% and 77% of the total in *Q. prinus* and *Q. alba,* respectively). R_g in stems of both *Quercus* species averaged 6% of the total respiration. Branch R_m averaged 36% of the total in *Q. prinus,* but only 7% in *Q. alba.* Stem R_m contributed about 10% of the total in both *Quercus* species. In *A. rubrum* and *L. tulipifera,* R_m was the largest contributor to total respiration in both stems and branches. Stem R_m contributed 30% of the total respiration in *A. rubrum* and 23% of the total in *L. tulipifera,* with branch R_m averaging 26% of the total in *A. rubrum* and 30% of the total in *L. tulipifera.*

Foliar dark-R_m data are presented in Table 4.3. Respiration rates in the overstory species (normalized to 20°C with published Q_{10} values for each species) ranch from 0.34 μmol CO_2 m^{-2} s^{-1} in blackgum to 0.43 in *A. rubrum.* Rates in the understory saplings averaged less than half the rates in the overstory species. Leaf R_m at the stand level (averaged over 8 years) for those species with a known leaf-

Table 4.3. Leaf maintenance respiration rates measured at the TDE compared to measurements from the same species in the southern Appalachian mountains (USA)

	Maintenance respiration (μmol CO_2 m^{-2} leaf area s^{-1} at 20°C)				
Species	TDE study	Bolstad et al. (1999)	Q_{10}[a]	LAI (m^2 m^{-2})	Dry mass (g m^{-2})
A. rubrum	0.43	0.47	2.3	1.72	117
Q. prinus		0.48	2.3	0.99	98
Q. alba	0.42	0.41	2.3	1.11	106
Nyssa sylvatica[b]	0.34	0.33	2.4	1.08	67
C. florida	0.12	0.31	2.7	0.05	3
C. glabra[c]	0.20	0.27	2.5	ND	ND
A. rubrum saplings[d]	0.16	ND	2.7[d]	ND	ND

Note: Bolstad's et al. (1999) measurements were on *in situ* or detached leaves at night, whereas TDE measurements were on *in situ* or detached leaves in a darkened chamber during the day. Comparisons from overstory leaves are from midcanopy positions. Leaf-area index (LAI) and dry mass were calculated from leaves collected in litterfall baskets at the TDE. An entry of ND indicates that no data were available.

[a] All Q_{10} values are from Bolstad et al. (1999).

[b] Respiration data are for *Nyssa sylvatica* only. LAI and dry mass data include *Oxydendrum arboreum.*

[c] *Cornus florida* and *Carya glabra* TDE data are from saplings. Bolstad et al. (1999) data are from the lower branches of *C. florida* and *C. glabra* trees.

[d] Q_{10} for *A. rubrum* saplings is based on lower-canopy *A. rubrum* overstory.

area index (LAI) totaled 389 g C m^{-2} soil surface y^{-1}. Of this total, *A. rubrum* contributed 37%, *Q. prinus* 22%, *Q. alba* 23%, *O. arboreum* L. (sourwood) plus blackgum 18%, and *Cornus* 0.3%. The LAI for all these species combined totals 4.95 out of a total-stand LAI of about 6. Leaf R_g ranged from 0.4 g C m^{-2} ground area y^{-1} in *Cornus* to 14.5 in *A. rubrum,* with a total-stand average of 49.

Foliar dark-respiration data in each treatment plot are insufficient to demonstrate any treatment effects except in *Q. alba* and *A. rubrum* overstory during the drought of 1995 (Fig. 4.6). During that period, R_m rates in *A. rubrum* were a factor of 1.4 greater ($p \leq 0.05$) in the wet treatment than in the ambient and dry treatments. In *Q. alba,* R_m was a factor of 1.9 greater ($p \leq 0.05$) in the wet treatment than in the ambient and dry treatments.

Comparing woody-tissue and dark foliar respiration at the stand level (Tables 4.2 and 4.4) reveals that foliar respiration accounted for 77% and woody-tissue accounted for 23% of the aboveground autotrophic respiration in *A. rubrum.* For the other species, the comparison was 51% foliar and 49% woody in *Q. prinus,* 52% foliar and 48% woody in *Q. alba,* and 42% foliar and 58% woody in *Cornus.*

Discussion

The Q_{10} values and base rates provided in Table 4.1 reflect the high degree of variability in responses of different species and organs to changes in temperature and underscore the fact that many other variables influence respiration rates. Therefore, arbitrarily assigning an average Q_{10} value for modeling purposes can potentially cause large errors in model predictions. Respiration rates predicted from stem temperatures varied less from year to year than did measured rates.

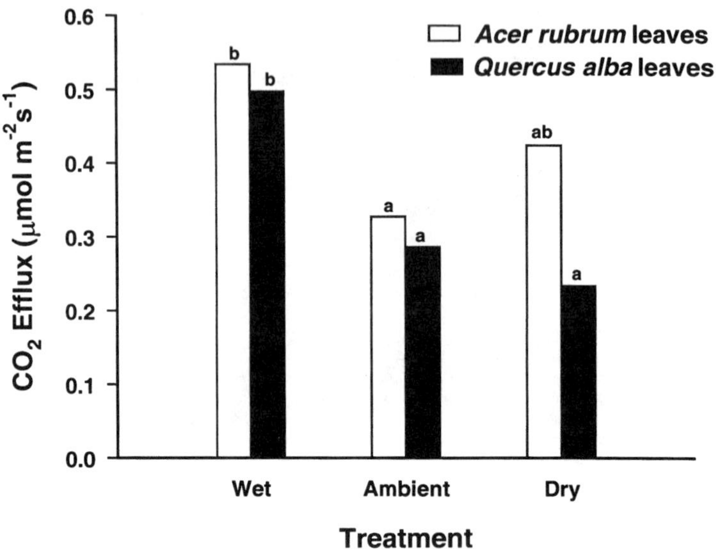

Figure 4.6. Dark respiration in midcanopy foliage of *A. rubrum* and *Q. alba* trees on September 5, 1995. Respiration was measured on darkened leaves attached to branches immediately after shotgun harvesting. Rates were normalized to 20°C with published Q_{10} values (Table 4.3). Bars designated by different letters indicate statistical treatment differences ($p \leq 0.05$, $n = 18$).

Table 4.4. Summary of yearly total leaf R_m and R_g by species

	Respiration (g C m^{-2} ground area y^{-1})					
Species	*A. rubrum*[a]	*Q. prinus*[b]	*Q. alba*	*Oxydendrum* and *Nyssa*[c]	*C. florida*	Totals
			Leaf-Maintenance Respiration			
Year						
1993	159	97	102	78	1.5	437
1994	139	85	89	67	1.2	381
1995	143	87	92	70	1.3	393
1996	131	80	84	63	1.2	360
1997	132	81	85	64	1.2	362
1998	154	94	99	75	1.4	424
1999	136	83	87	66	1.2	373
2000	139	85	89	67	1.2	381
Average	142	87	91	69	1.3	389
			Leaf-Growth Respiration			
Average	14.0	12.4	13.2	8.3	0.4	49

Note: R_m values were calculated from respiration rates and the LAI data in Table 4.3. R_g values were calculated by multiplying leaf dry mass by 0.125 (g C used to produce 1 g dry matter).

[a] Respiration rates were measured in *A. rubrum* only. The LAI combined the *A. rubrum* and *A. saccharum*.

[b] Our LAI values were applied to respiration data from Bolstad et al. (1999).

[c] *Nyssa sylvatica* respiration rates were applied to the LAI of *N. sylvatica* plus *O. arboreum*.

Whereas annual rates predicted from stem temperatures were highest during the dry years, measured rates (both from CO_2 efflux measurements and from diameter-increment extrapolations) were highest during the wet years. Growing-season rates based on diameter increments tended to exceed measured rates in the spring and early summer, whereas measured rates tended to be higher than calculated rates late in the growing season, especially in the *Quercus* species. This lag is in agreement with the conclusion by Sprugel and Benecke (1991) that R_g in tree stems trails diameter growth by about 1 month because wood synthesis occurs after cell expansion. This fact explains the relatively high number of positive R_g values in Fig. 4.2 at zero volume growth (i.e., positive respiration values were recorded in the late part of the growing season after volume growth had ended).

Annual R_g based both on diameter increments and measured rates between 1993 and 1996 was in line with what we would expect (i.e., higher growth rates in wet years and lower rates in dry years). Nevertheless, the lack of overall agreement between measured rates and those calculated from diameter increments in *A. rubrum* and *L. tulipifera* suggests problems with one or both of the methods. Rates based on diameter increments assume a given amount of CO_2 produced per unit mass of wood produced. Our calculations included species differences in wood density (based on estimates from the literature), but the assumption that 36 g CO_2 is produced for each 100 g of wood produced is based on the false assumption that all tree species contain equal proportions of various compounds (e.g., lignins, proteins, and cellulose). This faulty assumption could explain some or even all of this incongruity. On the other hand, there are also reasons to suspect problems with calculations using measured rates. Although measurements were performed on all species with identical methods, the measurements were point-in-time measurements, whereas the diameter-increment calculations were integrated over time periods that spanned the point-in-time measurements. Furthermore, the lack of quantification of living tissues in the sapwood volume used in the calculations is problematic, possibly more for some species than for others.

The relatively low rates of stem respiration in *L. tulipifera* as compared to the other dominant overstory species (Fig. 4.1) is possibly caused by the fact that sapwood thickness in *L. tulipifera* is more than twice that of the other species. It is possible that *L. tulipifera* has a lower proportion of living cells in its sapwood than do the other species. This uncertainty can only be addressed by quantifying the proportion of living tissue versus nonliving tissue in the sapwood as per Ryan (1990). Another possible reason for low respiration rates in *L. tulipifera* is that their growth rates were drastically reduced during much of the study period because of foliar damage from insect grazing (Hanson et al. 2001). However, the trees appeared to have recovered by the time our respiration measurements were taken.

The trend of higher stem-respiration rates in the ambient treatment in *Quercus* and *A. rubrum* apparently reflects differences in site characteristics not associated with the water manipulations. In the case of *L. tulipifera,* where stem-respiration rates were highest in the wet treatment, differences were probably caused more by slope position than by treatments. Because of the limited number of *L. tulipifera* on all plots, trees could not be randomly selected across slope positions. In fact, most of the trees measured in the wet treatment were located near the very bottom of the slope, whereas measured trees were located farther up slope in the ambient and dry treatment plots. This fact suggests that the reduced respiration in *L. tulipifera* may be caused by less favorable growing conditions (including moisture) for *L. tulipifera* at higher positions on the slope.

Unlike the growing-season stem R_m, dormant-season stem R_m was equal to or greater than (depending on species) branch R_m in this study. This finding suggests that the higher respiration rates in branches during the growing season were primarily caused by growth processes occurring in the cambium and by higher temperatures in the branches rather than differences in living cells in the sapwood. It also suggests that R_m in the cambial tissue is negligible, as reported by Sprugel (1990). The actively growing cambial area is much higher relative to sapwood in branches as opposed to stems. Unlike with our findings, Maier et al. (1998) found that R_m was higher in branches than in stems of 9-year-old *Pinus taeda* pine.

Sapling-respiration rates were lower than rates in canopy species except for *L. tulipifera.* The low rates in saplings reflects a very slow growth rate, which was expected based on their light-suppressed position in the forest. For example, rates in *A. rubrum* saplings were only about half the rates in mature *A. rubrum* despite the saplings having a higher percentage of living tissue per unit wood volume. However, this conditions held true even during the dormant season, suggesting more efficient use of food sources resulting from lower R_m costs. *Cornus* had an even lower R_m demand during the dormant season than did *A. rubrum* (Table 4.1).

The large contribution of branch respiration in the *Quercus* species at the stand level (Table 4.2) was caused by higher specific branch-respiration rates and to higher sapwood volume in branches. In *A. rubrum* and *L. tulipifera,* the higher contribution by branches was caused solely by the higher specific respiration rates in branches than in stems because stem-sapwood volume in these two species exceeded branch-sapwood volume. The importance of including branch respiration when scaling to tree and stand level is evident in this study. By including branch respiration, the 4-year average R_t was a factor of 2.4, 2.8, 1.3, and 1.4 greater in *Q. prinus, Q. alba, A. rubrum,* and *L. tulipifera,* respectively, than if the branch-respiration and stem-respiration rates were assumed to be equal. Ideally, measurements would be taken at several vertical locations of the stem and on a range of branch sizes. This level of exactitude was not logistically possible in this study. Given that our stem-respiration measurements were taken on a stem section with relatively low rates, our estimate of stand-level stem respiration is probably an underestimate. Conversely, our branch respiration rate probably represents an overestimate because measurements were taken on relatively young (1-year-old) branches. Thus, these discrepancies partially nullify each other.

Foliar dark-respiration rates (normalized to 20°C) on the TDE were in reasonably close agreement with rates reported by Bolstad et al. (1999) for the same species in the southern Appalachian Mountains of South Carolina. However, Bolstad et al. (1999) found that temperature–respiration response functions varied significantly with canopy position. Mitchell et al. (1999) found that nitrogen content per unit leaf area decreases by 41% from upper-canopy to lower-canopy leaves and that foliar respiration was linearly related to foliar nitrogen content per unit area. Therefore, because our extrapolations to the stand level are based only on mid-canopy respiration rates, they should be considered only an approximation of the true stand-level foliar respiration. Given the logistical difficulty of sampling foliage from enough canopy positions to accurately predict stand-level foliar respiration, the mid-canopy position was perhaps the best predictor for this study.

This study demonstrated no conclusive evidence of treatment effects on aboveground autotrophic respiration. However, higher rates of R_g occurred during the wetter years than during the drier years, indicating higher growth rates during the wetter years. This conclusion agreed with diameter-growth increments. The study also demonstrates the importance of several key factors when scaling from point-in-time, small-area respiration measurements to the whole tree or stand. For example, large differences between branch- and stem-respiration rates, the effect of the differences in temperatures in different parts of a tree on respiration rates, and variations in proportions of living tissue to total woody-tissue in different parts of the tree all contribute to the complexity of scaling up. In a mature deciduous forest, it is likely that our estimates of whole-tree autotrophic respiration are too low given the fact that our measurements are concentrated in the parts of the trees that have the lowest respiration rates.

References

Amthor JS (1989) *Respiration and crop productivity.* Springer-Verlag, New York.

Amthor JS (2000) The McCree-de Wit-Penning de Vries-Thornley respiration paradigms: 30 years later. Ann Bot 86:1–20.

Bolstad PV, Mitchell K, Vose MV (1999) Foliar temperature-respiration response functions for broad-leaved tree species in the southern Appalachians. Tree Physiol 19:871–878.

Carey EV, Callaway RM, DeLucia EH (1997) Stem respiration of ponderosa pines grown in contrasting climates: Implications for global climate change. Oecologia 111(1):19–25.

Edwards NT, Hanson PJ (1996) Stem respiration in a closed-canopy upland oak forest. Tree Physiol 16:433–439.

Edwards NT, McLaughlin SB (1978) Temperature-independent diel variations of respiration rates in *Quercus alba* and *Liriodendron tulipifera.* Oikos 31:201–206.

Edwards NT, Shugart HH Jr., McLaughlin SB, Harris WF, Reichle DE (1981) Carbon metabolism in terrestrial ecosystems. In Reichle DE (Ed) *Dynamic properties of forest ecosystems.* Cambridge University Press, London, pp 499–536.

Feist WC, Springer EL, Hajny GJ (1971) Viability of parenchyma cells in stored green wood. Tappi 54:1295–1297.

Hanson PJ, Todd DE, Amthor JS (2001) A six-year study of sapling and large-tree growth and mortality responses to natural and induced variability in precipitation and throughfall. Tree Physiol 21:345–358.

Lavigne MB, Ryan MG (1997) Growth and maintenance respiration rates of aspen, black spruce and jack pine stems at northern and southern BOREAS sites. Tree Physiol 17:543–551.

Lavigne MB, Ryan MG, Anderson DE, Baldocchi DD, Crill PM, Fitzjarrald DR, Goulden ML, Gower ST, Massheder JM, McCaughey JH, Rayment M, Strieg RG (1997) Comparing nocturnal eddy covariance measurements to estimates of ecosystem respiration made by scaling chamber measurements at six coniferous boreal sites. J Geophys Res 102(D24):28977–29985.

Maier CA (2001) Stem growth and respiration in loblolly pine plantations differing in soil resource availability. Tree Physiol 21:1183–1193.

Maier CA, Zarnoch SJ, Dougherty PM (1998) Effects of temperature and tissue nitrogen on dormant season stem and branch respiration in a young loblolly pine (*Pinus taeda*) plantation. Tree Physiol 18:11–20.

Mitchell KA, Bolstad PV, Vose JM (1999) Interspecific and environmentally induced variation in foliar dark respiration among eighteen southeastern deciduous tree species. Tree Physiol 19:861–870.

McCree KJ (1970) An equation for the rate of respiration of white clover plants grown under controlled conditions. In Setlik I (Ed) *Prediction and measurement of photosynthetic productivity.* PUDOC, Wageningen, The Netherlands, pp 221–229.

Penning De Vries FWT (1975) The cost of maintenance processes in plant cells. Ann Bot 39:77–92.
Penning De Vries FWT, Brunsting AHM, van Laar HH (1974) Products, requirements and efficiency of biosynthesis: A quantitative approach. J Theor Biol 45:339–377.
Reich PB, Walters MB, Elllsworth DS, Vose JM, Volin JC, Gresham C, Bowman WD (1998) Relationships of leaf dark respiration to leaf nitrogen, specific leaf area, and leaf life-span: A test across biomes and functional groups. Oecologia 114:471–482.
Roberts BR (1964) Effects of water stress on the translocation of photosynthetically assimilated carbon-14 in yellow poplar. In Zimmerman MH (Ed) *The formation of wood in forest trees.* Academic Press, New York, pp 273–287.
Ryan MG (1990) Growth and maintenance respiration in stems of *Pinus contorta* and *Picea engelmannii.* Can J For Res 20:48–57.
Ryan MG (1995) Foliar maintenance respiration of subalpine and boreal trees and shrubs in relation to nitrogen concentration. Plant Cell Environ 18:765–772.
Sprugel DG (1990) Components of woody-tissue respiration in young *Abies amabilis* (Dougl.) Forbes trees. Trees 4:88–98.
Sprugel DG, Benecke U (1991) Measuring woody-tissue respiration and photosynthesis. In Lassoie JP, Hinckley TM (Eds) *Techniques and approaches in forest tree ecophysiology.* CRC Press, Boca Raton, Florida, pp. 329–355.
Stockfors J (2000) Temperature variations and distribution of living cells within tree stem: Implications for stem respiration modeling and scale-up. Tree Physiol 20:1057–1062.
US Forest Products Laboratory (1974) *Wood handbook: Wood as an engineering material.* Agricultural Handbook No. 72. USGPO, Washington, District of Columbia, pp 1–55.
Waisel Y (1991) Adaptation to salinity. In Raghavendra A (Ed) *Physiology of trees.* John Wiley and Sons, New York, pp 359–383.

5. Dormant-Season Nonstructural Carbohydrate Storage

Timothy J. Tschaplinski and Paul J. Hanson

Like tree growth, dormant-season carbon storage is an important, integrative measure of tree health. Carbon is stored throughout trees in the form of carbohydrates for later use as an energy resource and substrate for synthetic products (Kozlowski et al. 1991). Trees rely on carbon fixed during the latter part of the growing season, and stored through the winter, to produce new organs for gathering light, water, and nutrients (Gholz and Cropper 1991; Kozlowski et al. 1991). Effects of altered throughfall levels leading to changes in carbon-exchange rates should be cumulative over time (Chapin et al. 1990), resulting in either accumulation or degradation of stored nonstructural carbohydrate pools in the wetter or drier environments, respectively. Dunn et al. (1987) and Parker and Patton (1975) have shown evidence of stress-induced carbohydrate reductions in *Quercus*. Stored carbohydrate concentrations of stems are typically similar to that of coarse roots, with both tissues responding similarly to environmental changes. However, given that stems are much more easily accessible, they are most often sampled to assess carbon storage.

Stored nonstructural carbohydrates represent a requisite predictive variable in a number of models of tree growth operating at the individual tree (Weinstein et al. 1991) or the forest stand level (Running and Coughlan 1988; Hunt and Running 1992; Amthor 1994). In these models, feedbacks between stored levels of carbohydrates are primarily or partially responsible for dictating growth rates or capacities of individuals or tissues within a plant. Given such a pivotal role of stored carbohydrates in plant or forest stand growth models, it is crucial that the relationships proposed in these models be evaluated against observed field data. Such evaluations are especially critical (Addiscott et al. 1995) considering that model interpretations of a limited number of experimental manipulation studies are to be extrapolated to assessments of global change impacts on forest ecosystems.

The Throughfall Displacement Experiment (TDE) near Oak Ridge, Tennessee is a large-scale manipulation study where canopy-dominant and understory deciduous species of a mature, southern upland forest, are exposed to ambient throughfall precipitation, ambient minus 33% throughfall precipitation, and ambient plus 33% throughfall precipitation. Although growth of plants or plant organs is the ultimate integrator of a plant's cumulative response to the external environment, a goal of the TDE experiment was to evaluate dormant-season carbon storage as a sensitive early sign of cumulative plant response to altered precipitation environments. It was hypothesized that the concentration of total nonstructural carbohydrates (TNC) in the dormant season would reflect the availability of water the previous growing season (short term), decreasing in a drier environment and increasing in a wetter environment, relative to the ambient condition. It was anticipated that species considered the most drought sensitive would have a lower TNC concentration [TNC] and be more responsive to changes in throughfall. The impact of chronic drought over multiple growing seasons (long term) on trees will be evident in an overall decline in the storage pools in all tissues (stem and branch). Drought will have a greater impact on the storage reserves of understory versus overstory trees of the same species, given that understory trees have limited root volumes and a shaded light environment, limiting carbon assimilation.

Methods

Study Site

The study was conducted on the TDE, located on a southern facing, uniformly sloped, upland oak site at Walker Branch Watershed, on the U.S. Department of Energy's National Environmental Research Park near Oak Ridge, Tennessee. The site characteristics, including soil type, aspect, history, species composition, climate, and system performance of the manipulation of throughfall precipitation have been described in detail elsewhere (Hanson et al. 2001; Chapter 2 this volume). The major canopy-dominant species on the site are *Quercus prinus* L., *Quercus alba* L., and *Acer rubrum* L. The understory consists largely of *A. rubrum, Cornus florida* L., and *Nyssa sylvatica* Marsh. Throughfall precipitation levels are manipulated by ~1900 subcanopy troughs suspended above the forest floor of the dry plot such that ~33% of the throughfall precipitation is trapped and passively transferred by gravity into piping and then across an ambient area to a wet plot. The three treatments, dry (−33% throughfall precipitation), ambient, and wet (+33% throughfall precipitation), each cover an 80 × 80 m area, and have a similar slope with the upper half of the all treatments typically drier than the lower half.

Sample Collection

Dormant-season TNC within tree boles of the major tree species on the TDE site were sampled February 20, 1993 to establish the pretreatment level of stored carbohydrates present. Paired, 3-cm increment cores (measured from the cambium inward) were removed from the north (i.e., shaded) side of all tree boles. Initial sampling showed little difference due to the sampling aspect on the tree. However, only the north side was sampled to control variability. Species sampled included *A. rubrum, A. saccharum* Marsh., *Q. alba,* and *Q. prinus.* To minimize damage to the tree boles, resampling was not done every year but was repeated in February/March of 1996, 1998, 1999, and 2000. In 1996, the same species were sampled on February 14–15, and additional species, including *N. sylvatica, Oxydendrum arboreum* L., and *Liriodendron tulipifera* L., were collected on February 26–27, 1996. All seven of these canopy-dominant and intermediate species were resampled February 26–27, 1998. Only *Q. alba* and *Q. prinus* were sampled on March 8, 1999, after the dry 1998 growing season. All seven species were again sampled on February 8, 2000. Flowering of *A. rubrum,* an early indicator of the initiation of physiological activity, was typically noted February 24–29 in each of the years of plant sampling.

Additional measurements of TNC storage were determined for canopy branches and branches from understory saplings. Tissue samples from understory/sapling-sized trees [<10 cm diameter at breast height (dbh)] were limited to the removal of terminal branch portions. Branch sampling was conducted during February of 1996, 1998, and 2000. Species sampled on March 12–13, 1996 included overstory *A. rubrum, Q. alba,* and *Q. prinus,* and understory *A. rubrum* and *C. florida.* The same species and canopy positions were again sampled on March 12, 1998. The same overstory canopy-dominant species were sampled on February 25, 2000. Overstory sampling was conducted with a shotgun, whereas a pole pruner was used to collect the understory samples.

After sample collection, the cores and branches were placed on ice, and then frozen in a deep freezer at −80°C, freeze-dried, ground, and finally stored at room temperature until analyses were preformed. For each overstory or understory species and each tissue type, 4–10 individuals per treatment were sampled for nonstructural carbohydrate analyses.

Soluble Carbohydrate Analysis

Total nonstructured caarbohydrate analyses included measurements of water-soluble carbohydrates and starch. Soluble carbohydrates were extracted from 50–75-mg subsamples of the dried and ground (1-mm pore size) increment cores and branch samples with three successive 5-ml extracts of 80% aqueous ethanol. Each extraction was conducted for 24 h and the extracts for each tree or sapling were combined. The residual material containing starch was dried overnight in an airstream in a flow hood. The eluant was analyzed for total water-soluble carbohydrates by using the phenol-sulfuric acid colorimetric reaction (Dubois et al. 1956), as described below for the starch digests.

Starch Analysis

Starch concentrations of the residue were determined using the perchloric acid digestion (Sutton et al. 1981; Tissue and Wright 1995). The dried residue (25–50 mg) from which soluble carbohydrates had been previously extracted three times was digested with 5 ml of perchloric acid (35%) for 1 h at room temperature. A 200-μl aliquot of the digest was added to 800 μl of distilled water, followed by the addition of 1 ml phenol (5% in water) reagent and 5 ml concentrated sulfuric acid (96% v/v). After 20 min, the absorbance at 490 nm was determined with a Spectronic 1001 spectrophotometer (Milton Roy Co., Rochester, NY). A glucose calibration curve was similarly prepared with concentrations ranging from 0 to 100 μg/ml. All soluble carbohydrate concentration [CHO], starch concentration [starch], and [TNC] results were thus reported in glucose equivalents.

Results

Pretreatment Differences in Dormant-Season Bole Carbohydrates

The only significant pretreatment effect was with *Q. alba* in the wet treatment, where bole wood [TNC] was higher than ambient and dry treatments, because of higher soluble [CHO] (particularly relative to ambient) (Table 5.1). Treatment differences in soluble [CHO] and [starch] were marginal and likely inconsequential (Tables 5.2 and 5.3). Otherwise, [TNC] of *Q. alba* did not differ between treatments over the 7-year study. Total soluble [CHO] of the ambient treatment was slightly reduced in the most recent 2000 sampling, but [TNC] was not different.

Main Effects of TDE on Dormant-Season Bole Carbohydrates

There were relatively few statistically significant effects of the TDE treatments on [TNC] of dormant-season bole wood samples over the 7 years of the throughfall manipulation (Table 5.1). Some notable exceptions included a 26% decline in [TNC] of *N. sylvatica* in the dry treatment in 1998, following a year of above-average precipitation. Additionally, soluble [CHO] of *N. sylvatica* in the dry and ambient treatments were lower than the wet treatment in 2000. Again, the 2000 growing season had a normal amount of precipitation, although the total annual precipitation was below average because of an end-of-year drought. In contrast, [TNC] of *L. tulipifera* in the dry and ambient treatments increased in both 1998 (+38% dry versus wet) and 2000 (+53% dry versus wet), relative to wet treatment, primarily the result of changes in [starch]. In years with normal to above-average precipitation, there was a negative correlation between bole [TNC] and

Table 5.1. Total nonstructural carbohydrate concentrations (%; g/100 g dw) of stems of overstory trees

	Year				
	1993	1996	1998	1999	2000
Q. prinus					
Wet	7.2 (1.1)	12.9 (1.4)	15.5 (0.6)	14.6 (1.3)	18.6 (1.0)
Ambient	6.7 (1.1)	10.3 (1.2)	15.5 (1.3)	12.9 (0.5)	21.3 (2.3)
Dry	6.8 (0.7)	9.8 (1.1)	17.3 (1.1)	13.1 (0.6)	18.7 (1.2)
Q. alba					
Wet	9.7 (0.4) a	21.3 (0.7)	14.1 (0.6) ab	14.0 (0.7)	15.7 (1.1)
Ambient	7.4 (0.5) b	19.1 (1.4)	12.2 (0.9) b	12.3 (1.2)	16.0 (0.7)
Dry	8.0 (0.5) b	21.3 (1.9)	16.0 (0.7) a	12.3 (0.6)	16.3 (1.5)
	0.006		0.014		
A. rubrum					
Wet	5.3 (0.7)	8.5 (0.7)	11.8 (0.6)		11.0 (0.4)
Ambient	7.4 (2.0)	9.7 (1.2)	13.4 (1.0)		11.9 (0.7)
Dry	8.0 (1.2)	8.9 (1.5)	13.1 (0.9)		11.1 (0.6)
A. saccharum					
Wet	4.8 (0.5)	5.3 (1.2)	9.3 (1.6)		7.3 (0.6)
Ambient	3.4 (0.4)	5.0 (0.9)	10.7 (0.2)		7.3 (0.6)
Dry	4.3 (0.5)	4.6 (0.1)	10.3 (0.3)		7.7 (0.5)
N. sylvatica					
Wet		8.3 (0.3)	13.6 (0.6) a		12.3 (0.3)
Ambient		9.4 (1.0)	13.7 (0.8) a		12.2 (0.6)
Dry		10.1 (0.9)	10.2 (0.4) b		11.5 (0.3)
			0.002		
O. arboreum					
Wet		10.0 (1.5)	16.8 (0.8)		4.6 (0.5)
Ambient		10.4 (1.0)	13.7 (0.4)		5.2 (0.2)
Dry		9.7 (0.8)	15.3 (1.0)		4.9 (0.4)
			0.069		
L. tulipifera					
Wet		5.7 (0.6)	10.7 (1.0) b		4.0 (0.3) b
Ambient		6.4 (0.5)	16.5 (0.7) a		7.1 (0.8) a
Dry		5.1 (0.3)	14.8 (0.8) a		6.1 (0.6) ab
		0.057	0.001		0.021

Note: Data represent the mean ± SEM. Treatments that share a letter are not significantly different ($p > 0.05$). Probabilities of statistical significance ($p < 0.10$) are shown.

basal-area growth the previous growing season regression (Fig. 5.1A, $r^2 = 0.84$, $n = 6$). Similarly, there was a negative correlation between bole [TNC] and basal-area growth the subsequent growing season regression (Fig. 5.1B, $r^2 = 0.84$, $n = 6$). Such correlations were not evident in any of the other species studied (data not shown).

Although [TNC] was not significantly impacted in other species, the components of TNC did significantly respond to the throughfall manipulation in the dormant season following years with the lowest growing-season precipitation (Tables 5.2 and 5.3). [Starch] of *Q. prinus* in the dry treatment was lower than in the wet and ambient treatments in 1999, with the same trend evident in 1996, but

Table 5.2. Total soluble carbohydrate concentrations (%; g/100 g dw) of stems of overstory trees

	Year				
	1993	1996	1998	1999	2000
Q. prinus					
Wet	2.4 (0.2)	3.6 (0.2)	7.1 (0.5)	11.2 (1.1)	7.6 (0.2)
Ambient	2.7 (0.1)	4.1 (0.2)	7.4 (0.7)	9.8 (0.4)	8.5 (0.8)
Dry	2.9 (0.4)	3.9 (0.1)	8.2 (0.6)	11.1 (0.7)	8.3 (0.4)
Q. alba					
Wet	4.3 (0.3) a	15.6 (0.7)	7.7 (0.2)	11.6 (0.6)	9.8 (0.5) a
Ambient	3.3 (0.3) b	12.8 (0.9)	6.6 (0.8)	10.1 (1.0)	7.9 (0.4) b
Dry	3.6 (0.2) ab	14.6 (1.3)	7.8 (0.5)	9.9 (0.7)	8.4 (0.7) ab
	0.048				0.029
A. rubrum					
Wet	1.9 (0.1)	3.1 (0.1)	6.4 (0.4)		7.7 (0.3)
Ambient	2.1 (0.1)	3.8 (0.4)	7.7 (0.9)		8.9 (0.6)
Dry	2.1 (0.3)	2.9 (0.1)	6.9 (0.7)		8.1 (0.6)
A. saccharum					
Wet	1.3 (0.1)	1.4 (0.1) b	6.6 (1.3)		3.1 (0.5)
Ambient	1.3 (0.3)	2.0 (0.1) a	6.7 (0.5)		3.4 (0.5)
Dry	1.3 (0.1)	1.7 (0.0) ab	6.9 (0.7)		3.5 (0.0)
		0.017			
N. sylvatica					
Wet		7.3 (0.4)	5.8 (0.6)		3.0 (0.2) a
Ambient		7.0 (0.5)	6.0 (0.7)		2.5 (0.1) b
Dry		7.6 (0.7)	5.1 (0.1)		2.4 (0.1) b
					0.019
O. arboreum					
Wet		7.7 (1.0)	9.1 (0.7)		2.5 (0.3)
Ambient		7.9 (0.9)	8.1 (0.3)		3.2 (0.2)
Dry		7.1 (0.5)	7.9 (0.3)		2.7 (0.3)
L. tulipifera					
Wet		4.4 (0.5) ab	3.7 (0.2)		2.0 (0.1)
Ambient		4.7 (0.3) a	4.2 (0.4)		2.4 (0.2)
Dry		3.5 (0.3) b	4.0 (0.6)		2.1 (0.1)
		0.034			

Note: Data represent the mean ± SEM. Treatments that share a letter are not significantly different ($p > 0.05$).

these did not translate into a significant effect on [TNC] (Table 5.3). The 1995 and 1998 seasons were the driest growing seasons, with soil-water potentials in 1998 the lowest over the whole experiment to date (Hanson et al. 2001).

Another statistically significant effect that was of lesser consequence, but still informative, included higher soluble [CHO] of *A. saccharum* in the ambient treatment versus the wet treatment in 1996 (Table 5.3). This response may be explained by the size of the trees in the different treatments. *A. saccharum* in the ambient treatment were considerably larger in diameter at breast height in 1999 (37.1 cm) than trees in the wet (26.5 cm) or dry (24.8 cm) treatments.

Table 5.3. Total starch concentrations (%; g/100 g dw) of stems of overstory trees

	Year				
	1993	1996	1998	1999	2000
Q. prinus					
Wet	4.7 (1.1)	9.2 (1.3)	8.4 (0.3)	3.4 (0.4) a	11.0 (1.0)
Ambient	3.9 (0.8)	6.2 (1.1)	8.1 (1.1)	3.1 (0.3) a	12.9 (2.7)
Dry	4.0 (1.4)	5.9 (1.0)	9.1 (0.6)	2.0 (0.3) b	10.4 (1.0)
		0.091		0.027	
Q. alba					
Wet	5.4 (0.3)	5.8 (0.4)	6.4 (0.5) ab	2.4 (0.3)	5.9 (0.8)
Ambient	4.1 (0.3)	6.3 (1.0)	5.6 (0.4) b	2.1 (0.4)	7.1 (0.4)
Dry	4.4 (0.4)	6.7 (0.9)	8.1 (0.8) a	2.4 (0.2)	7.9 (1.0)
	0.055		0.035		
A. rubrum					
Wet	3.4 (1.0)	5.5 (0.7)	5.4 (0.4)		3.3 (0.2)
Ambient	5.8 (1.1)	5.9 (1.0)	5.7 (0.3)		3.0 (0.2)
Dry	5.3 (1.9)	6.0 (1.9)	6.2 (0.4)		3.0 (0.3)
A. saccharum					
Wet	3.5 (1.1)	3.7 (1.3)	2.5 (0.3)		3.9 (0.1)
Ambient	2.2 (0.2)	3.0 (0.6)	3.4 (0.2)		4.3 (0.4)
Dry	3.0 (0.6)	2.9 (0.1)	3.4 (0.4)		4.1 (0.3)
N. sylvatica					
Wet		1.5 (0.5)	7.8 (0.1) a		9.2 (0.4)
Ambient		2.3 (0.7)	7.7 (0.4) a		9.7 (0.6)
Dry		2.5 (0.4)	5.2 (0.4) b		9.1 (0.3)
			0.001		
O. arboreum					
Wet		2.3 (0.6)	7.7 (0.5)		2.1 (0.3)
Ambient		2.4 (0.3)	5.6 (0.2)		2.0 (0.2)
Dry		2.5 (0.5)	7.4 (0.8)		2.0 (0.1)
			0.090		
L. tulipifera					
Wet		1.3 (0.2)	7.0 (0.8) b		2.0 (0.3) b
Ambient		1.7 (0.2)	12.3 (0.6) a		4.8 (0.8) a
Dry		1.6 (0.1)	10.8 (0.8) a		4.1 (0.5) a
			0.001		0.034

Note: Data represent the mean ± SEM. Treatments that share a letter are not significantly different ($p > 0.05$). Probabilities of statistical significance ($p < 0.10$) are shown.

Main Effects of TDE on Dormant-Season Branch Carbohydrates

Overstory

Acer rubrum in the dry treatment had higher [TNC] than trees in the wet (+21%) and ambient (+16%) treatments in 1996, because of an increase in [starch] (+28% dry versus wet) (Table 5.4). Additionally, *Q. prinus* in the dry treatment had a 42% increase in soluble [CHO] versus branches of trees in the wet treatment in 1996 that did not translate in significant [TNC] ($p = 0.11$, n was low) (Table 5.5). Another inconsequential effect was lower [TNC] of *Q. alba* in the ambient versus the wet and dry treatments, due to a decline in soluble [CHO] ([starch] also tended to be reduced, but variability was high) (Table 5.4). Again, this effect was likely due to the smaller stature of trees in the ambient treatment, resulting

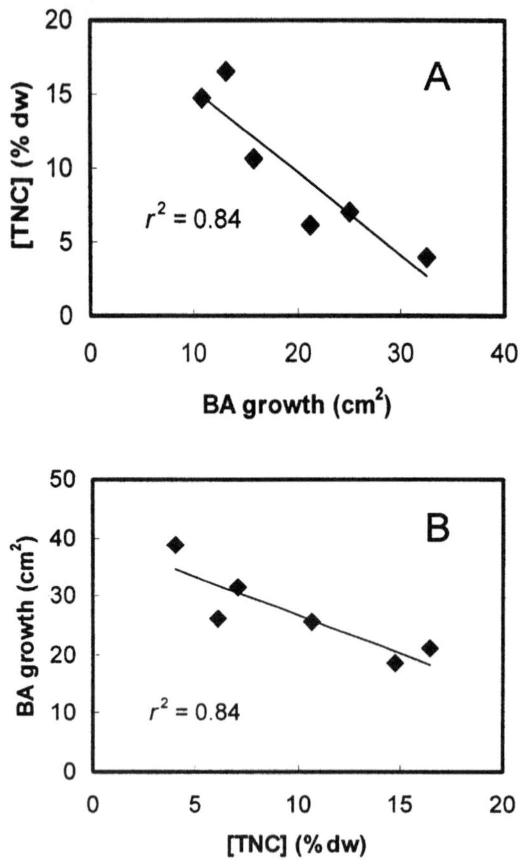

Figure 5.1. Regression of dormant-season total nonstructural carbohydrate concentrations (%; g/100 g dw) of branches with basal-area growth of *L. tulipifera* for (A) the preceding normal to above-normal precipitation years (1997, 1999) and (B) the subsequent growing seasons.

in more lateral shading from canopy-dominant neighbors and a lowering of carbon assimilation and soluble [CHO], a response similar to that observed in the boles of these trees. Similarly, a borderline significant decline in [starch] of *Q. alba* in the ambient versus wet treatment was also evident in 1998 (Table 5.6).

Understory

Branch [TNC] of *A. rubrum* saplings in the dry and ambient treatments was lower (−17% in dry; −9% in ambient) than in the wet treatment in 1996 (Table 5.4), resulting from a decline in [starch] (−30% in dry, −16% in ambient) (Table 5.6). Whereas overstory branches of *A. rubrum* indicated an increase in starch reserves in response to drought, the branches indicated a decline in reserves, likely reflecting the limited carbon assimilation of the shaded understory saplings. An-

Table 5.4. Total nonstructural carbohydrate concentrations (%; g/100 g dw) of branches

	Year		
	1996	1998	2000
Q. prinus—overstory			
Wet	21.4 (1.1)	34.1 (0.4)	14.2 (0.8)
Ambient	22.4 (0.9)	27.3 (5.7)	14.1 (0.5)
Dry	26.0 (1.3)	36.7 (3.6)	14.1 (0.3)
Q. alba—overstory			
Wet	18.5 (0.7) a	30.4 (0.6)	13.2 (0.6)
Ambient	13.4 (0.4) b	22.9 (5.5)	12.1 (0.3)
Dry	17.0 (0.7) a	20.0 (5.0)	13.5 (0.8)
	0.010		
A. rubrum—overstory			
Wet	15.2 (0.2) b	23.1 (2.3)	13.9 (0.6)
Ambient	15.8 (0.6) b	21.5 (2.6)	13.3 (0.7)
Dry	18.4 (0.8) a	23.8 (1.8)	13.9 (0.8)
	0.018		
A. rubrum—understory			
Wet	22.9 (0.7) a	21.4 (0.9)	
Ambient	20.8 (0.4) b	14.7 (0.7)	
Dry	19.0 (0.6) c	19.6 (1.8)	
	0.001	0.051	
C. florida—understory			
Wet	18.1 (2.1)	18.8 (1.6)	
Ambient	18.0 (0.9)	18.2 (2.0)	
Dry	19.4 (0.9)	20.2 (1.6)	

Note: Data represent the mean ± SEM. Treatments that share a letter are not significantly different ($p > 0.05$). Probabilities of statistical significance ($p < 0.10$) are shown.

other inconsequential effect was a lower soluble [CHO] of *A. rubrum* in the ambient versus the wet and dry treatments in 1998 (Table 5.5).

Interannual Differences in Tissue Carbohydrates

Branches

The [TNC] of branches from understory plants were consistent between years (Table 5.4). The [starch] and soluble [CHO] concentrations of understory *A. rubrum* and *C. florida* were similar between the two years sampled (1996 and 1998) (Tables 5.5 and 5.6). In contrast, there was more variation in [TNC] of branches from overstory plants (i.e., differences between sampling years ranging 50–100% from minimum to maximum). The [TNC] tended to be intermediate in 1996, highest in 1998, and lowest in 2000. The largest difference between 2000 and 1998 was caused by starch in *A. rubrum*, whereas the reductions in soluble [CHO] and [starch] were similar in *Q. prinus* and *Q. alba*. Overall, [starch] and soluble [CHO] paralleled each other in branches. The highest [TNC] occurred during 1997 following a normal amount of throughfall. Regressions of [TNC] with total annual precipitation yielded fairly high correlations for three species; *Q. prinus* ($r^2 = 0.58$), *A. rubrum* ($r^2 = 0.88$), and *Q. alba* ($r^2 = 0.98$), but data were

Table 5.5. Total soluble carbohydrate concentrations (%; g/100 g dw) of branches

	Year		
	1996	1998	2000
Q. prinus—overstory			
Wet	11.3 (0.3) b	18.9 (1.2)	8.3 (0.7)
Ambient	12.5 (0.6) b	15.3 (4.0)	7.4 (0.3)
Dry	16.0 (0.8) a	22.7 (3.6)	8.4 (0.4)
	0.009		
Q. alba—overstory			
Wet	9.6 (0.2) b	16.5 (0.7)	7.2 (0.4)
Ambient	7.5 (0.0) a	14.0 (5.2)	6.6 (0.3)
Dry	8.9 (0.5) b	10.1 (3.4)	7.4 (0.6)
	0.025		
A. rubrum—overstory			
Wet	8.7 (0.4)	9.5 (0.5)	9.5 (0.6)
Ambient	8.7 (0.6)	10.6 (1.9)	8.4 (0.6)
Dry	10.1 (0.6)	12.3 (1.4)	8.4 (0.8)
A. rubrum—understory			
Wet	8.5 (0.4)	10.7 (0.9) a	
Ambient	8.7 (0.5)	5.7 (0.6) b	
Dry	8.9 (0.6)	8.6 (0.8) a	
		0.007	
C. florida—understory			
Wet	9.3 (1.8)	10.3 (1.1)	
Ambient	9.0 (0.5)	8.7 (1.1)	
Dry	10.4 (0.6)	11.3 (1.6)	

Note: Data represent the mean ± SEM. Treatments that share a letter are not significantly different ($p > 0.05$).

collected in only 3 years of the study and two of those years tended to have similar annual precipitation (Fig. 5.2A).

Boles

As in the branches, the boles of overstory trees had the highest [TNC] in February 1998 in all species monitored, except *Q. prinus* and *N. sylvatica,* for which the [TNC] were the same in 1998 and 2000. Linear regressions of bole wood [TNC] of trees in the ambient plot against the total annual precipitation indicated that there were strong positive correlations in the case of *L. tulipifera* ($r^2 = 1.00$, $n = 3$), *A. rubrum* ($r^2 = 0.963$, $n = 4$), and *A. saccharum* ($r^2 = 0.913$, $n = 4$) (Fig. 5.2B). The TNC regressions against growing-season precipitation yielded coefficients of determination that were not quite as high. In contrast, the best fits of the regression of [starch] against precipitation were with growing-season precipitation. The best fit of the linear regressions was for *L. tulipifera* ($r^2 = 0.947$, $n = 3$) (Fig. 5.3). The correlation of soluble carbohydrate concentrations against annual precipitation was only strong for *A. saccharum* ($r^2 = 0.959$, $n = 4$). In summary, some of the species that are considered to be somewhat drought sensitive tended to have higher [TNC] in the dormant-season following years with

Table 5.6. Starch concentrations (%; g/100 g dw) of branches

	Year		
	1996	1998	2000
Q. prinus—overstory			
Wet	10.2 (1.4)	15.1 (0.9)	5.9 (0.2)
Ambient	9.9 (0.5)	12.1 (1.8)	6.8 (0.3)
Dry	10.1 (0.6)	14.0 (0.9)	5.7 (0.2)
Q. alba—overstory			
Wet	8.9 (2.4)	13.9 (1.2)	6.0 (0.4)
Ambient	5.9 (0.4)	8.9 (0.8)	5.6 (0.2)
Dry	8.1 (0.3)	9.9 (1.6)	6.1 (0.5)
		0.053	
A. rubrum—overstory			
Wet	6.5 (0.2) b	13.6 (2.3)	4.4 (0.8)
Ambient	7.1 (0.4) ab	10.9 (0.9)	4.8 (0.7)
Dry	8.3 (0.5) a	11.5 (0.7)	5.2 (0.6)
	0.047		
A. rubrum—understory			
Wet	14.4 (0.4) a	10.8 (0.4)	
Ambient	12.1 (0.7) b	9.0 (0.6)	
Dry	10.1 (0.4) c	11.0 (1.1)	
	0.001		
C. florida—understory			
Wet	8.7 (0.4)	8.5 (0.5)	
Ambient	8.8 (0.4)	9.5 (1.2)	
Dry	9.0 (0.5)	8.9 (0.6)	

Note: Data represent the mean ± SEM. Treatments that share a letter are not significantly different ($p > 0.05$). Probabilities of statistical significance ($p < 0.10$) are shown.

higher precipitation. Overall, using data from 1996 to 2000, branch [TNC] and stem [TNC], were poorly correlated (data not shown).

Species Differences

Averaging bole wood [TNC] across treatments and over time produces an approximate ranking of the major species studied, with *Q. prinus* having the highest [TNC] (13.4%), followed by *Q. alba* (12.3%), *N. sylvatica* (11.3%), *O. arboreum* (10.1%), *A. rubrum* (10.0%), *L. tulipifera* (8.5%), and *A. saccharum* (6.8%). Given that year-to-year variation was high, as described earlier, and that not all species were sampled at all times, these rankings should be considered approximations. The species ranking by soluble [CHO] was similar to that of total [TNC], with *Q. alba* (8.9%), followed by *Q. prinus* (6.6%), *O. arboreum* (6.2%), *N. sylvatica* (5.2%), *A. rubrum* (5.1%), *L. tulipifera* (3.5%), and *A. saccharum* (3.3%). The range in species' [starch] was narrower than soluble [CHO], with *Q. prinus* (6.1%), followed by *Q. alba* (5.4%), *L. tulipifera* (5.0%), *A. rubrum* (4.9%), *O. arboreum* (3.8%), and *A. saccharum* (3.3%). Although *L. tulipifera* and *A. saccharum* were at the lower end of the range with regard to storage carbohydrates and the concentrations tended to be lower after drought years,

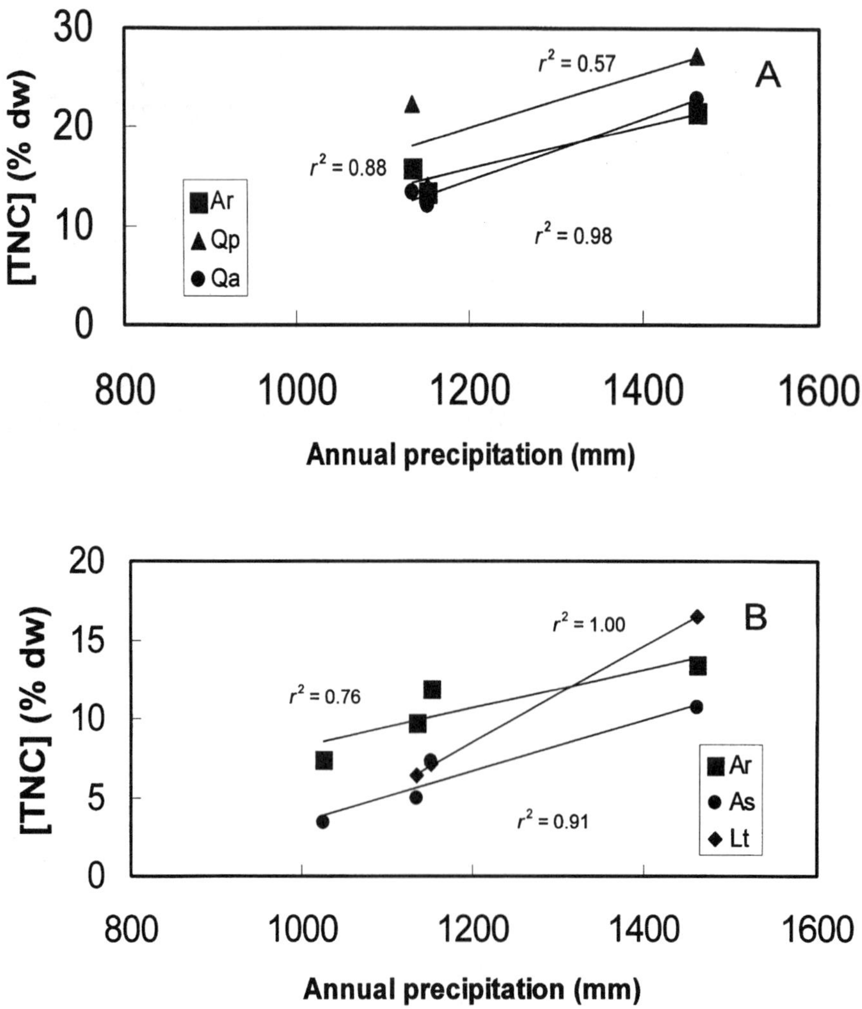

Figure 5.2. Regression of dormant-season [TNC] (%; g/100 g dw) of branches (A) and bole wood (B) with annual precipitation. Branch data are shown for *Q. prinus, Q. alba,* and *A. rubrum,* and bole data are for *A. rubrum, A. saccharum* and *L. tulipifera.*

[starch] and soluble [CHO] never dropped to negligible levels (<1%). In contrast, [TNC] for the *Quercus* species at the upper end of the range peaked at 21%.

Discussion

Species Differences

The trees in this study were sampled in late winter when their storage reserves should be at or near their annual peak. McLaughlin et al. (1980) reported that

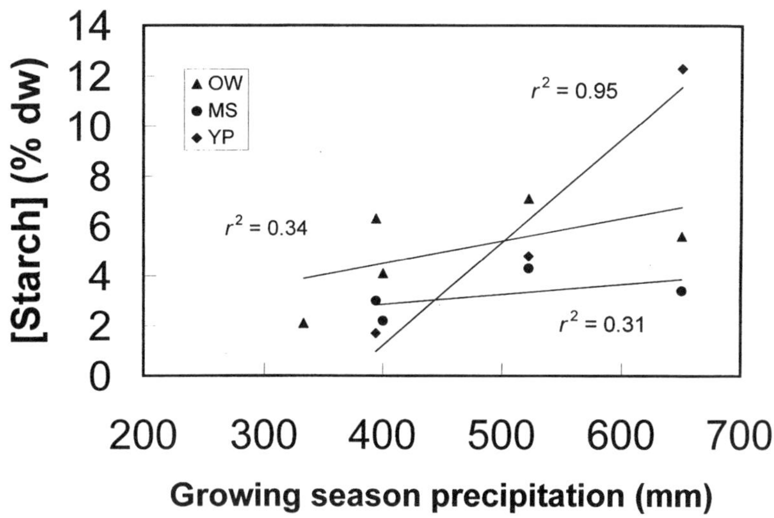

Figure 5.3. Regression of dormant-season starch concentrations (%; g/100 g dw) of bole wood with growing season precipitation (May–September) for *Q. alba, A. saccharum,* and *L. tulipifera.*

upper boles of *Q. alba* trees had peak [starch] and soluble [CHO] of 8% and 1.8%, respectively, which occurred late winter. Similarly, Gholz and Cropper (1991) reported a late-winter maximum in mature (>20 years old) *Pinus elliottii* Engelm., with [starch] and soluble [CHO] peaking at 0.8% and 0.7%, respectively, in stems, but, in contrast with coarse roots, stems showed less seasonal variation. [Starch] and soluble [CHO] that we observed ranged from 1.3% to 12.9% and 1.3% to 15.6%, respectively, depending on the species and year sampled. Although [starch] and soluble [CHO] varied inversely in red osier dogwood (*Cornus sericea* L.) over the winter months, [starch] of stems peaked at ~11% (Ashworth et al. 1993). Thus, our results are in general agreement with other studies.

Short-Term Treatment Effects on Stem Carbohydrate Storage

Treatment-induced changes in dormant-season TNC of stems and branches can be detected in some, but not all species, with [starch] impacted the most. Among overstory species, the dry treatment decreased the dormant-season stem [TNC] of *N. sylvatica* by 33% relative to the wet treatment, but increased [TNC] of *L. tulipifera* by 38% in 1998, following a growing season with above-average precipitation, but a late-season drought. Severe drought has been qualitatively shown to reduce [starch], but increase soluble [CHO], especially reducing sugars, in roots of *Q. velutina* Lam. (black oak) seedlings (Parker and Patton 1975). The differential response of stem [TNC] in our study, particularly [starch], in these drought-sensitive species may be caused by the degree of drought-induced growth suppression displayed, with *L. tulipifera* impacted the most. The higher sink demand of the *L. tulipifera* trees in the wet treatment likely sustained carbon utilization and thus depleted reserves. The negative correlation between bole [TNC] and basal-area growth the previous season supports the contention that high sink demand caused by growth reduces stored reserves in the following dormant season. Similarly, the significant negative correlation between rainfall (and relative humidity) and stem soluble sugar concentration observed in five hardwood species in a seasonally dry tropical climate (Latt et al. 2001) was also likely the result of sink demand depleting soluble sugars when environmental conditions favored growth and reproduction. Stem [starch] was not correlated with rainfall (Latt et al. 2001). In our study, stem [starch] of *Q. prinus* in the dry treatment was also lower than the wet and ambient treatments in 1999, with the same trend evident in 1996, but these did not translate into a significant effect on [TNC]. These changes in dormant-season carbohydrate storage should be considered responses that are short term, responding to precipitation and temperatures of the previous growing season, rather than to chronic, long-term treatment. Consistent trends on any storage carbohydrate pool in response to the treatments are not evident.

The differential short-term response of *L. tulipifera* and *N. sylvatica* to the late-season drought may indicate the degree to which growth may have been suppressed. *L. tulipifera* accumulated unused assimilate in the form of starch, whereas *N. sylvatica* depleted its reserves in response to the late-season drought. It is not surprising that the TDE manipulation would have its greatest effect when throughfall distribution over a growing season was near normal, as evident in 1994, 1997, and 1999, allowing the TDE infrastructure to redistribute the throughfall. In contrast, the TDE infrastructure would only have a short-term, transient effects if the drought was abrupt and long term, because all of the treatment plots would eventually approach the same level of water stress (i.e., there would be little to no throughfall to redistribute). This may explain why the greatest treatment effects were evident in the dormant season after above-average growing-season precipitation.

Branch Versus Stem Wood Carbohydrate Storage

Budbreak in relation to flowering for mature *A. rubrum* trees tended to occur no later than the last week of February and was in progress from mid-February onward. Therefore, some degree of physiological activity (albeit low) was in progress in *A. rubrum* when branches were sampled (February 25 to March 13). It was our intention to sample when physiological activity was minimal and this assumption likely holds for most of the branch samples collected. Although boles were typically sampled earlier than branches, the year-to-year responses were similar. The year that had the highest stem [TNC] (1998) followed two wet years and was also the peak year for branch [TNC], but the correlation between branch and stem [TNC] was not high for any given species. *A. rubrum, Q. alba,* and *Q. prinus* trees all tended to store higher [TNC] in branches than stems. *Q. prinus* tended to have higher stem and branch storage of TNC than the other species, reflecting its high drought tolerance capability, being able to accumulate storage carbohydrates even during stressful years.

There was an interesting difference between canopy-dominant *A. rubrum* versus understory saplings in 1996, following a dry year. Branches of overstory *A. rubrum* indicated a 21% increase in [TNC] in the dry treatment, whereas understory saplings had a 17% decline in [TNC], both largely the result of changes in [starch]. Such a differential response indicates the greater vulnerability of the understory saplings to drought, given the limitations of low light availability and limited root systems to access additional soil-water reserves. With respect to the latter response, drought tolerance is determined by a number of mechanisms that can be classified as those that postpone dehydration, including stomatal closure, leaf rolling, and deep rooting (Kramer 1980). By virtue of their small stature, understory species are not likely to exploit large soil volumes. With respect to the former response, understory species generally have lower net photosynthetic rate than overstory species, which may be related to their differences in irradiance levels they received (Sullivan et al. 1996), as well as reduced photosynthetic capacity (Boardman 1977). Overall, year-to-year variation in branch [TNC] of the understory species was less than that observed for the canopy dominants. Despite the potential susceptibility of the understory species, [TNC] of *C. florida* and *A. rubrum* (15 to 23%) never declined to levels that would be considered limiting with respect to growth or mortality (e.g., <1%). The greater mortality of *C. florida* relative to *A. rubrum* (Hanson et al. 2001) could not be attributed to factors related to carbon storage.

Carbohydrate Storage as an Indicator of Species Sensitivity to Chronic Drought

The best predictor of the dormant-season [TNC] in stems of some of the more sensitive species, including *A. rubrum, A. saccharum,* and *L. tulipifera,* was the previous year's precipitation (growing season or annual). Similarly, the best predictor of the dormant-season [starch] in stems of *L. tulipifera* was precipitation the previous growing season. As such, the data suggest that [TNC] is an integrator of the previous precipitation history and its resulting effects on physiological activity. But again, the response has not been observed beyond a single season to suggest that a chronic decline (or increase) is occurring. This indicates that although [TNC] can approach low levels following dry years, there was no carryover significantly limiting growth or survival in subsequent years. Some of the lowest [TNC] were observed in normal to above-normal precipitation years, where high growth depressed [TNC] in boles of *L. tulipifera,* but with no carryover to the following growing seasons, where growth was even higher than the preceding year. Current photosynthesis, in a subsequent growing season, evidently makes up for the previous year's deficit, whether induced by drought or high growth rates. The lack of carryover following drought years is applicable for hardwood ecosystems in the random drought region of the eastern United States, as defined by Hanson and Weltzin (2000), where resaturation of the soil in the dormant season and the absence of early season droughts is typical. This has been the pattern over the 7 years of the study (1993 to 2000). A carryover effect of declining [TNC] over multiple years might be observed in some eastern forest ecosystems, such as those on the prairie–forest border, where dormant-season resaturation may not occur.

Some of the variation between treatments, such as the significant pretreatment effect on soluble [CHO] in *Q. alba,* as well as the similar response in year 2000, could be explained, in part, by the size of the trees sampled. The treatment differences in *Q. alba* with the ambient treatment being lower than the other treatments, was most likely driven by these trees being somewhat smaller in stature. The regression of treatment means of soluble [CHO] on diameter at breast height (dbh) yielded a high coefficient of determination ($r^2 = 0.63$). Tree size also likely explained the occasional treatment differences in *A. saccharum.* The regression of treatment means of soluble [CHO] on dbh yielded a high coefficient of determination ($r^2 = 0.83$). The greater diameter of the ambient *A. saccharum* trees suggests that these trees were likely canopy dominants and had more of their crowns exposed to high light availability that would contribute to greater carbon assimilation than would be evident in trees that were more likely to be overtopped. Therefore, it is not surprising that the ambient *A. saccharum* would tend to have higher soluble [CHO] after a dry growing season. The concentrations were low in all treatments in 1996, following a dry year, but they were lowest in the smaller trees of the wet treatment. The CO_2 assimilation rates of the *Acer* species are comparatively low, exacerbating conditions that further limit carbon assimilation, including lateral shading and drought. In both cases, when treatment differences were evident, the treatment with the smallest trees of *A. saccharum* and *Q. alba* tended to have the lowest soluble [CHO]. Any consideration of the use of carbohydrate storage as an indicator of species sensitivity to chronic drought needs to consider tree size as a potential confounding factor.

Aside from such a confounding factor, some of the species that are considered to be somewhat drought sensitive tended to have higher [TNC] in years with higher precipitation. Starch concentration tended to be higher in the more drought-sensitive *A. saccharum* and *L. tulipifera.* It is interesting to note that the interannual differences, when evident, did not necessarily parallel the treatment trends. The treatment effects on [TNC] in *L. tulipifera* are even in the opposite direction as would be predicted from the interannual TNC responses. Other factors such as cooler temperatures in years with higher precipitation may cause this uncoupling by facilitating greater carbon gain as well as lower respiratory carbon loss in wetter years. Trees accumulate soluble carbohydrates and starch reserves in roots and shoots when carbon assimilation exceeds that consumed in respiration and anabolic events (Kozlowski et al. 1991). The depletion of [TNC] in the boles of *L. tulipifera* in the wet treatment was largely the result of a growth-driven decline in starch reserves in the years of high water availability. A similar growth-induced decline in stem [TNC] was reported following the application of irrigation to three tropical forest shrubs (*Psychotria* sp.) during the dry season (Tissue and Wright 1995).

Conclusion

Although treatment-induced changes in dormant-season [TNC] of stems and branches can be detected in some, but not all species, with [starch] impacted the most, these changes should be considered responses that are short term, reflecting the cumulative physiological activity in response to precipitation of the previous growing season, rather than to chronic, long-term drought. The difference between short- and long-term responses may simply be one of degree. The lack of cumulative impacts of the chronic TDE treatments on stored dormant season [TNC] of the mature components of upland deciduous forests support the current hypothesis that the mature trees are largely insensitive to low-level chronic precipitation changes. The differential response of *A. rubrum* at different canopy positions reflected the greater potential vulnerability of the understory to drought. The best predictor of the dormant-season [TNC] in stems of some of the more sensitive species, including *A. rubrum, A. saccharum,* and *L. tulipifera,* was precipitation (growing season or annual) the previous year, but there was no evidence of carryover to subsequent growing seasons that would indicate a chronic decline in TNC reserves. The annual variation in [TNC] as driven by precipitation was typically similar between stems and branches. Species considered drought sensitive, tended to have higher [TNC] in years with higher precipitation. In such normal to above-normal precipitation years, growth depresses [TNC] of *L. tulipifera,* but there is no carryover effect to the following growing season. Despite the evident short-term responses, we found little evidence to suggest that chronic changes in throughfall result in long-term effects on TNC reserves that would be responsible for driving changes in growth or mortality of the species studied at this location.

References

Addiscott TJ, Smith J, Bradbury N (1995) Critical evaluation of models and their parameters. J Environ Qual 24: 803–807.

Amthor JS (1994) Scaling CO2-photosynthesis relationships from the leaf to the canopy. Photosyn Res 39:321–350.

Ashworth EN, Stirm VE, Volenec JJ (1993) Seasonal variations in soluble sugars and starch within woody stems of Cornus sericea L. Tree Physiol 13:379–388.

Boardman NK (1977) Comparative photosynthesis of sun and shade plants. Ann Rev Plant Physiol 28:355–377.

Chapin FS III, Schulze E-D, Mooney HA (1990) The ecology and economics of storage in plants. Ann Rev Eco Syst 21:423–447.

Dubois M, Gilles KA, Hamilton JK, Rebers PA, Smith F (1956) Colorimetric method for determination of sugars and related substances. Anal Chem 28:350–356.

Dunn JP, Kimmerer TW, Potter DA (1987) Winter starch reserves of white oak as a predictor of attack by the two-lined chestnut borer, *Agrilus bilineatus* (Weber) (*Coleoptera:Buprestidae*). Oecologia 74:352–355.

Gholz HL, Cropper WP, Jr. (1991) Carbohydrate dynamics in mature *Pinus elliottii* var. *elliottii* trees. Can J For Res 21:1742–1747.

Hanson PJ, Weltzin JF (2000) Drought disturbance from climate change: Response of United States forests. Sci Total Environ 262:205–220.

Hanson PJ, Todd DE, Amthor JS (2001) A six-year study of sapling and large-tree growth and mortality responses to natural and induced variability in precipitation and throughfall. Tree Physiol 21:345–358.

Hunt ER Jr., Running SW (1992) Simulated dry matter yields for aspen and spruce stands in the North American boreal forest. Can J Remote Sensing 16:126–134

Kozlowski TT, Kramer PJ, Pallardy SG (1991) *The physiological ecology of woody plants.* Academic Press, San Diego, California.

Kramer PJ (1980) Drought stress, and the origin of adaptations. In Turner NC, Kramer PJ (Eds.). *Adaptation of plants to water and high temperature stress.* Wiley, New York, pp 7–22.

Latt CR, Nair PKR, Kang BT (2001) Reserve carbohydrate levels in the boles and structural roots of five multipurpose tree species in a seasonally dry tropical forest. For Ecol Manage 146:145–158.

McLaughlin SB, McConathy RK, Barnes RL, Edwards NT (1980) Seasonal changes in energy allocation by white oak (*Quercus alba*). Can J For Res 10:379–388.

Parker J, Patton RL (1975) Effects of drought and defoliation on some metabolites in roots of black oak seedlings. Can J For Res 5:457–463.

Running SW, Coughlan JC (1988) A general model of forest ecosystem processes for regional applications. Ecol Model 42:125–154.

Sullivan NH, Bolstad PV, Vose JM (1996) Estimates of net photosynthetic parameters for twelve tree species in mature forests of the southern Appalachians. Tree Physiol 16:397–406.

Sutton BG, Ting IP, Sutton R (1981) Carbohydrate metabolism of cactus in a desert environment. Plant Physiol 68: 784–787.

Tissue DT, Wright SJ (1995) Effect of seasonal water availability on phenology and the annual shoot carbohydrate cycle of tropical forest shrubs. Functional Ecol 9:518–527.

Weinstein DA, Beloin RM, Yanai RD (1991) Modeling changes in red spruce carbon balance and allocation in response to interacting ozone and nutrient stresses. Tree Physiol 9:127–146.

3. Water-Cycle Processes

6. Sensitivity of Sapling and Mature-Tree Water Use to Altered Precipitation Regimes

Stan D. Wullschleger and Paul J. Hanson

Long-term shifts in precipitation caused by changes in regional or global climate could exert a profound influence on the hydrologic regimes of terrestrial ecosystems. Increases and decreases in precipitation could result in altered patterns of runoff, evaporation, and soil-water content, and, in turn, these changes could have serious implications for catchment water yield, hydrologic budgets across broad spatial scales, impacts on terrestrial vegetation, and consequences for ecosystem goods and services upon which society depends. Because of their spatial extent, forests are especially important in this regard, and studies have implied that forests may be particularly vulnerable to changes in soil-water content as evidenced by results from manipulative studies on large trees (Cermák et al. 1993; Cienciala et al. 1994; Irvine et al. 1998; Phillips et al. 2001), field investigations on seedlings and saplings (Abrams et al. 1990, Kubiske and Abrams 1994), and computer simulations (Pastor and Post 1988; Running and Nemani 1991; Lüdeke et al. 1995). However, few studies have identified the relative sensitivity of multiple species to altered precipitation regimes and then integrated those sensitivities both for the forest overstory and understory into a picture depicting how species and climatic change may potentially impact forest water use in the future.

Although uncertainties surrounding the direction and magnitude of regional and global climatic change are high, the hydrologic cycle is generally expected to become more intense in a future climate (Mitchell et al. 1990; Loaiciga et al. 1996). As a result, some regions of the globe will experience an increased frequency of precipitation, and others will face a greater likelihood of drought because of decreased precipitation or increased evaporative demand (Rind et al. 1990). Because the consequences associated with such changes in precipitation are so great for terrestrial ecosystems, we addressed the impacts of altered precipitation on the whole-plant water demands of multiple species within an eastern deciduous forest in eastern Tennessee. Specifically, we were interested in addressing (1) the water-use requirements of forest understory and overstory species during natural and imposed drought, (2) the relative impact of drought on seedling, sapling, and mature-tree water use, and (3) whether species of the forest overstory and understory differed in their overall response to drought. This chapter primarily focuses on individual species, whereas in Chapter 21 we consider the manner in which components of the system are integrated to determine the overall response of forest stands to changes in precipitation.

Methods

Techniques for Measuring Sapling and Mature-Tree Water Use

Sap-flow gauges, which operate on the theory of heat balance, were used to estimate rates of whole-plant water use for *Acer rubrum* saplings of different stem diameters during 1995. These gauges apply a constant input of heat to the stem, and the resulting heat fluxes in the radial and vertical directions are measured with a thermopile and a series of thermocouples (Baker and van Bavel 1987). Gauge sizes were selected on the basis of stem diameter, which varied between 0.51 and 13.0 cm for the 53 saplings measured with this technique (Wullschleger et al. 1998). Closed-cell foam was used to insulate the gauge and stem, and the entire assembly was wrapped in aluminum foil. A data logger was located in each Throughfall Displacement Experiment (TDE) plot to record data every minute and store estimates of average sap flow every 30 min. Minimum sheath conductance was determined as the average value between 0100 and 0600 h, when whole-plant sap flow was assumed to be zero.

Sap flow rates for *Cornus florida* saplings in 1998 and mature trees in 2000 were determined with thermal-dissipation probes. These probes operate on the constant-power principle (Granier 1987) and consist of two probes, each 2.0 mm in diameter, that are inserted into the sapwood. The upper probe was installed at a height of 1.3 m and contained a heating element; the lower probe served as an unheated reference. Each probe contained a thermocouple, and the temperature difference between the two probes was influenced by the rate of sap velocity in the vicinity of the heated probe. All probes were installed on the north side of trees to avoid direct solar heating and were shielded to minimize diurnal temperature fluctuations.

Additional Plant and Soil Measurements

Leaf photosynthesis and stomatal conductance were measured on four *C. florida* saplings per TDE treatment plot during 1998 with a Li-Cor LI-6400 gas analyzer.

All measurements were made at a photosynthetically active radiation (PAR) of 500 μmol m^{-2} s^{-1} and a leaf temperature of 25°C. Leaf water potentials were measured on fully expanded leaves with a pressure chamber. Bark and sapwood thickness for each sapling and/or tree were measured with an electronic caliper on a 0.5-cm core extracted with an increment borer. Color differences were used to identify the boundary between sapwood and heartwood. Cross-sectional sapwood area was calculated from stem diameter and bark and sapwood thickness. Soil-water content measurements for the top 35 cm of soil were made with a time-domain reflectometer. Soil water contents were converted to water potential based on soil-water release curves for the AE horizon (Chapter 2, this volume; Hanson et al. 1998).

Results

Sapling Water Use and Physiology: *A. rubrum*

Sap flow rates for *A. rubrum* saplings measured in 1995 varied by size class (Table 6.1). Saplings growing in the forest understory with a stem diameter larger than 10 cm used water at an average rate almost 50 times that of saplings with a stem diameter of 1–2 cm. The maximum rate of sap flow for saplings greater than 10 cm in diameter was ~5900 g $tree^{-1}$ d^{-1} and occurred during midsummer (Wullschleger et al. 1998). These size-dependent differences in whole-tree water use were driven almost exclusively by leaf and/or cross-sectional sapwood area; water-use rates were not different among size classes when expressed on either a leaf-area or sapwood-area basis (Table 6.1).

Water-use rates for all *A. rubrum* saplings expressed on the basis of leaf area varied both throughout the season and among the three treatment plots of the TDE (Fig. 6.1A). Rates of water use were greatest for understory saplings in the early summer, decreasing to minimum rates in late July, with rates recovering by the October sampling date. Soil-water potentials also varied throughout much of the season, with water potentials generally declining below −1.2 MPa on July 16 and below −1.7 MPa on September 10 (Fig. 6.1B).

Table 6.1. Water-use rates for understory *A. rubrum* saplings of different diameter classes expressed per unit tree, per unit leaf area, and per unit cross-sectional sapwood area

	Water-use rates		
Diameter class (cm)	(g $tree^{-1}$ d^{-1})	(g m^{-2} d^{-1})	(g dm^{-2} d^{-1})
< 1	16 ± 2 b	121 ± 15 a	5796 ± 702 a
1–2	78 ± 16 c	105 ± 21 a	4662 ± 918 a
2–5	596 ± 155 c	134 ± 27 a	5472 ± 1098 a
5–10	1771 ± 484 b	162 ± 39 a	6282 ± 1548 a
> 10	3922 ± 1044 a	124 ± 34 a	4572 ± 1242 a

Note: Values (mean ± SE) are means of all sampling dates in 1995 (seven dates). Data within a column followed by the same letter are not significantly different at $p = 0.05$.

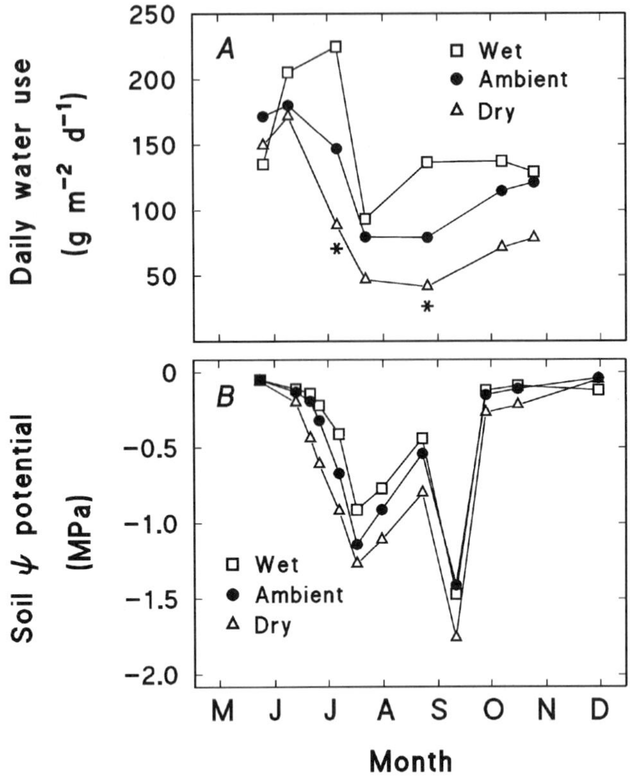

Figure 6.1. Seasonal patterns of (A) daily water use (expressed per unit leaf area) for *A. rubrum* saplings and (B) soil water potential for the three TDE treatment plots in 1995. The two asterisks indicate that significant differences in transpiration between the wet, ambient, and dry plots were observed during the July 6 and August 27 sampling sessions.

Treatment effects were observed for saplings across the TDE plots, with significant differences in water use occurring among the wet, ambient, and dry plots on the July 6–7 ($p = 0.01$) and August 27–28 ($p = 0.04$) sampling dates. Rates of water use on these dates were 61–70% lower for saplings in the dry plot compared to saplings in the wet plot and 40–48% lower compared to saplings in the ambient plot. Soil-water potentials on July 6 were −0.4 MPa in the wet plot, −0.7 MPa in the ambient plot, and −0.9 MPa in the dry plot, whereas on August 23, soil-water potentials were −0.4 MPa in the wet plot, −0.5 MPa in the ambient plot, and −0.8 MPa in the dry plot (Fig. 6.1B). Treatment-imposed differences in soil-water potential were sufficient to drive changes in leaf water potential for saplings on the three TDE plots. On July 6, leaf water potentials were −1.6 MPa for saplings in the wet plot, −1.7 MPa in the ambient plot, and −2.1 MPa in the dry plot, whereas on August 30, leaf water potentials were −1.9 MPa in the wet plot, −2.1 MPa in the ambient plot, and −2.4 MPa in the dry plot.

No TDE-imposed effects on sapling water use, soil-water potential, or leaf water potential were observed at or near the mid-July sampling period.

Sapling Water Use and Physiology: *C. florida*

Sapling water use varied throughout the 1998 season for understory *C. florida* (Fig. 6.2). Rates were highly variable and depended on weather (radiation and vapor pressure deficit), but whole-sapling water use on all plots peaked at just over 4 kg d^{-1} during late June and early July. Thereafter, rates of daily maximum water use for saplings in the ambient and dry plots began to decline, whereas rates of water use for saplings in the wet plot remained fairly stable (Fig. 6.2). Treatment differences in water use were most noticeable for saplings in the dry plot, where measurements showed a 23–54% reduction in daily water use compared to saplings in the ambient and wet plots. Low rates of sapling water use in

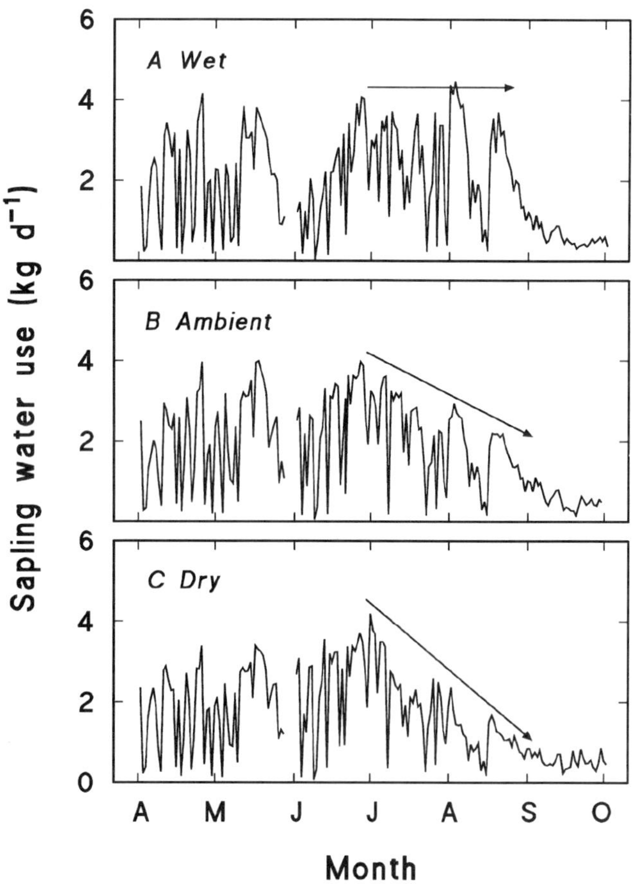

Figure 6.2. Seasonal pattern of daily water use (expressed per tree) for flowering *C. florida* saplings growing on each of the three TDE treatments during 1998.

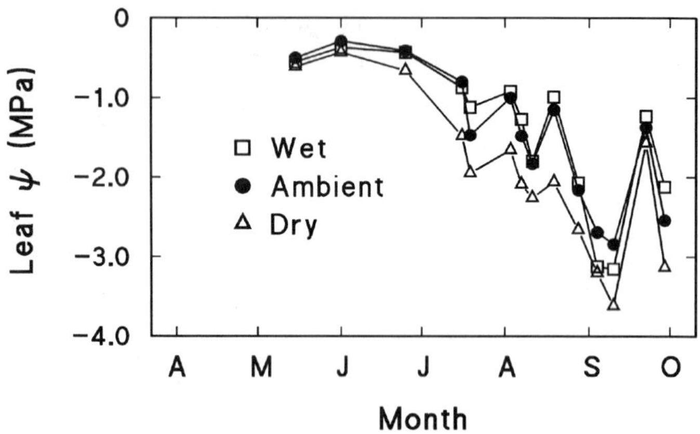

Figure 6.3. Leaf water potential as measured during the 1998 season for *C. florida* saplings growing on the wet, ambient, and dry TDE plots.

the dry plot coincided with extremely dry soil-water conditions, which contributed to low leaf water potentials. Midday leaf water potentials in the dry plot declined from −0.6 MPa in late June to below −2.3 MPa in mid-August (Fig. 6.3). Leaf water potentials were, in general, more negative for saplings on the dry plot compared to either the ambient or wet plots. On two sampling dates (in particular, July 7 and July 19), differences in leaf water potential between TDE treatments were significant (Table 6.2). Such differences in leaf water potential were most noticeable between the dry plot and those of the ambient and wet plots. Differences were observed among plots not only for leaf water potential but also for measurements of photosynthesis and stomatal conductance (Table 6.2).

Treatment-imposed differences in sapling water use and leaf water potential continued into mid-August, at which time a moderate precipitation event occurred and, as a result, water use by saplings increased on all TDE treatments (Fig. 6.2). However, the greatest increase in water use occurred for saplings on the wet plot,

Table 6.2. Measured rates of leaf photosynthesis, stomatal conductance, and leaf water potential (mean ± SE) for *C. florida* saplings growing on the three TDE treatment plots

Date/treatment	Photosynthesis (μmol m^{-2} s^{-1})	Stomatal conductance (mmol m^{-2} s^{-1})	Leaf water potential (MPa)
July 7			
Wet	2.9 ± 0.4 a	37 ± 6 a	−1.3 ± 0.06 a
Ambient	1.5 ± 0.4 b	26 ± 1 ab	−1.5 ± 0.13 a
Dry	0.8 ± 0.4 b	14 ± 2 b	−2.1 ± 0.05 b
July 19			
Wet	3.8 ± 0.4 a	70 ± 9 a	−1.0 ± 0.07 a
Ambient	3.2 ± 0.1 b	59 ± 3 a	−1.2 ± 0.05 a
Dry	2.2 ± 0.2 b	27 ± 3 b	−2.1 ± 0.12 b

Note: Data were taken on two dates in 1998 during a midsummer period when soil water potential was declining because of a lack of precipitation and imposed TDE treatments. Data within a column followed by the same letter are not significantly different at $p = 0.05$.

followed by saplings on the ambient plot, and then saplings on the dry plot. Rates of water use on August 18 for the dry treatment were only 47% of those observed on the wet plot and about 75% of those observed on the ambient plot. Thereafter, water use declined precipitously as leaf water potentials fell below -3.0 MPa for saplings growing in the three TDE treatments (Fig. 6.3). This period of severe drought induced premature leaf senescence in all plots, such that late-season precipitation on September 22 elicited essentially no noticeable response in sapling water use. The lack of response in water use occurred even though the leaf water potential of leaves remaining on the saplings did show a modest increase following the precipitation event (Fig. 6.3). However, total leaf area per sapling was so reduced from that of a few weeks earlier that water use remained low despite increases in soil and leaf water potential.

Diurnal patterns of water use for *C. florida* saplings were also strongly affected by treatments imposed by the TDE (Fig. 6.4). Midday (1300 h) rates of water use on July 18 were 0.53 kg h^{-1} for saplings in the wet plot, 0.37 kg h^{-1} for saplings in the ambient plot, and 0.26 kg h^{-1} for saplings in the dry plot. Water use rates decreased by the August 18–19 sampling date, although treatment differences were still observed. Rates of water use during the early afternoon of August 19

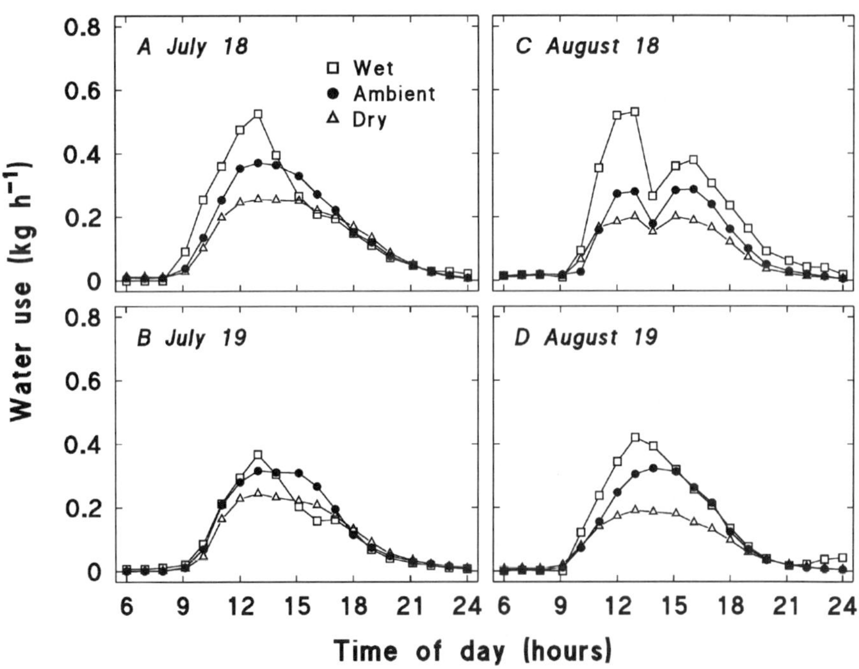

Figure 6.4. Diurnal rates of water use for *C. florida* saplings measured on the three TDE treatment plots during the (A, B) July 18–19 and (C, D) August 18–19, 1998, sampling dates. It was on these dates that significant differences in water use were observed among sapling water use for the wet, ambient, and dry treatments.

were almost 55% lower for saplings in the dry plot than for those in the wet plot and 37% lower compared to saplings in the ambient plot (Fig. 6.4).

Integration of daily rates of water use for saplings on the three TDE plots yielded long-term estimates of water use. Over the 225-day season, for example, the total amount of water used by *C. florida* was ~344 kg for saplings on the wet plot, 325 kg on the ambient plot, and 295 kg on the dry plot. A similar analysis applied only to data collected after July 1 indicated that water-use rates were ~175 kg for saplings on the wet plot, 143 kg for the ambient plot, and 124 kg for the dry plot. In either case, TDE-imposed reductions in water use between the ambient and dry plots were on the order of −9 to −13%.

Mature-Tree Water Use and Physiology

Large trees that occupied the forest overstory on the TDE had peak rates of daily water use that ranged from 40 to 325 kg d^{-1} and higher. The greatest rates of whole-tree water use were recorded for a 45-cm diameter *Liriodendron tulipifera* (340 kg d^{-1}) tree, followed by 250 kg d^{-1} for a 38-cm diameter *Carya* spp., and ~90 kg d^{-1} for a 48-cm diameter *Quercus alba* tree. Water-use rates for species that dominated the overstory were fairly uniform compared to rates recorded for codominant and/or suppressed individuals within the canopy. This nonuniformity was especially noticeable in the *A. rubrum* trees, where rates of water use exhibited considerable variation, ranging from 35 kg d^{-1} for one particularly suppressed individual to 180 kg d^{-1} for trees that occupied more-dominant canopy positions.

Seasonal patterns of whole-tree water use showed considerable day-to-day variation associated with changing daily radiation and vapor pressure deficit (Fig. 6.5). Furthermore, water-use rates for three species measured on the ambient TDE plot showed noticeable seasonal differences when compared to trees on the dry plot. These treatment-imposed differences in water use began in mid to late July and persisted throughout the remainder of the season. Although trees on the ambient and dry plots are not directly comparable because of differences in size and canopy position, the magnitude of a treatment effect can, as a first approximation, be examined by integrating daily water use for each tree over the season. In the case of *Q. prinus,* water use for the tree in the ambient plot was about 14% lower than that of the tree in the dry plot (3285 kg versus 2825 kg); whereas for *Q. alba,* the difference amounted to a 12% reduction for the tree in the ambient compared to the dry plot (5505 kg versus 4825 kg). Treatment differences in the codominant *A. rubrum* are not so obvious, caused, in part, by the gradual decline in water use associated with progressive canopy shading during the season, but treatment differences were still on the order of 10–15% (Fig. 6.5).

The seven overstory species examined on the TDE exhibited considerable variation in their apparent response to soil-water potential (i.e., drought tolerance) as reflected in species-specific measurements of water use. All trees had rates of water use relative to an observed maximum (T/T_{max}) close to 1.0 at soil-water potentials between 0 and −0.5 MPa (Table 6.3). However, as soil-water avail-

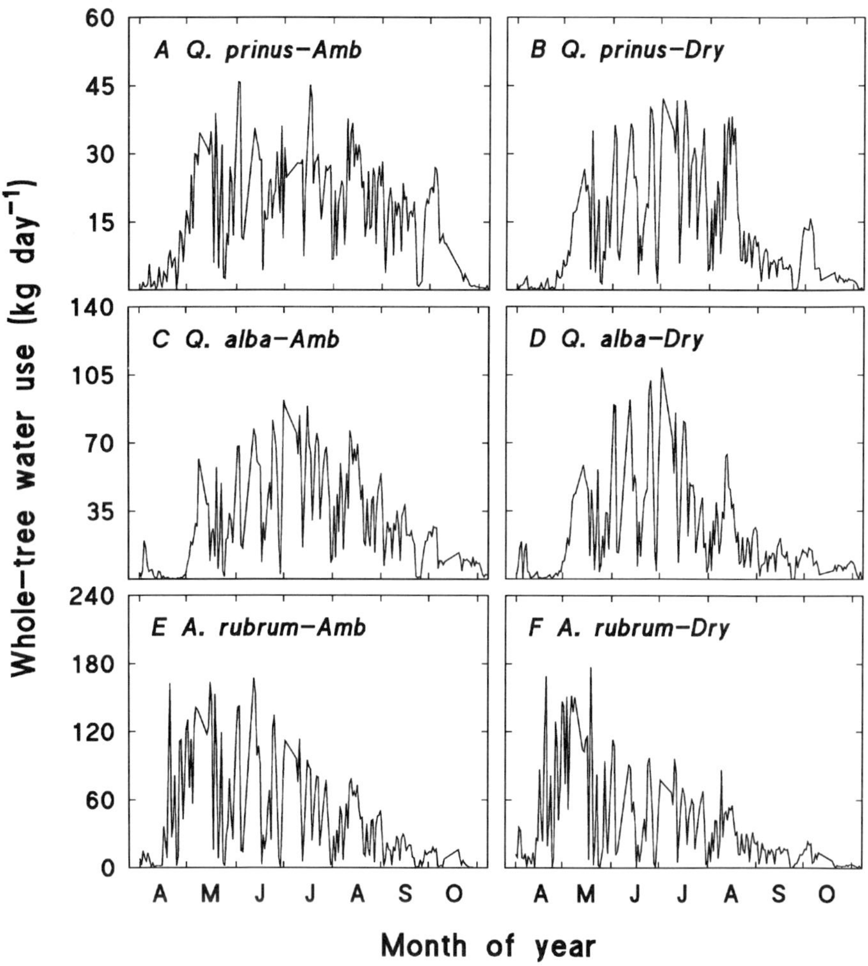

Figure 6.5. Whole-tree estimates of water use for *Q. prinus* (A, B), *Q. alba* (C, D), and *A. rubrum* (E, F) trees of the forest overstory. Data are shown for trees from both the ambient and dry plots. Note decline in late-season water use caused by the TDE-imposed drought.

ability declined below −0.5 MPa and then below −1.0 MPa, a general separation of responses occurred, indicating that some species were able to maintain water use to progressively lower and lower soil-water potentials. *Q. rubra* appeared most sensitive to declining soil-water potentials, whereas *Q. prinus, Carya* spp., and *Q. stellata* were less sensitive (Table 6.3). A similar response of relative water use to soil-water potential was also observed for three understory species, and *A. rubrum* was considerably different from *C. florida* and *Prunus serotina,* two other common understory species on the TDE (Table 6.3).

Table 6.3. Species-specific estimates of relative water use (T/T_{max}) as measured using sap-flow techniques across a range of soil-water potentials for trees and saplings growing on the TDE

	Soil-water potential (MPa)		
Species	0 to −0.5	−0.5 to −1.0	< −1.0
Overstory species			
A. saccharum	0.92	0.50	0.23
Carya spp.	0.94	0.79	0.48
N. sylvatica	0.97	0.67	0.37
Q. alba	0.96	0.52	0.23
Q. prinus	0.98	0.74	0.34
Q. rubra	0.97	0.23	0.14
Q. stellata	0.98	0.56	0.56
Understory species			
A. rubrum	0.97	0.59	0.21
C. florida	0.98	0.35	0.15
P. serotina	0.91	0.22	0.09

Discussion

Seedlings and saplings of the forest understory have been shown to be affected by modest soil water deficits during dry periods in normal years and serious drought in unusual years (Federer 1980; Abrams and Mostoller 1985; Abrams et al. 1990; Kubiske and Abrams 1994; Caspersen and Kobe 2001). Most often the impact of drought on understory trees has been characterized by leaf-level measurements of photosynthesis, stomatal conductance, midday and predawn leaf water potential, and components of leaf water potential including leaf osmotic and turgor potentials (Tschaplinski et al. 1998; Gebre et al. 1998). These studies, although focusing almost exclusively on leaf-specific processes, have shown that seedlings and saplings are sensitive to drought. At the same time, it has often been observed that species exhibit a range of characteristics that make them more or less able to withstand the many impacts of drought. In a study with particular relevance to the results from the TDE, Bahari et al. (1985) examined the response of six woody species to seasonal drought using saplings growing in an *Quercus–Carya* forest of central Missouri. The species examined (*Q. velutina, Q. alba, Q. rubra, C. florida, A. saccharum,* and *Juniperus virginiana*), displayed a wide range of observed tolerance to seasonal drought. Although the *Quercus* species were generally tolerant of reduced soil-water availability, *C. florida* exhibited an accentuated response to drought, as evidenced by predawn water potentials that fell below −3.4 MPa during an especially dry period in 1980. By comparison, predawn water potentials for the other species seldom decreased below −2.0 MPa. Such an inability of *C. florida* to avoid low predawn leaf water potentials was, according to Bahari et al. (1985), attributable to the comparatively shallow rooting habit of this species.

Although our data cannot attest to the rooting characteristics of *C. florida,* measurements of sap flow and leaf water potential made during the 1998 season do support the general finding that among saplings of the forest understory, *C.*

florida stands out as a drought-intolerant species. This characteristic was seen in the diurnal pattern of water use for saplings on the three TDE plots, where imposed reductions in water availability decreased midday water use on the dry plot by 37–55% compared to that for saplings on the ambient and wet plots. Manipulation of throughfall on the TDE decreased total seasonal water use by 9–13% for saplings on the dry compared to ambient and wet plots. Furthermore, ancillary measurements of leaf water potential made on August 19, 1998 indicated that midday leaf water potentials for *C. florida* (−2.0 MPa) were far more negative than those observed for either *A. rubrum* (−1.3 MPa), *P. serotina* (−1.3 MPa), or *Fagus grandifolia* (−1.4 MPa). Interestingly, even though all saplings were growing in conditions of similar soil-water potential, *C. florida* was unique in not being able to access available soil water because of presumed limitations in root distribution or to otherwise effectively regulate water loss via stomatal closure. This inability to moderate leaf and whole-plant water balance resulted in low leaf water potentials, severe leaf wilting, and (at least for the saplings studied here) high rates of mortality as observed in the spring of the following year. Evidence for the apparent sensitivity of *C. florida* to drought was also seen in the pattern of comparative sapling mortality on the wet, ambient, and dry plots during 6 years of the TDE (Chapter 15, this volume; Hanson et al. 2001).

Although previous research has focused considerable attention on tree seedling and sapling responses to reduced soil-water availability, little research has been conducted from which to describe the response of larger trees to drought (Cermák et al. 1993; Lu et al. 1995; Irvine et al. 1998). This lack of information is especially true for trees in mesic temperate deciduous forests, presumably because soil moisture limitations have traditionally been thought to be less significant in these ecosystems. Our findings show, however, that a 33% reduction in throughfall, as implemented on the TDE, was sufficient to reduce whole-tree water use in several species of the forest overstory by 12–14% during a single season. Such reductions were observed by comparing seasonal patterns of water use for individual species from the dry, ambient, and wet plots.

Slightly more dramatic differences between individual trees on the dry and ambient (or wet) plots could be observed by focusing on specific periods during the season and/or restricting our analysis of water use to specific ranges of soil-water potential. For example, natural and TDE-imposed droughts were most common during the mid to late summer, with treatment-induced reductions in soil water potential becoming evident in late June and early July, with considerable differences between TDE plots often persisting into September and October. In the case of *Q. prinus* and *Q. alba* (see Fig. 6.5), water use during the month of September for trees on the dry plot were 51–61% lower than rates measured on trees in the ambient plot. Treatment differences in water use were most evident during the early stages of drought development because, as soils dried, all plots tended to achieve similar levels of soil water potential.

An interesting outcome from our whole-tree studies was that not all species responded similarly to imposed drought. Rather, a range of responses was observed, some severe and others not so severe, leading to the conclusion that tolerance to drought differs among species. For saplings of the forest understory, *C. florida* was clearly not able to prevent deleterious effects from restricted soil-water supplies. For overstory trees, a range of sensitivities was also observed. At soil-water potentials between -0.5 and -1.0 MPa, *Carya* spp. and *Q. prinus* were able to maintain modest rates of relative water use (0.79 and 0.74, respectively), whereas *Q. rubra* (0.23) could not. The fact that *Q. rubra* typically occupies mesic sites has been emphasized in earlier physiological studies with saplings (Abrams et al. 1990) and this species clearly stands apart from other *Quercus* species like *Q. stellata* and *Q. prinus* in this regard. More recently, Oren and Pataki (2001) quantified water use for several species in an upland hardwood forest in North Carolina and showed that *C. tomentosa* was able to maintain rather high rates of water use during a late-July period of soil-water deficit compared to other *Quercus* and *Acer* species in which water use declined 50% or more from predrought levels. This observation is indeed consistent with our sap-flow measurements, which suggest that, for all the species studied on the TDE, *Carya* ranked as one of the more drought tolerant components of this upland *Quercus* forest ecosystem. As such, it deserves more investigation.

Summary

A range of sap-flow and leaf-level physiological techniques were used to assess the water use characteristics of seedlings, saplings, and large trees in response to both seasonal and TDE-imposed droughts. In relation to the response of trees to alterations in water availability, our data on rates of water use for a variety of overstory and understory species are sufficient to speculate that (1) mature trees are less likely to be impacted by drought than are seedlings and saplings, (2) the relative ranking of drought tolerance among understory saplings is *A. rubrum* > *C. florida* = *P. serotina*, and (3) the relative ranking of drought tolerance for trees of the forest overstory is *Carya* spp. > *Q. prinus* = *Q. stellata* > *Nyssa sylvatica* > *Q. alba* > *Q. rubra* = *A. rubrum*. The exact mechanisms that contribute to observed differences between saplings and mature trees and to differences between species in the forest overstory and understory cannot be addressed solely on the data presented here. Explanations perhaps lie in leaf-level physiology, rooting characteristics, and whole-tree considerations reflecting species-specific differences in leaf area, cross-sectional sapwood area, and hydraulic architecture.

References

Abrams MD, Mostoller SA (1995) Gas-exchange, leaf structure and nitrogen in contrasting successional tree species growing in open and understory sites during a drought. Tree Physiol 15:361–370.

Abrams MD, Schulze JC, Kleiner KW (1990) Ecophysiological responses in mesic versus xeric hardwood species to an early-season drought in central Pennsylvania. For Sci 36:970–981.

Bahari ZA, Pallardy SG, Parker WC (1985) Photosynthesis, water relations, and drought adaptation in six woody species of oak-hickory forests in Central Missouri. For Sci 31:557–569.

Baker JM, van Bavel CHM (1987) Measurement of the mass flow of water in stems of herbaceous plants. Plant Cell Environ 10:777–782.

Caspersen JP, Kobe RK (2001) Interspecific variation in sapling mortality in relation to growth and soil moisture. Oikos 92:160–168.

Cermák J, Matyssek R, Kučera J (1993) Rapid response of large, drought-stressed beech trees to irrigation. Tree Physiol 12:281–290.

Cienciala E, Lindroth A, Cermák J, Hallgren JE, Kučera J (1994) The effects of water availability on transpiration, water potential and growth of *Picea abies* during a growing season. J Hydrol 155:57–71.

Federer CA (1980) Paper birch and white oak saplings differ in responses to drought. For Sci 26:313–324.

Gebre GM, Tschaplinski TJ, Shirshac TL (1998) Water relations of several hardwood species in response to throughfall manipulation in an upland forest during a wet year. Tree Physiol 18:299–305.

Granier A (1987) Evaluation of transpiration in a Douglas-fir stand by means of sap flow measurements. Tree Physiol 3:309–320.

Hanson PJ, Todd DE, Amthor JS (2001) A six-year study of sapling and large-tree growth and mortality responses to natural and induced variability in precipitation and throughfall. Tree Physiol 21:345–358.

Hanson PJ, Todd DE, Huston MA, Joslin JD, Croker J, Augé RM (1998) *Description and field performance of the Walker Branch Throughfall Displacement Experiment: 1993–1996.* ORNL-TM-13586. Oak Ridge National Laboratory, Oak Ridge, Tennessee.

Irvine J, Perks MP, Magnani F, Grace J (1998) The response of *Pinus sylvestris* to drought: stomatal control of transpiration and hydraulic conductance. Tree Physiol 18:393–402.

Kubiske ME, Abrams MD (1994) Ecophysiological analysis of woody species in contrasting temperate communities during wet and dry years. Oecologia 98:303–312.

Loaiciga HA, Valdes JB, Vogel R, Garvey J, Schwarz H (1996) Global warming and the hydrologic cycle. J Hydrol 174:83–127.

Lu P, Biron P, Bréda N, Granier A (1995) Water relations of adult Norway spruce (*Picea abies* (L) Karst) under soil drought in the Vosges mountains: water potential, stomatal conductance and transpiration. Ann Sci For 52: 117–129.

Lüdeke MKB, Dönges S, Otto RD, Kindermann J, Badeck FW, Ramge P, Jäkel U, Kohlmaier GH (1995) Responses in NPP and carbon stores of the northern biomes to a CO_2-induced climatic change, as evidenced by the Frankfurt Biosphere Model (FBM). Tellus 47B:191–205.

Mitchell JFB, Manabe S, Meleshko V, Tokioka T (1990) Equilibrium climate change and its implications for the future. In Houghton JT, Jenkins GT, Ephraums JJ (Eds) *Climate change, The IPCC second assessment.* Cambridge University Press, Cambridge, England, pp 131–172.

Oren R, Pataki DE (2001) Transpiration in response to variation in microclimate and soil moisture in southeastern deciduous forests. Oecologia 127:549–559.

Pastor J, Post WM (1988) Response of northern forests to CO_2-induced climate change. Nature 334:55–58.

Phillips N, Bergh J, Oren R, Linder S (2001) Effects of nutrition and soil water availability on water use in a Norway spruce stand. Tree Physiol 21:851–860.

Rind D, Goldberg R, Hansen J, Rosenzweig C, Ruedy R (1990) Potential evaporation and the likelihood of future drought. J Geophys Res 95:9983–10004.

Running SW, Nemani RR (1991) Regional hydrologic and carbon balance responses resulting from potential climatic change. Clim Change 19:342–368.

Tschaplinski TJ, Gebre GM, Shirshac TL (1998) Osmotic potential of several hardwoods species as affected by manipulation of throughfall precipitation in an upland oak forest during a dry year. Tree Physiol 18:291–298.

Wullschleger SD, Hanson PJ, Tschaplinski TJ (1998) Whole-plant water flux in understory red maple exposed to altered precipitation regimes. Tree Physiol 18:71–79.

7. Stomatal Behavior of Forest Trees in Relation to Hydraulic, Chemical, and Environmental Factors

Robert M. Augé

Stomata regulate plant carbon gain, water loss, and other physiological determinants of forest productivity. Our ability to assess impacts of environmental changes on forest ecosystems relies heavily, therefore, on understanding stomatal function and control. The recent discovery of nonhydraulic, root-sourced stress signals is changing our understanding of how plants "sense" and how stomata respond to fluctuations in soil moisture. Formerly, it had been widely held that stomatal conductance (g_s) was hydraulically regulated by leaf water potential (Ψ) or turgor potential (Ψ_p) Kramer and Boyer 1995), at least in anisohydric plants (Tardieu et al. 1996; Tardieu and Simmoneau 1998). However, there are several instances in which g_s was inhibited in drying soils even in the absence of perturbations in leaf water status (Davies et al. 1994). Such studies suggest that stomatal closure resulting from soil-water depletion can be mediated by changes in root water status through effects on the chemical flow of information from root to shoot.

This behavior has been explained as follows. Soil dehydration alters root metabolism, which results in the production of a nonhydraulic chemical signal, probably a positive inhibitor (Gowing et al. 1990) that moves via the transpiration stream to leaves, where it affects various physiological responses. A drop in the water status of part of the root system (e.g., roots near the soil surface) may trigger or elevate the signal, which then reduces g_s, even though other portions of the root system are exposed to sufficient moisture to fully supply shoot needs. Nonhydraulic inhibition of stomatal opening can be substantial. Declines in g_s of up to 50% or more (relative to well-watered controls) have been reported for both woody (Khalil and Grace 1993; Croker et al. 1998) (Fig. 7.1) and herbaceous species (Zhang and Davies 1989a, 1990a).

Concentrations or rates of flow of several components of xylem sap have been suggested to play a role in chemical root-to-shoot signaling of soil drying: abscisic acid (ABA) (Davies et al. 1994), pH (Hartung et al. 1998), malate (Patonnier et al. 1999), cation/anion balance (Hartung and Radin 1989), and cytokinins (Neuman et al. 1990; Fußeder et al. 1992). Most evidence implicates the involvement of ABA. ABA is synthesized in increasing quantities in dehydrating root tips (Liu et al. 2001a, 2001b), where it or its conjugates (Hansen and Dorffling 1999; Sauter and Hartung 2000) move into xylem, to shoots, and to the leaf apoplast surrounding guard cells. It has long been recognized that ABA is a potent stomatal-closing agent [e.g., Jones and Mansfield (1970)]. Roots produce ABA in increasing amounts as soil water declines [e.g., Cornish and Zeevaart (1985); Zhang and Davies (1987)], and these increases have been correlated with subsequent increases in ABA content of stem and petiole xylem sap (Zhang and Davies 1990a, 1990b; Tardieu et al. 1993). Stomatal conductance is often more closely correlated with xylem ABA concentrations ([ABA]) than with leaf or soil-water status.

Stomata may also be responding to root-sourced changes in xylem-sap pH that occur in response to changes in soil moisture (Hartung et al. 1998; Holbrook et al. 2002). Small changes in the flux of H^+ to leaves via xylem can create large

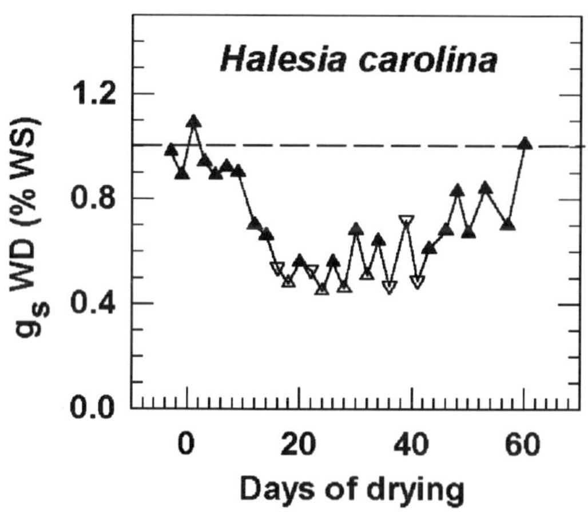

Figure 7.1. Stomatal conductance (g_s) during the drying period of WD trees (about half the root system unwatered) relative to WS control trees (about half the root system severed) of *Halesia carolina* (see page 103 for definitions of WS and WD). Daily relative g_s of each of the eight WD trees was calculated as absolute g_s of an individual WD tree divided by the average absolute g_s of the four WS trees for that day, expressed as percent: WD relative g_s = (WD absolute g_s)/(WS average absolute g_s) $\times$ 100. Each individual tree's absolute or relative g_s value represents the average of four leaves. Symbols show when average g_s of WD trees (n = 8) were significantly different from WS controls (solid triangle, NS; ▲, $p \leq 0.05$; ∇, $p \leq 0.1$), as determined by analysis of variance. [Redrawn from Croker et al. (1998), by permission of Oxford University Press.]

changes in apoplastic pH, with alkalization of leaf apoplast, in turn, enhancing the release of ABA from leaf mesophyll cells into the apoplast surrounding guard cells (Hartung et al. 1988; Netting 2000). pH gradients in leaf tissues control ABA distribution in the leaf and ABA concentrations at the primary site(s) of action at guard-cell complexes and hence influence stomatal aperture and transpirational water loss (Hartung et al. 1998). Much evidence implicates both ABA and xylem-sap pH in the regulation of g_s [e.g., Jia and Zhang (1997); Zhang et al. (1997); Hartung et al. (1998); Wilkinson et al. (1998)].

Hydraulic signals, whether originating in response to changes in atmospheric or soil-water content, probably act in concert with chemical signals to regulate g_s (Davies et al. 1994; Saliendra et al. 1995; Thomas and Eamus 1999; Netting 2000). Whether the ABA moving to guard-cell complexes is predominantly of leaf or root origin (Saliendra et al. 1995; Liang et al. 1997; Thompson et al. 1997; Hartung et al. 1998), changes in g_s have been correlated with foliar ABA levels (Thomas and Eamus 1999) and more specifically with [ABA] in leaf epidermis (Zhang et al. 1987) and xylem sap (Davies et al. 1994). Stomatal sensitivity to xylem [ABA] can be altered by leaf Ψ, temperature, cytokinin activity, or light intensity, and changes in stomatal sensitivity to xylem [ABA] may play as important a role as actual xylem [ABA] in regulating g_s during drought (Tardieu et al. 1993; Tardieu and Davies 1993; Heckenberger et al. 1996; Correia et al. 1997; Socias et al. 1997). Often, g_s during diurnal time-course studies is not correlated with leaf Ψ or xylem [ABA] but is better correlated with these hydraulic and chemical parameters from day to day [e.g., Socias et al. (1997)].

Although the mechanism of action is still being elucidated, nonhydraulic, root-to-shoot signaling appears to be an important component of plant response to drought conditions (Davis et al. 1994). The ability to respond dynamically to changes in available soil moisture usually improves a plant's long-term water-use efficiency and survival (Ludlow et al. 1989; Mansfield and McAinsh 1995). We conducted a series of experiments to (1) characterize stomatal response of temperate, deciduous tree species to nonhydraulic root-to-shoot signals of soil drying, (2) measure the foliar dehydration tolerance of these tree species, (3) determine whether seasonal g_s of forest trees is better correlated with hydraulic, chemical, or environmental variables, and (4) determine if g_s of forest trees is inhibited during soil-drying episodes before leaf water status is affected.

Stomatal Response of Deciduous Trees to the Nonhydraulic Signaling Process

Although nonhydraulic, root-to-shoot signaling of soil drying has been well documented for herbaceous species, it has been argued that this process may not figure prominently in regulating responses to soil drought in woody species; it would be an inefficient mechanism for rapid stomatal control because of lengthy transport time (Saliendra et al. 1995). If nonhydraulic signaling does not play a role in regulating drought responses of mature woody species, it may be important in seedling and sapling development, where it has, in fact, been observed several times [e.g., *Malus* × *domestica* (Gowing et al. 1990), *Prunus dulcis* (Fußeder et al. 1992), *Rosa hybrida* (Augé and Duan 1991), *Acer pseudoplatanus* (Khalil and Grace 1993), *Prunus avium* × *pseudocerasus* (Gowing et al. 1993), and some coniferous species (Jackson et al. 1995)].

Methods

To quantify the extent of response to strictly nonhydraulic root-to-shoot signaling of soil drying among temperate forest trees, we examined 12 species under standard conditions in a rain shelter (Croker et al. 1998; Augé and Moore 2002). Two- to three-year-old saplings of the following species were planted with the root system divided between two 11-L plastic pots: *Acer rubrum* L., *A. saccharum* Marsh., *Cornus florida* L., *Chionanthus virginicus* L., *Halesia carolina* L., *Liriodendron tulipifera* L., *N. sylvatica* Marsh., *Oxydendrum arboreum* [L.] DC., *Q. prinus* L., *Q. rubra* L., *Q. alba* L., and *Q. acutissima* Carruthers.

Each species had three split-root treatments. Water was withheld from one pot for a sustained drying period for one group of seedlings (WD). A two-control system was used as follows. In one group, one-half of the root system was left intact and watered and the other half was severed prior to the drought episode (WS). In the second group, both halves of the root system were left intact and were fully watered throughout the experiment (WW). The WS individuals provided a control for potential effects on leaf Ψ caused by reducing the water supply of the WD plants by half, relative to WW plants. Half-dried and WS trees had largely similar amounts of roots receiving water regularly, so decreases in the g_s of WD relative to WS trees were not caused by a direct hydraulic signal linked to a difference in water-gathering capacity. Bulk-soil matric potential (Ψ_M) was recorded every 4 h throughout the drying period for the drying pots of WD plants, with heat dissipation sensors as described by Augé et al. (1994). Stomatal conductance was measured with diffusion porometers about every other day before and during the 60 to 90-day drying periods. Root dry weights were determined at the end of the experiment to gauge how evenly roots were divided between the two pots and to compare the size of the root systems among treatments. Full experimental details are provided in Croker et al. (1998).

Results

Drying reduced the relative g_s of WD trees (WD as a percentage of WS controls) of all genera except *Quercus* for most days during the drying episode. The largest single-day declines in average relative g_s were to about 40% in *L. tulipifera, H. carolina,* and *A. rubrum,* to about 50% in *C. florida, N. sylvatica,* and *A. saccharum,* to about 60% in *C. virginicus* and *O. arboreum,* and to about 80% in each of the four *Quercus* species (Fig. 7.2). Relative g_s declined and mostly remained low for some species (*C. florida, C. virginicus, L. tulipifera, A. saccharum,* and *Q. rubra*). In other species (*H. carolina, A. rubrum, N. sylvatica,* and *O. arboreum*), relative g_s declined for several days and then recovered during

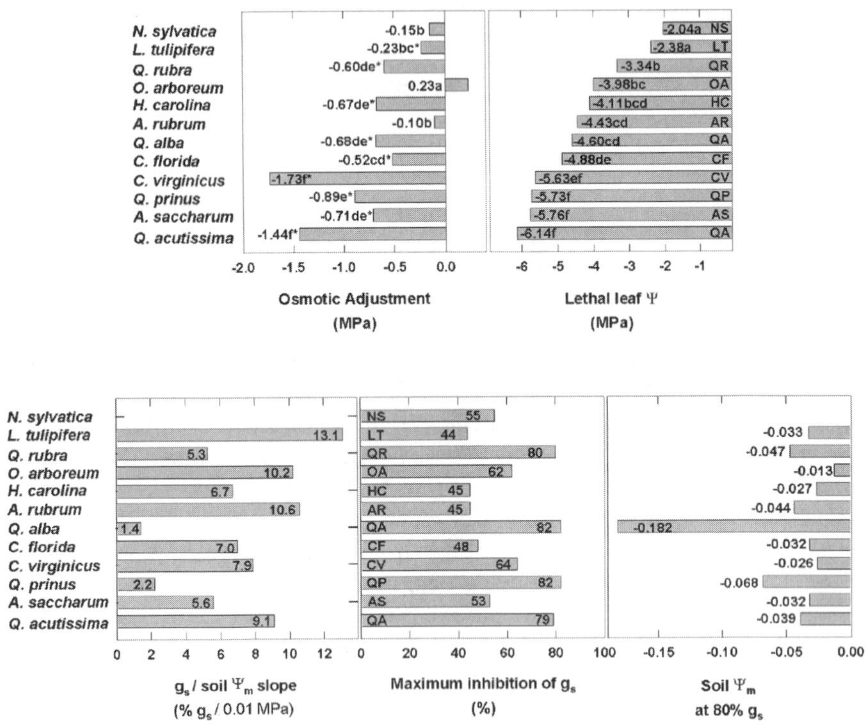

Figure 7.2. Foliar dehydration tolerance and stomatal response to nonhydraulic signals of soil drying in 12 temperate, deciduous tree species. Foliar dehydration tolerance was characterized as leaf Ψ at the end of and osmotic adjustment during a lethal drying episode. Species are listed in order of decreasing lethal leaf Ψ. Osmotic adjustment was measured as full turgor Ψ_π at the lethal point $-$ prestress full turgor Ψ_π. Larger absolute numbers (more negative numbers) represent larger active osmotic adjustments. Stomatal response to nonhydraulic signals was characterized in three ways: maximum declines in daily average relative g_s of WD trees; soil Ψ_M required to depress relative g_s by 20% (a measure of how much soil drying was necessary to initiate inhibition of g_s), and slope of the relative g_s/soil-Ψ_M regression, which indicates how quickly g_s declined as soil Ψ_M changed in the one drying pot.

the latter part of the drying period (e.g., Fig. 7.1). As a group, the *Quercus* species displayed little stomatal inhibition to partial soil drying.

Stomatal conductance began to decrease in all species before soil Ψ_M decreased measurably. The moisture release plot for this soil indicated that it lost about 60% of its water before its Ψ_M began to decline (Croker et al. 1998). For all genera except *Quercus,* the extent of inhibition in g_s was associated with the extent of decline in soil moisture within the one drying pot of WD trees. In *H. carolina,* for example, the average g_s of WD trees was 114% that of WS trees before the drying episode began. Their g_s declined to 81% of WS trees with some drying (but soil Ψ_M still at -0.01 MPa). With more drying, g_s dropped to 64% of that of controls. Further drying of soil, to -0.02 MPa and below, did not result in further decreases in g_s; relative g_s held mostly steady, at an average of 70% within the −0.02 to −0.10-MPa soil Ψ_M range, 58% within the −0.1 to −1.0-MPa soil Ψ_M range, and 56% when soil Ψ_M dropped below −1.0 MPa. For all species, maximum inhibition of g_s of WD trees occurred before the soil in the one drying pot reached −1.0 MPa, and additional soil drying did not depress g_s further.

In most species, including *Quercus,* relative g_s was positively correlated with soil Ψ_M between −0.01 and −0.10 MPa. Below −0.10 MPa, relative g_s and soil Ψ_M were not correlated. Rates of decline in relative g_s as a function of soil Ψ_M varied among species, with *L. tulipifera* (largest slope) showing a nine-times-steeper decline than *Q. alba* (smallest slope). As another way of describing and comparing stomatal sensitivity among species, we calculated the soil Ψ_M at 20% inhibition of g_s (i.e., the soil Ψ_M required to bring the average g_s of WD trees to 80% of the WS controls). We chose 80% because it likely represents the first real nonhydraulically regulated declines in g_s; above 80% relative g_s, stomata may just have been showing biological variation in g_s unrelated to drought. *O. arboreum* responded to the least soil drying, reaching 80% of WS g_s at an average soil Ψ_M of −0.013. *Q. alba* required the most soil drying (−0.18 MPa) to reach 80%.

Discussion

Carbon gain appears to be maximized in woody species at the expense of water loss, whereas water is conserved in herbaceous species at the expense of carbon gain (Knapp and Smith 1990). One might therefore conclude that herbaceous species would be more likely than woody plants to use a nonhydraulic signal to control g_s. Trees generally have relatively low photosynthetic and transpiration rates, deep and extensive root systems, and large amounts of tissue for water storage. These factors may lessen the impact of water stress; therefore, the physiological status of mature trees may not be as dependent on environmental variation as in the case for herbaceous plants (Knapp and Smith 1990). However, because of low boundary-layer resistance, canopy transpiration is expected to be regulated more by stomatal responses in trees than in herbaceous canopies, which typically have much higher boundary-layer resistances (Schulze 1986). Because of their smaller stature, tree seedlings and saplings may rely on a root-sourced, nonhydraulic mechanism for regulating stomata and may be more susceptible to drought than older, larger trees.

Foliar Dehydration Tolerance of Deciduous Trees

Plants have evolved many physiological, morphological, anatomical, and phenological characteristics for responding to and resisting drought stress. In the spectrum of responses ranging from extreme drought avoidance to extreme tolerance, perhaps the single most important determinant of drought-resistance strategy is the dehydration tolerance of the species, which is a measure of tissue capacity for withstanding desiccation (Ludlow 1989). Dehydration tolerance has been operationally defined as the Ψ or relative water content of the last surviving leaves (called the lethal value) on a plant subjected to a slow, continuous soil-drying episode (Ludlow 1989). Drought avoidance, a strategy possessed by both annuals and perennials, is characterized by relatively high lethal leaf Ψ and relatively little osmotic adjustment during drought. In contrast, drought tolerators, mostly perennials, have lower values of lethal leaf water status and typically display much more active accumulation of solutes when subjected to drought.

The division of plant responses into categories or strategies can be a valuable aid to understanding the ecological consequences for a species (e.g., potential for carbon acquisition and growth during drought, metabolic costs of resistance mechanisms, long-term plant survival). Although lethal leaf Ψ have been measured in several herbaceous species, such characterizations had not been made of tree species. We compared the potential for foliar dehydration tolerance and maximum capacity for osmotic adjustment in 11 tree species native to the southeastern United States (representing a range of natural habitat and, hence, probable leaf-dehydration tolerance) and an additional *Quercus* species.

Methods

Two- to three-year-old saplings were subjected to a continuous soil-drying episode by withholding water from pots. Each tree was checked daily after beginning the drying episode, and lethal measurements were begun when fewer than eight live leaves with minimal necrotic areas (less than 25% of total leaf area) remained. Leaf Ψ and osmotic potential (Ψ_π) at full turgor were measured with thermocouple psychrometers. Many experimental factors can influence foliar responses to severe drought stress: plant status (nutrition, age, overall vigor), edaphic conditions (aeration, ion content, microbes, soil-drying rates), and atmospheric factors [irradiance, air temperature, vapor-pressure deficit (VPD)]. The approach was to standardize these conditions as much as possible, growing trees of similar age in soils that would maximize vigor and promote adequate nutrition and conducting the experiment in a rain shelter (20 m × 7.7 m × 5 m high) constructed to simulate the ambient light and atmospheric conditions common to the understory of deciduous forests of the southeastern United States. The soil-drying rate was not controlled, either among or within species, but did not materially affect lethal leaf Ψ or changes in Ψ_π. Other experimental details are summarized in Augé et al. (1998).

Results

These 12 tree species represented a fairly wide range of intrinsic drought tolerance/avoidance, with lethal leaf Ψ ranging from −2 to −6 MPa and osmotic adjustment from essentially 0 to −1.7 MPa (Fig. 7.2). Decreasing lethal leaf Ψ was significantly correlated with increasing capacity for osmotic adjustment. This relationship was as strong or stronger for these deciduous trees as has been observed before for other kinds of plants (e.g., 22 forage legumes) [Ludlow (1989)]. Drying time to the lethal point ranged from 26 to 87 days and averaged 44 days across species. Drying time was not correlated with lethal leaf Ψ or osmotic adjustment, so estimates of relative species foliar dehydration tolerance did not appear to be confounded by differences in rates of soil drying.

Discussion

Ranking of species according to foliar dehydration tolerance under these standardized conditions reveals some interesting relationships among species (Fig. 7.2). For instance, previous works have ranked the drought resistance of *Q. alba, Q. rubra, A. saccharum,* and *C. florida* in that order as sites moved from xeric to mesic (Bahari et al. 1985). Yet, leaves of *A. saccharum* were able to withstand more desiccation (lower lethal leaf Ψ) than either of the two *Quercus* species, despite the fact that the soil ultimately became as dry (*Q. alba*) or much drier (*Q. rubra*) in the *Quercus* species than in *A. saccharum* [data for soil moisture provided in Augé et al. (1998)]. The similar foliar dehydration tolerance and osmotic adjustment of *A. saccharum,* considered a drought-sensitive species in nature (Graves 1994; Berkowitz et al. 1995), and *Q. alba,* a drought-tolerant species, reveals the value of determining lethal leaf Ψ. The implication is that, in nature, *Q. alba* must have some other means of more successfully avoiding leaf dehydration than does *A. saccharum.* One possibility is a deeper root system, and this has been indicated by root experiments comparing the two species (Pallardy and Rhoads 1993). In nature, *C. florida* usually exhibits relatively little drought tolerance, perhaps because of its inability to avoid low Ψ [e.g., Bahari et al. (1985)], but its leaves withstood as much or more dehydration than two of the native *Quercus* species studied. This observation suggests that *C. florida* probably has poorly developed mechanisms for drought avoidance, such as deep rooting, and, in fact, it is considered to have relatively shallow root systems. *Q. rubra* had perhaps surprisingly high lethal leaf Ψ, but others have observed mortality in this species when predawn leaf Ψ fell to -3.6 MPa (Viven et al. 1993). Among the *Quercus, Q. rubra*'s relatively high lethal leaf Ψ is consistent with its usual ranking as a mesic *Quercus* species [e.g., Timbal and Lefebvre (1995)]. Obviously, *Q. rubra* and *Q. alba* possess mechanisms other than leaf-dehydration tolerance that allow them to grow on drier sites than do *A. saccharum* and *C. florida.* Not surprisingly, foliage of the cove hardwoods *N. sylvatica, L. tulipifera, O. arboreum, H. carolina,* and *A. rubrum* (species tending to inhabit moist or bottomland sites) were most sensitive to dehydration in this study. *Acer rubrum* is usually considered to be less drought resistant than *A. saccharum,* and this is reflected by the leaf-dehydration tolerances of the two species observed here. *L. tulipifera* and *Q. rubra* had relatively similar lethal leaf Ψ; these two species previously showed similar leaf hydraulic properties in a study comparing drought response of diffuse and ring-porous tree species (Shumway et al. 1993).

Correlations in the Forest of g_s with Chemical, Hydraulic, and Environmental Factors

The theory of whole-plant stomatal regulation by root-sourced, nonhydraulic factors has been advanced on the basis of a few field studies (Loveys 1984; Wartinger et al. 1990; Tardieu et al. 1991), but mostly by work with potted plants in controlled conditions. There are indications that g_s and xylem-sap [ABA] may be correlated in trees in field conditions (Loewenstein and Pallardy 1998b), but additional work is needed to corroborate or refute the idea that the g_s of woody species in the field is typically better correlated with chemical than with hydraulic plant variables. Some evidence indicates that modeling of stomatal behavior requires consideration of both chemical and hydraulic aspects of root-to-shoot communication, with g_s regulated by xylem [ABA] with a sensitivity dependent upon leaf Ψ (Tardieu et al. 1993; Tardieu and Davies 1993).

Our objectives were to (1) characterize seasonal g_s, shoot water relations, xylem-sap [ABA], and xylem-sap pH of several deciduous tree species, (2) determine if the g_s of tree species is better correlated with plant chemical factors ([ABA] or pH of xylem fluid) or with plant hydraulic factors (shoot Ψ components), (3) determine if the g_s of tree species is better correlated with plant or with environmental factors [photosynthetic photon flux density (PPFD), VPD, air temperature, and air velocity], (4) test a published model that incorporates both leaf Ψ and xylem-sap [ABA] in describing g_s, and (5) search for other, empirical models that account for variation in g_s.

Methods

Field data were collected between May and September 1997 on saplings and trees of 11 species growing outdoors at the University of Tennessee Forestry Experiment Station and Arboretum (36°1′ N and 84°13′ W; ~890 ha) in Oak Ridge, Tennessee. The long-term mean annual precipitation (30-year normals, 1967 to 1996) at the arboretum is 1385 mm of rainfall and the mean temperature is 14.0°C. We measured *A. rubrum, A. saccharum, C. florida, L. tulipifera, N. sylvatica, O. arboreum, Q. alba, Q. prinus, Q. rubra, Liquidambar styraciflua* L., and *Q. muehlenbergii* Willd.

Six to eight replicates of each species were sampled weekly throughout most of the growing season. All measurements for each replicate were made on one terminal shoot, consisting of about 8–12 leaves. The following data were collected for each shoot before moving on to the next tree: g_s, leaf temperature, leaf dimension (four leaves per shoot); leaf Ψ_π and Ψ_p (two leaves per shoot); shoot Ψ, xylem [ABA], xylem pH, transpiration (E), and ABA flux; and wind speed, relative humidity, PPFD, and air temperature within each shoot. ABA flux was computed as [ABA] $\times$ E. Predawn shoot Ψ was measured with a pressure chamber. Full site, measurement, and statistical details are provided in Augé et al. (2000).

Various models were fitted to each species subset of the g_s data, using single and multiple independent variables (X). Linear models, with C_i representing various parameters, included simple linear regression:

$$g_s = C_1 + C_2X + e, \tag{1}$$

and response surface:

$$g_s = C_1 + C_2X_1 + C_3X_2 + C_4X_{12} + C_5X_{22} + C_6X_1X_2 + e, \tag{2}$$

where e is the residual error. Nonlinear models included the double exponential explored by Tardieu and Davies (1993) and Tardieu et al. (1993):

$$g_s = C_2 \exp(C_3X_1 \exp(C_4X_2)) + e, \tag{3}$$

a sum of exponentials:

$$g_s = C_1 \exp(C_2X_1) + C_3 \exp(C_4X_2) + e, \tag{4}$$

exponential response surface:

$$g_s = C_7 + C_8 \exp(C_1 + C_2X_1 + C_3X_2 + C_4X_{12} + C_5X_{22} + C_6X_1X_2) + e, \tag{5}$$

and the hyperbolic:

$$g_s = \frac{C_1 + C_2X_1}{C_1C_2X_1} + \frac{1}{C_3X_2} + e. \tag{6}$$

Seasonal behaviors of these trees (plant hydraulic, plant chemical, and environmental variables) and results of single- and multiple-regression analyses are detailed and illustrated in Augé et al. (2000). There were no dramatic climatic extremes, yet there was sufficient variability in precipitation and soil moisture to provide a range in values of g_s and other plant and environmental variables.

Results

The correlation coefficients for single- and multiple-variable models explaining g_s are provided in Tables 7.1 and 7.2 respectivley. Xylem-sap [ABA] was not strongly associated with g_s in this study. Stomatal conductance/[ABA] correlation coefficients were mostly higher than those observed before in large *Quercus* trees under natural conditions (Triboulet et al. 1996) and lower than those observed before in *Juglans nigra, A. saccharum,* and *Q. alba* (Loewenstein and Pallardy 1998b). Averaged across species, g_s correlation coefficients were slightly higher with shoot Ψ than with xylem [ABA], but effectively g_s appeared to be similarly linked with shoot Ψ and xylem [ABA] throughout the season. Moreover, the same number of species showed significant correlations for g_s/shoot Ψ and g_s/[ABA]. Shoot Ψ_π better explained g_s than either shoot Ψ or xylem-sap [ABA], and xylem pH generally most poorly explained g_s relative to other plant and environmental variables.

The more easily measured environmental variables were generally better indicators of g_s than either the hydraulic or chemical plant variables. For instance, g_s was significantly correlated with the air temperature surrounding leaves for each of the four species having significant g_s/shoot Ψ correlations as well as for an additional three species. More species showed significant g_s correlations with

Table 7.1. Correlation coefficients (R) for single-variable models explaining g_s

Species	Shoot Ψ	Shoot Ψ_π	Shoot Ψ_P	[ABA]	Xylem pH	VPD	PPFD	Air temperature	Air velocity
A. rubrum	0.27**	0.16	—	−0.08	−0.03	−0.18	0.24*	0.28**	0.19
A. saccharum	0.01	0.17	—	−0.06	0.12	0.25**	0.23*	0.17	0.01
C. florida	−0.17	−0.17	−0.06	0.01	−0.05	0.20	0.17	0.13	0.24*
L. styraciflua	0.37**	0.46**	0.21	−0.04	−0.15	0.39**	−0.11	0.39**	0.17
L. tulipifera	−0.07	−0.12	−0.00	0.31*	−0.15	0.43***	0.41***	0.33***	0.18
N. sylvatica	0.14	−0.27*	0.27*	−0.25*	−0.02	0.05	0.28*	0.05	0.24*
O. arboreum	−0.21*	−0.23*	−0.06	0.17	−0.30**	0.14	0.19	0.37***	0.06
Q. alba	−0.02	0.07	−0.16	−0.27*	0.14	0.44***	0.13	0.49***	0.06
Q. muehlenbergii	−0.01	−0.23	0.10	0.02	−0.12	0.01	0.28*	0.23	0.23
Q. prinus	−0.27	0.31*	−0.34	−0.34**	0.09	0.32**	−0.18	0.52***	0.06
Q. rubra	−0.33**	−0.17	−0.26*	−0.02	0.04	0.29**	0.01	0.27*	0.11
Average, g_s	0.17	0.21	0.16	0.14	0.11	0.25	0.20	0.29	0.14
Average, transpiration	—	0.22 (5)	—	0.15 (3)	0.16 (4)	—	0.54 (10)	—	—

Note: Models were fit to each species separately with all the data collected during the season. Asterisks indicate significance of regression (R^2) at $p < 0.05$ (*), 0.01 (**), or 0.001 (***). Average correlation coefficients for g_s across species for each variable were computed from absolute values of species correlation coefficients. For comparative purposes, average correlation coefficients across species are also provided for transpiration. Numbers in parentheses (transpiration) indicate how many species showed significant correlation, $p = 0.05$. Blanks indicate regression was not computed because the variable was used in calculating transpiration (or, in the case of shoot Ψ_p, shared a variable).

Table 7.2. Multiple correlation coefficients (R; always positive) for various two-variable models explaining stomatal conductance

Species	Tardieu	[ABA]/ shoot Ψ	[ABA]/ shoot Ψ_π	pH/shoot Ψ	Air temp./ PPFD	Air temp./ VPD	VPD/PPFD	Shoot Ψ_π/ air temp.
A. rubrum	0.14	0.52***	0.47***	0.36*	0.53***	0.44**	0.44***	0.37*
A. saccharum	0.21	0.22	0.28	0.30	0.42**	0.28	0.43***	0.35*
C. florida	0.24	0.24	0.36	0.34	0.25	0.33	0.30	0.37
L. styraciflua	0.45***	0.47*	0.46*	0.39	0.49**	0.62***	0.61***	0.60***
L. tulipifera	0.24	0.32	0.28	0.30	0.55***	0.48***	0.52***	0.38*
N. sylvatica	0.30*	0.34	0.48**	0.22	0.35	0.26	0.34	0.43*
O. arboreum	0.22	0.29	0.36	0.56***	0.42**	0.44**	0.29	0.42**
Q. alba	0.32*	0.42**	0.41**	0.31	0.61***	0.65***	0.57***	0.54***
Q. muehlenbergii	0.04	0.30	0.43*	0.35	0.43*	0.49**	0.47***	0.41*
Q. prinus	0.45***	0.46**	0.45**	0.23	0.62***	0.56***	0.48*	0.61***
Q. rubra	0.17	0.41*	0.31	0.40*	0.36	0.39*	0.38*	0.39*
Average, g_s	0.25	0.36	0.39	0.34	0.46	0.45	0.44	0.44

Note: Tardieu represents the double-exponential model developed for maize (Tardieu and Davies 1993; Tardieu et al. 1993). All other models are response surfaces. Models were fitted to each species separately with all the data collected during the season. Asterisks indicate significance of regression (R^2) at $p < 0.05$ (*), 0.01 (**) or 0.001 (***). Average correlation coefficients across all species for each variable were computed from absolute values of species correlation coefficients.

PPFD and VPD than with either [ABA] or shoot Ψ. Stomatal conductance was significantly correlated with air temperature for more species than with any other single plant or environmental variable. Moreover, the nature of the g_s correlation was more predictable for environmental than for plant variables. Unlike [ABA] or shoot Ψ components, higher PPFD, VPD, air temperature, and air velocity were always correlated with higher g_s when correlations were significant. We found that g_s was somewhat better related to changes in VPD than in PPFD.

The mechanism by which stomata perceive an ABA signal has yet to be clearly defined, and others have addressed the question of whether the stomatal response appears better linked to xylem [ABA] or, alternatively, to a measure of the amount of ABA entering the leaf. We evaluated xylem-sap [ABA] because most investigators have found that g_s has been better correlated with [ABA] than with flux of ABA to leaves in both herbaceous plants (Tardieu et al. 1992; Tardieu and Davies 1993; Tardieu et al. 1993; Jia et al. 1996) and trees (Gowing et al. 1993; Liang et al. (1996); Loewenstein and Pallardy 1998a). Mechanistically, it is difficult to envision how xylem [ABA] could affect the guard-cell plasmalemma directly because so many factors act to modify this concentration in the transpiration stream on its path from the xylem vessel to the guard cell (Gowing et al. 1993). However, the complex pattern of ABA balance in the apoplast and of stomatal response to apoplastic ABA has been adequately portrayed with a simple relationship between g_s and xylem [ABA], and strong correlations between g_s and xylem-sap [ABA] have been reported under many conditions for several species, including trees [e.g., Liang et al. (1996); Loewenstein and Pallardy (1998a, 1998b)]. ABA fluxes were measured to compare their seasonal trends with [ABA] (Augé et al. 2000).

We tested numerous models incorporating many environmental and plant variables, both chemical and hydraulic, for the best empirical correlation with g_s and found that response surfaces provided the best fits (Table 7.2). The xylem [ABA]/shoot Ψ response surface was significantly correlated with g_s in more species than was the Tardieu model, and the average correlation coefficient across species was higher. However, a response surface that modified stomatal sensitivity to xylem [ABA] by shoot Ψ_π rather than by shoot Ψ accounted for even more variation. Of all models incorporating a plant variable, the shoot Ψ_π/air temperature response surface accounted for the most variation in tree g_s. The shoot Ψ_π/air temperature response surface also resulted in significant g_s correlations in 10 of the 11 species, more than for any other model. As in the double-exponential models, using environmental variables in the response surfaces explained as much or more variation in g_s than using plant variables. Two variable response surfaces incorporating any two of these three environmental variables (air temperature, PPFD, and VPD) were equally robust in describing seasonal g_s and better described g_s than any two-variable combination of [ABA], pH, shoot Ψ, shoot Ψ_π, or shoot Ψ_p.

Stomatal conductance of more than half of the tree species studied here showed significant seasonal correlations with VPD, and VPD was one of the best describers of seasonal g_s, better than any of the plant variables (Tables 7.1 and 7.2).

Stomatal conductance of large *Q. alba* trees was previously seen to be better correlated with VPD than with xylem [ABA] (Loewenstein and Pallardy 1998b), as in our study, and we found that among the 11 species, *Q. alba* had the highest g_s/VPD correlation coefficient. When g_s is related to VPD [not always the case in woody species; e.g., Turner et al. (1984)], the relationship is typically a negative one, with g_s decreasing as VPD increases [e.g., Turner et al. (1984); Ball et al. (1987); Thomas and Eamus (1999)]. However, in every case in our study of significant correlation between seasonal g_s and VPD (*A. saccharum, Liquidambar styraciflua, Liriodendron tulipifera, Q. alba, Q. prinus,* and *Q. rubra*), g_s increased as VPD increased. Vapor-pressure deficit is greatly affected by temperature, and temperature also affects g_s. Sometimes the g_s/air temperature relation is negative, and sometimes it is positive, depending on species and temperature range (Wilson 1948; Meyer and Anderson 1952; Wuenscher and Kozlowski 1971). Stomatal conductance has previously been positively correlated with temperature in *A. saccharum* in the air temperature range of 26–30°C (Pereira and Kozlowski 1977), and we also observed positive correlations between g_s and air temperature in each of the seven instances for which individual species correlations were significant. It is likely, therefore, that the positive g_s/VPD relationships were driven by the positive g_s/air temperature relationships. Air temperature was the single strongest describer of g_s among all plant and environmental variables examined.

Tardieu et al. (1993) also tested a model involving [ABA] and VPD and determined that, in their maize system, leaf Ψ better described stomatal sensitivity to ABA than did VPD. We found that, overall, VPD better described stomatal sensitivity to [ABA] than did shoot Ψ. Nine tree species showed significant correlations of g_s with the [ABA]/VPD response surface, and only five species with the [ABA]/shoot Ψ response surface. Average correlation coefficients across all 11 species were 0.42 and 0.36 for the [ABA]/VPD and [ABA]/shoot Ψ response surfaces, respectively.

Investigating Nonhydraulic Signaling of Soil-Drying in the TDE

Throughfall Displacement Experiment researchers are interested in effects of the throughfall displacement treatments on plants of all ages (Hanson et al. 1995; Hanson et al. 1998; Hanson et al. 2001). Seedlings and saplings may be particularly sensitive to decreases in yearly rainfall because their root systems are not as large as or as deep as those of larger trees. The ability of saplings to cope with changed precipitational inputs will dictate their long-term competitive position within the forest ecosystem. Our goal was to investigate the significance of the nonhydraulic signaling process in saplings on the TDE. Our specific objectives were to (1) determine whether g_s of native *in situ* saplings was better correlated with environmental parameters (e.g., air temperature, VPD, and PPFD), plant hydraulic factors (shoot Ψ and Ψ_π), or plant chemical factors (ABA fluxes, xylem [ABA], and xylem-fluid pH), (2) determine whether g_s was better correlated with root-to-shoot fluxes of ABA or with [ABA] in shoot-xylem fluid, and (3) determine if nonhydraulic inhibition of g_s occurred in the forest.

Methods

Four species were present on the TDE site in sufficient abundance for the sampling required: *A. rubrum, C. florida, N. sylvatica,* and *Rhamnus caroliniana.* We selected the first three because of their importance in the understory and because they were examined in other TDE work [e.g., Hanson et al. (2001)]. Studies were conducted during the growing seasons of 1996 and 1997. Plant water relations, xylem [ABA], xylem pH (1997 only), and environmental measurements were made throughout the season as described in the previous section, between 10 AM and 3 PM during most summer weeks of 1996 and about every other week during 1997. During the first 2.5 h of each sampling day, nine trees were sampled, one replicate of each species from each treatment plot. One species was sampled on dry, ambient, and wet plots before moving on to the next species. This sequence was repeated during the next 2.5 h. Species sequence was alternated each day. Measurements were made each sampling day on saplings from the same transect or from two adjacent transects to minimize bias from site heterogeneity. It took 3 days to sample six replicates of each species on each treatment plot; weekly means portrayed in Fig. 7.3 represent 3-day averages. Soil-water content was measured biweekly during the growing season as described in Chapter 2.

The study was designed as a 3 (species) $\times$ 3 (precipitation treatments) factorial, with six replicates of each combination, for a total of 54 sampled trees for each week that measurements were made. In 1997, we sampled only black gum and red maple on wet and dry plots; there were no longer sufficient numbers of *C. florida* saplings on the site for the required measurements because of mortality in 1996, possibly as a result of infections by dogwood anthracnose, *Discula destructiva,* and powdery mildew, *Microspherae pulchra.* Correlation analyses were performed as described in previous section.

Results

The years 1996 and 1997 were relatively moist years of the 8-year (1993–2000) TDE (Figs. 2.6 and 2.11 of Chapter 2, this volume). Average soil Ψ of the upper 35 cm in the dry plot dropped to -2 MPa in 1993, 1995, and 1999 and below -2 MPa in 1998 and 2000, whereas the low points in 1996 and 1997 were about -0.5 and -1.0 MPa, respectively. Soil Ψ of wet and ambient plots were similar during the 1996 and 1997 seasons and stayed mostly above -0.1 MPa in 1996 and -0.5 MPa in 1997. The lowest soil Ψ during the 1996 and 1997 3-month sampling periods was about -0.2 MPa, recorded on the dry plot during the first week in July in 1996 and during the first 3 weeks of August in 1997 (Fig. 7.3e).

Other environmental variables did not differ much among the treatment plots in 1996 and 1997. Air temperature, PPFD, and VPD during porometry measurements were similar among treatments during both seasons (data not shown). In 1996, air temperature and VPD increased during May and June, then remained

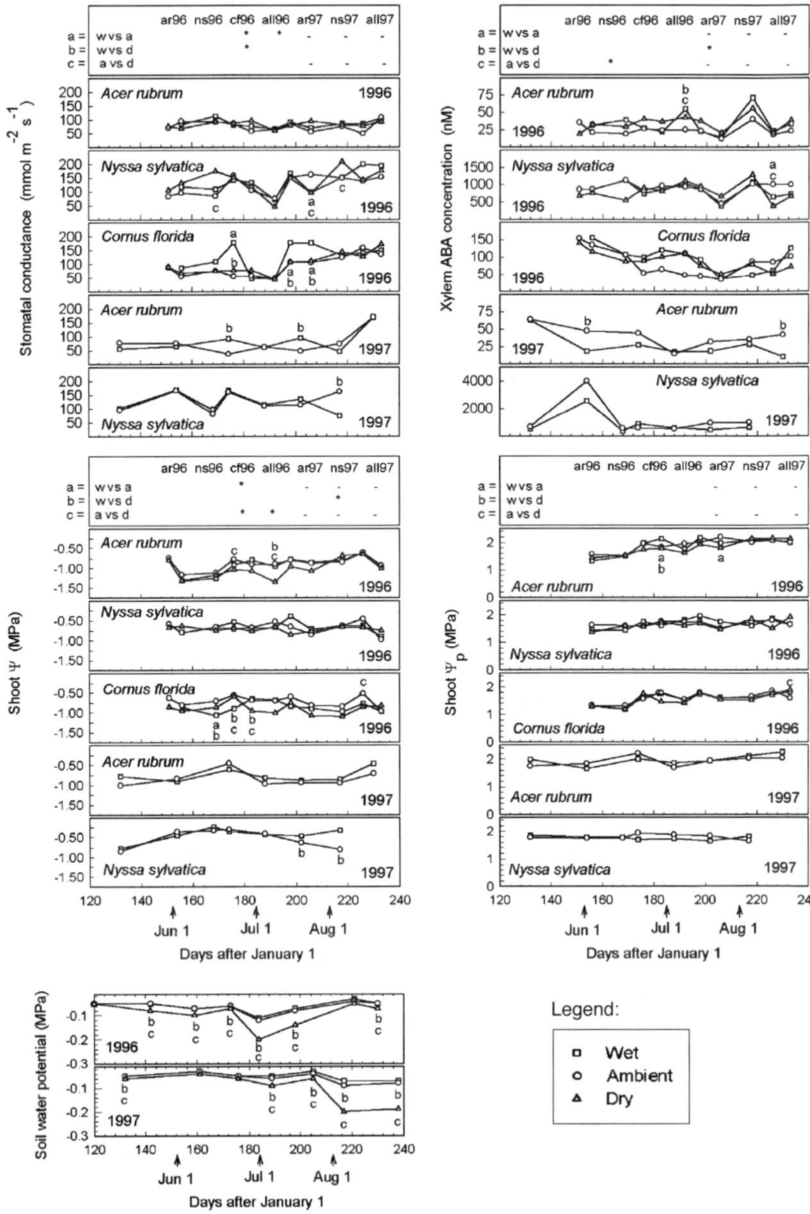

Figure 7.3. Plant and soil variables on the wet (w), ambient (a), and dry (d) plots on the TDE during the summers of 1996 (96) and 1997 (97). Stomatal conductance, xylem [ABA], shoot Ψ, and shoot Ψ_p in *Nyssa sylvatica, Acer rubrum,* and *Cornus florida* saplings, and soil Ψ. Because of high mortality, *C. florida* saplings were not measured in 1997. Shoot Ψ_p was computed as shoot $\Psi - \Psi_\pi$, with psychrometric Ψ_π estimates corrected for apoplastic dilution with apoplastic water fractions determined by pressure–volume analysis, as described in Augé and Stodola (1990). The table at the top of each of the first four groups of panels summarizes linear contrasts (abbreviated a, b, or c) that compare treatment means over all measurement days for 1996 or 1997.

mostly constant for the remainder of the sampling period. In 1997, air temperature and VPD increased during the second half of June and the first half of July, then declined for the remaining sampling days. In 1996, air velocity tended to be consistently higher on the wet plots; daily means were not significantly different, but seasonal means were significantly different among treatments.

Correlation analysis involving all data collected over the 1996 season on individual trees showed that the g_s of saplings was better correlated with shoot Ψ than with xylem [ABA] or ABA flux. Stomatal conductance was also better correlated with some environmental parameters (VPD and PPFD) than with nonhydraulic factors (ABA fluxes, and xylem [ABA]). Stomatal conductance was not correlated with xylem [ABA] in any of the nine treatment/species combinations, except for *A. rubrum* on the wet plot, where g_s increased as [ABA] increased. Stomatal conductance was correlated with shoot Ψ in all of the treatment/species combinations, except for *N. sylvatica* on the dry plot. Each of these correlations were negative, indicating that g_s increased as shoot Ψ decreased. The negative correlations between g_s and shoot Ψ suggest that g_s was controlling shoot Ψ rather than the reverse. If shoot Ψ was controlling g_s, declines in Ψ would have been expected to result in declines in g_s. It appears that shoot Ψ dropped when environmental conditions favored higher g_s. Others have shown that although the influence of shoot Ψ and turgor on g_s can be important, the influence of g_s on shoot Ψ and turgor may be equally and sometimes more important (Jones 1990a, 1990b). Among all environmental and plant parameters measured, g_s was mostly strongly correlated with shoot Ψ, VPD, and PPFD. However, as is often observed in field and forest work, correlations of g_s with any single parameter were not particularly strong.

Stomatal conductance was better correlated with root-to-shoot fluxes of ABA than with [ABA] in shoot-xylem fluid. However, the g_s/ABA flux correlations were positive, rather than negative as the theory predicted. Hence, it appeared that g_s was not being regulated by ABA fluxes; if it was, greater fluxes would have resulted in inhibition rather than promotion of stomatal opening. Rather, it appeared that ABA fluxes were being regulated by g_s, as with shoot Ψ.

The relatively slight soil drying of the dry plot in 1996 and 1997 provided a reasonable system for detecting nonhydraulic inhibition of g_s. Nonhydraulic inhibition of g_s begins before soil has dried enough to cause declines in soil Ψ, and much nonhydraulic inhibition occurs under controlled conditions when part of the root system is exposed to soil Ψ in the range of ~0 to -0.10 MPa, as discussed earlier. Mild soil-drying may trigger increased root-signal activity, with soil moist enough that roots still contribute to the transpiration stream. For the signal to affect g_s, its production must not only increase but also be loaded into root xylem and move to leaves. During the course of the season, if the g_s of dry plot trees was lower than the g_s of wet or ambient plot trees in the absence of treatment effects on shoot Ψ or Ψ_p, this would suggest that the g_s of dry plots was being depressed by some nonhydraulic process, especially if the lower g_s coincided with increased xylem [ABA] on the dry plot.

A few declines in g_s were observed in the absence of differences in shoot Ψ or Ψ_p among treatments. During the weeks of July 16 and July 24, 1996, the g_s

of dry and ambient plot dogwoods was substantially lower than the g_s of wet plot dogwoods (Fig. 7.3a), despite the fact that shoot Ψ and Ψ_p were similar among treatments (Figs. 7.3c and 7.3d). Soil Ψ in the 0–35-cm range was lower in the dry plot than in ambient and wet plots during these 2 weeks (Fig. 7.3e). No environmental circumstances would explain this; VPD, PPFD, and air and leaf temperatures were similar among moisture treatments during these weeks (data not shown). It is possible, therefore, that g_s was inhibited by some nonhydraulic, possibly drought-induced factor. However, it is also possible that the lower dry plot g_s was hydraulically induced because shoot Ψ was significantly lower in dry plot than wet plot dogwoods during the 3 weeks prior to July 17. Xylem [ABA] (Fig. 7.3b) and ABA fluxes to leaves (not shown) were not different among treatments during the latter half of July, so if a nonhydraulic factor was responsible for the g_s decline, it was not directly related to a change in xylem [ABA].

Nonhydraulic inhibition of g_s also possibly occurred in *A. rubrum* on two occasions during the 1997 season. Stomatal conductance was significantly lower in dry plot trees than in wet plot trees on June 23 and July 21 (Fig. 7.3a). Shoot Ψ and Ψ_p were similar in wet and dry plot red maple on these dates (Fig. 7.3c and 7.3d), whereas xylem [ABA] tended to be higher in dry plot trees (Fig. 7.3b). However, g_s was sometimes higher in trees of one treatment even when shoot Ψ was lower in trees of that treatment. Examples include *N. sylvatica* on August 18, 1997 and *C. florida* on June 25, 1996. For the most part, fluctuations in g_s in 1996 and 1997 did not appear to be obviously related to nonhydraulic, root-sourced signals.

Conclusions

The most sensitive indicator of a plant's overall physiological state is often its stomatal behavior (Smith and Hollinger 1991). Stomata are probably the most dynamic organ on a plant, responding rapidly to changes in external environmental and internal plant conditions, such as wind speed, PPFD, VPD, leaf temperature, CO_2 concentration, and water content. Additionally, stomatal sensitivity to one environmental or internal physiological variable can vary with changes in one or more of the other variables, as well as with more long-term conditions, such as stress history and nutritional status.

We conclusively demonstrated that, under controlled conditions, the g_s of temperate, deciduous tree saplings can be markedly inhibited by root-to-shoot signals that are strictly nonhydraulic in nature. However, we did not observe direct effects of these signals in the forest. This does not mean that they are not operating there to influence g_s. It probably means we were not able to discern them within the complexity of a natural system, in which g_s is continually reacting to so many fluctuating variables.

We can say with confidence that stomata of temperate tree saplings are more sensitive to small changes in water content of forest soils than previously thought. In all species examined, inhibition of g_s occurred under controlled conditions when only part of a root system was dried. Further, in all species except the

Quercus species, nonhydraulic inhibition of g_s had begun to occur before soil had dried to the extent that its Ψ_M changed (i.e., before soil Ψ_M declined below control values of -0.01 MPa). In most genera, stomatal inhibition was extensive by the time the soil around roughly half of the root system had dried to -0.10 MPa, with soil around the other half of the root system kept moist enough to fully supply shoot needs. Surface forest soils frequently undergo drying episodes of this and greater magnitude, even in relatively moist regions and even in wet years.

References

Augé RM, Duan X (1991) Mycorrhizal fungi and nonhydraulic root signals of soil drying. Plant Physiol 97:821–824.

Augé RM, Duan X, Croker JL, Witte WT, Green CD (1998) Foliar dehydration tolerance of twelve deciduous tree species. J Exp Bot 49:753–759.

Augé RM, Duan X, Ebel C, Stodola AJW (1994) Nonhydraulic signaling of soil drying in mycorrhizal maize. Planta 193:74–82.

Augé RM, Green CD, Johnson TB, Stodola AJW, Saxton AM, Olinick JB, Evans RM (2000) Correlations of stomatal conductance with hydraulic and chemical factors in several deciduous tree species in a natural habitat. New Phytol 145:483–500.

Augé RM, Moore JL (2002) Stomatal response to nonhydraulic root-to-shoot communication of partial soil drying in relation to foliar dehydration tolerance. Exp Environ Bot 47:217–229.

Augé RM, Stodola AJW (1990) An apparent increase in symplastic water contributes to greater turgor in mycorrhizal roots of droughted *Rosa* plants. New Phytol 115:285–295.

Bahari ZA, Pallardy SG, Parker WC (1985) Photosynthesis, water relations, and drought adaptation in six woody species of oak-hickory forests in central Missouri. Forest Sci 31:557–569.

Ball J, Woodrow IE, Berry JA (1987) A model predicting stomatal conductance and its contribution to the control of photosynthesis under different environmental conditions. In Biggins J (Ed) *Progress in photosynthesis research, Vol. 4.* Proceedings of the VIIth International Congress on Photosynthesis, Providence, Rhode Island, Aug 10–15, 1986. Martinus Nijhoff Publishers, Dordrecht, Netherlands.

Berkowitz AR, Canham CD, Kelly VR (1995) Competition vs facilitation of tree seedling growth and survival in early successional communities. Ecology 76:1156–1168.

Cornish K, Zeevaart JAD (1985) Abscisic acid accumulation by roots of Xanthium strumarium L. and Lycopersicon esculentum Mill. in relation to water stress. Plant Physiol 79:653–658.

Correia M, Rodrigues M, Ferreira M, Pereira J (1997) Diurnal change in the relationship between stomatal conductance and abscisic acid in the xylem sap of field-grown peach trees. J Exp Bot 48:1727–1736.

Croker JL, Witte WT, Augé RM (1998) Stomatal sensitivity to nonhydraulic root-to-shoot signals of partial soil drying in six temperate, deciduous forest trees. J Exp Bot 49:761–774.

Davies WJ, Tardieu F, Trejo CL (1994) How do chemical signals work in plants that grow in drying soil? Plant Physiol 104:309–314.

Fußeder A, Wartinger A, Hartung W, Schulze ED, Heilmeier H (1992) Cytokinins in the xylem sap of desert-grown almond (*Prunus dulcis*) trees—Daily courses and their possible interactions with abscisic acid and leaf conductance. New Phytol 122:45–52.

Gowing DJG, Jones HG, Davies WJ (1993b) Xylem-transported abscisic acid—The relative importance of its mass and its concentration in the control of stomatal aperture. Plant Cell Environ 16:453–459.

Gowing DJG, Davies WJ, Jones HG (1990) A positive root-sourced signal as an indicator of soil drying in apple, Malus X domestica Borkh. J Exp Bot 41:1535–1540.

Graves WR (1994) Seedling development of sugar maple and black maple irrigated at various frequencies. Hort Sci 119:1292–1294.

Hansen H, Dorffling K (1999) Changes of free and conjugated abscisic acid and phaseic acid in xylem sap of drought-stressed sunflower plants. J Exp Bot 50:1599–1605.

Hanson PJ, Todd DE, Amithor JS (2001) A six-year study of sapling and large-tree growth and mortality responses to natural and induced variability in precipitation and throughfall. Tree Physiol 21:345–358.

Hanson PJ, Todd DE, Huston MA, Joslin JD, Croker JL, Augé RM (1998) *Description and field performance of the Walker Branch Throughfall Displacement Experiment: 1993–1996.* ORNL/TM-13586. Oak Ridge National Laboratory, Oak Ridge, Tennessee.

Hanson PJ, Todd DE, Edwards NT, Huston MA (1995) Field performance of the Walker Branch Throughfall Displacement Experiment. In Jenkins A, Ferrier RC, Kirby C (Eds) *Ecosystem manipulation experiments: Scientific approaches, experimental design and relevant results.* Ecosystem Research Report #20. Commission of the European Communities, Brussels, Belgium, pp 307–313.

Hartung W, Radin JW (1989) Abscisic acid in the mesophyll apoplast and in the root xylem sap of water-stressed plants: The significance of pH gradients. In Randall DD, Blevins DG (Eds) *Current topics in plant biochemistry and physiology.* University of Missouri, Columbia, Missouri.

Hartung W, Radin JW, Hendrix DL (1988) Abscisic acid movement into the apoplastic solution of water-stressed cotton leaves. Plant Physiol 86:908–913.

Hartung W, Wilkinson S, Davies W (1998) Factors that regulate abscisic acid concentrations at the primary site of action at the guard cell. J Exp Bot 49:361–367.
Heckenberger U, Schurr U, Schulze E (1996) Stomatal response to abscisic acid fed into the xylem of intact *Helianthus annus* (L.) plants. J Exp Bot 47:1405–1412.
Holbrook NM, Shashidhar VR, James RA, Munns R (2002) Stomatal control in tomato with ABA-deficient roots: response of grafted plants to soil drying. J Exp Bot 2002:1503–1514.
Jackson GE, Irvine J, Grace J, Khalil AMM (1995) Abscisic acid concentrations and fluxes in droughted conifer saplings. Plant Cell Environ 18:13–22.
Jia W, Zhang J (1997) Comparison of exportation and metabolism of xylem-delivered ABA in maize leaves at different water status and xylem sap pH. Plant Growth Reg 21:43–49.
Jia W, Zhang J, Zhang DP (1996) Metabolism of xylem-delivered ABA in relation to ABA flux and concentration in leaves of maize on *Commelina communis.* J Exp Bot 47:1085–1091.
Jones HG (1990a) Control of growth and stomatal behavior at the whole plant level: Effects of soil drying. In Davies WJ, Jeffcoat B (Eds) *Importance of root to shoot communication in the responses to environmental stress.* Mongraph No.1. British Society for Plant Growth Regulation, University of Bristol, England.
Jones HG (1990b) Physiological aspects of the control of water status in horticultural crops. HortScience 25:19–26.
Jones RJ, Mansfield TA (1970) Suppression of stomatal opening in leaves treated with abscisic acid. J Exp Bot 21: 714–719.
Khalil AAM, Grace J (1993) Does xylem sap ABA control the stomatal behavior of water stressed sycamore (*Acer pseudoplatanus* L.) seedlings? J Exp Bot 44:1127–1134.
Knapp AK, Smith WK (1990) Stomatal and photosynthetic responses to variable sunlight. Physiol Plant 78:160–165.
Kramer PJ, Boyer JS (1995) *Water relations of plants and soils.* Academic Press, San Diego, California.
Liang J, Zhang J, Wong M (1996) Stomatal conductance in relation to xylem sap ABA concentrations in two tropical trees, *Acacia confusa* and *Litsea glutinosa.* Plant Cell Environ 19:93–100.
Liang J, Zhang J, Wong M (1997) How do roots control xylem sap ABA concentration in response to soil drying? Plant Cell Physiol 38:10–16.
Liu L, McDonald AJS, Stadenberg I, Davies WJ (2001a) Abscisic acid in leaves and roots of willow: Significance for stomatal conductance. Tree Physiol 21:759–764.
Liu L, McDonald AJS, Stadenberg I, Davies WJ (2001b) Stomatal and leaf growth responses to partial drying of root tips in willow. Tree Physiol 21:765–770.
Loewenstein NJ, Pallardy SG (1998a) Drought tolerance, xylem sap abscisic acid and stomatal conductance during soil drying: A comparison of young plants of four temperate deciduous angiosperms. Tree Physiol 18:421–430.
Loewenstein NJ, Pallardy SG (1998b) Drought tolerance, xylem sap abscisic acid and stomatal conductance during soil drying: A comparison of canopy trees of three temperate deciduous angiosperms. Tree Physiol 18:431–439.
Loveys BR (1984) Diurnal changes in water relations and abscisic acid in field-grown *Vitis vinifera* cultivars. III. The influence of xylem-derived abscisic acid on leaf gas exchange. New Phytol 98:563–573.
Ludlow MM (1989) Strategies in response to water stress. In Kreeb HK, Richter H, Hinckley TM (Eds) *Structural and functional responses to environmental stresses: Water shortage.* SBP Academic Publishing, The Hague, The Netherlands.
Ludlow MM, Sommer KJ, Flower DJ, Ferraris R, So HB (1989) Influence of root signals resulting from soil dehydration and high soil strength on the growth of crop plants. Curr Top Plant Biochem Physiol 8:81–99.
Mansfield TA, McAinsh MR (1995) Hormones as regulators of water balance. In Davies PJ (Ed) *Plant hormones.* Kluwer Academic Publishers, The Hague, The Netherlands.
Meyer BS, Anderson DB (1952) *Plant physiology.* D. Van Nostrand & Co., New York.
Neuman DS, Rood SB, Smit BA (1990) Does cytokinin transport from root-to-shoot in the xylem sap regulate leaf responses to root hypoxia? J Exp Bot. 41:1325–1333.
Netting AG (2000) pH, abscisic acid and the integration of metabolism in plants under stressed and non-stressed conditions: Cellular responses to stress and their implication for plant water relations. J Exp Bot 51:147–158.
Pallardy SG, Rhoads JJ (1993) Morphological adaptations to drought in seedlings of deciduous angiosperms. Can J Forest Res 23:1766–1774.
Patonnier M, Peltier J, Marigo G (1999) Drought-induced increase in xylem malate and mannitol concentrations and closure of *Fraxinus excelsior* L. stomata. J Exp Bot 50:1223–1231.
Pereira JS, Kozlowski TT (1977) Influence of light intensity, temperature and leaf area on stomatal aperture and water potential of woody plants. Can J Forest Res 7:145–153.
Saliendra NZ, Sperry JS, Comstock JP (1995) Influence of leaf eater status on stomatal response to humidity, hydraulic conductance, and soil drought in Betula occidentalis. Planta 196:357–366.
Sauter A, Hartung W (2000) Radial transport of abscisic acid conjugates in maize roots: Its implication for long distance stress signals. J Exp Bot 51:929–935.
Schulze ED (1986) Whole-plant responses to drought. Aust J Plant Physiol 13:127–141.
Shumway DL, Steiner KC, Kolb TE (1993) Variation in seedling hydraulic architecture as a function of species and environment. Tree Physiol 12:41–54.
Smith WK, Hollinger DY (1991) Measuring stomatal behavior. In Lassoie JP, Hinckley TM (Eds) *Techniques and approaches in forest tree ecophysiology.* CRC Press, Boca Raton, Florida.
Socias X, Correia M, Chaves M, Medrano H (1997) The role of abscisic acid and water relations in drought responses of subterranean clover. J Exp Bot 48:1281–1288.
Tardieu F, Davies WJ (1993) Integration of hydraulic and chemical signalling in the control of stomatal conductance and water status of droughted plants. Plant Cell Environ 16:341–349.

Tardieu F, Simonneau Th (1998) Variability among species of stomatal control under fluctuating soil water status and evaporative demand: Modelling isohydric and anisohydric behaviors. J Exp Bot 49:419–432.
Tardieu F, Katerji N, Benthenod O, Zhang J, Davies WJ (1991) Maize stomatal conductance in the field: Its relationship with soil and plant water potentials, mechanical constraints and ABA concentration in the xylem sap. Plant Cell Environ 14:121–126.
Tardieu F, Zhang J, Katerji N, Bethenod O, Palmer S, Davies WJ (1992) Xylem ABA controls the stomatal conductance of field-grown maize subjected to soil compaction or soil drying. Plant Cell Environ 15:193–197.
Tardieu F, Zhang J, Gowing DJG (1993) Stomatal control by both [ABA] in the xylem sap and leaf water status: A test of a model for droughted or ABA-fed field-grown maize. Plant Cell Environ 16:413–420.
Tardieu F, LaFarge T, Simonneau T (1996) Stomatal control by fed or endogenous ABA in sunflower: Interpretation of correlations between leaf water potential and stomatal conductance in anisohydric species. Plant Cell Environ 19:75–84.
Thomas D, Eamus D (1999) The influence of predawn leaf water potential on stomatal responses to atmospheric water content at constant Ci and on stem hydraulic conductance and foliar ABA concentrations. J Exp Bot 50: 243–251.
Thompson DS, Wilkinson S, Bacon MA, Davies WJ (1997) Multiple signals and mechanisms that regulate leaf growth and stomatal behavior during water deficit. Physiol Plant 100:303–313.
Timbal J, Lefebvre C (1995) Seasonal changes in water potential and the growth of young *Quercus rubra* and *Quercus palustris* plants during soil drought. Ann Sci For 52:67–79.
Triboulot MB, Fauveau ML, Bréda N, Label P, Dreyer E (1996) Stomatal conductance and xylem-sap abscisic acid (ABA) in adult oak trees during a gradually imposed drought. Ann Sci For 53:207–220.
Turner NC, Schulze ED, Gollan T (1984) The responses of stomata to vapour pressure deficits and soil water content. I. Species comparison at high soil water contents. Oecologia 63:338–342.
Viven P, Aussen G, Levy G (1993) Difference in drought resistance among three deciduous oak species grown in large boxes. Ann Sci For 50:221–233.
Wartinger A, Heilmeier H, Hartung W, Schulze D (1990) Daily and seasonal courses of leaf conductance and abscisic acid in the xylem sap of almond trees [*Prunus dulcis* (Miller) DA Webb] under desert conditions. New Phytol 116:581–587.
Wilkinson S, Corlett JE, Oger L, Davies WJ (1998) Effects of xylem pH on transpiration from wild-type and *flacca* mutant tomato leaves: A vital role for abscisic acid in preventing excessive water loss even from well-watered plants. Plant Physiol 117:703–709.
Wilson CC (1948) The effect of some environmental factors on the movements of guard cells. Plant Physiol 23: 5–37.
Wuenscher JE, Kozlowski TT (1971) The response of transpiration resistance to leaf temperature as a desiccation resistance mechanism in tree seedlings. Physiol Plant 24:254–259.
Zhang J, Davies WJ (1987) Increased synthesis of ABA in drying root tips and root-shoot communication via the transpiration stream. J Exp Bot 38:2015–2023.
Zhang J, Davies WJ (1989a) Abscisic acid produced in dehydrating roots may enable the plant to measure the water status of the soil. Plant Cell Environ 12:73–81.
Zhang J, Davies WJ (1989b) Sequential response of whole plant water relations to prolonged soil drying and the involvement of xylem sap ABA in the regulation of stomatal behavior of sunflower plants. New Phytol 113: 167–174.
Zhang J, Davies WJ (1990a) Changes in the concentration of ABA in xylem sap as a function of changing soil water status can account for changes in leaf conductance and growth. Plant Cell Environ 13:277–285.
Zhang J, Davies WJ (1990b) Does ABA in the xylem control the rate of leaf growth in soil-dried maize and sunflower plants? J Exp Bot 41:1125–1132.
Zhang J, Schurr U, Davies WJ (1987) Control of stomatal behavior by abscisic acid which apparently originates in the roots. J Exp Bot 38:1174–1181.
Zhang J, Jia W, Zhang D-P (1997) Effect of leaf water status and xylem pH on metabolism of xylem-transported abscisic acid. Plant Growth Reg 21:51–58.

8. Leaf Water Potential, Osmotic Potential, and Solute Accumulation of Several Hardwood Species as Affected by Manipulation of Throughfall Precipitation in an Upland *Quercus* Forest

Timothy J. Tschaplinski and G. Michael Gebre

Plant growth is more often limited by water stress in the temperate region than by any other environmental variable (Kramer 1983). The response of forests to increased occurrence of drought, associated with increased atmospheric concentrations of CO_2 and other greenhouse gases, has emerged as a key issue in the global climate change scenarios (Wigley et al. 1984). The forests of the southeastern United States have been identified as potentially sensitive to climate change because present rainfall levels are barely sufficient to meet current potential evaporation and transpiration demands of vegetation (Nielson et al. 1989). An increase in the frequency and severity of drought associated with the greenhouse effect (Rind et al. 1990) has the potential to greatly alter the distribution and productivity of tree species (Pastor and Post 1988), with an increase in the proportion of the growing season during which soil moisture is below wilting point inevitably increasing (Pastor and Post 1988). Drought tolerance of the individual species will dictate how ecosystems respond to such climate-change scenarios.

A mechanistic understanding of forest ecosystem adjustment to climate change requires an understanding of how component species respond to environmental perturbation at their predominant canopy position. Drought-tolerance mechanisms include those that postpone dehydration (e.g., stomatal closure, leaf rolling, and deep rooting) or those that increase dehydration tolerance (Kramer 1980). By maintaining low osmotic potentials and/or by accumulating solutes [osmotic adjustment (Morgan 1984)], dehydration-tolerant species function under low plant water potentials, in part, by enhancing the capacity of plants to take up water from dry soils and facilitating recovery after drought. The range of osmotic potentials of several North American tree species measured at full turgor was reported to be between −0.72 and −2.00 MPa under moist conditions and between −1.51 and −3.25 MPa under dry conditions (Abrams 1988), with the drought-tolerant species exhibiting generally lower osmotic potentials. Whereas much of the research on dehydration tolerance of woody species has been conducted on seedlings or saplings (reviewed in Abrams 1988), large trees have attributes that distinguish them from seedlings, including a deeper rooting habit, a greater proportion of canopy exposed to high light levels, and lower relative growth rates. Comparative studies are therefore needed. It has been hypothesized that light and nutrient regimes that promote high net carbon assimilation rates should promote the accumulation of osmotica (Abrams 1988). Given that osmotic adjustment and the maintenance of dehydration tolerance may require high light availability, overstory trees are likely to display greater dehydration tolerance than understory trees.

Mixed forests have assemblages of plant species with different drought-tolerance mechanisms, and species that maintain lower water potentials are able to compete better under dry conditions than those that do not. Therefore, species with low osmotic potentials may be more tolerant than species with high potentials, even if the former do not display osmotic adjustment. The objective of this study was to assess dehydration tolerance, as determined by osmotic potential and osmotic adjustment, of several major overstory and understory deciduous species in a mature forest growing under ambient throughfall precipitation, ambient minus 33% throughfall precipitation, and ambient plus 33% throughfall precipitation during both wet and dry years. Ultimately, the goal was to identify potentially sensitive species, assess the role that canopy position has in determining dehydration tolerance, and determine the nature of the solutes that are correlated with osmotic adjustment to drought.

Methods

Site Description and Experimental Design

The study was conducted on the Throughfall Displacement Experiment (TDE), where the mean annual precipitation is 1360 mm and the mean daily temperature is 14.2°C. The site characteristics, including soil type, aspect, history, species composition, climate, and system performance of the manipulation of throughfall precipitation, have been described in detail elsewhere (Hanson et al. 2001; Chapter 2 this volume). The major canopy-dominant species on the site are *Quercus prinus* L., *Quercus alba* L., and *Acer rubrum* L.. The understory consists largely of *A. rubrum, Cornus florida* L., and *Nyssa sylvatica* Marsh. Throughfall precipitation levels are manipulated by ~1900 subcanopy troughs suspended above the forest floor of the dry plot such that ~33% of the throughfall precipitation is trapped and passively transferred by gravity into piping and then across an ambient area to a wet plot. The three treatments, dry (-33% throughfall precipitation), ambient, and wet ($+33\%$ throughfall precipitation), each cover an 80×80-m area and have a similar slope with the upper half of the all treatments typically drier than the lower half. During the drought of 1995, Ψ_s declined on the drier upper slope to -2.0 MPa or below in all treatments by day of year (DOY) 254 (Hanson et al. 2001).

Sample Collection

Midcanopy branches were collected periodically during the growing season for trees of all three throughfall precipitation levels. Trees that were sampled were generally selected from the drier upper- and middle-slope positions of the study site. Sampling was typically conducted 10:00 AM through 4:00 PM on sunny days, with all the samples from a given species collected within 2 h. Species were selected for study largely based on their abundance on the site. Typically, the same trees/saplings were used throughout the season (May to September). In 1994, seedlings of *Q. prinus* and *Fagus grandifolia* Ehrh. were sampled only once, on July 29.

The following species were sampled in the 1995 growing season: *Q. prinus* (overstory and understory), *A. rubrum* (overstory and understory), *C. florida* (understory), *F. grandifolia* (understory), *Q. alba* (overstory), and *Acer saccharum* Marsh. (overstory). Branch samples of understory trees were harvested with a pole pruner, whereas overstory trees were sampled with a 12-gauge shotgun. Except for *A. saccharum,* six trees or saplings were selected at random from the top- and middle-slope positions of each plot. For *A. saccharum,* there were fewer trees available, and $n = 3$, 4, and 5 for the dry, wet, and ambient plots, respectively.

In 1995, sampling commenced June 8 (DOY 159) and continued through September 8, 1995 (DOY 251). Sampling occurred in the middle of the growing season to limit phenological/seasonal changes in Ψ_s, which were generally <-0.1 MPa for most species during the period sampled the previous, wetter growing season (Gebre et al. 1998a). Sampling occurred after spring leaf out and was completed without rainfall during sample collection. The three sampling periods for the major species in 1995 corresponded to when there was little or no stress, moderate stress, and severe stress. Virtually all understory *C. florida* were wilted in the dry treatment at the last sampling, compared to only a few trees displaying foliar wilting in the ambient treatment.

Plant Water Relations

Leaf water potential (Ψ_l) was determined with a Scholander-type pressure chamber immediately after the branch was collected (to minimize leaf dehydration) at midday, following the precautions of Ritchie and Hinckley (1975). Measurements were typically completed within 30–90 s of sampling, whether the plants were sampled by a pole pruner or by a shotgun. Sample variability was generally low for all species within the sampling time frame, suggesting that continued transpirational losses after excision were not problematic. Given the difficulty in negotiating the terrain and collecting samples at night, predawn Ψ_l was only determined for *C. florida* on Aug. 30, 1995 (DOY 242) when Ψ_s was low. Osmotic potential at full turgor ($\Psi_{\pi o}$) was used to assess the degree of dehydration tolerance and evidence of osmotic adjustment to drought. The branch was recut under water, rehydrated overnight in the dark, and a leaf sampled the following day to determine $\Psi_{\pi o}$ with a vapor-pressure osmometer. The protocol used to determine $\Psi_{\pi o}$ of leaf extracts was modified from Markhart and Lin (1985), as described by Gebre et al. (1998a). Briefly, the leaf samples were frozen on dry ice or in a −70°C freezer until they were analyzed. Samples were allowed to thaw for 15–18 min in plastic bags before the tissue sap was extracted with a leaf press and collected on a filter disk, which was then removed from the press and placed in the osmometer to obtain a reading (mmol kg^{-1} = mol m^{-3}), which was converted to MPa with the van't Hoff relation at 25°C (Nobel 1991). Data were not corrected for possible apoplastic dilution of symplast $\Psi_{\pi o}$, which was determined to be small (~0.2 MPa) for the species tested (unpublished data). Osmotic adjustment was calculated as the difference in $\Psi_{\pi o}$ of the dry treatment and that of the ambient or wet treatments.

Solute Analysis

At the time of leaf sapling in the field, one to two leaves per shoot were immediately collected and frozen under dry ice for metabolite analyses. The frozen leaf samples were lyophilized and ground before extraction. The extraction procedure for solutes followed that of Tschaplinski and Tuskan (1994). Briefly, 50 mg of dry ground leaf sample was twice extracted with 5 ml of 80% ethanol, and the extracts were combined prior to analyses. After air-drying 1 ml of the 10-ml extract, soluble carbohydrates, organic acids, and phenolic compounds were analyzed by gas chromatography–mass spectrometry (GC-MS) as trimethylsilyl derivatives (Gebre et al. 1998b). Extracts were silylated with 0.5 ml of *N*-methyl-*N*-trimethylsilyltri-fluoroacetamide (MSTFA) + 1% trimethyl-chlorosilane (TMCS) and 0.5 ml of acetonitrile and heated for 45 min at 70°C. The following day, 1 μl was injected into an HP 5890 Series II GC and 5972 MS. Sorbitol was used as an internal standard to develop response factors for all identified compounds; concentrations of quercitol and some unknown compounds were estimated in glucose equivalents. The concentration of scylloinositol was estimated in myoinositol equivalents.

Data Analyses

Given that individual soil-water contents recorded on the TDE were not correlated with each other and, therefore, could be treated as independent measurements (Hanson et al. 2001), spatially separated trees and saplings were treated as independent samples with respect to soil-water content. Treatment differences in measured variables, within a given species at a specific time, were determined by one-way analysis of variance (ANOVA) with *t*-tests used for mean comparisons.

Species differences in water relations variables and solute concentrations within a given treatment at a specific sampling time were determined by one-way ANOVA and ranked by Waller–Duncan multiple-range tests. Typically, six replications were collected for each species and treatment combination, unless specified otherwise.

Results

General Water Relations Responses

The 1994 through 1996, growing seasons can be characterized as wet with a short-term, early-season drought that was alleviated midseason in 1994, followed by a severe drought throughout 1995 and a return to a wetter-than-normal 1996 (Fig. 2.11 of Chapter 2 this volume). Combining data collected over the three growing seasons permits determination of the relationships of plant water relations variables over the full range of Ψ_s. Regressions of leaf $\Psi_{\pi o}$ versus Ψ_s indicate a series of lines with *F. grandifolia* having higher $\Psi_{\pi o}$ at a given Ψ_s than *C. florida,* followed by *Q. prinus* (Fig. 8.1). These regressions demonstrate that the $\Psi_{\pi o}$ data combined across years were relatively constant until Ψ_s declined below approximately -1.0 MPa, which can be considered the threshold for severe-stress ini-

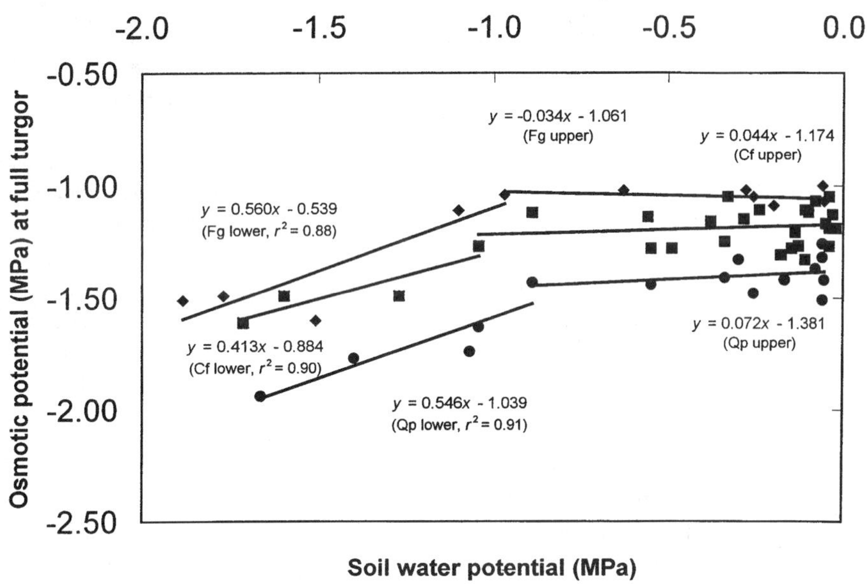

Figure 8.1. Linear regression of soil-water potential of the treatment plots at depth of 0–35 cm with leaf osmotic potential at full turgor of three understory species, including all data collected in 1994 through 1996 for *F. grandifolia* (Fg; diamonds), *C. florida* (Cf, squares), and *Q. prinus* (Qp, circles). Data points represent treatment means for each species at a point in time. Regressions of the data collected at high soil water potentials were separated from those collected at low water potentials (< -1.0 MPa).

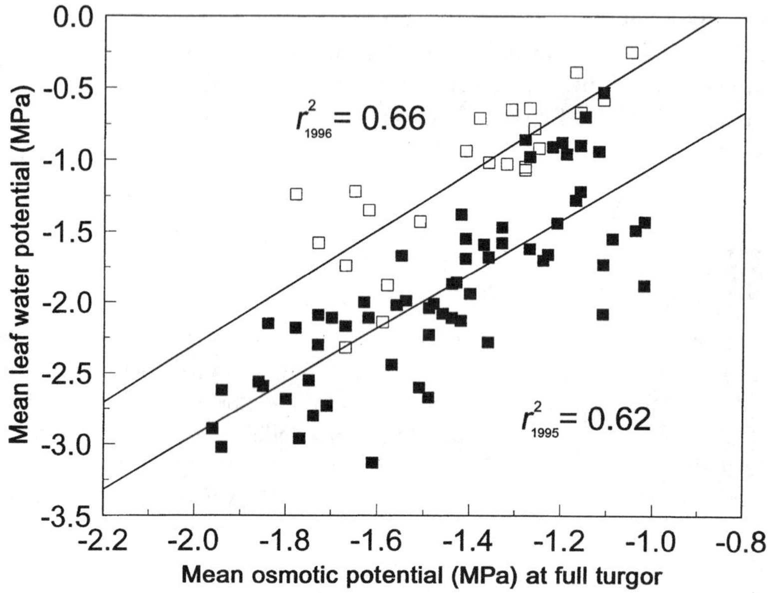

Figure 8.2. Regression of leaf water potential against leaf osmotic potential (determined between 10 AM and 4 PM) for data collected in the dry 1995 (■) and wet 1996 (□) growing seasons, including all species sampled at all times in those growing seasons. Data points represent treatment means for each species at a point in time.

tiation. The curves also indicate that the general species' differences in dehydration tolerance were maintained over the whole observed range of Ψ_s (Fig. 8.1). When data were pooled across species, a high correlation was observed between Ψ_s and Ψ_l in both the dry 1995 and wet 1996 growing season, with the regression lines paralleling each other (Fig. 8.2), and with lower Ψ_l coupled with lower $\Psi_{\pi o}$. Trees sampled in the dry 1995 growing season had a lower Ψ_l at a given $\Psi_{\pi o}$, and those trees achieved $\Psi_{\pi o}$ values that were up to 0.9 MPa lower than those of trees sampled during the wet year, indicating that a high degree of stress was experienced and dehydration tolerance was attained.

Species Differences

Given that the April to July precipitation in 1994 was 45% more than the long-term average, midday Ψ_l were generally high in all treatments. *Q. alba* maintained high midday Ψ_l (−0.9 to −1.3 MPa) compared to that of *Q. prinus* (−1.5 to −2.1 MPa) in the dry plot from July to September. A much larger range of values was observed the following dry growing season. Among the understory species sampled in the ambient treatment in 1995, *C. florida* had the highest midday Ψ_l (−0.53 MPa) early in the growing season, when soil water availability was high, followed by *A. rubrum* (−0.88 MPa), *Q. prinus* (−1.47 MPa), and *F. grandifolia* (−1.55 MPa). Among the overstory species, *A. saccharum* (−1.22 MPa) had the highest Ψ_l in the ambient treatment when water availability was high, followed by *A. rubrum* (−1.69 MPa), *Q. alba* (−2.11 MPa), and *Q. prinus* (−2.18 MPa). With the exception of overstory *A. rubrum,* Ψ_l declined in all species during the 1995 growing season, coincident with the decline in Ψ_s.

Pronounced species differences were observed in $\Psi_{\pi o}$, with species rankings consistent over time within a growing season and year after year (Fig. 8.3). The largest declines in $\Psi_{\pi o}$ occurred during the drought of the 1995 growing season (Fig. 8.3) in all species except *A. saccharum,* which was only measured early in the growing season. Based on $\Psi_{\pi o}$ measurements of the ambient treatment, a general trend in the ranking of species from low to high $\Psi_{\pi o}$ were *Q. prinus* < *Q. alba* < *A. rubrum* < *A. saccharum* = *N. sylvatica* = *C. florida* < *F. grandifolia.* This hierarchy was first evident in 1994 and was sustained through 1996, being a robust indicator of dehydration tolerance (Fig. 8.3).

The regressions of mean Ψ_l on mean $\Psi_{\pi o}$ for the combined data from the dry 1995 growing season demonstrated that, except for overstory *A. rubrum,* there were strong, positive correlations within a given species at a particular canopy position (Fig. 8.4). The coefficients of determination (r^2) for each species at a given canopy position were as follows: understory *A. rubrum* (0.96), understory *Q. prinus* (0.89), *C. florida* (0.89), understory *F. grandifolia* (0.85), overstory

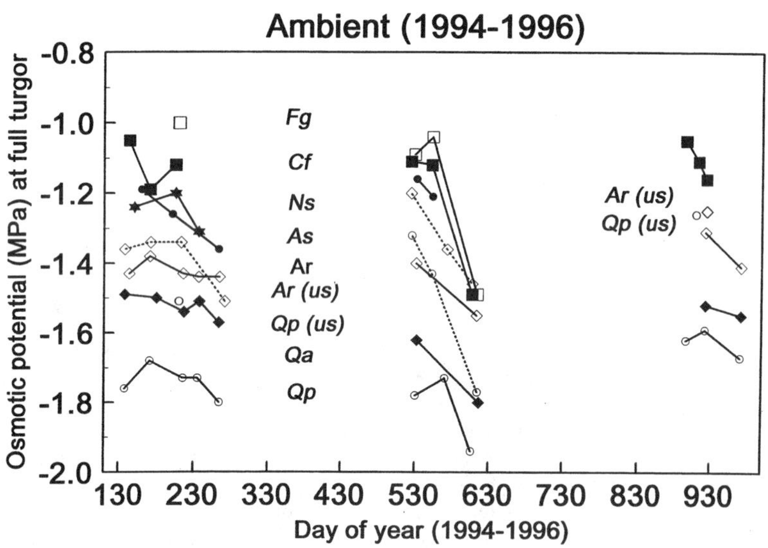

Figure 8.3. Leaf osmotic potential over time of all species sampled from 1994 through 1996. The symbols of the last 1994 data point match the species' abbreviations. Species that were sampled in both the understory (us) and overstory have the same symbol designations, but understory seedlings/saplings are additionally labeled and lines are dashed rather than solid. For Figure 8.3 the symbols are: ○, *Q. prinus* (Qp); ♦, *Q. alba* (Qa); ◇ – ◇, *A. rubrum* (Ar; US); ◇—◇, *A. rubrum* (Ar); ●, *A. saccharum* (As); ✱, *N. sylvatica* (Ns); ■, *C. florida* (Cf); and □, *F. grandifolia* (Fg).

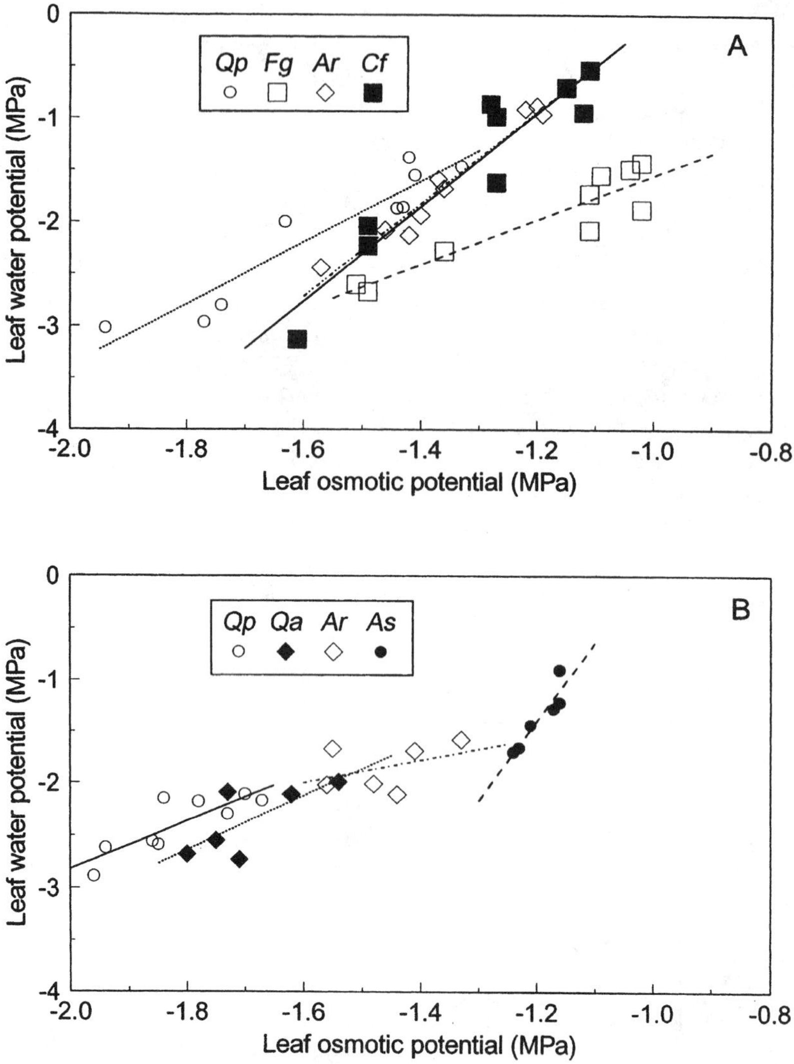

Figure 8.4. Regression of midday leaf water potential of tree species against leaf osmotic potential at full turgor for (A) understory saplings and (B) overstory trees growing in the wet, ambient, or dry treatments during the dry 1995 growing season. Data points are the means of six replicate trees (four for *A. saccharum*), which were pooled across treatment and over time. Monitored species included *Q. prinus* (Qp), *Q. alba* (Qa), *A. rubrum* (Ar), *A. saccharum* (As), *C. florida* (Cf), and *F. grandifolia* (Fg). [Graphs modified from Tschaplinski et al. (1998).]

A. saccharum (0.85), overstory *Q. prinus* (0.71), and overstory *Q. alba* (0.55, not significant). Low $\Psi_{\pi o}$ permits leaves to achieve and withstand low Ψ_l even within species that did not display significant treatment effects on $\Psi_{\pi o}$. Whereas the range of values of the understory species overlapped considerably, the overstory species occupied discrete ranges on the total distribution of values, with *A. saccharum* at the highest values of $\Psi_{\pi o}$ and Ψ_l, followed by *A. rubrum, Q. alba,* and *Q. prinus* (with the latter two species overlapping somewhat). The ranges of $\Psi_{\pi o}$ of each of the overstory species are related to the capacity for osmotic adjustment in response to declining Ψ_l. *Q. prinus* had the greatest range among the canopy-dominant species, followed by *Q. alba, A. rubrum,* and *A. saccharum.* Among the understory species, all of the species had similar ranges in $\Psi_{\pi o}$, but *F. grandifolia* had a lower Ψ_l at a given $\Psi_{\pi o}$. *A. rubrum* and *C. florida* occupied identical ranges, whereas *Q. prinus* again achieved the lowest $\Psi_{\pi o}$ of all the understory species measured.

Treatment Differences

All tree species measured in 1994 had lower Ψ_l in the dry plot than in the wet plot at some time during the growing season, despite the high amount of precipitation (Gebre et al. 1998a). The dry treatment had significantly lower Ψ_s than did the wet and ambient treatments from DOY 116 (April 26) onward (Fig. 2.11 of Chapter 2, this volume). *C. florida* exhibited lower $\Psi_{\pi o}$ and midday Ψ_l in the dry plot than the ambient or wet plots corresponding to the driest period in June ($\Psi_s = -0.37$ MPa). The only tree species to display osmotic adjustment in 1994 were understory *C. florida* (0.07–0.12 MPa) and overstory *Q. prinus* (0.19–0.25 MPa). As early as the second observation period in 1995, the dry treatment lowered the Ψ_l of understory *C. florida, F. grandifolia,* and *A. rubrum,* relative to the ambient treatment, but the dry treatment had no effect on the Ψ_l of understory *Q. prinus.* Although all treatments were under some degree of drought stress by August 30, 1995 (DOY 242), the predawn Ψ_l of *C. florida* in the wet treatment was -0.98 MPa, compared to -1.92 MPa for the ambient treatment and -2.54 MPa for the dry treatment. Among the overstory species, *Q. prinus* was the only species to demonstrate a decline in Ψ_l in the dry treatment relative to the other treatments during the last sampling period. Early in the growing season, when the degree of drought stress was moderate, *C. florida* and *Q. prinus* were the only understory species to display osmotic adjustment (0.15–0.20 MPa). Among the overstory tree species, only *Q. prinus* displayed osmotic adjustment (0.19 MPa) in the dry treatment (Tschaplinski et al. 1998). Under severe drought, overstory trees had similarly low $\Psi_{\pi o}$ in all treatments in late August and early September, as the $\Psi_{\pi o}$ of all species except *A. rubrum* declined in all treatments (Tschaplinski et al. 1998). In contrast with the overstory trees, all understory species, including *A. rubrum, C. florida, F. grandifolia,* and *Q. prinus,* displayed osmotic adjustment (0.12–0.20 MPa) in the dry treatment during the severe drought relative to the wet treatment (Tschaplinski et al. 1998).

Canopy Differences

There was no difference in $\Psi_{\pi o}$ between the overstory and understory *A. rubrum* trees from May to September 1994. In contrast, overstory *Q. prinus* had significantly lower $\Psi_{\pi o}$ (-1.91 MPa; dry plot) than the seedlings (-1.48 MPa; dry plot) during the same period (DOY 210 and 216). A comparison of tree species growing as both canopy dominants and as understory saplings in the drier 1995 growing season indicated that both understory *A. rubrum* and *Q. prinus* had higher Ψ_l and $\Psi_{\pi o}$ than did overstory trees when water availability was high. As severe drought was established, understory *A. rubrum* achieved lower Ψ_l (-1.9 to -2.4 MPa) than did canopy-dominant trees (-1.7 to -2.0 MPa), but the $\Psi_{\pi o}$ was the same. In contrast, the Ψ_l and $\Psi_{\pi o}$ of *Q. prinus* seedlings and mature trees were the same in the dry treatment under severe drought. Both seedlings and mature *Q. prinus* trees displayed the capacity for osmotic adjustment to drought, whereas

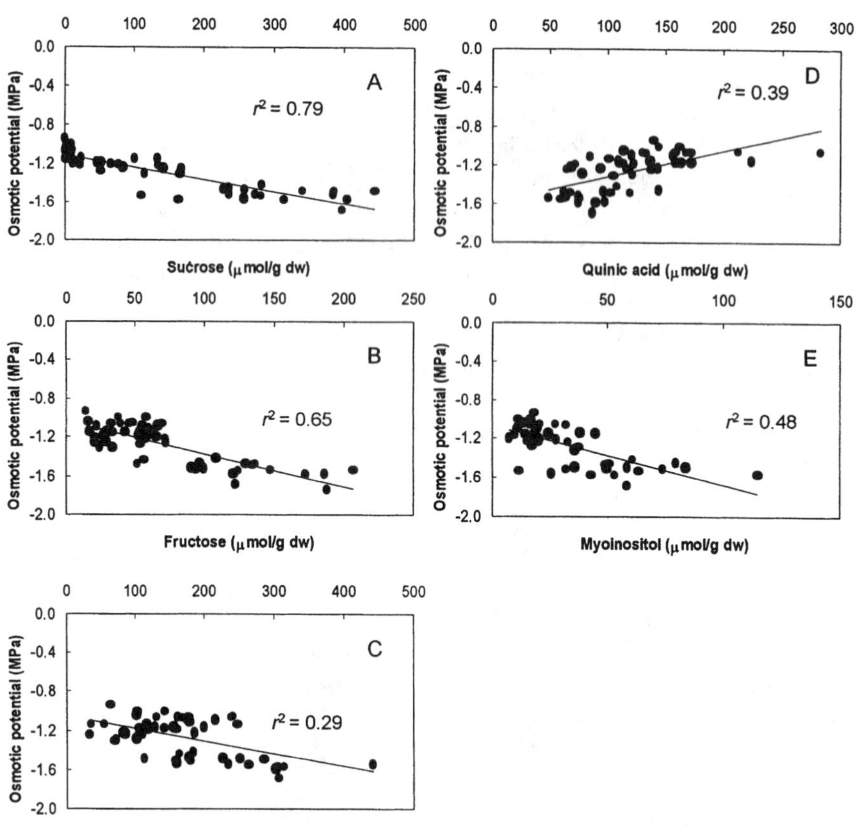

Figure 8.5. Correlation of organic solute concentrations with osmotic potential at full turgor in leaves of *C. florida* sampled when osmotic adjustment was greatest, including May 1994, June 1994, and August 1995. Data points are the means of treatment plots.

Table 8.1. Correlation coefficients (r) between leaf osmotic potential at full turgor ($\Psi_{\pi o}$) and specific solute concentrations of *C. florida* and *Q. prinus* leaves analyzed when osmotic adjustment was greatest during the 1994 and 1995 growing seasons

C. florida: Understory

	Fructose	Quinic acid	Glucose	Myoinositol	Sucrose	Potassium
Fructose						
Quinic acid	−0.51					
Glucose	0.82	−0.21				
Myoinositol	0.52	−0.30	0.44			
Sucrose	0.64	−0.53	0.42	0.81		
Potassium	0.06	0.63	−0.54	−0.70	−0.89	
$\Psi_{\pi o}$	−0.81	−0.20	−0.21	−0.11	0.21	−0.17

Q. prinus: Understory

	Quercitol	Fructose	Quinic acid	Glucose	Scylloinositol	Sucrose	Potassium
Quercitol							
Fructose	0.13						
Quinic acid	−0.03	−0.31					
Glucose	0.41	0.85	−0.24				
Myoinositol	0.52	−0.03	−0.02	0.27			
Sucrose	0.79	0.18	−0.18	0.39	0.56		
Potassium	−0.33	−0.67	0.33	−0.65	−0.24	−0.58	
$\Psi_{\pi o}$	−0.13	0.09	0.14	−0.01	−0.10	−0.13	0.03

Q. prinus: Overstory

	Quercitol	Fructose	Quinic acid	Glucose	Scylloinositol	Sucrose	Potassium
Quercitol							
Fructose	−0.30						
Quinic acid	0.07	−0.30					
Glucose	0.32	0.68	−0.13				
Myoinositol	−0.58	0.43	0.16	−0.13			
Sucrose	0.67	−0.34	0.04	0.00	−0.50		
Potassium	−0.28	−0.28	0.08	−0.35	0.18	−0.30	
$\Psi_{\pi o}$	0.27	0.22	−0.34	0.33	−0.16	0.00	−0.34

A. rubrum seedlings only displayed osmotic adjustment under severe drought conditions.

Solutes That Constitute Osmotic Potential and Adjustment

Extensive analyses of the of the solutes that accumulated as $\Psi_{\pi o}$ declined were conducted on *C. florida* and *Q. prinus,* the only species to consistently demonstrate osmotic adjustment to the dry treatment. Combining data from 1994 to 1995 indicated that the $\Psi_{\pi o}$ of *C. florida* was positively correlated with the concentrations of specific soluble carbohydrates, including sucrose ($r^2 = 0.79$), fructose ($r^2 = 0.65$), myoinositol ($r^2 = 0.48$), and glucose ($r^2 = 0.29$), and negatively correlated with the concentration of quinic acid ($r^2 = 0.39$) (Table 8.1 and Fig. 8.5). Such correlations were also evident in understory *Q. prinus* with $\Psi_{\pi o}$

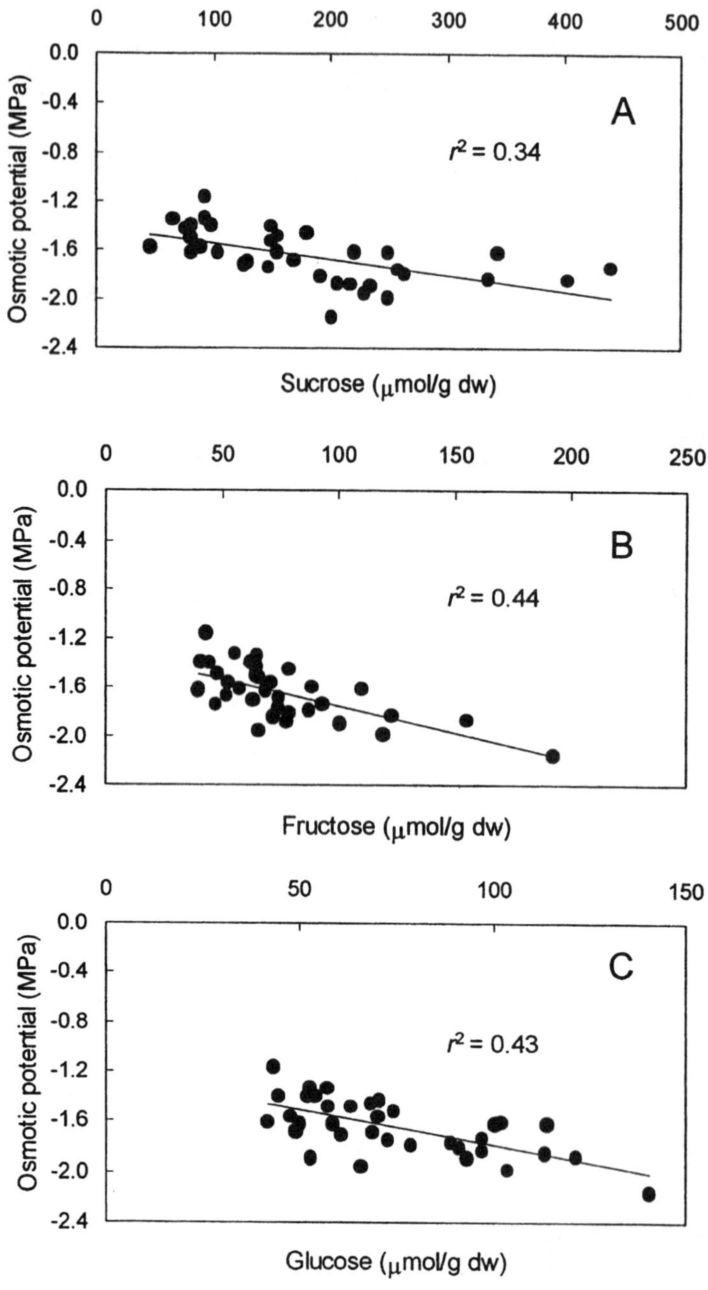

Figure 8.6. Correlation of organic solute concentrations with osmotic potential at full turgor in leaves of understory *Q. prinus* sampled when osmotic adjustment was greatest, including July and September 1995. Data points are the means of treatment plots.

correlated with sucrose (r^2 = 0.34), fructose (r^2 = 0.44), and glucose (r^2 = 0.43) (Table 8.1 and Fig. 8.6). Although K concentrations were high, they were not highly correlated with $\Psi_{\pi o}$ (Table 8.1). Similar correlations that used data from overstory *Q. prinus* yielded low coefficients of determination (r^2). These solutes represent a broad range, including soluble carbohydrates, sugar alcohols, organic acids, and inorganic ions that together constitute the bulk of the foliar $\Psi_{\pi o}$. In addition to these compounds, quercitol, a methylated sugar alcohol, is the dominant sugar alcohol in *Q. prinus* leaves and constitutes a significant fraction of their foliar organic solute pool. Given that quercitol concentrations were stable over time, quercitol's correlation with $\Psi_{\pi o}$ was not high (r^2 = 0.11). Additionally, scylloinositol was the most abundant nonmethylated sugar alcohol in *Q. prinus,* which contrasts with *C. florida,* which had myoinositol as the most abundant sugar alcohol.

Discussion

Species Differences

Species variation in $\Psi_{\pi o}$ was high, indicating considerable differences in dehydration tolerance among species of the eastern deciduous forest, with species differences consistent year to year (Fig. 8.3). The species differed in baseline $\Psi_{\pi o}$ when water availability was high, and the differences were maintained as $\Psi_{\pi o}$ declined with the decline in Ψ_s (Fig. 8.1). Given that the rate of decline in $\Psi_{\pi o}$ increased when Ψ_s was ≤ -1.0 MPa, it can be considered as a general threshold for the initiation of a severe stress response. This is not to suggest that trees, especially shallow-rooted, understory species, do not experience stress above this threshold. Within a given year, there were strong correlations between $\Psi_{\pi o}$ and Ψ_l (with data pooled across all species), indicating that lower $\Psi_{\pi o}$ facilitates functioning at lower Ψ_l (Fig. 8.2), which was especially evident when analyses were confined to a particular species (Fig. 8.4). Additionally, the relationship between $\Psi_{\pi o}$ and Ψ_l may vary on a year-to-year basis to some degree, because trees would experience differing degrees of adjustment to the prevailing conditions (Gebre et al. 1998a; Tschaplinski et al. 1998). Such a response was indicated in the relationship of Ψ_l versus $\Psi_{\pi o}$, which had a lower Ψ_l at a given $\Psi_{\pi o}$ in the drier year. Other drought tolerance mechanisms, such as stomatal closure that functions to maintain higher Ψ_l, may modify the relationship, contributing to some of the variation observed. Higher carbon assimilation rates when water availability is high likely leads to the accumulation of the carbohydrates and organic acids required for solute accumulation and the production of low $\Psi_{\pi o}$ at a correspondingly high Ψ_l. Conversely, long-term drought can restrict carbon assimilation, leading to relatively higher $\Psi_{\pi o}$ at a given Ψ_l (Kwon and Pallardy 1989; Gebre and Kuhns 1993).

Species differences in carbon assimilation rates under drought conditions can be used as an indicator of dehydration tolerance. For example, Hinckley et al. (1979) ranked some of the species in a central Missouri study, with *Q. alba* > *A.*

saccharum > *C. florida* saplings, from best to least able to photosynthesize during drought. Our species ranking in $\Psi_{\pi o}$ in 1994 is similar to other rankings in the drought tolerance of tree species using other methods (Hinckley et al. 1979; Martin et al. 1987). Oak species are generally considered more dehydration tolerant than mesic species, such as *A. saccharum, C. florida,* and *Juglans nigra* (Martin et al. 1987). These findings also agree with those of Bahari et al. (1985), who reported that *C. florida* was more sensitive to drought than were *A. saccharum* and *Q. alba.* Thus, the high $\Psi_{\pi o}$ of *C. florida,* a species recognized for its sensitivity to water stress (Roberts et al. 1980), and the low $\Psi_{\pi o}$ of the *Quercus* species, known for their adaptation to xeric sites (Abrams 1990), supports the use of $\Psi_{\pi o}$ to rank the degree of species dehydration tolerance. Overall, the species rankings in $\Psi_{\pi o}$ parallel their putative dehydration tolerance. Osmotic potential provides one assessment of dehydration tolerance of a given species. Species rankings in the ambient treatment were maintained when water availability was high or low and from year to year. Among the overstory trees, *Q. prinus* coincidently occupied the driest (pretreatment) ridge-top sites, maintained lower $\Psi_{\pi o}$ than *Q. alba,* achieved lower Ψ_l, and hence should be considered more dehydration tolerant. *Q. prinus* consistently maintained the lowest $\Psi_{\pi o}$, followed by *Q. alba, A. rubrum,* and *A. saccharum.* Of the understory species, *Q. prinus* again was the most dehydration tolerant, followed by *A. rubrum, C. florida,* and *F. grandifolia.*

Treatment Differences

Under moderate stress, understory *C. florida* and *Q. prinus* were the only two species that displayed osmotic adjustment to drought, having a lower $\Psi_{\pi o}$ in the dry treatment. Given that *C. florida* displayed osmotic adjustment under dry conditions and that it and *F. grandifolia* had the highest $\Psi_{\pi o}$ of the major species monitored, the capacity for osmotic adjustment was not only restricted to species considered dehydration tolerant. Subjecting the ambient and wet treatments to drying during severe drought resulted in underestimated capacities for osmotic adjustment (defined as the difference in $\Psi_{\pi o}$ between treatments) among the understory species. Overstory *Q. prinus,* the most dehydration-tolerant species of those studied, was the only canopy-dominant species to display osmotic adjustment under moderate stress. The absence of osmotic adjustment of canopy-dominant *Q. prinus* under severe drought was also indicative of all treatments being subjected to lower Ψ_s later in the growing season. Interannual variation in precipitation induced greater responses to drought among species than did the treatments, which tended to establish transient differences dependent on the precipitation pattern. The absence of osmotic adjustment in overstory *A. rubrum* and *A. saccharum* was expected, given that *Acer* plants are considered mesic species (Bahari et al. 1985). However, the lack of adjustment in *Q. alba* was unexpected, given studies that report the contrary (Parker et al. 1982; Parker and Pallardy 1988), and the low $\Psi_{\pi o}$ typically reported for *Quercus* species growing under xeric conditions (cf. Abrams 1990). With *Q. alba* being a deep-rooted species

(Hinckley et al. 1981), it can maintain higher predawn Ψ_l during drought than *C. florida,* a shallow-rooted species, can (Bahari et al. 1985). Deep rooting is a dehydration postponement mechanism that can delay *Q. alba* from reaching the low Ψ_l that initiates adjustment (Abrams 1990). In contrast, little to no change in predawn Ψ_l can induce adjustments of -0.2 to -0.3 MPa in *Q. prinus* (Abrams and Knapp 1986; Kubiske and Abrams 1991). However, severe drought in our study in the latter half of the summer of 1995 reduced the $\Psi_{\pi o}$ of all species to their lower limits (Tschaplinski et al. 1998).

It was only under severe drought that understory *A. rubrum* and *F. grandifolia* displayed osmotic adjustment to drought in addition to *C. florida* and *Q. prinus.* The prolonged drought of 1995 contrasted with the generally wetter 1994 summer. Whereas Ψ_l of seedlings/saplings did not decline below -1.5 MPa in 1994 (Gebre et al. 1998a), Ψ_l declined to -2.0 to -3.0 MPa for most species from July to September. Leaf water potentials declined during the growing season in all species measured except overstory *A. rubrum,* with many of the Ψ_l below the minimum values previously reported for field-grown saplings and trees (Tschaplinski et al. 1998). In contrast with overstory *A. rubrum,* which exhibited a minimum Ψ_l of -2.0 MPa, similar to that reported by Roberts et al. (1979), Ψ_l of saplings declined to -2.44 MPa in the dry treatment. The decline in Ψ_l of understory *C. florida* to -3.13 MPa was well below the minimum of -2.2 MPa reported for trees growing in North Carolina (Roberts et al. 1979) but similar to the predawn Ψ_l reported by Bahari et al. (1985). Similarly, both understory and overstory *Q. prinus* declined to ≤ -2.89 MPa, >1.0 MPa below the minimum Ψ_l reported for trees growing in West Virginia (Kochenderfer and Lee 1973). It can be concluded from these comparisons that the 1995 growing season was extremely dry in eastern Tennessee, with the drought further exacerbated by throughfall manipulation (Tschaplinski et al. 1998).

Canopy Differences

Understory trees maintained higher Ψ_l than overstory trees of the same species when water availability was high, but as Ψ_s declined, the Ψ_l of understory trees declined to below that observed for canopy dominants (Tschaplinski et al. 1998). Given that the lowest Ψ_l were achieved by understory trees, it was concluded that they exploit primarily shallow surface layers and are influenced sooner and to a greater extent than overstory trees. It was hypothesized that overstory trees were more likely to exhibit osmotic adjustment than understory trees because of differences in net photosynthetic rate. There were no canopy differences in $\Psi_{\pi o}$ of *A. rubrum,* a species that did not readily exhibit osmotic adjustment, except under severe drought and only in saplings. In contrast, overstory *Q. prinus* had lower $\Psi_{\pi o}$ than did seedlings in this study. Although shaded versus exposed trees of *C. florida* were reported to exhibit differences in $\Psi_{\pi o}$ (Knippling 1967), differences between overstory and understory plants may be observed only in species capable of osmotic adjustment. How a particular species responded to drought depended on the severity of drought, the species' relative position within the canopy structure, and its ability to respond in different light environments. The hypothesis that osmotic adjustment and maintenance of dehydration tolerance requires high light availability was supported by data from *Q. prinus* in the wet 1994 but not from that of *A. rubrum* obtained the same year (Gebre et al. 1998a) nor from the data of either species the following year (Tschaplinski et al. 1998). Understory *A. rubrum* and *Q. prinus* had higher $\Psi_{\pi o}$ and Ψ_l than overstory trees when water availability was high, but Ψ_l of understory saplings declined below that of overstory trees, and $\Psi_{\pi o}$ of understory saplings approached that of the overstory trees. The lower rates of carbon assimilation of *A. rubrum,* a more shade-tolerant species than *Q. prinus,* may have contributed to the lack of an effect of canopy position on $\Psi_{\pi o}$. In contrast, during July of the wetter 1994, the $\Psi_{\pi o}$ of understory *Q. prinus* seedlings remained higher than that of the overstory trees. The difference between understory and overstory trees increased in the dry treatment as overstory trees displayed osmotic adjustment to a short-duration drought whereas understory trees did not (Gebre et al. 1998a).

Role of Soluble Carbohydrates in Osmotic Potential

Although there are notable exceptions, soluble carbohydrates typically constitute the bulk of the solutes that determine the $\Psi_{\pi o}$ of leaves of deciduous hardwood species (Gebre and Tschaplinski 2002; Tschaplinski, unpublished data). In species that displayed osmotic adjustment, sucrose, fructose, and glucose were correlated with $\Psi_{\pi o}$, especially in understory seedlings/saplings. In the case of overstory *Q. prinus,* a larger array of solutes constituted $\Psi_{\pi o}$, which weakened the correlations with specific soluble carbohydrates (Gebre and Tschaplinski 2002). Given that soluble carbohydrate concentrations can fluctuate, other solutes are required to maintain $\Psi_{\pi o}$. Sugar alcohols, methylated sugar alcohols, and organic acids may serve to buffer the fluctuating concentrations of primary soluble carbohydrates, with their concentrations not necessarily correlated with $\Psi_{\pi o}$. For example, in addition to soluble carbohydrates, *C. florida* had high concentrations of myoinositol, which was positively correlated with $\Psi_{\pi o}$, and quinic acid, which was negatively correlated with $\Psi_{\pi o}$. Quinic acid concentrations are typically high early in the growing season, when growth is most active, and declines later in the growing season, at the same time when drought is more likely to occur, inducing osmotic adjustment (Gebre and Tschaplinski 2002; Tschaplinski, unpublished data). These responses together resulted in the negative correlation. In the case of *Q. prinus,* quercitiol, a methylated sugar alcohol, and scylloinositol, another sugar alcohol, together constitute a large fraction of the solute pool with concentrations stable over time (Gebre and Tschaplinski 2002). Effective maintenance of $\Psi_{\pi o}$ requires that concentrations of some solutes remain stable, whereas others, such as soluble carbohydrates, can respond to the prevailing conditions and are, therefore, more likely to be involved in osmotic adjustment.

Significance of Osmotic Potential

The Ψ_l observed for both understory and overstory trees generally paralleled $\Psi_{\pi o}$, indicating that lower $\Psi_{\pi o}$ facilitates plants surviving lower Ψ_l. Such correlations of midday, predawn, or Ψ_s with $\Psi_{\pi o}$ have been reported, but often a correlation is not observed (Bahari et al. 1985; Abrams 1990; Parker et al. 1982; Gebre et al. 1998a). A significant correlation is less likely to occur if the data are collected only when soil-water availability is generally high or if the species studied do not readily lower $\Psi_{\pi o}$ in response to a decline in Ψ_l, as is the case for overstory *A. rubrum* and *Q. alba* (Gebre et al. 1998a; Tschaplinski et al. 1998). A significant correlation is more likely to occur if data are pooled across species that differ in dehydration tolerance and across a wide range of soil-water availability, as in the present study. *F. grandifolia* had a unique response, maintaining a high $\Psi_{\pi o}$ but at low Ψ_l, a combination of responses that are likely to keep it suppressed in the understory during dry conditions.

Using a single water relations variable to characterize a species' drought tolerance capability may be problematic, as highlighted by the contrast between *A. rubrum* and *C. florida.* Both species demonstrate the same relationship between $\Psi_{\pi o}$ and Ψ_l. Although *C. florida* has a greater range in values, facilitated by osmotic adjustment, this does not result in greater survival. Under the driest conditions, Ψ_l of *C. florida* declined to values that were 0.6 MPa lower than those of *A. rubrum,* and its mortality was four times greater in the dry treatment than in the wet treatment, with a mean rate of 13% over the dry 1995 summer, compared to a mean rate of 2.2% for *A. rubrum* (Hanson et al. 2001; Chapter 15, this volume). Both species had similar mortality rates of ~1% across all treatments during the wet 1994 growing season. Other drought-tolerance mechanisms and characteristics, such as stomatal regulation and root distribution, additionally determine Ψ_l and its impact on the plant (i.e., whether lethal Ψ_l has been exceeded). The greater drought sensitivity of *C. florida* has been attributed to its shallow root system (Hinckley et al. 1981). Understory species that also have high $\Psi_{\pi o}$, such as *F. grandifolia, N. sylvatica,* and *A. saccharum,* also have to be considered vulnerable to increased mortality in response to an increased frequency and severity of droughts, potentially altering the species composition of the remaining understory.

Conclusions

Species displayed significant differences in $\Psi_{\pi o}$, with drought-tolerant species, such as *Q. prinus,* exhibiting lower $\Psi_{\pi o}$ and greater osmotic adjustment than other species. Given that *C. florida* had high $\Psi_{\pi o}$ and exhibited osmotic adjustment in response to drought, osmotic adjustment was not limited to species with a low baseline $\Psi_{\pi o}$. Whereas such a capability aids survival under mild water stress, species with a high baseline $\Psi_{\pi o}$, such as *C. florida,* will be affected more than other species under severe stress in a mixed forest, where *Quercus* or other species with a lower $\Psi_{\pi o}$ predominate by functioning at a lower leaf Ψ_l facilitated by a low $\Psi_{\pi o}$. Prolonged drought in the latter half of the 1995 growing season reduced the Ψ_l of most species to below that previously reported for field-grown trees. Throughfall manipulation (−33%) was effective in further exacerbating drought, but interannual variation in precipitation produced greater drought responses than did the treatments. The shallow-rooted understory saplings/trees were influenced sooner and to a greater extent than were the overstory trees, highlighting the greater vulnerability of understory species with a high $\Psi_{\pi o}$. The Ψ_l observed for both understory and overstory trees generally paralleled $\Psi_{\pi o}$, suggesting that a low $\Psi_{\pi o}$ facilitates leaf functioning at a lower Ψ_l. The leaf $\Psi_{\pi o}$ was correlated with concentrations of specific soluble carbohydrates, especially in understory *Q. prinus* and *C. florida,* where fructose, glucose, and sucrose accounted for much of the solute accumulation (osmotic adjustment). Whether determined under wet or dry years, the ranking among species in $\Psi_{\pi o}$ was similar, with *Q. prinus* and *Q. alba* being the most dehydration tolerant, as defined by $\Psi_{\pi o}$, followed by *A. rubrum,* which was intermediate, and *A. saccharum, C. florida, N. sylvatica,* and *F. grandifolia,* which were least tolerant. The greater drought sensitivity of *C. florida,* resulting in greater mortality than for *A. rubrum,* despite similar water relations responses, is likely attributable to its shallow root system, which resulted in a lower Ψ_l. Species, canopy position, and Ψ_s in the rooting zone interact to determine $\Psi_{\pi o}$, osmotic adjustment, and the Ψ_l that can be tolerated before mortality occurs. Understory seedlings/saplings are more vulnerable to droughts than overstory trees, with species differences in dehydration tolerance likely to dictate the species composition of the regenerating forest following sustained, chronic drought.

References

Abrams MD (1988) Sources of variation in osmotic potentials with special reference to North American tree species. For Sci 34:1030–1046.

Abrams MD (1990) Adaptations and responses to drought in *Quercus* species of North America. Tree Physiol 7: 227–238.

Abrams MD, Knapp AK (1986) Seasonal water relations of three gallery forest hardwood species in northeast Kansas. For Sci 32:687–696.

Bahari ZA, Pallardy SG, Parker WC (1985) Photosynthesis, water relations, and drought adaptation in six woody species of oak-hickory forests in central Missouri. For Sci 31:557–569.

Gebre GM, Kuhns MR (1993) Effects of water stress preconditioning on gas exchange and water relations of *Populus deltoides* clones. Can J For Res 23:1291–1297.

Gebre GM, Tschaplinski TJ (2002) Solute accumulation of chestnut oak and dogwood leaves in response to throughfall manipulation of an upland oak forest. Tree Physiol 22:251–260.

Gebre GM, Tschaplinski TJ, Shirshac, TL (1998a) Response in water relations of several hardwood species to throughfall manipulation in an upland oak forest during a wet year. Tree Physiol 18:299–305.

Gebre GM, Tschaplinski TJ, Tuskan GA, Todd DE (1998b) Clonal and seasonal differences in leaf osmotic potentials and organic solutes of five hybrid poplar clones grown under field conditions. Tree Physiol 18:645–652.

Hanson PJ, Todd DE, Amthor JS (2001) A six-year study of sapling and large-tree growth and mortalility responses to natural and induced variability in precipitation and throughfall. Tree Physiol 21:345–358.

Hinckley TM, Dougherty PM, Lassoie JP, Roberts JE, Teskey RO (1979) A severe drought: Impact on tree growth, phenology, net photosynthetic rate and water relations. Am Midl Nat 102:307–316.

Hinckley TM, Teskey RO, Duhme F, Richter H (1981) Temperate hardwood forests. In Kozlowski TT (ed) *Water deficits and plant growth.* Academic Press, New York, pp 153–208.

Knippling EB (1967) Effect of leaf aging on water deficit-water potential relationships of dogwood leaves growing in two environments. Physiol Plant 20:65–72.

Kramer PJ (1980) Drought stress, and the origin of adaptations. In Turner NC, Kramer PJ (Eds). *Adaptation of plants to water and high temperature stress.* Wiley, New York, pp 7–22.

Kramer PJ (1983) *Water relations of plants.* Academic Press, New York.

Kochenderfer J, Lee R (1973) Indexes to transpiration by forest trees. Oecol Plant 8:175–184.

Kubiske ME, Abrams MD (1991) Rehydration effects on pressure-volume relationships in four temperate woody species: Variability with site, time of season and drought conditions. Oecologia 85:537–542.

Kwon KW, Pallardy SG (1989) Temporal changes in tissue water relations of *Quercus acutissima, Q. alba,* and *Q. stellata* subjected to chronic water stress. Can J For Res 19:622–626.

Markhart AH III, Lin TY (1985) New hand-operated press for the extraction of tissue sap for the measurement of osmotic potential. Agron J 77:182–185.

Morgan JM (1984) Osmoregulation and water stress in higher plants. Annu Rev Plant Physiol 35:299–319.
Martin U, Pallardy SG, Bahari ZA (1987) Dehydration tolerance of leaf tissues of six woody angiosperm species. Physiol Plant 69:182–186.
Nielson RP, King GA, DeVelice RL, Lenihan J, Marks D, Dolph J, Campbell B, Glick G (1989) *Sensitivity of ecological landscapes and regions to global climate change.* US Environmental Protection Agency, Environmental Research Laboratory, Corvallis, Oregon.
Nobel PS (1991) *Physicochemical and environmental plant physiology.* Academic Press, San Diego, California.
Parker WC, Pallardy SG (1988) Leaf and root osmotic adjustment in drought-stressed *Quercus alba, Q. macrocarpa,* and *Q. stellata* seedlings. Can J For Res 18:1–5.
Parker WC, Pallardy SG, Hinckley TM, Teskey RO (1982) Seasonal changes in tissue water relations of three woody species of the *Quercus–Carya* forest type. Ecology 63:1259–1267.
Pastor J, Post WM (1988) Response of northern forests to CO2-induced climate change. Nature 334:55–58.
Rind D, Goldberg R, Hansen J, Rosenzweig C, Ruedy R (1990) Potential evapotranspiration and the likelihood of future drought. J Geophys Res 95:9983–10004.
Ritchie GA, Hinckley TM (1975) The pressure bomb as an instrument for ecological research. Adv Ecol Res 9: 165–254.
Roberts SW, Knoerr KR (1977) Components of water potential estimated from xylem pressure measurements in five tree species. Oecologia 28:191–202.
Roberts SW, Knoerr KR, Strain BR (1979) Comparative field water relations of four co-occurring forest tree species. Can J Bot 57:1876–1882.
Roberts SW, Strain BR, Knoerr KR (1980) Seasonal patterns of leaf water relations in four co-occurring forest tree species: parameters from pressure-volume curves. Oecologia 46:330–337.
Tschaplinski TJ, Tuskan GA (1994) Water-stress tolerance of black cottonwood and eastern cottonwood clones and four of their hybrid progeny. II. Metabolites and inorganic ions that constitute osmotic adjustment. Can J For Res 24:681–687.
Tschaplinski TJ, Gebre GM, Shirshac TL (1998) Osmotic potential of several hardwood species as affected by throughfall manipulation of an upland oak forest during a dry year. Tree Physiol 18:291–298.
Wigley TML, Briffa KR, Jones PD (1984) Predicting plant productivity and water resources. Nature 312:102–103.

9. ^{18}O and ^{13}C in Leaf Litter Versus Tree-ring Cellulose as Proxy Isotopic Indicators of Climate Change

Lee W. Cooper and Corina Solis

The Throughfall Displacement Experiment (TDE) has resulted in a number of arboreal responses, examples of which range from water potential differences in treatments (Gebre et al. 1998) to whole-tree mortality (Hanson et al. 2001). Many of these direct ecosystem adjustments have been readily measurable and are discussed in more detail throughout this volume. Nevertheless there are compelling reasons to measure more indirect physiological responses, such as stable-isotope ratios, within trees of this eastern North American deciduous forest subjected to a throughfall displacement disturbance. Stable-isotope ratios of plant organic material have been shown to have utility for integrating plant water status in the case of carbon isotopes [e.g., Berninger et al. (2000)], and hydrogen and oxygen isotopes of organic matter also show potential for recording climatic variations (Edwards 1993; Farquhar et al. 1998). This has led to widespread interest in the climatic and physiological information that is potentially recorded in the isotopic composition (oxygen, carbon, and hydrogen) of organic matter in tree rings. However, interpreting isotope data from recent tree rings occurs in the context of industrial-era increases in CO_2, decreasing $^{13}C/^{12}C$ ratios in the atmospheric fraction of the gas, and incomplete or unavailable weather records for older tree-ring sequences. These complexities impair the use of isotopic variations for interpreting prehistoric and historic tree-ring records.

At the initiation of the TDE, we anticipated that the 33% reduction and the 33% increase in precipitation throughfall would give us the opportunity to validate the responses observed in the distribution of stable-isotope ratios in plant organic matter synthesized during the experiment. Therefore, we saw this experiment as an opportunity to assess and improve the applicability of the isotopic content of organic matter to infer past climate; for example, for trees in the dry treatment:

- Would greater water-use efficiency be observed in the molar ratio of carbon dioxide fixed relative to water transpired? For stable carbon isotopes, less discrimination against ^{13}C would be expected during photosynthesis (Farquhar et al. 1982).
- If this water-efficiency marker is manifested as less negative $\delta^{13}C$ values in leaf tissue, is that isotopic discrimination conserved through the biochemical changes that are required prior to cellulose synthesis in tree rings?
- Are changes in the drought status of trees, or the natural annual variation in humidity reflected in the oxygen (and hydrogen) isotopic composition of leaf cellulose?

For oxygen (and hydrogen) isotopes, the composition of water in leaves and other plant tissues subject to stomatal transpiration are significantly enriched in heavy isotopes relative to stem and soil waters [reviewed by Flanagan and Ehleringer (1991), and Buhay et al. (1996)]. The enrichment in ^{18}O in leaf water relative to meteoric water can reach as high as 20‰ in some arid-zone plants that have long water-residence times (Cooper and DeNiro 1989; Roden and Ehleringer 2000a). The oxygen isotopic composition of leaf organic matter is functionally related to the degree of enrichment of heavy isotopes in leaf water (Farquhar et al. 1998). As a result, it has been assumed that at least some heavy-isotope enrichment expressed in leaf water and organic material will be transferred and preserved during the translocation of photosynthate to other portions of trees, including the synthesis of tree-ring tissue (Burk and Stuiver 1981; Edwards 1993).

This assumption has served as the basis for empirically based assessments of past climates that use stable-isotope analyses of preserved wood [e.g., Clague et al. (1992); Buhay and Edwards (1995); Lipp et al. (1996); Saurer et al. (1997b)]. Despite the apparent success of this approach, there are indications that significant isotopic exchange can occur between the oxygen atoms in translocated photosynthates and the oxygen atoms of *in situ* water at sites of cellulose synthesis, such as within tree trunks. The significance of this isotopic exchange is that climatic information associated with the oxygen isotopic composition of leaf water, which would have been affected by evapotranspiration and photosynthesis, will be diluted by the oxygen isotopic composition of waters available at the sites of cellulose synthesis. These indications were first demonstrated with cell cultures (Sternberg et al. 1986) and sprouting potato tubers (DeNiro and Cooper 1989), and (1) patterns consistent with this hypothesis and (2) known biochemical transformations have also been observed in trees (Hill et al. 1995). If these isotopic exchange processes are variable in different species, or under different physiological or physical conditions, there can be no generalized, practical model for interpreting the isotopic composition of organic material preserved in wood cellulose that would also include isotopic fractionation during photosynthesis (DeNiro and Cooper 1989).

Field and experimental approaches contemporary with our TDE experimental work have contributed new insights to the problem of relating the isotopic composition of leaf cellulose to the wood cellulose that is more commonly preserved and analyzed. White et al. (1994) found that deuterium/hydrogen (D/H) ratios in the wood of eastern white pine (*Pinus strobus* L.) could be related to its source water D/H ratios by a model for isotopic fractionation that included isotopic fractionation within pine needle water during photosynthesis. The importance of leaf water isotopic fractionation was sensitive, however to the isotopic composition of the water vapor that exchanged with the leaf water, which was not necessarily in isotopic equilibrium with the leaf water. White et al. also pointed out that the relationship between the isotopic composition of water vapor and relative humidity also varies by location, which complicates developing a predictive, general model for the isotopic composition of wood cellulose. They also suggested that an alternative model for the isotopic composition of wood cellulose, advanced by Terwilliger and DeNiro (1995), which used avocado seedlings as an experimental system, remained consistent with their findings. Terwilliger and DeNiro (1995) showed that the hydrogen isotopic composition of carbon-bonded hydrogen in cellulose (analyzed as cellulose nitrate) was primarily a function of source water and that carefully controlled experimental shifts in relative humidity had little or no influence on the hydrogen isotopic composition of wood synthesized by avocados. This work with tree seedlings represented a more sophisticated approach than the simple experiments with potatoes (DeNiro and Cooper 1989), in which relative humidity was held constant, humidity variations were not specifically investigated, and only source water in tubers was varied.

Many empirical examples remain, however, where relative humidity and/or water availability are at least correlated with the isotopic composition of wood cellulose. Saurer et al. (1997a), for example, observed higher $\delta^{18}O$ values in wood cellulose of several tree species associated with drier soil moisture within a small Swiss valley. They also observed increasing $\delta^{13}C$ values in these same trees, relating these trends to greater stomatal resistance under drier conditions, which, in turn, leads to less isotopic discrimination against ^{13}C during photosynthesis. Lipp et al. (1996) also found that oxygen isotopic composition of wood cellulose in *Tamarix jordanis* over an aridity gradient in Israel could be related to the degree of heavy-isotope enrichment observed in leaf water, which was, in turn, controlled by relative humidity over the gradient.

Most recently, Roden and Ehleringer (1999, 2000a, 2000b) have used riparian trees growing under both greenhouse and field conditions and evaluated the hydrogen and oxygen isotopic composition of the wood cellulose formed. They found that the oxygen and hydrogen isotopic composition of wood cellulose were functions of both source water and heavy-isotope enrichment that occurs within the leaves. These studies observed that some of the heavy-isotope enrichment in leaves of the trees studied is partially and ultimately incorporated into wood cellulose. This work criticized previous experiments that demonstrated postphotosynthetic changes in the isotopic content of synthesized cellulose [e.g., DeNiro and Cooper (1989); Terwilliger and DeNiro (1995)] while constructing model adjustments to account for these postphotosynthetic changes. These efforts (Roden and Ehleringer 1999, 2000a, 2000b) are, no doubt, subject to their own practical limitations for field research. It is not straightforward, for example, to estimate what fraction of water within a leaf is actively available for incorporation into organic material during photosynthesis (Yakir et al. 1989). Models that incorporate the oxygen and hydrogen isotopic compositions of leaf water into predictions of the isotopic composition of wood cellulose are also dependent on finite, laborious sampling of leaf water that varies continuously. Similarly, the degree of heavy-isotope enrichment in leaves is also dependent on the isotopic composition of atmospheric water vapor that exchanges with leaf water, which may or may not be in equilibrium, is laborious to measure and varies continuously. Leaf boundary-layer differences and physiological differences among species (Buhay et al. 1996) may also lead to different isotopic compositions in leaf and wood cellulose for trees growing under identical climatic conditions.

Comparisons with Natural "Throughfall Displacement" in a Dry Tropical Forest

The fidelity of stable isotopic ratios to record predictable physiological responses in organic matter is therefore an important element for validating generalized models of the relationship between stable-isotope ratios and meteorological variables. Although our work in the TDE met some of these requirements through the study of separate species growing in both the dry and wet treatments, the responses observed in the isotopic content of organic matter of deciduous trees in the humid temperate conditions of eastern North America are probably not universally applicable in other ecological settings. In order to assess the context of our findings at the TDE site, we also measured the isotopic content of organic matter produced at a dry tropical forest site that in many respects represents an extreme case of throughfall displacement on an annual climatic cycle. On the Pacific Coast of México, at Chamela (19°29' N, 105°01' W), some species of trees are "wet deciduous" (i.e., these trees bear leaves during the dry season, taking advantage of abundant light when most other trees have shed their leaves). These leaves are then shed during the wet season, when light levels are lower due to competition with the majority of species that are dry deciduous. This adaptation to annual physiological drought stress is achieved through stomatal control (Fanjul and Barradas 1985, 1987; Holbrook et al., 1995). As part of this companion sampling to our work on the TDE site, we determined the carbon and oxygen isotope composition of whole organic matter and cellulose, respectively, in leaves of two wet deciduous trees (*Jaquinia pungens* and *Coccoloba liebmannii*) and a drought deciduous (*Thouinia paucidentata*) species during the beginning of the dry season, when leaves are present on all trees.

Throughfall Displacement Experimental Design

Despite the complexities that relate to the species and ecological settings studied, it is clear that, in many instances, the hydrogen and oxygen isotopic composition of both leaf and wood cellulose can be related to climatic and physiological processes that affect the isotopic composition of source and leaf water and to atmospheric water vapor. Carbon isotopic ratios are similarly influenced by physiological responses to stomatal conductance and to the ambient isotopic composition of carbon dioxide. We designed our TDE work to address a series of questions:

- Could we identify trees showing a response to the throughfall displacement in the isotopic composition of wood or leaf cellulose?
- Given the high species diversity, were some trees showing a response and others not?
- Is the choice of species analyzed a determinant of the isotopic variation observed?
- Was interannual variation in precipitation and relative humidity a significant factor in governing isotopic responses relative to the disturbance ($\pm 33\%$ throughfall) itself?
- Was there a relationship between the physiological responses of the trees, as studied by other investigators, and the isotopic variation we observed over the wet and dry plots?
- Is the expected heavy-isotope enrichment in cellulose of leaf tissue observed in the dry treatment or in drier years?
- What is the relationship between the heavy-isotope content of leaf tissue and the heavy-isotope content of tree-ring cellulose synthesized in the same growing season or the following growing season?

Oxygen Isotopes

Because of financial constraints, we chose (1) not to pursue work on carbon-bound (nonexchangeable) hydrogen isotopes in order to develop more complete oxygen and carbon isotopic datasets and (2) not to sample on the ambient plot that was unaffected by the throughfall disturbance. We chose to sample the five most common species of trees that were available on both the wet and dry plots. Our sampling revolved around a central assumption: that cellulose synthesized within the leaf is more sensitive to climatic factors, such as relative humidity, temperature, and isotopic exchange with water vapor than the cellulose ultimately incorporated into wood cellulose in the trunk. We hypothesized that any differences in the oxygen isotopic composition of cellulose between the wet and dry treatments would be most readily observed in leaf cellulose and that there would be less significant differences in the oxygen isotopic composition of wood cellulose between wet and dry treatments. Our sampling was accomplished on an annual basis with leaf cellulose derived from random leaf-litter samples, and we ultimately compared the oxygen and carbon isotopic compositions of wood cellulose produced in each of four years to that of leaf cellulose produced each year.

Other Stable Isotope Datasets

Other Oak Ridge Reservation datasets were used to constrain the isotopic composition of local soil water, organic matter and atmospheric carbon dioxide, and ranges of heavy-isotope enrichment in leaf water (Garten and Taylor, 1992; Cooper and Norby 1994; Halleran 1995; Garten et al. 2000; Bowling et al. 1999). For example, the annual range in oxygen isotopic variability observed in water stored in all but surface soil and groundwater sources is small and close to analytical precision, [~0.2‰ (Halleran 1995)]. We assumed, therefore, that differences in the ^{18}O content of deeper waters directly used by trees between the wet and dry treatments would be relatively small over the course of the sampling period. We also assumed that changes in soil moisture would not have significant effects on the oxygen isotopic content of water used, based on the year-round measurements made by Halleran (1995) over many different soil moisture conditions. Based on previous work (Cooper and Norby, 1994), we also assumed that adjustments in long-term stomatal conductance, such as would also occur with reduced soil-water availability, could result in a significantly higher ^{18}O content in leaf water and cellulose even if humidity and source water remained constant.

Carbon Isotopes

We hypothesized that incorporation of ^{13}C in organic matter would follow established models [e.g., Farquhar et al. (1982)], in which higher ^{13}C content in leaf organic matter results from lower stomatal conductance under drought stress, whether on a species or treatment basis. Other effects that we anticipated included (1) more negative $\delta^{13}C$ values in leaves of light-limited, understory species as a consequence of slower rates of photosynthesis under lower light conditions or (2) exposure to heavy-isotope-depleted CO_2 released from the soil sources that are present on Walker Branch Watershed (Bowling et al. 1999).

Methods

Incoming rainfall, air temperature, and relative humidity data at the TDE site were collected using data loggers (Hanson et al. 1998). Additional information on physical conditions at the TDE site are described in more detail in Hanson et al. (1998).

Leaf cellulose was extracted from leaves of five of the more dominant species of trees that are growing on both the wet and dry treatments of the TDE: *Quercus prinus, Q. alba, Acer rubrum, Liriodendron tulipifera*, and *Cornus florida*. Leaves were sorted from bulk leaf-litter collections made in the month of September for each year in the period from 1993 to 1996. Because the autumn period is when the bulk of leaf-fall occurs, we assumed that each individual leaf sample was derived from random heights in the forest canopy and from random individual trees within the treatment plot. Cellulose was extracted from three individual leaves of each species from each of the wet and dry treatments for each year of analysis. Thus, for each of the 4 years analyzed, 1993–1996, 30 samples were analyzed (3 leaves $\times$ 5 species $\times$ 2 treatments).

Leaf organic matter sampling and processing at Chamela was similar to that at the TDE site. Leaves of two wet deciduous trees (*Jaquinia pungens* and *Coccoloba liebmannii*) and a drought deciduous species (*Thouinia paucidentata*) were sampled in December 1992, at the beginning of the annual dry season, from sites with different soil moisture: arroyos (wet) and ridge (dry).

For the tree-ring measurements, 12 cores, 5 mm in diameter, were obtained from 3 of the 5 species at trunk heights $\sim$1 m above ground level on the west and east sides of specific trees. The sampling was accomplished on both the wet and dry treatment plots of the TDE in early April 1997, several weeks before the onset of leaf growth. The wood recovered consisted of two cores each from both the wet and dry plots (four total) for *A. rubrum, L. tulipifera,* and *C. florida.* Cores were sectioned with the aid of a dissecting microscope to separate each of the last four full-ring increments, corresponding to 1993 to 1996.

Briefly, the cellulose extraction used for both individual leaves and tree-ring increments is derived from Wise (1944); additional details are given by DeNiro and Cooper (1989). For oxygen-isotope measurements of cellulose, the standard pyrolysis method was used during sample preparation (Epstein et al. 1977) except that isoquinoline was used instead of benzoquinoline (DeNiro and Cooper, 1989). Reproducibility of the oxygen isotope measurement, based on repeated measurements of an internal cellulose standard during the 3-year period in which samples were analyzed, was $\pm 0.51‰$ (standard deviation, $n = 22$). Carbon isotope sample preparation also used off-line, standard methods (Lajtha and Michener 1994). An internal cellulose standard was measured to assess the standard deviation associated with all sample preparation and analytical procedures ($\pm 0.05‰$, $n = 10$).

Results

Climatic Variables

During the first 4 years of our sampling program at the TDE site, meteorological conditions showed considerable interannual variability. Total rainfall was nearly twice as high during the April 1 to October 1 growing season in 1994 and 1996, compared to 1993 and 1995 (Table 9.1). Mean daily air temperatures during the growing season were also slightly higher in the dry intervening years (1993 and 1995), although analysis of variance did not indicate that the differences were significant (Table 9.2). Mean daily relative humidity was highest in 1994, lowest

Table 9.1. Selected meteorological conditions at the TDE site, 1993–1996

Year	Annual rainfall (mm year^{-1})	Rainfall, April 1–October 1 (mm)	Daily mean relative humidity, ± SE, April 1–October 1	Daily mean air temperature ± SE, April 1–October 1
1993	1139	488	76.1 ± 0.8	21.42 ± 0.37
1994	1675	795	81.4 ± 0.9	20.74 ± 0.29
1995	1135	458	76.2 ± 0.9	21.59 ± 0.33
1996	1638	886	72.3 ± 1.0	20.55 ± 0.36

Note: The relative humidities and daily air temperatures are based upon measurements made on an hourly basis each day, with each date assigned a mean value, which was then summed and averaged.
Source: Data adapted in part from Hanson et al. (1998).

Table 9.2. Results of tests for significant differences among years, in daily mean relative humidity and daily mean air temperature, April 1–October 1 (growing season)

Means comparison among years	Daily mean relative humidity, April 1–October 1	Daily mean air temperature, April 1–October 1
1993–1994	$p < 0.0001$	$p = 0.1670$
1993–1995	$p = 0.9523$	$p = 0.7328$
1993–1996	$p < 0.0048$	$p = 0.0752$
1994–1995	$p < 0.0001$	$p = 0.0850$
1995–1996	$p = 0.0039$	$p = 0.0340$

in 1996, and intermediate in the 2 years with lowest rainfall, with significant differences among years (Table 9.2). Month-by-month analysis of the low mean relative humidity observed in 1996 indicates that dry air conditions in April, May, and June of that year were responsible for the low overall mean rather than arid conditions throughout the growing season.

Oxygen-Isotope Data

Each year, the $\delta^{18}O$ values observed in tree-ring cellulose varied over a smaller range than leaf cellulose collected (Fig. 9.1). For leaf cellulose, analysis of variance indicates that the oxygen isotopic ratios observed in the dry and wet treatments were not significantly different except in the case ($p = 0.007$) of the oxygen isotopic composition of leaf cellulose of *A. rubrum.* For all other species, there was no significant difference between the mean oxygen isotopic composition of leaf cellulose of the same species between the wet and dry treatments. Analysis of variance indicates that differences in the mean oxygen isotopic compositions of leaf cellulose can be significantly explained by species (Table 9.3) and year of sampling (Table 9.4). Based on the finding that the wet and dry treatments in the TDE are not significant in influencing the oxygen isotopic composition of leaf cellulose, when data from both treatments are combined, significant year-to-year and species-to-species differences can be identified (Table 9.4). For the most part, species, treatment, and interannual differences among the two *Quercus* species, *Q. alba* and *Q. prinus,* were relatively small; therefore, tree-ring cores for these trees were not collected, and only the leaf-cellulose data are tabulated in Table 9.5.

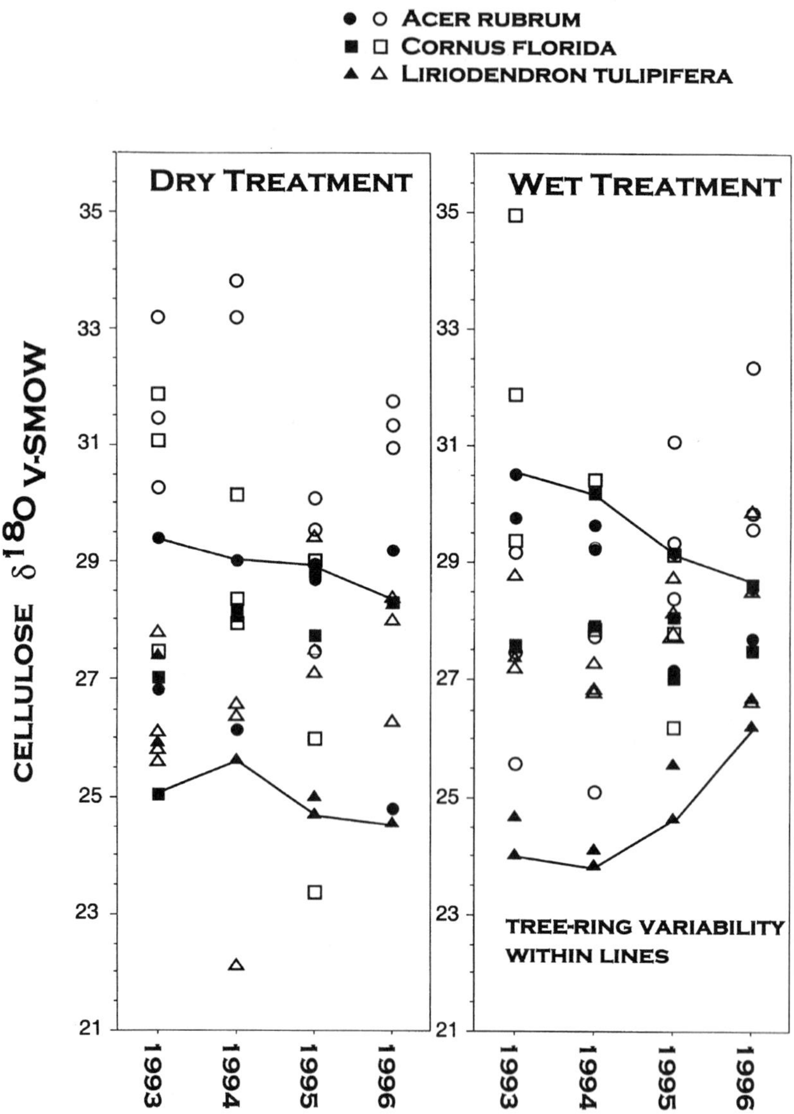

Figure 9.1. Oxygen-isotope composition of cellulose in the leaf litter and tree rings of three species of trees at the Walker Branch Watershed, Throughfall Displacement Experiment site, 1993–1996. Lines are provided to illustrate the lower degree of oxygen-isotope variability in tree rings relative to leaf litter. Tree rings are filled symbols; leaf litter are open symbols.

Table 9.3. Analysis of variance (ANOVA) for the oxygen- and carbon-isotope composition of leaf cellulose

Variable	Degrees of freedom	Sum of squares	Mean square	*F*-Value	*p*-Value
$\delta^{18}O_{V\text{-}SMOW}$					
Treatment	1	3.04	3.04	0.54	0.4613
Species	4	176.07	44.02	7.91	<0.0001
Treatment × species	4	56.07	14.02	2.52	0.0458
Residual	99	550.52	5.56		
$\delta^{13}C_{V\text{-}PDB}$					
Treatment	1	0.30	0.30	0.21	0.6491
Species	4	32.62	8.16	5.75	<0.0003
Treatment × species	4	15.94	3.99	2.81	0.0291
Residual	108	153.30	1.42		

Note: All years (1993–1996) of analysis are combined.
V-SMOW = Vienna Standard Mean Ocean Water.

Table 9.4. ANOVA for the oxygen-isotope composition of leaf cellulose

Variable	Degrees of freedom	Sum of squares	Mean square	*F*-Value	*p*-Value
Species	3	159.168	53.056	12.463	<0.0001
Year	2	93.042	46.521	10.928	<0.0001
Species × year	11	132.578	12.053	2.831	0.032
Residual	90	383.137	4.257		

Fisher's post hoc differences among species

Species compared	Mean difference	Critical difference	*p*-Value
A. rubrum, Q. alba	−0.7	1.2	0.26
A. rubrum, Q. prinus	−0.3	1.3	0.59
A. rubrum, C. florida	0.9	1.3	0.19
A. rubrum, L. tulipifera	2.7	1.2	<0.0001
Q. alba, Q. prinus	0.3	1.3	0.59
Q. alba, C. florida	1.5	1.3	0.02
Q. alba, L. tulipifera	3.4	1.2	<0.0001
Q. prinus, C. florida	1.2	1.3	0.08
Q. prinus, L. tulipifera	3.0	1.2	<0.0001
C. florida, L. tulipifera	1.8	1.3	0.005

Fisher's post hoc differences among years

Years compared	Mean difference	Critical difference	*p*-Value
1993, 1994	2.3	1.1	<0.0001
1993, 1995	1.7	1.1	0.002
1993, 1996	0.9	1.2	0.14
1994, 1995	−0.6	1.1	0.24
1994, 1996	−1.4	1.2	0.02
1995, 1996	−0.8	1.2	0.18

Note: Both treatments (wet and dry) are combined. Fisher's protected least significant difference (PLSD) was used to identify significant differences among species and years.

Table 9.5. Oxygen- and carbon-isotope composition of leaf litter from *Q. alba* and *Q. prinus,* collected in September for the years 1993–1996

	1993		1994		1995		1996	
Treatment	Dry	Wet	Dry	Wet	Dry	Wet	Dry	Wet
				Q. alba				
$\delta^{18}O$	33.0	33.3	27.3	28.9	27.7	33.1	30.2	28.2
V-SMOW	31.3	30.8	25.1	28.0	35.6	29.4	30.0	31.2
	34.2	35.1	29.8	30.6	25.9	30.5	31.9	33.8
$\delta^{13}C$	−26.3	−28.2	−26.2	−26.5	−27.3	−25.6	−28.4	−28.6
V-PDB	−29.5	−26.9	−26.9	−26.9	−27.7	−28.2	−28.0	−27.7
	−27.6	−27.4	−26.0	−27.2	−28.4	−26.8	−26.8	−26.7
				Q. prinus				
$\delta^{18}O$	30.5	34.2	31.0	29.2	31.0	32.6	28.1	no data
V-SMOW	32.6	32.1	28.6	27.0	28.9	25.5	27.5	
	34.6	34.8	26.0		31.0	30.1		
$\delta^{13}C$	−29.2	−27.5	−28.0	−26.1	−26.9	−26.1	−26.0	−25.8
V-PDB	−27.5	−27.6	−25.7	−26.4	−27.8	−25.4	−29.2	−25.9
	−27.9	−28.3	−26.6	−25.0	−26.1	−28.4	−29.3	−26.1

Note: Each data point corresponds to an independent leaf-cellulose measurement.

For comparing the oxygen isotopic composition of leaf cellulose with the oxygen isotopic composition of tree-ring cellulose, two approaches were used. When data from all five species, both treatments (six leaves total in most cases), and each of the 4 years of study are combined, the mean $\delta^{18}O$ value for leaf cellulose ($+29.4 \pm 2.7$‰ SD, $n = 109$) is significantly higher (Mann–Whitney test, $p < 0.0001$) than the mean observed in individual tree rings ($+26.2 \pm 1.8$‰ SD, $n = 46$) sampled during the same period (1993 to 1996). Because species effects appear to be significant and we did not analyze tree-ring cellulose from the two *Quercus* species, we also employed a comparison using a Wilcoxon signed-rank test. This test was used to compare linked leaf and tree-ring data: the oxygen isotopic composition of leaf cellulose relative to tree-ring cellulose synthesized in trees of the same species for each year of study and for both wet and dry treatments. In other words, the mean oxygen isotopic composition of leaf cellulose for both the wet and dry treatment of each species for each year was compared to the corresponding oxygen isotopic composition of tree-ring cellulose synthesized in that year in both the wet and dry treatments for each of the three species for which cores were analyzed. The mean $\delta^{18}O$ values in individual tree-ring cellulose linked with the mean $\delta^{18}O$ values of the corresponding wet or dry treatment leaf cellulose synthesized the same year were significantly lower ($p < 0.0004$; Table 9.6). Because of the possibility that some photosynthate produced in 1 year would not be synthesized into secondary wood cellulose until the following growing season, we also compared the oxygen isotopic composition of leaf cellulose for each species and treatment against the oxygen isotope composition of tree-ring cellulose observed the following year. The mean $\delta^{18}O$ values in cellulose in these individual tree rings were also significantly lower than species- and treatment-linked leaf cellulose that had been produced the previous

Table 9.6. Wilcoxon signed-rank tests for comparison among the mean $\delta^{18}O$ value observed for leaf cellulose and the mean $\delta^{18}O$ value observed in tree-ring cellulose synthesized in the same year and in the following year

Comparison of $\delta^{18}O$ values of leaf cellulose with matching (species- and treatment-specific) $\delta^{18}O$ values of wood cellulose synthesized the same year			
Number with no differences		0	
Number of ties		0	
Z-Value		−3.52	
p-Value		0.0004	
Count	Sum	Ranks	Mean rank
Ranks <0	3	18.0	6.0
Ranks >0	19	235.0	12.4
Comparison of $\delta^{18}O$ values of leaf cellulose with matching (species- and treatment-specific) $\delta^{18}O$ values of wood cellulose synthesized the following year			
Number with no differences		0	
Number of ties		0	
Z-Value		−3.36	
p-Value		0.0008	
Count	Sum	Ranks	Mean rank
Ranks <0	1	3.0	3.0
Ranks >0	15	133.0	8.9

Note: Test results indicate that the $\delta^{18}O$ values observed for leaf cellulose are significantly higher than the mean $\delta^{18}O$ value observed in tree-ring cellulose synthesized the same year, or in the following year.

year ($p < 0.0008$; Table 9.6). Mean leaf-litter $\delta^{18}O$ values for all species and all treatments ranged from a low of +28.3 in 1994, an unusually wet year, to +30.6 in 1993, which had warm and dry conditions. However, the mean tree-ring cellulose ^{18}O (grouping both treatments and the three species cored) varied only within a narrow range for each of the 4 years [+26.1 (1995) to +26.4‰ (1994)]. In tree rings, as with leaf cellulose, differences among species appear to have a larger influence on the oxygen isotopic composition of cellulose than do year of growth or treatment. When split by species (combining both treatments and all years), the mean $\delta^{18}O$ value of tree-ring cellulose was 24.4‰ ± 1.3 SD for *L. tulipifera*, 26.8‰ ± 1.1 SD for *C. florida*, and 27.4‰ ± 1.5 SD for *A. rubrum*.

Carbon-Isotope Data

The $\delta^{13}C$ values in leaf cellulose were also significantly less negative than $\delta^{13}C$ values in tree-ring cellulose (Fig. 9.2, Table 9.3). Significantly higher leaf-cellulose $\delta^{13}C$ values in the dry treatment were only observed in *A. rubrum*. Grouping all leaf-cellulose data, regardless of treatment or year, indicates that the mean $\delta^{13}C$ values of *C. florida* are significantly more negative ($p < 0.05$) than any other species.

Tropical Forest Sampling

At Chamela, all leaves sampled from the drier ridge site had consistently higher ^{13}C content than leaves collected at the wetter arroyo sites, which reflects higher water-use efficiency at the dry site. In addition, wet deciduous species showed

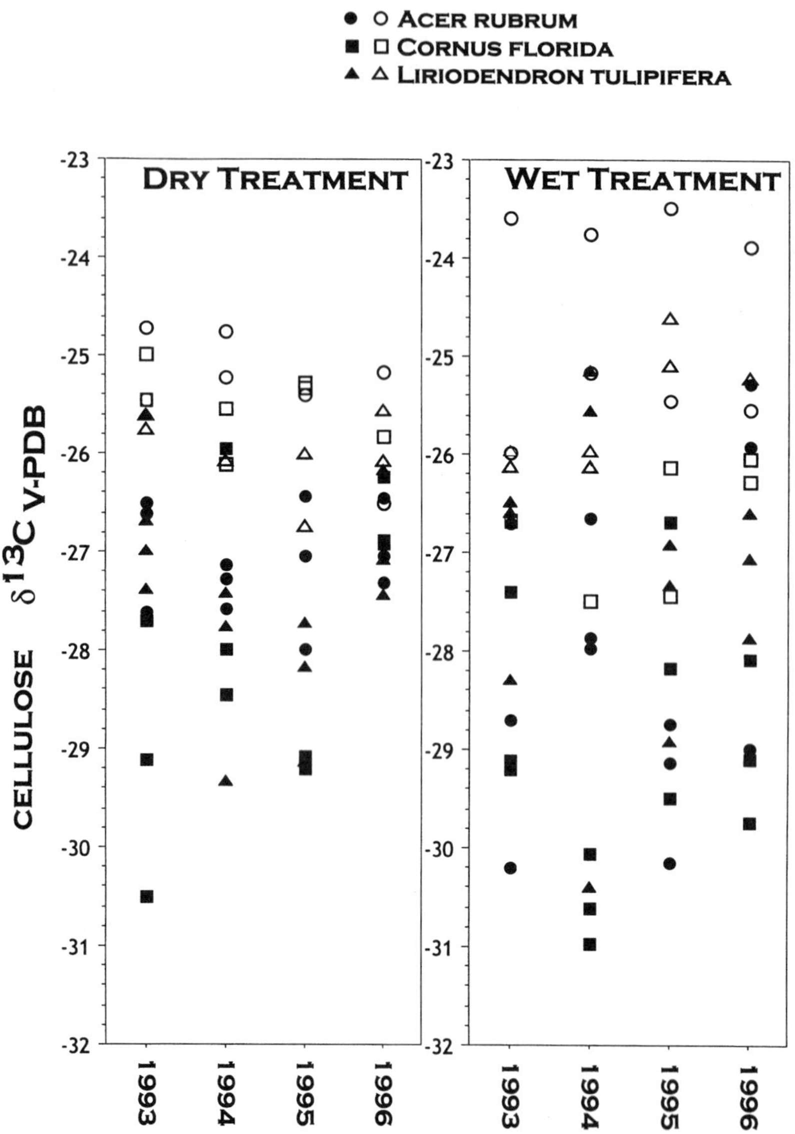

Figure 9.2. Carbon-isotope composition of cellulose in the leaf litter and tree rings of three species of trees at the Walker Branch Watershed Throughfall Displacement Experiment site, 1993–1996. Tree rings are filled symbols; leaf litter are open symbols.

higher water-use efficiency relative to the dry deciduous tree (Fig. 9.3, top). However, cellulose ^{18}O content did not reflect moisture differences between growth sites. Both wet deciduous trees growing on different sites showed similar cellulose $\delta^{18}O$ values (around +28‰) with the exception of *J. pungens* from the arroyo site 1, which showed a significantly higher ^{18}O content. However, we observed similar trends in the relative ^{18}O content in cellulose among species [i.e., wet deciduous species showed higher ^{18}O contents relative to drought deciduous (Figure 9.3, bottom)]. The difference in leaf heavy-isotope enrichment is not likely to be the result of soil water with a different isotopic signature taken up by the plants because no major difference was observed in the isotopic composition of stem water of plants occurring on the same site (data not shown).

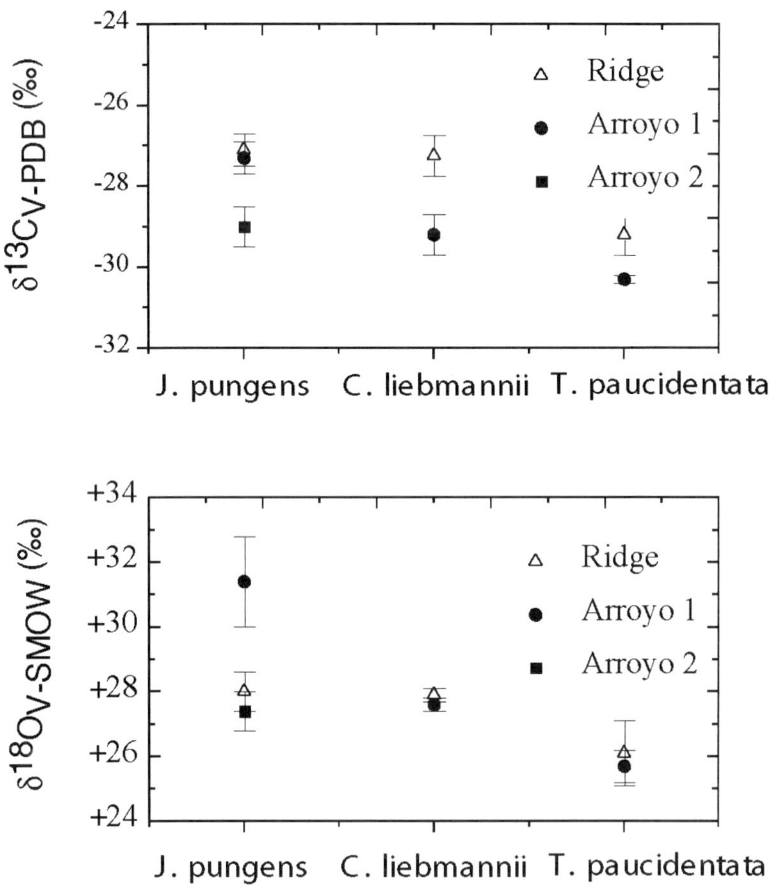

Figure 9.3. Top: $\delta^{13}C$ values; and bottom: $\delta^{18}O$ values of leaf cellulose of *J. pungens, C. liebmannii,* and *T. paucidentata* collected in December 1992 (early dry season) at a dry (ridge) and two wet sites (arroyo 1 and arroyo 2). Each point represent the average of at least three samples.

Discussion

Leaf Versus Wood Cellulose

Our results are consistent with the integration of heavy-oxygen-isotope enrichment in leaf water within leaf cellulose but suggest that this heavy-oxygen-isotope enrichment may not be effectively recorded in secondary wood cellulose. We do not exclude the possibility that some of the ^{18}O originating from leaf cellulose is incorporated into wood cellulose to the contrary, it appears that some isotopic-fractionation effects caused by evapotranspiration are preserved, as we describe below. However, the variability in $\delta^{18}O$ values within wood cellulose is much smaller than in leaf cellulose (Fig. 9.1), indicating that much of the variation in leaf cellulose presumably caused by meteorological variation is lost before incorporation into wood cellulose. For example, for all species and treatments, the mean $\delta^{18}O$ content of *leaf* cellulose (+28.3‰) was lowest in 1994, the wettest and most humid year (Fig. 9.1; Tables 9.1 and 9.2), which is consistent with expectations. However, the mean $\delta^{18}O$ content of *wood* cellulose synthesized in 1994 (+26.4‰) was analytically indistinguishable from the narrow range of mean $\delta^{18}O$ values for wood cellulose, combining all species and treatments, that have been observed during the 4 years of this study (+26.1‰ to +26.4‰). The commonly observed oxygen isotopic fractionation between cellulose and its source water in submerged, nontranspiring organisms is 27 ± 3‰, (DeNiro and Epstein 1981). This result implies that the source water for this wood cellulose (in 1994) had a $\delta^{18}O$ value as high as +2 or as low as −4‰ if there was no contribution of ^{18}O from leaf cellulose and only this biochemical fractionation was present. Available soil-water data from the Walker Branch Watershed (Halleran 1995) indicates that isotopically unfractionated soil water has a $\delta^{18}O$ value of approximately −6‰ with no significant seasonal variation. This indicates that isotopically unfractionated soil or groundwater could not have been the sole determinant of oxygen isotopes at the point of wood cellulose synthesis in these trees. At 26.1‰, wood-cellulose $\delta^{18}O$ values are 2–8‰ higher than would be expected if biochemical fractionation of soil water [−6‰ + (27 ± 3‰)] was the sole source of oxygen. Despite this conclusion, the variations in mean wood-cellulose $\delta^{18}O$ values that occurred during the 4 years of our work were no larger than analytical error and were therefore insensitive to the significant climatic and drought fluctuations that occurred, whether measured as mean daily humidity, temperature, precipitation, or throughfall displacement.

Early Wood Versus Late Wood

In other work, it is has been pointed out that early wood may be isotopically distinct from late wood because the early wood is synthesized partly from the previous year's photosynthetic production (Hill et al. 1995). We were unable to separate early wood from late wood in the species we sectioned. In the slowest growing species, *C. florida,* too little late wood was synthesized to sample, and in the two other species, *A. rubrum* and *L. tulipifera,* the boundaries between early and late wood were relatively indistinct. That notwithstanding, the small range of oxygen-isotope variability for the mean tree-ring tissue provides no evidence that the presumably higher ^{18}O content of photosynthate is transferred effectively to the secondary wood cellulose in tree-ring tissue.

Species Differences

Analysis of variance indicates that the TDE dry treatment had no significant effect on the oxygen or carbon isotopic compositions of either leaf or tree-ring cellulose for four of the five species during the 4 years experimental sampling was under way. Species effects in the trees studied are significant variables influencing the leaf-cellulose ^{18}O content (Tables 9.3 and 9.4). Generally, higher $\delta^{18}O$ values were observed in *Quercus* and *A. rubrum* and consistently lower leaf and tree-ring cellulose $\delta^{18}O$ values in *L. tulipifera* (Table 9.4). High variability in $\delta^{18}O$ values were observed in *C. florida,* which also had the most negative mean $\delta^{13}C$ values. Interestingly, *C. florida* appears to have been the species that was perhaps most affected by mortality on the dry plot as a result of the TDE manipulation (Hanson et al. 2001). Tschaplinski et al. (1998) and Gebre et al. (1998) also report that *C. florida* is among the least drought-tolerant of the trees present on the TDE site. These factors might otherwise be expected to result in less negative $\delta^{13}C$ values, resulting from reduced stomatal conductance under drought stress. The understory growth form of *C. florida,* however, suggests that within-canopy variations in atmospheric carbon dioxide were probably more important than drought stress in influencing the carbon-isotope ratios observed. Bowling et al. (1999) report that carbon dioxide concentrations in summer at ground level (0.75 m) on the Walker Branch watershed can be significantly higher (440 μmol mol^{-1} at 9: 00 AM, local daylight time) than at the canopy top at 26 m (390 μmol mol^{-1}). This difference was also reflected during this sampling event by differences in carbon-isotope values observed, with $\delta^{13}C$ values of $-12‰$ at 0.75m and $-9.5‰$ at the canopy top; significant differences were maintained throughout daylight hours (Bowling et al. 1999). This difference in atmospheric carbon dioxide concentrations and isotopic composition is probably driven by respired carbon from leaf litter and soil organic matter on the watershed, which has a $\delta^{13}C$ value of approximately -27 to $-29‰$ (Garten et al. 2000). These contributions of soil-respired carbon to materials preserved in organic matter have some implications for the use of carbon-isotope ratios in wood cellulose as an integrated water-status indicator. Clearly, at least in the case of this subcanopy species, the potential response to drought stress in the carbon-isotope composition of leaf cellulose was probably masked by the more negative $\delta^{13}C$ values of carbon dioxide at the lower canopy levels.

Relative Isotopic Sensitivity to Drought Stress

The results of our study suggest that the oxygen isotopic content of tree-ring cellulose at the Walker Branch site is relatively insensitive to precipitation throughfall volume and humidity and is, in some cases, more influenced by species differences. This result is probably a consequence of the relatively low degree of oxygen-isotope variability in Walker Branch soil and groundwaters (Halleran 1995) that serve as the water source available to the trees studied. The low degree of isotopic variability in soil and groundwaters is a consequence of the recharge of groundwater on Walker Branch, which occurs only during the winter (Mulholland et al. 1990; Mulholland 1993), despite a nearly year-round pattern of consistent precipitation. Consequently, our findings may not be generalizable to hydrologic settings where there is significant seasonal or annual variation in the oxygen-isotope composition of soil waters used by trees. In those settings, variation in the oxygen-isotope composition of soil waters and associated proxy climatic information may be recorded within wood cellulose. Seasonal or interannual isotopic variation in waters used by trees may explain some of the results reported elsewhere that correlate the oxygen-isotope composition of wood cellulose over small spatial scales with climatic variables, such as soil moisture availability [e.g., Saurer et al. (1997a)].

The ^{13}C content of leaf litter has also not been significantly influenced by large interannual differences in precipitation or treatment effects during the 4 years of the study period. This result is somewhat surprising because of the lower soil moisture that has been observed in the drier years, particularly 1993 and 1995, on the dry treatment of the TDE (Hanson et al. 1995, 1998). Because the carbon isotopic composition of leaf cellulose is established during leaf growth, which is more important in the spring, it is possible that the carbon isotopic composition of leaf litter collected in September predominantly reflects heavy-isotope discrimination during carboxylation in the spring, which may or may not coincide with soil moisture conditions that develop later during the summer. This situation is particularly true for the *Quercus* species studied; *L. tulipifera* and *A. rubrum* continue to produce leaves at a lower rate throughout the summer. Consistent with this leaf phenology, when all years are combined, *A. rubrum* leaf cellulose had significantly more positive $\delta^{13}C$ values in the dry treatment, which was the only species for which a significant treatment effect was observed. The largest changes in soil moisture have been observed on the TDE site in late summer in years when the highest annual temperatures and multiweek-long periods without rainfall have coincided (Hanson, et al. 1995, 1998).

Comparison with Tropical Forest Results

The results obtained from trees growing in dry deciduous forests show that physiological adaptation to drought can lead to significant differences in the carbon isotopic composition of leaf organic matter among plants with different life-forms. By comparison, oxygen isotopes in organic matter appear be less sensitive to soil moisture than do carbon isotopes. Compared to the TDE site, soil moisture effects were more pronounced on the carbon isotopic content of leaves at Chamela, where all sampled trees exhibited higher ^{13}C levels at dry sites relative to wet sites. At the TDE site, only *A. rubrum* showed an increased water-use efficiency, as implied by less negative $\delta^{13}C$ values, in the dry treatment.

As observed in trees from the Walker Branch experiment, interspecific variablity in the mean oxygen isotopic compositions of leaf cellulose from Chamela trees are small and do not appear to be a function of site humidity or soil moisture (Table 9.3 and Fig. 9.3). One explanation for the wet deciduous plants such as *J. pungens* is that these trees utilize stored starch during the wet season and synthesize new leaves at the start of the dry season (Janzen and Wilson 1974). As long as the original structural cellulose synthesized at the beginning of the dry season remains in the leaves, the oxygen and hydrogen isotope of that cellulose should remain largely constant and independent of the leaf water fractionation that can occurs as the dry season continues. As a result, the oxygen-isotope composition of cellulose likely reflects only the isotopic composition of water available at the time of leaf initiation at the wet season–dry season boundary. These results are similar to studies in the Sonoran desert that show that plants with long water-residence times have associated oxygen and hydrogen isotopic plant water compositions that bear little relationship to contemporaneous meteorological conditions (Cooper and DeNiro 1989).

Conclusions

Our study indicates that carbon- and oxygen-isotope-fractionation differences among species are at least as significant as differences caused by variations in throughfall and relative humidity for this humid, temperate site and for the species studied. If the TDE manipulation is typical of what can be expected for precipitation-pattern variation in the deciduous forests of eastern North America, stable-isotope variability recorded in wood cellulose does not appear to be sensitive enough to record that variation. It also appears that it would be unwise to combine oxygen-isotope data from more than one species during the extraction of proxy climatic data. At other locations, stable-isotope proxy data may more sensitively record climatic information in wood cellulose, particularly where there are seasonal or interannual variations in the oxygen-isotope composition of water available to plants. In our study, the low degree of variation in the $\delta^{18}O$ values of soil water available to the trees we studied appears to be responsible for the homogeneous isotopic composition of wood cellulose. Because of these observations, the goal of evaluating stable-isotope data from wood cellulose on a generally applicable basis to interpret paleoclimate may remain impractical, at best, and even mechanistically incorrect in some instances. The cautions expressed in Terwelliger and DeNiro (1995) and DeNiro and Cooper (1989) concerning the variable degree of isotopic exchange that is likely to occur following photosynthesis and during cellulose synthesis remain barriers to the practical use of stable isotopes as proxy indicators in wood cellulose.

References

Berninger F, Sonninen E, Aalto T, Lloyd J (2000) Modeling ^{13}C discrimination in tree-rings. Global Biogeochem Cycles 14:213–223.

Bowling DR, Baldocchi DD, Monson RK (1999) Dynamics of isotopic exchange of carbon dioxide in a Tennessee deciduous forest. Global Biogeochem Cycles 13:903–922

Buhay WM, Edwards TWD (1995) Climate in southwestern Ontario, Canada, between AD 1610 and 1885 inferred from oxygen and hydrogen isotopic measurements of wood cellulose from trees in different hydrologic settings. Quat Res 44:438–446.

Buhay WM, Edwards TWD, Aravena R (1996) Evaluating kinetic fractionation factors used for ecologic and paleoclimatic reconstructions from oxygen and hydrogen isotope ratios in plant water and cellulose. Geochim Cosmochim Acta 60:2209–2218.

Burk RL, Stuiver M (1981) Oxygen isotope ratios in tree cellulose reflect mean annual temperature and humidity. Science 211:1417–1419.

Clague JJ, Mathewes R, Buhay WM, Edwards TWD (1992) Early Holocene climate at Castle Peak, southern Coast Mountains, British Columbia, Canada. Palaeogeogr Palaeoclimatol Palaeoecol 95:153–167.

Cooper LW, DeNiro MJ (1989) Covariance of oxygen and hydrogen isotopic compositions in plant water: Species effects. Ecology 70:1619–1628.

Cooper LW, Norby RJ (1994) Atmospheric CO_2 enrichment can increase the ^{18}O content of leaf water and cellulose: paleoclimatic and ecophysiological implications. Clim Res 4:1–11.

DeNiro MJ, Cooper LW (1989) Post-photosynthetic modification of oxygen isotope ratios of carbohydrates in the potato: Implications for paleoclimatic reconstruction based upon isotopic analysis of wood cellulose. Geochim Cosmochim Acta 53:2573–2580.

DeNiro MJ, Epstein S (1981) Isotopic composition of cellulose from aquatic organisms. Geochim Cosmochim Acta 45:1885–1894.

Edwards TWD (1993) Interpreting past climate from stable isotopes in continental organic matter. In Swart PK, Lohmann KC, McKenzie J, Savin S (Eds) *Climate Change in Continental Isotopic Records.* American Geophysical Union, Washington, District of Columbia, pp 333–341.

Epstein S, Thompson P, Yapp CJ (1977) Oxygen and hydrogen isotopic ratios in plant cellulose. Science 218:1209–1215.

Fanjul L, Barradas VL (1985) Stomatal behaviour of two heliophile understory species of a tropical deciduous forest in México. J Appl Ecol 22:943–954.

Fanjul L, Barradas VL (1987) Diurnal and seasonal variation in the water relations of some deciduous and evergreen trees of a deciduous dry forest of the western coast of México. J Appl Ecol 24:289–303.

Farquhar GD, Leary MO, Berry JA (1982) On the relationship between carbon isotope discrimination and the intercellular carbon dioxide concentration in leaves. Aust J Plant Physiol 9:121–137.

Farquhar GD, Barbour MM, Henry BK (1998) Interpretation of oxygen isotope composition of leaf material. In Griffiths H (Ed) *Stable Isotopes,* BIOS Scientific Publishers, Oxford, England, pp 27–62.

Flanagan LB, Ehleringer JR (1991) Stable isotope composition of stem and leaf water: Applications to the study of plant water use. Functional Ecol 5:270–277.

Garten CT, Taylor GE (1992) Foliar $\delta^{13}C$ within a temperate deciduous forest: Spatial, temporal and species sources of variation. Oecologia 90:1–7.

Garten CT, Cooper LW, Post WM, Hanson PJ (2000) Climate controls on forest soil C isotope ratios in the Southern Appalachian Mountains. Ecology 81:1108–1119.

Gebre GM, Tschaplinski TJ, Shirshac TL (1998) Response in water relations of several hardwood species to throughfall manipulation in an upland oak forest during a wet year. Tree Physiol 18:299–305.

Halleran AL (1995) *Comparison of hydrologic cycle components for pre-storm and storm conditions, Walker Branch Watershed, Oak Ridge Reservation.* M.S. Thesis. The University of Tennessee, Knoxville, Tennessee.

Hanson PJ, Todd DE, Edwards NT, Huston MA (1995) Field performance of the Walker Branch Throughfall Displacement Experiment. In Jenkins A, Ferrier RC, Kirby C (Eds) *Ecosystem manipulation experiments: Scientific approaches, experimental design and relevant results.* Commission of the European Communities, Brussels, Belgium, pp 307–313.

Hanson PJ, Todd DE, Huston MA, Joslin JD, Croker JL, Augé RM (1998) *Description and field performance of the Walker Branch Throughfall Displacement Experiment: 1993–1996.* ORNL/TM–13586. Oak Ridge National Laboratory, Oak Ridge, Tennessee.

Hanson PJ, Todd DE, Amthor JS (2001) A six year study of sapling and large-tree growth and mortality responses to natural and induced variability in precipitation and throughfall. Tree Physiol 21:345–358.

Holbrook NM, Whitbeck JL, Mooney HA (1995) Drought responses of neotropical dry forest trees. In Bullock SH, Mooney HA, Medina E (Eds). *Seasonally dry tropical forests.* Cambridge Press, Cambridge, England, pp 243–270.

Hill SA, Waterhouse JS, Field EM, Swisur VR, ap Rees T (1995) Rapid recycling of triose phosphates in oak stem tissue. Plant Cell Environ 18:931–936.

Janzen DH, Wilson DE (1974) The cost of being dormant in the tropics. Biotropica 6:260–262

Lajtha K, Michener RH (1994) *Stable isotopes in ecology and environmental science.* Blackwell Scientific Publications, London.

Lipp J, Trimborn P, Edwards T, Waisel Y, Yakir D (1996) Climatic effects on the $\delta^{18}O$ and $\delta^{13}C$ of cellulose in the desert tree *Tamarix jordanis.* Geochim Cosmochim Acta 60:3305–3309.

Mulholland P (1993) Hydrometric and stream chemistry evidence of three storm flowpaths in Walker Branch Watershed. J Hydrol 151:291–316.

Mulholland PJ, Wilson GV, Jardine PM (1990) Hydrogeochemical response of a forested watershed to storms: Effects of preferential flow along shallow and deep pathways. Water Resour Res 26:3021–3036.

Roden JS, Ehleringer JR (1999) Hydrogen and oxygen isotope ratios of tree-ring cellulose for riparian trees grown long-term under hydroponically controlled environments. Oecologia 121:467–477.

Roden JS, Ehleringer JR (2000a) Hydrogen and oxygen isotope ratios of tree-ring cellulose for field-grown riparian trees. Oecologia 123:481–489

Roden JS, Lin GG, Ehleringer JR. (2000b) A mechanistic model for interpretation of hydrogen and oxygen isotope ratios in tree-ring cellulose. Geochim Cosmochim Acta 64:21–35

Saurer M, Aellen K, Siegwolf R (1997a) Correlating $\delta^{13}C$ and $\delta^{18}O$ of trees. Plant Cell Environ 20:1543–1550.

Saurer M, Borella S, Leunberger M (1997b) $\delta^{18}O$ of tree-rings of beech (*Fagus silvatica*) as a record of $\delta\ ^{18}O$ of the growing season precipitation. Tellus 49B:80–92.

Sternberg LDSL, DeNiro MJ, Savidge RA (1986) Oxygen isotope exchange between metabolites and water during biochemical reactions leading to cellulose synthesis. Plant Physiol 82:423–427.

Terwilliger VJ, DeNiro MJ (1995) Hydrogen isotope fractionation in wood-producing avocado seedlings: Biological constraints to paleoclimatic interpretations of delta D values in tree ring cellulose. Geochim Cosmochim Acta 59:5199–5207

Tschaplinski TJ, Gebre GM, Shirshac TL (1998) Osmotic potential of several hardwood species as affected by manipulation of throughfall precipitation in an upland oak forest during a dry year. Tree Physiol 18:291–298

White JWC, Lawrence JR, Broecker WS (1994) Modeling and interpreting D/H ratios in tree rings: A test case of White Pine in the Northeastern United States. Geochim Cosmochim Acta 58:851–862

Wise LE (1944) *Wood chemistry.* Reinhold, New York.

Yakir D, DeNiro MJ, Rundel PW (1989) Isotopic inhomogeneity of leaf water: Evidence and implications for the use of isotopic signals transduced by plants. Geochim Cosmochim Acta 53:2769–2773.

4. Decomposition and Soil Carbon Turnover

10. Soil Respiration and Litter Decomposition

Paul J. Hanson, Elizabeth G. O'Neill, M. Lala S. Chambers, Jeffery S. Riggs, J. Devereux Joslin, and Mark H. Wolfe

Emissions of CO_2 from the organic and mineral soil horizons into the atmosphere are considered a good representation of soil respiration (R_{soil}) when the diffusion-driven losses to the atmosphere are at equilibrium with the biological rates of CO_2 production (Hanson et al. 2000). R_{soil} is the net effect of the biological activity of autotrophic roots and associated rhizosphere organisms plus the mineral soil heterotrophic activity of bacteria, fungi, and soil fauna. Whereas the activity of soil heterotrophic organisms is proportionate to the decomposition of soil carbon, CO_2 lost from root and rhizosphere activity is a function of the consumption of organic compounds supplied by aboveground organs of plants (Horwath et al. 1994). Notwithstanding the complex nature of the sources of CO_2 contributing to it, R_{soil} represents a good measure of the collective response of root and soil biological activity to environmental conditions (Edwards et al. 1970; Hanson et al. 2000). Because soil respiration is a very large fraction of gross primary productivity (Curtis et al., 2002), its quantification must be a high priority in any attempt to establish carbon budgets for ecosystems (see Chapter 22, this volume).

This chapter describes seasonal patterns of R_{soil} for the Walker Branch temperate deciduous forest, and characterizes the factors responsible for controlling the observed seasonal dynamics, including temperature, soil-water availability, root growth, and the wetting and drying of the soil organic horizons. Direct measurements of litter decomposition using litter decay bags are also described. A model capable of capturing the seasonal dynamics of the litter layer, mineral-soil, and root-growth components of R_{soil} is presented, and annual estimates of R_{soil} and its components are provided and discussed. This chapter concludes by comparing R_{soil} for the Throughfall Displacement Experiment (TDE) (Hanson et al. 2001a) forest with published data for similar temperate forests throughout the eastern United States.

Methods

Observations of soil respiration and its components, litter layer respiration, and litter decomposition were conducted on the Walker Branch Watershed (35°58' N and 84°17' W) near Oak Ridge, Tennessee, as a part of the Throughfall Displacement Experiment (TDE; Hanson et al. 2001a). As described in Chapter 2 (Table 2.1), mean annual precipitation was 1329 mm and mean temperature was 14.5°C. The acidic forest soils (pH 3.5–4.6) are Typic Paleudults, and the depth to bedrock is ~30 m. The site is dominated by *Quercus alba, Quercus prinus,* and *Acer rubrum.* Stand basal area averages 20–25 $m^2\ ha^{-1}$. Annual leaf-litter production has been nearly constant from 1993 to 2000 at about 225 $g\ C\ m^{-2}\ y^{-1}$ with a corresponding leaf area index of 6.2 $m^2\ m^{-2}$.

Soil Respiration

Soil respiration or net forest floor CO_2 efflux (R_{soil}) was measured with the LiCor 6200 closed-gas-exchange system and a custom-designed aluminum chamber having a surface area of 594 cm^2 and a volume of 5227 cm^3 (Hanson et al. 1993). The open bottom of the chamber was pressed into the litter layer to form the closed cuvette for measurements that lasted from 1.5 to 3 min depending on the season (i.e., shorter observations in the summer when rates were high). Permanent soil collars were not used to avoid inappropriate trapping of precipitation on the litter layers. At periodic weekly or biweekly intervals, 30 observations per treatment (wet, ambient, and dry) were conducted at fixed locations along three transects of the TDE site representing upper, middle, and lower slope positions. Soil temperature and soil-water content were monitored simultaneously with each observation. Soil moisture (%, $cm^3\ cm^{-3}$) was measured with a time-domain reflectometer (Soil Moisture Equipment Corporation, Santa Barbara, CA) to a depth of 35 cm (Topp and Davis 1985). Soil temperature (°C) was measured to ~10 cm with an OMEGA digital temperature probe. Soil-water-content data were converted to soil-water potential with the soil moisture release curves documented by Hanson et al. (1998; Chapter 2, this volume).

R_{soil} was measured from 1992 to 1999 to establish the seasonal patterns and evaluate potential TDE treatment effects. Data collected from 1994 to 1996 emphasized observations of seasonal patterns and TDE treatments, and R_{soil} observations made from 1997 to 1999 were done to obtain supplemental information for cold soils and dry conditions to fill out the response surface relating R_{soil} to temperature and soil-water potential. Wetting and drying periods associated with rain events were also targeted in 1998 and 1999 to test the influence of wet surface organic horizons on R_{soil}.

Measurements of Litter-Specific CO_2 Efflux

Independent measurements of litter specific respiration [nmol CO_2 (g dry matter)$^{-1}$ s^{-1}] were conducted on litter samples collected from the forest organic horizons, including the Oi layer and a combined Oe/Oa layer (SSSA 1997). These layers correspond to the L and combined F and H layers of the taxonomy proposed by Green et al. (1993). Samples were collected throughout an annual cycle to provide a seasonal sequence and a range of litter water contents. Organic-horizon samples were collected in the field with each horizon layer placed in a sealed plastic bag and transferred to the laboratory for respiration measurements. The collected litter samples were separated into two litter layers: Oi (recognizable litter) and Oe/Oa (fragmented litter and humus). All samples were allowed to equilibrate in their sealed bags to 22 ± 1.6°C, 17 ± 0.5°C, or 11 ± 0.4°C. These temperatures covered the range from growing season maximum to common forest floor temperatures during the dormant season. Following temperature equilibration, a subsample of the equilibrated sample was placed on an inert closed-cell foam background, and the soil respiration chamber, described above was inverted over this sample. Three replicate 15-s measurements of the CO_2 efflux rate from each sample were collected along with corresponding chamber blanks (i.e., identical observations without the litter sample). Immediately after the measurements, the temperature of each sample was verified with an infrared thermometer (Everest Interscience Inc., Fullerton, CA). The mass of all samples was determined before and after being oven-dried (70°C for 24 h) for dry mass determinations. The mean rate of efflux for each sample was corrected for the small blank effect and subsequently expressed on a per-unit-litter dry mass basis.

In Situ Litter Water Content and Litter Water Potential

Continuous mean hourly observations of the litter water content of the Oi/Oe horizons were logged on a data logger (CR10X, Campbell Scientific, Logan, UT) beginning March 20, 1996 with a new approach that uses real leaf litter and its characteristic electrical conductance with variable water content. The approach is analogous to the electrical-impedance grids reported on by Gillespie and Kidd (1978), but it uses intact leaves of the most recent year's cohort placed within the organic horizons as the sensing device. This method has been used to verify surface evaporation patterns of the forest floor (Wilson et al. 2000). Replicate *Quercus prinus* leaves (eight per treatment plot) were chosen because of their higher resistance to decomposition within one growing season. The leaves were attached to a voltage supply, external resistors, and datalogger as shown in Fig. 10.1 and returned to the forest floor between the Oi and Oe horizons for the duration of the growing season. Once each minute, a voltage measurement was made across the leaf, and the hourly averages were stored on the datalogger. Calibration of this approach was accomplished by periodic collections of separate

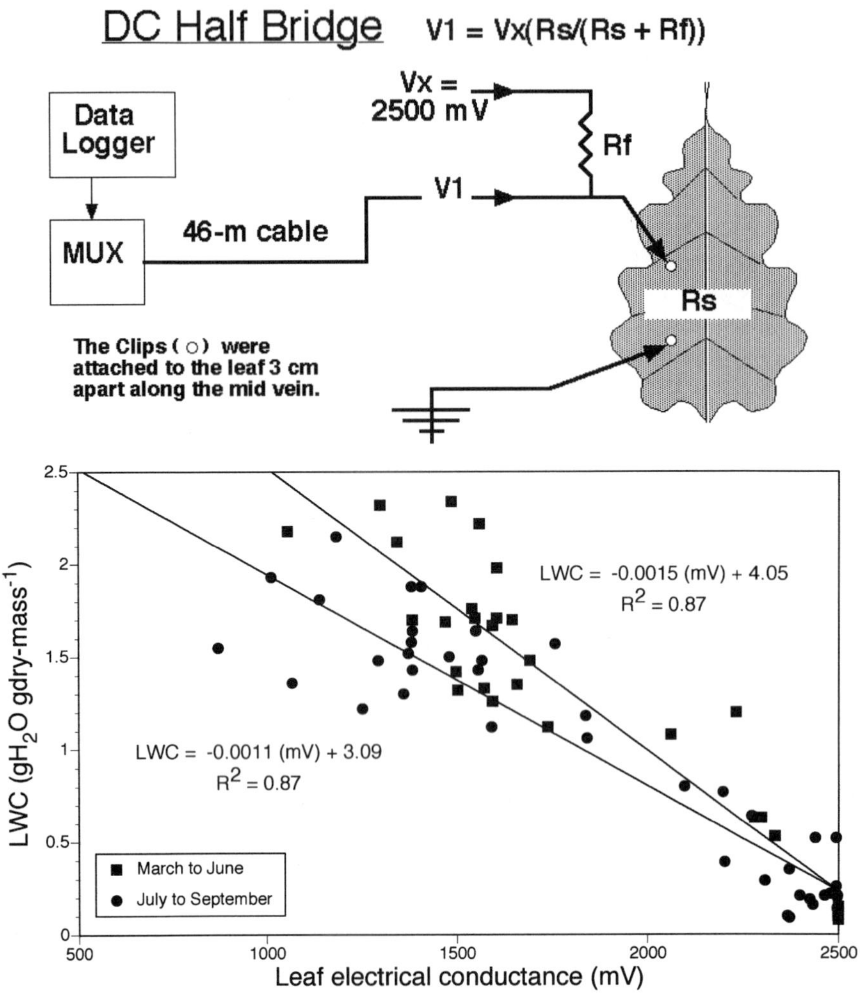

Figure 10.1. Diagram of the datalogger, voltage supply, and external resistors used to monitor the voltage across a leaf in the litter layer as a function of variable litter water content (upper graph). DC voltage of 2500 mV is applied across the half-bridge circuit. An output signal related to the resistance is measured as Rs. As the leaf becomes moist, the resistance Rs decreases, thus decreasing the voltage measured at V1. Rf is a fixed resistor; Rs is the chestnut oak leaf. A calibration curve between litter water content (LWC) and leaf electrical conductance (lower graph) shows the relationship between *Q. prinus* leaf electrical conductance (mV) and LWC. Data presented are for two different sets of leaves collected early versus late in the growing season. The two sets have different regression lines and represent different stages of foliar decomposition.

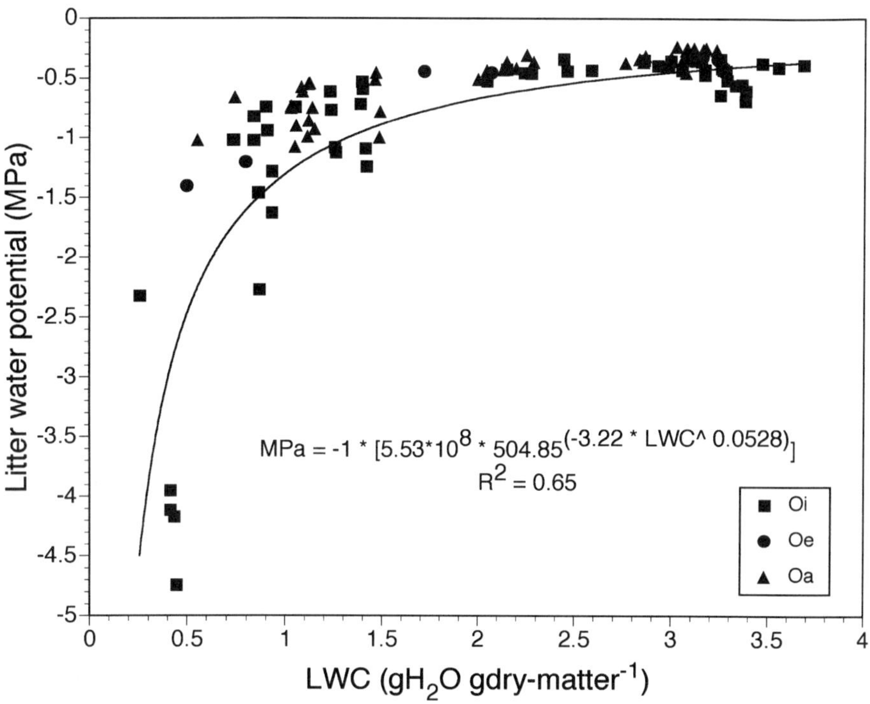

Figure 10.2. Relationship between LWC and litter water potential based on thermocouple psychrometry.

Q. prinus leaves over time and under a range of water contents. A calibration-leaf's ability to conduct current was determined in the field, the leaf was sealed in a plastic bag and transferred to the laboratory; and its wet and dry masses were determined as described earlier. The coefficients for the resulting linear relationship changed slightly over the course of the season, as shown in Fig. 10.1.

The relationship between litter water content (LWC) and litter water potential (LWP) was determined with psychrometric measurements (SC-10, Decagon Devices, Inc. Pullman, WA) of water potential and gravimetric measurements of litter water content. The curve-linear relationship and a fitted equation are provided in Fig. 10.2.

Litter Chemistry and the Decomposition

Foliar-litter chemistry was measured annually from 1993 to 1999 with *A. rubrum* and *Q. alba* as indicator species. Six litter collection baskets were chosen per treatment, three upslope and three downslope. Litter composed of a representative mixture of species was subsampled from these baskets during fall senescence and analyzed for total carbon (C) and nitrogen (N) (Carlo Erba C-N analyzer, Carlo Erba Strumentazione, Milan, Italy), and the C : N ratio calculated.

Table 10.1. Chemical composition of the 1995 litter cohort by species or mixture of species

		Species			
Chemical	Treatment	*A. rubrum*	*C. florida*	*Q. prinus*	"Mixture"
Nitrogen	Wet	1.27	1.35	1.05	1.01
	Ambient	1.35	1.38	1.06	1.02
	Dry	1.32	1.24	1.06	1.04
Lignin	Wet	13.8	17.8	20.9	16.0
	Ambient	13.1	19.2	20.1	16.4
	Dry	13.7	17.8	20.0	15.6
Lignin : N	Wet	10.9	13.2	20.0	16.0
	Ambient	9.7	14.0	19.0	16.0
	Dry	10.4	14.3	18.9	15.0

Note: Values reported in the "Mixture" column are for material used in the litterbag study and comprise a 45 : 35 : 20 mix of the three species (Qp : Ar : Cf) ground together before analysis.

In 1995, newly fallen litter from *A. rubrum, Cornus florida,* and *Q. prinus* was collected from all three plots. Litter was air-dried and then mixed within plots in a ratio of 45 parts *Q. prinus,* 35 parts *A. rubrum,* and 20 parts *C. florida.* This ratio is representative of the occurrence of the three species on the site (Hanson et al. 2001; Chapters 2 and 17, this volume). Approximately 10 g of this mixed litter were placed within mesh bags (2-mm nylon mesh on top and 1-mm fiberglass mesh on bottom) and weighed. A subsample of each species litter and one of the mixture were oven dried for 2 days at 65°C and analyzed for lignin and nitrogen by near-infrared spectroscopy (Wessman et al. 1988) to determine the initial litter quality and to convert air-dried weights to oven-dried weights. The mixed litter had lignin : nitrogen characteristics that were intermediate for the three species (Table 10.1).

All litterbags were placed in the field in March 1996. Litterbags prepared from wet, ambient, and dry plot litter ($n = 36$) were divided among the treatments and placed at random locations on each treatment plot just below the Oi layer so that interactions between litter source and decomposition environment could be evaluated. Bags placed in the dry treatment area were placed directly under the troughs, at the midpoint of the trough length. Bags in the wet treatment area were placed directly under drip holes in the delivery pipes. A subset of four bags from each source/treatment combination was retrieved in October 1996 (elapsed time: 217 days), October 1997 (elapsed time: 594 days), and October 2000 (elapsed time: 1689 days). At each retrieval, litter residue was oven-dried at 65°C for 2 days before weighing. Residue was ground in a Cyclotec 1093 Sample Mill (Techator, Hoganas, Sweden) and analyzed for carbon and nitrogen with a C–N analyzer (Carlo-Erba, Milan, Italy).

Results

Measurements of R_{soil} along upper, middle, and lower slope transects of the TDE study area at intervals throughout 1993 and 1994 allowed us to evaluate the spatial

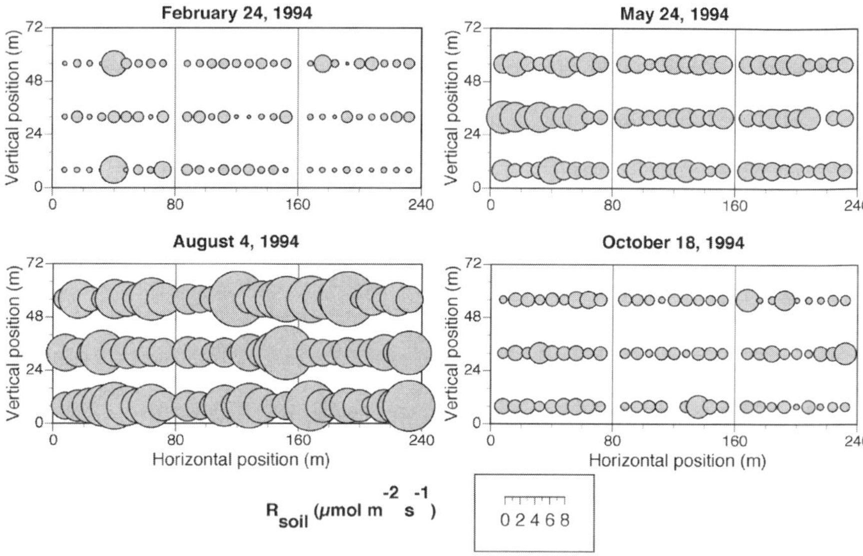

Figure 10.3. Plot of the variation of soil respiration (R_{soil}) across the TDE area for selected dormant and cool-season periods. The diameter of individual bubbles represents R_{soil} at a given location. Mean soil temperatures for the February, May, August, and October dates were 9.8°C, 15.9°C, 20.3°C, and 14.6°C, respectively. Soil-water deficits did not represent a significant limitation for R_{soil} in 1994.

variability of R_{soil}. Figure 10.3 shows data for measurement intervals in February, May, August, and October of 1994 corresponding to mean soil temperatures of 9.8°C, 15.9°C, 20.3°C, and 14.6°C, respectively. Visual inspection of these bubble plots showed minimal site-to-site variability across treatment plots. This uniformity notwithstanding, significant effects of the TDE treatments were seldom found for any of the measurement dates, suggesting that treatment-induced changes in the litter layer or soil-water budgets were transient in nature and difficult to capture with instantaneous measurements of R_{soil}.

Seasonal patterns of R_{soil} followed expected patterns (Fig. 10.4), with dormant-season values ranging from near zero to 1 μmol m^{-2} s^{-1} and typical summer maximum rates of R_{soil} ranging from 4 to 6 μmol m^{-2} s^{-1}. Much of the intra-annual variation in R_{soil} can be attributed to environmental changes in soil temperature, with maximum R_{soil} corresponding to soil temperatures of between 21°C and 23°C. Midseason reductions in R_{soil} were observed during drought periods in 1993, 1995, and 1998. Combined observations of R_{soil} from 1993 to 1999 (more than 2500 measurements) showed clear relationships among R_{soil}, soil temperature, and soil-water potential (Fig. 10.5A), but substantial variation was apparent. Plotting only the daily means by treatment (Fig. 10.5B) smoothed much of the variation in Fig. 10.5A and helped demonstrate the response pattern.

The first few years, R_{soil} measurements were typically collected under "nice" field conditions between rain events when the litter layer was dry (Hanson et al.

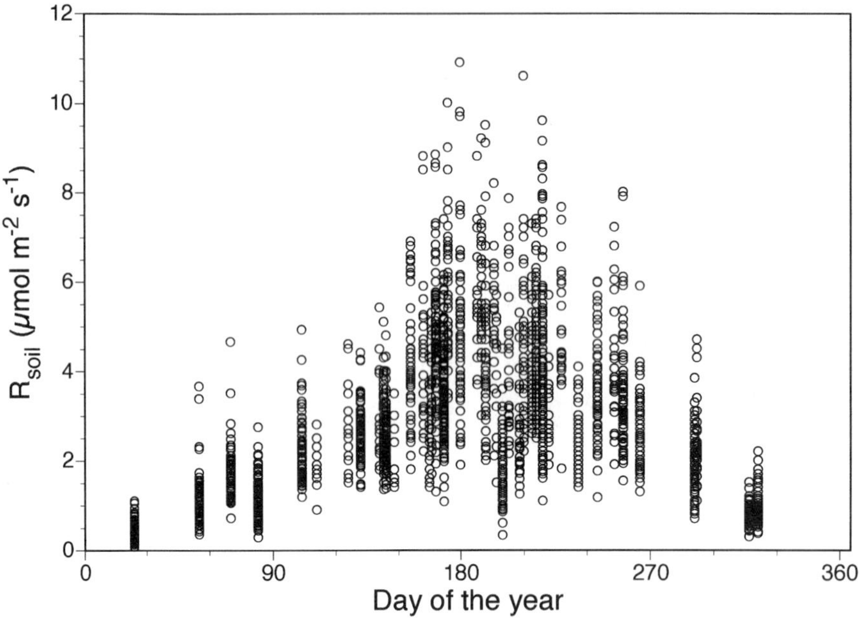

Figure 10.4. Seasonal pattern of soil respiration (R_{soil}) for the combined dataset from 1993 to 1999. Declining midseason R_{soil} was only observed at the depth of drought periods in 1993, 1995, and 1998.

1993). Under these conditions, R_{soil} was composed primarily of CO_2 losses from mineral-soil heterotrophic organisms and root activity (open symbols in Fig. 10.5B) and contributions from organic-layer decomposition were minimal. Small incremental increases in R_{soil} were observed during periods of active root growth (gray symbols in Fig. 10.5B), but the wetting of the surface litter layer increases R_{soil} as much as 1–2 μmol m^{-2} s^{-1} (black symbols in Fig. 10.5B). The elevation of R_{soil} through the stimulation of heterotrophic decomposition of the surface litter layer is a significant process that has often been overlooked in studies of R_{soil}.

Litter-specific respiration (L_{resp}) increased curve-linearly with increasing litter water potential and temperature (Fig. 10.6). Although we had expected L_{resp} under constant conditions to be higher for the Oi than for the Oe/Oa samples because of the Oi layer's larger C : N ratio (Pastor and Post 1985), that was not the case, and the relationship with temperature and LWP was the same (data not shown). *In situ* field observations of litter water content demonstrated that the water status of the Oi layer was highly dynamic, with wetting and drying cycles associated with rain events lasting 1–2 days (Fig. 10.7B). Applying the relationship among L_{resp}, LWP, and litter temperature from Fig. 10.6 to measured litter water potentials and temperatures provides an indirect method of estimating the contribution of Oi litter to R_{soil} (Fig. 10.7C). The decomposition of the Oi layer is tightly coupled to litter water status surrounding precipitation events, and often the litter will dry to levels inhibiting microbial activity in less than 2 days. The increasing patterns

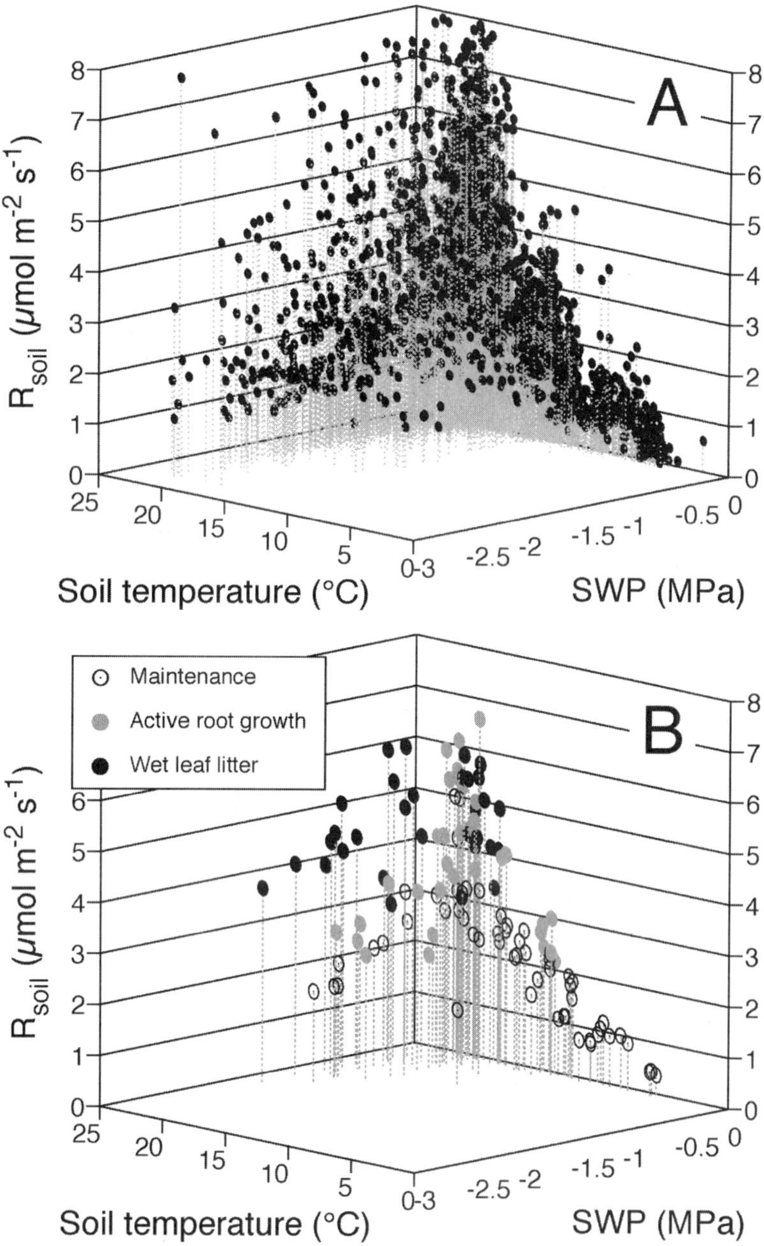

Figure 10.5. Soil respiration (R_{soil}) as a function of soil temperature at 10 cm and soil water potential (SWP) for 0–35 cm. The upper graph (A) shows the scatter plot for all data collected from 1993 to 1999 (>2500 observations). Data in the lower graph (B) are daily means by treatment for individual measurement dates (individual treatments are not indicated on this graph). Shading of the symbols in (B) delineates R_{soil} during maintenance conditions (open symbols), during periods of active root growth (gray symbols), and during periods when the litter layer was wet (black symbols).

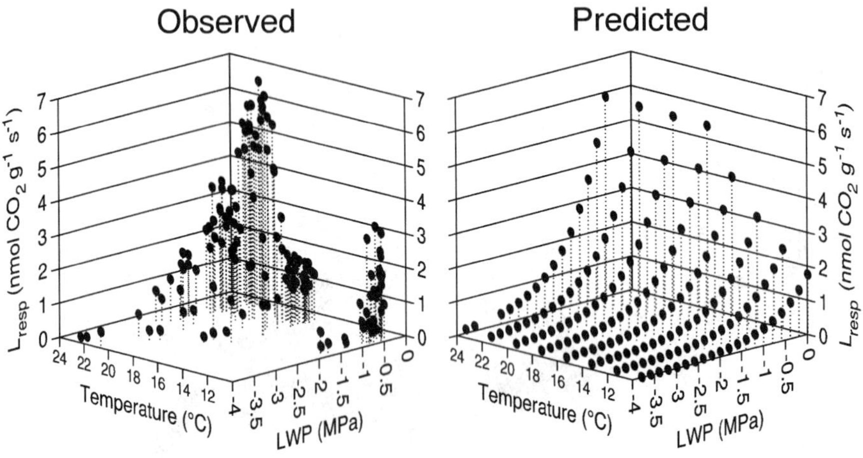

Figure 10.6. Observed and predicted relationships among litter-specific respiration (L_{resp}), litter temperature, and litter water potential (LWP). Data for samples taken from the Oi, Oe and Oa horizons are combined in this graph. Predicted litter respiration follows the formula $L_{resp} = (a\ b^{LWP})\ (Q^{(Temp-20)/10})$, where a, b, and Q are constants and 20 is the reference temperature (in °C). For the predicted relationship shown, $a = 7.1 \pm 0.4$, $b = 3.48 \pm 0.31$, and $Q = 4.05 \pm 0.39$, and the $R^2 = 0.80$.

of L_{resp} between day 80 and day 220 are related to temperature increases (compare Figs. 10.7A and 10.7C).

No significant differences in fresh-litter C, N, or C : N ratio from the TDE treatment were found for *A. rubrum* or *Q. prinus* litter collected from 1993 to 2000 (data not shown). Mean values for fresh-litter N concentration and the C : N ratio varied slightly from year to year for both species. *A. rubrum* fresh-litter N concentration ranged from 0.6% to 0.94% with a multiyear mean of 0.70%. *Q. prinus* fresh-litter N concentration ranged from 0.65% to 0.95%, with a multiyear mean of 0.77%. The mean C : N ratio for *A. rubrum* and *Q. prinus* was 70 and 64, respectively. The lack of treatment differences on fresh-litter chemistry suggests few long-term effects of altered throughfall regimes on litterfall chemistry as expressed by N and C : N ratios. Analyses of other constituents of the litterfall chemistry (e.g., K, Ca, and Mg) have not been completed, but are planned.

After 4.6 years, mass loss of litter as measured in litter decay bags on the dry plots was slightly greater than mass loss on the ambient or wet plots, however, because the trend was not significant ($p = 0.08$; Fig. 10.8), we conclude that the TDE treatment had little or no effect on the decomposition of the 1995 cohort of litter. There were no significant differences in mass loss or in residue N concentrations due to litter origins in the wet, ambient or dry plots. Decomposition of the contents of litterbags during the first year takes place in the Oi layer and is subject to high temperature and LWC variability. In subsequent years, the measured litter-cohort becomes buried under additional layers of litter input, and the temperature and water content variability become muted and eventually take on

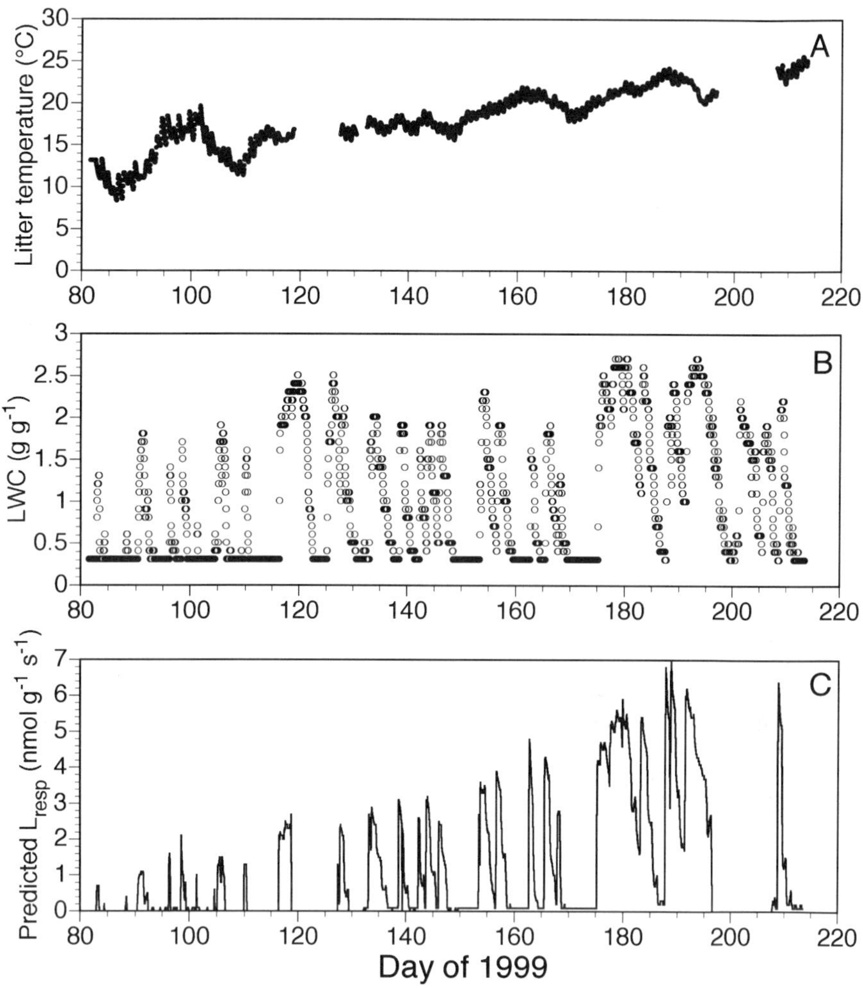

Figure 10.7. Temporal pattern of Oi litter layer temperature (A), litter water content (LWC) (B), and predicted litter-specific respiration (L_{resp}) (C) for days 80–220 of 1999. L_{resp} estimates were calculated using the equation in the caption to Fig. 10.6 after converting LWC data to leaf water potential with the relationship in Fig. 10.2.

the characteristics of the surface mineral horizon. The change in position and environment of the litterbags over the 5-year measurement period confounds comparisons with the litter-specific respiration data discussed previously. To model mass loss within a litterbag for multiyear periods, *in situ* information on the temperature and water potential environments would need to be collected for that cohort as it becomes incrementally buried by subsequent leaf-litter layers.

The C : N ratio of the decomposed litter, on the other hand, was significantly higher at every harvest of litter from the dry plots (Table 10.2). Nitrogen was significantly lower in litter decomposed on the dry plot for two out of three

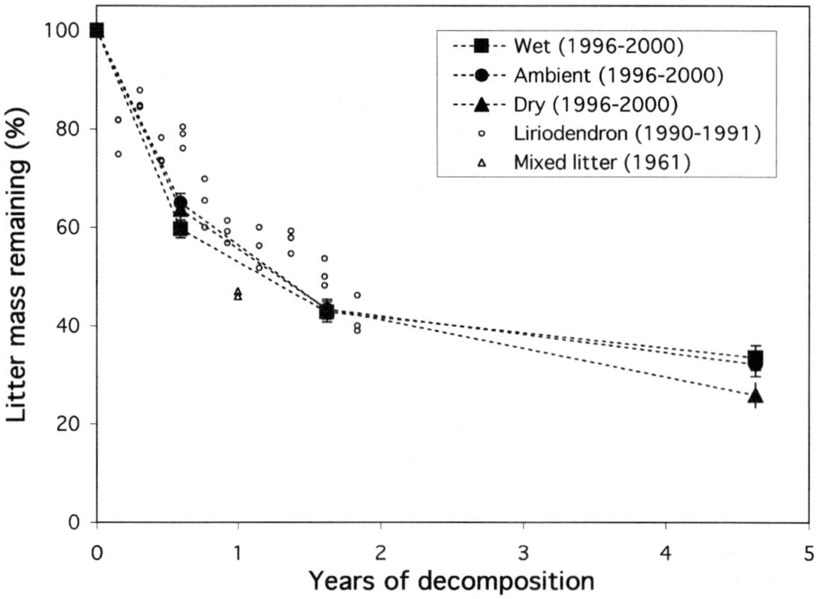

Figure 10.8. Pattern of litter decomposition over time expressed as the percent of original mass remaining in the litter bag. Solid symbols correspond to the TDE treatment means for data collected from 1996 to 2000. Open circles are data from O'Neill and Norby (1996) for *L. tulipifera* litter decomposed on an upland site of the Walker Branch Water shed from 1990 to 1991. Open triangles are data from Witkamp (1966) for mixed litter decomposed on the Oak Ridge Reservation between November 1960 and November 1961.

Table 10.2. Carbon and nitrogen concentrations in litter residue of 1995 mixed litter as a function of time in the organic horizons and TDE treatment

Elapsed time (year)	TDE treatment	Carbon	Nitrogen	C : N
0.6	Wet	47.1 ± 2.2 a	1.32 ± 0.14 a	36.2 ± 4.4 b
	Ambient	44.8 ± 5.2 b	1.26 ± 0.16 a	36.0 ± 5.1 b
	Dry	47.8 ± 1.1 a	1.09 ± 0.14 b	44.3 ± 5.6 a
1.6	Wet	38.4 ± 8.6 a	1.65 ± 0.35 a	23.1 ± 1.7 b
	Ambient	39.4 ± 2.7 a	1.65 ± 0.12 a	23.9 ± 1.5 b
	Dry	41.6 ± 4.0 a	1.54 ± 0.14 a	27.0 ± 2.0 a
4.6	Wet	38.4 ± 4.4 a	1.95 ± 0.24 a	19.2 ± 0.9 b
	Ambient	33.8 ± 4.4 b	1.79 ± 0.27 b	19.0 ± 1.6 b
	Dry	39.1 ± 3.7 a	1.86 ± 0.17 ab	21.1 ± 0.9 a

Note: Data given as means ± sd, $n = 24$. For each nutrient and within each harvest, means with the same letter are not significantly different ($p = 0.05$).

harvests. If residue chemistry is a better indicator of decomposition dynamics than is mass loss alone, then we might conclude that long-term reduced precipitation does retard decomposition. Calculation of the decomposition constant (k_d) from the negative exponential function proposed by Olson (1963) suggests that litter on the TDE is decomposing at the rate of ~0.44 % per year ($k_d = -0.44$). A similar constant ($k_d = -0.47$) was observed during a 2-year period from 1990 to 1991 for *L. tulipifera* litter placed in a similar upland location on Walker Branch Watershed (O'Neill and Norby 1996).

Modeling R_{soil}

Hanson et al. (1993) proposed a simple empirical model of R_{soil} based on soil temperature, soil-water content, and the soil's rock content for application to upland oak forests. Because the patterns of R_{soil} reported in Fig. 10.5 showed substantial change in R_{soil} associated with wetting of the organic horizons, the model was updated to account for root-growth respiration and to use soil-water potential in the place of soil-water content as follows:

$$R_{soil} = M_{resp} + G_{resp}, \tag{1}$$

$$M_{resp} = R_b \, Q^{(T_{soil}/10)} \left(1 - \frac{CF}{100}\right) \left(\frac{\Psi_{max} - \Psi_{35}}{\Psi_{max}}\right), \tag{2}$$

$$G_{resp} = RG \; G_{cost}, \tag{3}$$

where

M_{resp} = the maintenance respiration rate of roots and soil microbes
G_{resp} = the cost of growing new roots
R_b = the maintenance respiration of roots and soil microbes when temperatures approach 0°C (determined from data when active root growth was not occurring)
Q = the rate of change in M_{resp} for a 10°C increase in soil temperature (i.e., the Q_{10} value)
T_{soil} = the soil temperature at a depth of 10–15 cm (in °C)
CF = the percent coarse fraction of the soil
Ψ_{max} = the soil matric potential (in Mpa) corresponding to the complete inhibition of R_{soil}
Ψ_{35} = the observed soil matric potential of the upper 35 cm of soil (in MPa) (i.e., the zone of maximum rooting density)
RG = the rate of root growth (g dry matter as C) on a ground-area basis
G_{cost} = the carbon cost for growing roots (i.e., g C used for each g dry-matter carbon produced).

This model was successful for integrating seasonal patterns of R_{soil} when applied to nontransient R_{soil} between rain events (Hanson et al. 1993), and it has been used recently to estimate the soil-respiration components of eddy covariance measurements of net ecosystem exchange (Wilson et al. 2001). Fang and Moncrieff (2001) recently reviewed a wide range of empirical models for application to R_{soil} and found that a number of forms performed well. Our model is an expanded form of a first-order exponential model. We favor it over an Arrhenius model to avoid the pitfall of inappropriate interpretation of that model's "activation energy" constant. The complex nature of R_{soil} makes such an interpretation problematic at best.

The R_b variable of Eq. (2) is very important but purely empirical for this study. In reality, as suggested by Hanson et al. (1993) and Chambers (1998), R_b would be site-specific and a function of a variety of characteristics, including soil organic matter, root density, nutrient status, and associated microbial populations and activity. For this chapter, R_b characteristics are assumed constant for intra-annual periods, but such an assumption would need to be modified for multiyear simulations.

Hanson et al. (1993) proposed that quantification of the respiratory losses of CO_2 associated with the growth of new roots in the soil (i.e., growth respiration) would improve predictions of R_{soil}. To facilitate this calculation, seasonal patterns of root-growth from minirhizotrons and total annual root production (Joslin et al. 2000, 2001; Chapter 16, this volume) were tracked each year, and estimates of the root-growth respiration were derived from the product of the fractional pattern of new root growth, annual new root mass (as carbon), and the growth-respiration coefficient (Amthor 2000). The growth-respiration coefficient for roots of the Walker Branch forest was estimated to be 0.23 g C expended to produce 1 g of dry matter as carbon based on the estimation methods of Penning de Vries et al. (1974) and Vertregt and Penning de Vries (1987) as presented by Amthor (1996).

In the previous Results section, R_{soil} measurements were occasionally shown to increase following rain events. Such a phenomenon is unexplained by the R_{soil} model outlined in Eqs. (1)–(3). The dynamic response of surface litter layers to wetting and drying cycles (Figs. 10.6 and 10.7) suggests an explanation for the pattern and the need for another model component to explain the dynamics of litter-specific respiration. Therefore, a new term that captures the dynamics of the surface (Oi) litter layer decomposition (L_{resp}) was added to Eq. (1), consistent with the litter-decomposition characteristics shown in Figures 10.6 and 10.7:

$$R_{soil} = M_{resp} + G_{resp} + L_{resp}, \tag{4}$$

$$L_{resp} = (L_b \, b^{\mathrm{OiWP}}) \, [\mathrm{Q}_l^{(\mathrm{O_{itemp}} - 20)/10}] \text{ (Oi litter mass per m}^2\text{ of ground)}, \tag{5}$$

where

R_{soil} = total soil respiration from all contributing components as earlier

L_{resp} = the CO_2 respired from the surface litter layer (Oi horizon) as a function of the Oi litter water potential ($\mathrm{Oi_{wp}}$), Oi temperature ($\mathrm{Oi_{temp}}$), and Oi litter mass

L_b = the litter-specific respiration at 20°C and maximum litter water potential

b defines the shape of the relationship previously shown in Fig. 10.6

Q_l = the temperature response coefficient for a 10°C change in litter temperature.

The utility of this model for estimating R_{soil} for the period from 1993 to 1999 is demonstrated in Fig. 10.9. Inset graphs A and B in Fig. 10.9 show that the model has the flexibility to capture transient increases and decreases in R_{soil} associated with wetting and drying of the surface organic horizons. Estimates for the model parameters outlined in Eqs. (2), (3), and (5) are provided in Table 10.3.

Estimated Annual R_{soil} and Its Components

Application of the revised model of R_{soil} expressed by Eq. (4) to environmental data from the TDE study for the period from 1993 to 2000 provides a method for estimating annual ecosystem carbon losses from R_{soil} (Table 10.4). Estimated 7-year (1994–2000) annual ambient R_{soil} showed mean values of 929, 907, and 938 g C m^{-2} y^{-1} for the ambient, dry, and wet simulations, respectively. Interannual differences were not large, but the 1995 drought year produced noticeably lower R_{soil}. A similar reduction for the 1998 drought year was not predicted because higher than normal annual and growing-season temperatures (Table 2.1

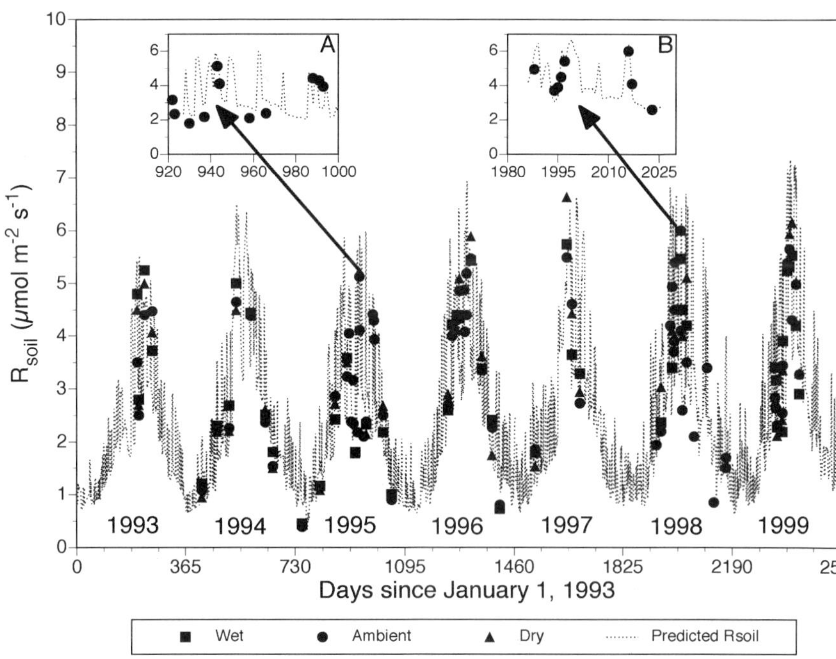

Figure 10.9. Mean values for observed soil respiration (R_{soil}) by TDE treatment plotted along with predicted R_{soil} for 1993 through 1999. Inset graphs for periods in 1995 (A) and 1998 (B) are provided to show the details of the dynamics of a wetting and drying litter layer.

Table 10.3. Parameter estimates for the variables needed to predict R_{soil} from soil and surface litter temperature and water potential

Equation variables	Value	Units
Equation (2)		
R_b	0.45 ± 0.03	$\mu mol\ m^{-2}\ s^{-1}$
Q	2.5 ± 0.1	Unitless
Ψ_{max}	−7.1 ± 0.8	MPa
CF	14.5	Unitless
Equation (3)		
G_{cost}	0.2333	gC gC biomass $^{-1}$
Equation (5)		
L_b	7.10 ± 0.38	nmol C $g^{-1}\ s^{-1}$
b	3.48 ± 0.31	Unitless
Q_l (litter)	4.05 ± 0.39	Unitless

Note: Estimates of error (±SE) for the parameter estimates are provided if they were derived from regression analysis of observed data.

of Chapter 2, this volume) increased M_{resp}, thus compensating for reduced R_{soil} under drought. The small differences in annual R_{soil} across treatments are consistent with our inability to measure direct effects of TDE treatments on R_{soil} in the field. Of the components in our model of $R_{soil,}$, M_{resp} contributed 72.9%, and G_{resp} only 3.9% of the total. At 23.4% of R_{soil}, L_{resp} is clearly a significant component of the total and should be included in future models of R_{soil} for other sites. The TDE treatments had little impact on the relative contribution of the components to R_{soil}.

Discussion

Soil Respiration and Its Components

The relationship between R_{soil} (or L_{resp}) and temperature and soil-water status is demonstrated and quantified for our aerobic soils. Such a pattern is not unexpected and many studies have reported observations of similar range and magnitude for upland temperate forests (Andersen 1973; Keith et al. 1997; Davidson et al. 1998). Some studies have found limited R_{soil} responses to soil-water deficits (Reiners 1968; Castelle and Galloway 1990), but they may not have measured R_{soil} over a sufficient range of conditions to observe the reduced rates associated with severe drought that we observed in 1993, 1995, and 1998. Andersen (1973) suggested that R_{soil} was only affected by changes in soil moisture when approaching "limiting conditions." Many past studies do not express soil-water availability as water potential, making direct comparisons to other studies difficult, if not impossible, because soil characteristics at any given experimental location can yield different values for soil-water potential for similar volumetric or gravimetric estimates of soil-water content. Future studies should express the response of R_{soil} to water deficits in terms of soil-water potential.

Recent studies have contrasted various methods for obtaining *in situ* observations of R_{soil} (e.g., Norman et al. 1997; Kelliher et al. 1999; Law et al. 1999) and

Table 10.4. Simulated components of annual soil respiration (R_{soil}) from an upland oak forest for 8 years according to the apparent source: maintenance of roots and microbes (M_{resp}), root-growth costs (G_{resp}), and decomposition of the surface litter layer (L_{resp})

Components of R_{soil} ($gC\ m^{-2}\ y^{-1}$)	Year								Mean flux ($gC\ m^{-2}\ y^{-1}$)	Percent
	1993	1994	1995	1996	1997	1998	1999	2000		
Ambient treatment										
M_{resp}	618	676	647	663	665	710	707	720	676 ± 12	72.9
G_{resp}	26	38	23	48	32	28	42	39	35 ± 3	3.9
L_{resp}	145	244	192	249	250	227	223	217	218 ± 13	23.4
Total	789	958	862	960	947	965	972	976	929 ± 24	100.0
Dry treatment										
M_{resp}	590	670	623	657	660	680	688	692	658 ± 12	72.5
G_{resp}	23	38	19	48	32	28	42	38	34 ± 3	3.7
L_{resp}	141	242	182	249	254	220	223	220	216 ± 13	23.8
Total	754	950	824	954	946	928	953	950	907 ± 27	100.0
Wet treatment										
M_{resp}	623	677	666	664	666	732	721	724	684 ± 13	72.9
G_{resp}	26	38	26	48	32	28	42	39	35 ± 3	3.7
L_{resp}	150	242	197	250	247	231	221	218	220 ± 12	23.4
Total	799	957	889	962	945	991	984	981	938 ± 23	100.0

Note: The 8-year means ± SE of the interannual variation and the components respective percent contribution to R_{soil} are provided. Simulations are provided for ambient, dry (−33.3% throughfall), and wet (+33.3% throughfall) scenarios.

concluded that the closed-system approach can provide acceptable quantitative information for scaling the CO_2 losses from soils over space and time. Similarly, Nay and Bormann (2000) demonstrated good agreement between chamber and mass-loss estimates of R_{soil}. All of these data suggest that interpolated data from chamber measurements should yield acceptable estimates of carbon exchange for the evaluation of stand carbon budgets or changes in soil carbon pools. Annual R_{soil} estimates for the TDE site on Walker Branch (Table 10.4) show a mean value of 929 g C m^{-2} y^{-1}, with insignificant differences resulting from the TDE treatments. Interannual variability from 1993 to 2000 driven by temperature differences or drought was not large. Lack of interannual variability, however, does not imply insensitivity to either factor. It simply suggests that mean annual temperatures are stable and drought occurrence is typically not sustained long enough to affect annual R_{soil} to a large degree.

Ambient R_{soil} data for our site are contrasted with data for other temperate hardwood studies in Table 10.5, which is an expansion of data for temperate broadleaf forests previously summarized by Raich and Schlesinger (1992). The data for the current study range from 789 to 976 g C m^{-2} y^{-1} and agree well with other data with the same mean annual temperature and precipitation. Across a latitudinal gradient from 52° N to 34° N (plus one data point at 35° S), corresponding to mean annual temperatures (MATs) ranging from 4°C to 15°C, annual R_{soil} showed a linearly increasing relationship with temperature [$R_{soil} = 35.9$ (MAT) + 407; $R^2 = 0.29$], but no relationship to mean annual precipitation. Raich and Schlesinger (1992) published similar relationships for a global database. Garten et al. (1999) also showed that the turnover times for unprotected soil carbon were linearly related to soil MAT, with the turnover times increasing as temperature decreased. Substantial variation around the relationship between R_{soil} and site temperatures is undoubtedly driven by site-specific differences that influence both the heterotrophic and autotrophic components of R_{soil}. The large contribution of root respiration to R_{soil} in forests (Hanson et al. 2000) together with a recognized relationship between stand net primary production and ecosystem respiration (Valentini et al. 2000; Janssens et al. 2001) may also account for the substantial variation in the positive correlation between R_{soil} and stand MAT.

Temperature Sensitivity of R_{soil}

The Q_{10} value of 2.5 reported for R_{soil} in this work applies only to the maintenance respiration of heterotrophic organisms and roots of the mineral soils [Eq. (2) and Table 10.3] but is well within the range of values published for other temperate deciduous forests (Table 10.5) (Kirschbaum 1995). The Q_{10} value for this study was estimated from only those data known to be collected when active root growth was minimal. Hanson et al. (1993) pointed out that midseason R_{soil} data included CO_2 losses from root-growth activity that might not follow normal temperature-driven patterns. Boone et al. (1998) also showed that the presence of roots increases Q_{10} values of R_{soil}. Davidson et al. (1998) provide a complete discussion

Table 10.5. Published estimates of annual R_{soil} for a variety of deciduous forests

Location	Species	Latitude	Longitude	MAT (°C)	MAP (mm)	Annual R_{soil} (gC m^{-2} y^{-1})	Q_{10}	Reference
Japan	*Fagus, Abies*	34° 05′ N	136° 05′ E	~7.0	na	494	na	Nakane 1980
Japan	*Quercus* mixed	34° 45′ N	135° 55′ E	~13.0	na	1414	na	Kirita 1971
Japan	*Quercus* mixed	34° 45′ N	135° 55′ E	~14.5	na	1098	na	Nakane 1975
North Carolina	*Quercus* plus	35° 03′ N	83° 25′ W	13.6	1628	857	na	Wade after Edwards et al. 1989
Australia	*Eucalyptus*	35° 23′ S	148° 48′ E	9.5	1200	711	na	Keith et al. 1997
Tennessee	*Quercus*	35° 38′ N	85° 18′ W	11.3	1450	674	na	Larkin and Kelly 1987
Tennessee	*Quercus*	35° 38′ N	85° 18′ W	13.0	1550	683	na	Larkin and Kelly 1987
Tennessee	*Liriodendron*	35° 55′ N	84° 17′ W	13.3	1265	1070	na	Edwards 1975
Tennessee	*Liriodendron*	35° 55′ N	84° 17′ W	13.3	1265	1065	na	Edwards and Harris 1977
Tennessee	*Liriodendron*	35° 55′ N	84° 17′ W	13.3	1265	1036	na	Edwards and Sollins 1973
Tennessee	*Quercus, Acer*	35° 57′ N	84° 17′ W	13.8	1352	950	2.5	Curtis et al. 2002
Tennessee	*Quercus, Carya*	35° 58′ N	84° 17′ W	13.3	1400	529	na	Edwards and Ross-Todd 1983
Tennessee	*Quercus, Carya*	35° 58′ N	84° 17′ W	13.3	1400	610	na	Edwards et al. 1989
Tennessee	*Quercus, Pinus*	35° 58′ N	84° 17′ W	14.5	1390	927	2.4	Hanson et al. 1993
Tennessee	*Quercus, Acer*	35° 58′ N	84° 17′ W	14.5	1390	845	2.0	Hanson et al. 1993
Tennessee	*Quercus, Liriodendron*	35° 58′ N	84° 17′ W	14.5	1390	818	2.4	Hanson et al. 1993
Tennessee	*Fagus, Acer*	35° 58′ N	84° 17′ W	14.5	1390	736	3.2	Hanson et al. 1993
Tennessee	*Quercus, Acer*	35° 58′ N	84° 17′ W	14.3	1139	789	2.5	This study
Tennessee	*Quercus, Acer*	35° 58′ N	84° 17′ W	14.4	1675	958	2.5	This study
Tennessee	*Quercus, Acer*	35° 58′ N	84° 17′ W	14.4	1135	862	2.5	This study
Tennessee	*Quercus, Acer*	35° 58′ N	84° 17′ W	13.6	1638	960	2.5	This study
Tennessee	*Quercus, Acer*	35° 58′ N	84° 17′ W	14.1	1462	947	2.5	This study
Tennessee	*Quercus, Acer*	35° 58′ N	84° 17′ W	15.8	1225	965	2.5	This study
Tennessee	*Quercus, Acer*	35° 58′ N	84° 17′ W	15.0	1152	972	2.5	This study
Tennessee	*Quercus, Acer*	35° 58′ N	84° 17′ W	14.5	1207	976	2.5	This study
Misouri	*Quercus, Carya*	38° 40′ N	92° 12′ W	13.0	939	1013	na	Garrett and Cox 1973
Indiana	*Acer, Liriodendron*	39° 53′ N	86° 25′ W	11.1	1012	1207	na	Curtis et al. 2002

Table 10.5. *Continued*

Location	Species	Latitude	Longitude	MAT (°C)	MAP (mm)	Annual R_{soil} (gC m^{-2} y^{-1})	Q_{10}	Reference
New York	*Quercus, Pinus*	40° 50′ N	72° 54′ W	9.8	1240	502	na	Woodwell and Botkin 1970
Massachusetts	*Quercus, Pinus*	41° 40′ N	70° 35′ W	~8.0	na	648	na	Raich and Schlesinger 1992
Italy	*Fagus*	41° 05′ N	12° 27′ E	7.9	1959	395	na	Virzo De Santo et al. 1976
Massachusetts	*Quercus, Tsuga*	42° 32′ N	72° 10′ W	7.1	1066	800	na	Curtis et al. 2002
Massachusetts	*Quercus, Acer*	42° 32′ N	72° 11′ W	7.9	na	710	3.5	Davidson et al. 1998
Massachusetts	*Quercus, Acer*	42° 32′ N	72° 11′ W	7.9	na	720	4.1	Davidson et al. 1998
Massachusetts	*Quercus, Acer*	42° 32′ N	72° 11′ W	7.9	na	850	3.4	Davidson et al. 1998
Michigan	*Acer, Quercus*	44° 48′ N	85° 48′ W	7.2	810	487	na	Toland and Zak 1994
Michigan	*Acer, Tilia*	44° 48′ N	85° 48′ W	7.2	810	469	na	Toland and Zak 1994
Minnesota	*Fraxinus* mixed	45° 30′ N	93° 10′ W	7.1	676	707	na	Reiners 1968
Minnesota	*Quercus*	45° 30′ N	93° 10′ W	7.1	676	794	na	Reiners 1968
Michigan	*Populus, Quercus*	45° 35′ N	84° 42′ W	6.2	750	1028	na	Curtis et al. 2002
Wisconsin	*Acer, Tilia*	45° 47′ N	90° 05′ W	4.8	776	810	na	Curtis et al. 2002
Quebec	*Populus*	46° 00′ N	77° 25′ W	~4.0	800	334	na	Weber 1990
France	*Fagus*	48° 40′ N	7° 05′ E	9.2	819	575	3.9	Epron et al. 1999b
France	*Fagus*	48° 40′ N	7° 05′ E	9.6	837	663	3.9	Epron et al. 1999b
Germany	*Fagus, Picea*	49° 20′ N	8° 40′ E	na	na	470	na	Dörr and Münnich 1987
Belgium	*Fagus*	50° 18′ N	6° 00′ E	7.5	792	870	na	Longdoz et al. 2000
Belgium	*Quercus* etc.	51° 18′ N	4° 31′ E	7.5	792	578	na	Janssens et al. 2000
Belgium	*Quercus* etc.	51° 18′ N	4° 31′ E	7.5	792	633	na	Janssens et al. 2000
United Kingdom	*Castanea*	51° 20′ N	1° 05′ E	na	na	630	3.2	Andersen 1973
United Kingdom	*Fagus*	51° 20′ N	1° 05′ E	na	na	575	na	Andersen 1973
Germany	*Fagus*	~52° N	~10° E	6.4	10	322	2.3	Brumme 1995
Germany	*Fagus*	~52° N	~10° E	6.4	10	439	2.3	Brumme 1995

Note: Abbreviations: MAT = mean annual temperature; MAP = mean annual precipitation, Q_{10} = change in R_{soil} for a 10°C change in soil temperature; na = not available.

[a] These are 6-month (i.e., growing season) estimates only.

of factors that might influence Q_{10} values for R_{soil} and pointed out that active root growth often produced elevated estimates of the Q_{10} for R_{soil} if not considered separately. Högberg et al. (2001) demonstrated that R_{soil} was directly affected by root activity driven by aboveground photosynthetic production in a tree-girdling experiment. Similar data were reported in a grass clipping study by Craine et al. (1999). R_{soil} driven by autotrophic activities associated with aboveground photosynthesis may not show a constant temperature response.

Hanson et al. (2000) reviewed a substantial amount of published work showing that roots contribute on average 46–60% of the CO_2 in R_{soil}, and a recent study for temperate deciduous forests showed similar results (Epron et al. 1999a). Because roots and heterotrophic organisms can function differentially under similar environmental conditions (Boone et al. 1998; Högberg et al. 2001), additional complexity may need to be added to models of R_{soil} before they can adequately capture the response of R_{soil} to long-term changes in climate conditions. On the basis of the model applied in this study, root-growth costs contributed only 3–4% to annual R_{soil} (Table 10.4), but such activity on a daily basis added nearly 0.5 μmol m^{-2} s^{-1} to R_{soil} (data not shown). Coleman et al. (2000) also reported a poor correlation between root-growth activity, as measured with minirhizotron cameras, and soil respiration.

Litter-Specific Respiration Dynamics

Few studies have attempted to quantify litter layer contributions to R_{soil}. Virzo De Santo et al. (1976) reported litter-specific respiration rates of *Fagus* litter to be between 2.4 and 4.3 nmol g^{-1} s^{-1}, which are in the range of those reported here. Unfortunately, they did not provide corresponding temperature and moisture data. Their graphs of seasonal patterns of R_{soil} with and without the litter layer showed only a small increase in the presence of litter, but it may not have been wet and actively respiring. Keith et al. (1997) measured litter layer water content in their study of R_{soil} in a *Eucalyptus pauciflora* forest and found an positive correlation between R_{soil} and litter water content only when air temperatures exceeded 10°C. Van Cleve and Sprague (1971) also found minimal litter layer respiration when litter water content fell below 0.7 g g^{-1} as reported here, but they found that temperature was the most important driver for litter respiration in their boreal *Betula* and *Populus* forests.

The model predictions for reduction in L_{resp} with increasing LWP in Figure 10.6 follow a continuous negative exponential function. An alternate model was proposed by Sommers et al. (1981), who showed that strong positive relationships with temperature and negative relationships with soil-water potential existed. They concluded that reductions in decomposition followed a two-phase pattern, with rapid declines occurring from -0.03 to -1 Mpa, followed by slower linear rates of declining decomposition for much lower water potentials.

When integrated over an entire growing season, the "spikes" of litter respiration shown in Fig. 10.9 sum to a mean annual L_{resp} of more than 200 g C m^{-2} y^{-1} and represent 23% of the total annual R_{soil} (Table 10.4). Tesarova and Gloser

(1976) studied the influence of variable temperature, soil, and litter water content on the CO_2 losses from alluvial soils covered by grassland communities. They found a strong linear relationship between litter decomposition and litter water content and estimated that between 20% and 30% of the R_{soil} from their sites was from the litter layer. Edwards and Sollins (1973) provided a small amount of data on litter layer contributions to R_{soil} and showed that litter wetness could increase R_{soil} by around 1 μmol m^{-2} s^{-1}. These estimates are similar to the changes reported for the TDE data. Edwards and Harris (1977) reported a lower contribution of the Oi litter layer to total R_{soil} of a *Liriodendron* forest (10%), but they may not have measured peak loss rates from wet litter directly following rain events. Future studies and models of R_{soil} in forest stands with a quantifiable litter layer must accommodate the phenomenon of carbon-efflux dynamics from a wetting and drying litter layer or they will underestimate respiratory losses.

Litterbag Mass Loss

The mixed-species litterbag study conducted on the TDE plots revealed the mass loss constant (k_d) (Olson 1966) to be 0.44. A similar constant (k_d = 0.47) was observed during a two-year period from 1990 to 1991 for *Liriodendron tulipifera* litter placed in another upland location on Walker Branch Watershed (O'Neill and Norby 1996). Whitkamp (1966) also studied the decomposition of forest litter in relation to the species type and stand location on Walker Branch watershed from November 1960 to 1961. Decomposition during that year varied by species according to the following pattern *Morus rubra* L. > *Cercis canadensis* L. > *Q. alba* > *Pinus taeda.* Consistent with the TDE observations for mixed species, mass remaining after 1 year averaged 46–47% (see open triangles in Figure 10.9), and mass loss for *Q. alba* was 45%.

Larkin and Kelly (1987) found k_d values for litter in another Tennessee oak forest to range between 0.35 and 0.39, and Cromack (1973) reported higher values (k_d = 0.61) for mixed hardwood litter in a wetter Coweeta, North Carolina site. Kelly and Beauchamp (1987) studied leaf-litter decomposition in an upland oak and mesic hardwood setting in Tennessee and found similar rates of decomposition, but concluded that an alternate model yielding a variable decomposition rate with time was most appropriate. Kuperman (1999) found litter-decay constants for *Quercus–Carya* forests in Illinois, Indiana, and Ohio ranging from 0.24 to 0.75. The higher k values were found to be associated with litter with a higher nitrogen concentration. The Long-Term Intersite Decomposition Experiment (LIDET) (Gholz et al. 2000) was designed to follow the decomposition of fixed litter types across a range of sites representing a range of environmental conditions. It showed leaf-decomposition constants ranging from 0.38 to 0.41 for *Drypetes glauca* litter and from 0.27 to 0.29 for *Pinus resinosa* litter. Gholz et al. (2000) found consistent decomposition constants for litter placed on three different temperate broadleaf forest sites; but across the full range of ecosystems studied, increasing mean annual temperature was associated with greater mass loss and larger k values. A similar pattern was observed for litter decomposition across a range of northern forests (Moore et al. 1999).

Changes in decomposition rates of litter were hypothesized at the beginning of this study to serve as an early indicator of change in carbon and nitrogen turnover produced by altered precipitation. Decomposition rates were expected to change through environmental (litter moisture) or chemical (tissue chemistry) means. The 5-year sequence of litterbag decomposition observations from 1995 to 2000 did not show a strong relationship between litter decomposition and the throughfall manipulations. However, dry plot litter retained a higher C : N ratio throughout the measurement. Litter quality is well documented as an important variable controlling decomposition (Meentemeyer 1978; Flanagan and Van Cleve 1983; McClaugherty et al. 1985), where decomposition is negatively related to C : N ratios. Therefore, high C : N ratios of dry plot leaf litter may indicate reduced decomposition rates even though mass-loss differences were not statistically significant. The leaf-chemistry data in Table 10.2 show increasing litter nitrogen over time and a decline in C : N ratios to values near 19 after 5 years of *in situ* decomposition. Similar increases in nitrogen with time have been reported for upland deciduous forests in Wisconsin (Fassnacht and Gower 1999). In their documentation of the Linked Forest Productivity–Soil Process model (LINKAGES), Pastor and Post (1985) summarize species-specific critical litter nitrogen levels, that identify its transition to the soil humus pool (Aber and Mellilo 1982). These values are 1.6%, 1.3%, and 1.3% for *Acer, Cornus,* and *Quercus,* respectively. These critical levels had nearly been reached at the second harvest (1.6 years) and had certainly been attained at the last litterbag harvest (4.6 years) when nitrogen concentrations of the mixed litter had reached 1.8–1.95% (Table 10.2). These data suggest a rapid (i.e., < 3 year) transfer of litter organic matter to the soil–humus pool. Berg and Matzner (1997) recognized the similarities between differences in initial litter quality and the chemical changes taking place in litter decomposition over time. They proposed a multiphase model of litter decomposition that appears to capture much of the pattern of mass loss and litter chemical changes observed on the TDE. Future work on litter decomposition and nitrogen turnover might well focus on their model structure.

Johnson et al. (2002; Chapter 23, this volume) showed an increase in the standing pool of organic matter on the TDE dry plot that is consistent with the hypothesis of reduced decomposition with lower throughfall inputs. These results, however, are not consistent with the litter decay bag study results that showed a trend towards more rapid decomposition with lower throughfall inputs (i.e., on the dry plot). O'Neill et al. (2003) conducted a parallel experiment to test whether short-term seasonal precipitation exclusions (months) would affect decomposition processes over longer periods (i.e., years). Rain was completely excluded from miniplots for four 3-month periods representing the seasons of 1999. Litterbags placed in each plot were collected immediately following each exclusion period and again 9 months later. Results were similar to litter-decomposition data from the TDE litterbag measurements reported previously. Mass loss proved to be a highly variable and unreliable indicator of exclusion effects, with significant differences seen only in the spring exclusion. Nitrogen concentration and the C : N ratio were significantly lower ($p < 0.05$) in the precipitation exclusion treatment after the 4-month rainfall exclusions for all seasons of the year. However, differences in nitrogen and C : N ratio were not subsequently present for litterbags left in the field unsheltered from precipitation for an additional 9 months following the initial exclusions. This result suggests that decomposition processes are well buffered against short-term disturbance (at least in the case of precipitation) and that seasonal droughts will have no long-term effects on carbon or nitrogen mineralization. On the other hand, the short-term litterbag experimental results do not explain the increase in the standing pool of organic matter on the TDE dry plot, as reported by Johnson et al. (2002; Chapter 23, this volume). Further detailed measurements and modeling of organic-layer water dynamics and associated decomposition will be required to resolve these two studies.

Extraplotion of R_{soil} Data

The short-term data in this chapter show clearly that R_{soil} or L_{resp} are both related to temperature and soil-water potential. Simple application of these data to long-term simulations of climatic warming would result in predictions of increasing soil and litter-layer carbon losses. Such predictions have been reported (Jenkinson et al. 1991; Andersen 1992; Raich and Schlesinger 1992; Kirschbaum 1995), but other analyses have tempered these predictions because of the feedbacks anticipated from carbon inputs from increased net primary production under increasing atmospheric CO_2 (Kirschbaum 2000).

Several recent articles, however, have questioned such relationships based on measured ecosystem respiration along latitudinal gradients (Valentini et al. 2000) or laboratory-based mineral soil decomposition rates (Giardina and Ryan 2000). Discrepancies among studies and appropriate concerns about the appropriate application of experimental data to long-term climatic-change questions may in a large part stem from the attempt to apply overly simplistic models of R_{soil} to complex systems. The model of R_{soil} described in this chapter represents an incremental improvement over simple exponential temperature functions commonly applied (Fang and Moncrieff 2001). This chapter focused primarily on factors controlling short-term rates of R_{soil}, but it is the long-term balance between carbon inputs and outputs from forest ecosystems that will determine net changes in soil carbon storage (Janssens et al. 2001). Improving our understanding of R_{soil} by accounting for the complexity of the contributing components, their individual characteristics, and the nonuniform environmental gradients existing in soils will go a long way to provide models of R_{soil} that can provide robust predictions across a range of temporal and spatial scales.

References

Aber JD, Mellilo JM (1982) Nitrogen immobilization in decaying hardwood leaf litter as a function of initial nitrogen and lignin content. Can J Bot 60:2263–2269.

Amthor JS (1996) *Models of plant growth costs and their application to trees in Walker Branch Watershed.* UCRL-AR-125547. Lawrence Livermore National Laboratory, Livermore, California.

Amthor JS (2000) The McCree-de Wit-Penning de Vries-Thornley Respiration paradigms: 30 years later. Ann Bot 86:1–20.

Andersen JM (1973) Carbon dioxide evolution from two temperate deciduous woodland soils. J Appl Ecol 10:361–378.

Andersen JM (1992) Responses of soils to climate change. Adv Ecol Res 22:163–210.

Berg B, Matzner E (1997) Effect of N deposition on decomposition of plant litter and soil organic matter in forest systems. Environ Rev 5:1–25.

Boone RD, Nadelhoffer KJ, Canary JD, Kaye JP (1998) Roots exert a strong influence on the temperature sensitivity of soil respiration. Nature 396:570–572.

Brumme R (1995) Mechanisms of carbon and nutrient release and retention in beech forest gaps III. Environmental regulation of soil respiration and nitrous oxide emissions along a microclimatic gradient. Plant Soil 168–169: 593–600.

Castelle AJ, Galloway JN (1990) Carbon dioxide dynamics in acid forest soils in Senandoah National Park, Virginia. Soil Sci Soc Amer J 54:252–257.

Chambers LS (1998) *Characterization of forest floor carbon dioxide efflux from three forest ecosystems in East Tennessee, USA.* Masters thesis. The University of Tennessee, Knoxville, Tennessee.

Coleman MD, Dickson RE, Isebrands JG (2000) Contrasting fine-root production, survival and soil CO_2 efflux in pine and poplar plantations. Plant Soil 225:129–139.

Craine JM, Wedin DA, Chapin FS III (1999) Predominance of ecophysiological controls on soil CO_2 flux in a Minnesota grassland. Plant Soil 207:77–86.

Cromak K (1973) *Litter production and decomposition in a mixed hardwood watershed and a white pine watershed at Coweeta Hydrologic Station, North Carolina.* Ph.D. dissertation. University of Georgia, Athens, Georgia.

Curtis PS, Hanson PJ, Bolstad P, Barford C, Randolph JC, Schmid HP, Wilson KB (2002) Biometric and eddy-covariance-based estimates of annual carbon storage in five eastern North American deciduous forests. Agric For Meteorol 113:3–19.

Davidson EA, Belk E, Boone RD (1998) Soil water content and temperature as independent or confounded factors controlling soil respiration in a temperate mixed hardwood forest. Global Change Biol 4:217–227.

Dörr H, Münnich KO (1987) Annual variation in soil respiration in selected areas of the temperate zone. Tellus 39B:114–121.

Edwards CA, Reichle DE, Crossley DA Jr. (1970) The role of soil invertebrates in turnover of organic matter and nutrients. In Reichle DE (Ed) *Analysis of temperate forest ecosystems.* Springer-Verlag, New York, pp 147–172.

Edwards NT (1975) Effects of temperature and moisture on carbon dioxide evolution in a mixed deciduous forest floor. Soil Sci Soc Amer J 39:361–365

Edwards NT, Harris WF (1977) Carbon cycling in a mixed deciduous forest floor. Ecology 58:431–437.

Edwards NT, Ross-Todd BM (1983) Soil carbon dynamics in a mixed deciduous forest following clear-cutting with and without residue removal. Soil Sci Soc Amer J 47:1014–1021.

Edwards NT, Sollins P (1973) Continuous measurement of carbon dioxide evolution from partioned forest floor components. Ecology 54:406–412.

Edwards NT, Johnson DW, McLaughlin SB, Harris WF (1989) Carbon dynamics and productivity. In Johnson DW, Van Hook RI (Eds), *Analysis of biogeochemical cycling processes in Walker Branch Watershed.* Springer-Verlag, New York, pp 197–232.

Epron D, Farque L, Lucot E, Badot P-M (1999a) Soil CO_2 efflux in a beech forest: the contribution of root respiration. Ann For Sci 56:289–295.

Epron D, Farque L, Lucot E, Badot P-M (1999b) Soil CO_2 efflux in a beech forest: dependence on soil temperature and soil water content. Ann For Sci 56:221–226.

Fang C, Moncrieff JB (2001) The dependence of soil CO_2 efflux on temperature. Soil Biol Biochem 33:155–165.

Fassnacht KS, Gower ST (1999) Comparison of the litterfall and forest floor organic matter and nitrogen dynamics of upland forest ecosystems in north central Wisconsin. Biogeochemistry 45:265–284.

Flanagan PW, VanCleve K (1983) Nutrient cycling in relation to decomposition and organic matter quality in taiga ecosystems. Can J For Res 13:795–817.

Garrett HE, Cox GS (1973) Carbon dioxide evolution from the floor of an oak-hickory forest. Soil Sci Soc Amer J 37:641–644.

Garten CT Jr., Post WM III, Hanson PJ, Cooper LW (1999) Forest soil carbon inventories and dynamics along an elevation gradient in the southern Appalachian Mountains. Biogeochemistry 45:115–145.

Gholz HL, Wedin DA, Smitherman SM, Harmon ME, Parton WJ (2000) Long-term dynamics of pine and hardwood litter in contrasting environments: Toward a global model of decomposition. Global Change Biol 6:751–765.

Giardina CP, Ryan MP (2000) Evidence that decomposition rates of organic carbon in mineral soil do not vary with temperature. Nature 404:858–861.

Gillespie TJ, Kidd GE (1978) Sensing duration of leaf moisture retention using electrical impedance grids. Can J Plant Sci 58:179–187.

Green RN, Trowbridge RL, Klinka K (1993) *Towards a taxonomic classification of humus forms.* Forest Science Monograph 29. Society of American Foresters, Bethesda, Maryland.

Hanson PJ, Wullschleger SD, Bohlman SA, Todd DE (1993) Seasonal and topographic patterns of forest floor CO_2 efflux from an upland oak forest. Tree Physiol 13:1–15.

Hanson PJ., Chambers MLS, Riggs JS, Misek TL, Stringfellow JB (1996) Litter CO_2 Efflux: Daily Dynamics from Field Measurements and Model Extrapolations. *Agronomy abstracts.* Agronomy Society of America, Madison, Wisconsin, p 300.

Hanson PJ, Todd DE, Huston MA, Joslin JD, Croker J, Augé RM (1998) *Description and field performance of the Walker Branch Throughfall Displacement Experiment: 1993–1996.* ORNL/TM-13586. Oak Ridge National Laboratory, Oak Ridge, Tennessee.

Hanson PJ, Edwards NT, Garten CT Jr., Andrews JA (2000) Separating root and soil microbial contributions to soil respiration: A review of methods and observations. Biogeochemistry 48:115–146.

Hanson PJ, Todd DE, Amthor JS (2001) A six year study of sapling and large-tree growth and mortality responses to natural and induced variability in precipitation and throughfall. Tree Physiol 21: 345–358.

Högberg P, Nordgren A, Buchmann N, Taylor AFS, Ekblad A, Högberg MN, Nyberg G, Ottosson-Löfvenius M, Read DJ (2001) Large-scale forest girdling shows that current photosynthesis drives soil respiration. Nature 411:789–792.

Horwath WR, Pregitzer KS, Paul EA (1994) ^{14}C allocation in tree-soil systems. Tree Physiol 14:1163–1176.

Janssens IA, Meiresonne L, Cuelemans R (2000) Mean soil CO_2 efflux from a mixed forest: temporal and spatial integration. In Ceulemans R, Veroustraete F, Gond V, Van Rensbergen JBHF (Eds) *Forest ecosystem modelling, upscaling, and remote sensing.* Academic Publishing, The Hague, The Netherlands, pp 19–31.

Jansssens IA, Lankreijer H, Matteucci G, Kowalski AS, Buchmann N, Epron D, Pilegaard K, Kutsch W, Longdoz B, Grunwald T, Montagnani L, Dore S, Rebmann C, Moors EJ, Grelle A, Rannik U, Morgenstern K, Oltchev S, Clement R, Gudmundsson J, Minerbi S, Berbigier P, Ibrom A, Moncrieff J, Aubinet M, Bernhofer C, Jensen NO, Vesala T, Granier A, Schulze ED, Lindroth A, Dolman AJ, Jarvis PG, Ceulemans R, Valentini R (2001) Productivity overshadows temperature in determining soil and ecosystem respiration across European forests. Global Change Biol 7:269–278.

Jenkinson DS, Adams DE, Wild A (1991) Model estimates of CO_2 emissions from soil in response to global warming. Nature 351:304–306.

Johnson DW, Hanson PJ, Todd DE (2002) The effects of throughfall manipulation on soil solution chemistry and leaching in a deciduous forest. J Environ Qual 31:204–216.

Joslin JD, Wolfe MH (1998) Impacts of long-term water input manipulations on fine root production and mortality in mature hardwood forests. Plant Soil 204:165–174.

Joslin JD, Wolfe MH, Hanson PJ (2000) Effects of shifting water regimes on forest root systems. New Phytol 147: 117–129.

Joslin JD, Wolfe MH, Hanson PJ (2001) Factors controlling the timing of root elongation intensity in a mature upland oak stand. Plant Soil 228:201–212.

Kelliher FM, Lloyd J, Arneth A, Lühker B, Byers JN, McSeveny TM, Milukova I, Grigoriev S, Panfyorov M, Soatchev A, Varlargin A, Ziegler W, Bauer G, Wong S-C, Schultze E-D (1999) Carbon dioxide efflux density from the floor of a central Siberian pine forest. Agric For Meteorol 94:217–232.

Kelly JM, Beauchamp JJ (1987) Mass loss and nutrient changes in decomposing upland oak and mesic-hardwood leaf litter. Soil Sci Soc Amer J 51:1616–1622.

Keith H, Jacobsen KL, Raison RJ (1997) Effects of soil phosphorus availability, temperature and moisture on soil respiration in *Eucalyptus pauciflora* forest. Plant Soil 190:127–141.

Kirita H (1971) Studies of soil respiration in warm-temperate evergreen broadleaf forests of southwestern Japan. Jpn J Ecol 21:230–244.

Kirschbaum MUF (1995) The temperature dependence of soil organic matter decomposition, and the effect of global warming on soil organic C storage. Soil Biol Biochem 27:753–760.

Kirschbaum MUF (2000) Will changes in soil organic carbon act as a positive or negative feedback on global warming? Biogeochemistry 48:21–51.

Kuperman RG (1999) Litter decomposition and nutrient dynamics in oak-hickory forests along a historic gradient of nitrogen and sulfur deposition. Soil Biol Biochem 31:237–244.

Larkin RP, Kelly JM (1987) Influence of elevated ecosystem S levels on litter decomposition and mineralization. Water Air Soil Pollut 34:415–428.

Law BE, Baldocchi DD, Anthoni PM (1999) Below-canopy and soil CO_2 fluxes in ponderosa pine forest. Agric For Meteorol 94:171–188.

Longdoz B, Yernaux M, Aubinet M (2000) Soil CO_2 efflux measurements in a mixed forest: Impact of chamber disturbances, spatial variability and seasonal evolution. Global Change Biol 6:907–917.

McClaugherty CA, Pastor J, Aber JD (1985) Forest litter decomposition in relation to soil nitrogen dynamics and litter quality. Ecology 66:266–275.

Meentemeyer V (1978) Macroclimatic and lignin control of litter decomposition rates. Ecology 59:465–472.

Moore TR, Trofymow JA, Taylor B, Prescott C, Camire C, Duschene L, Fyles J, Kozak L, Kranabetter M, Morrison I, Siltanen M, Smith S, Titus B, Visser S, Wein R, Zoltai S (1999) Litter decomposition rates in Canadian forests. Global Change Biol 5:75–82.

Nakane K (1975) Dynamics of soil organic matter in different parts on a slope under evergreen oak forest. Jpn J Ecol 25:206–216.

Nakane K (1980) Comparative studies of cycling of soil organic carbon in three primeval moist forests. Jpn J Ecol 30:155–172.

Nay SM, Bormann BT (2000) Soil carbon changes: Comparing flux monitoring and mass balance in a box lysimeter experiment. Soil Sci Soc Amer J 64:943–948.

Norman JM, Kucharik CJ, Gower ST, Baldocchi DD, Crill PM, Rayment M, Savage K, Striegl RG (1997) A comparison of six methods for measuring soil-surface carbon dioxide fluxes. J Geophys Res 102:28771–28777.

Olson JS (1963) Energy storage and the balance of producers and decomposers in ecological systems. Ecology 44: 322–331.

O'Neill EG, Norby RJ (1996) Litter quality and decomposition rates of foliar litter produced under CO_2 enrichment. In Koch GW, Mooney HA (Eds) *Carbon dioxide and terrestrial ecosystems.* Academic Press, San Diego, pp 87–103.

O'Neill EG, Johnson DW, Ledford J, Todd DE (2003) Acute seasonal drought does not permanently alter mass loss and nitrogen dynamics during decomposition of red maple (*Acer rubrum* L.) litter. Global Change Biol 9: 117–123.

Pastor J, Post WM (1985) *Development of a linked forest productivity-soil process model.* ORNL/TM-9519. Oak Ridge National Laboratory, Oak Ridge, Tennessee.

Penning de Vries FWT, Brunsting AHM, van Laar HH (1974) Products, requirements and efficiency of biosynthesis: A quantitative approach. J Theor Biol 45:339–377.

Raich JW, Schlesinger WH (1992) The global carbon dioxide flux in soil respiration and its relationship to vegetation and climate. Tellus 44B:81–99.

Reiners WA (1968) Carbon dioxide evolution from the floor of three Minnesota forests. Ecology 49:471–483.

Soil Science Society of America [SSSA] (1997) *Glossary of soil science terms 1996.* Soil Science Society of America, Madison, Wisconsin.

Sommers LE, Gilmour CM, Wildung RE, Beck SM (1981) The effect of water potential on decomposition processes in soils. In *Water potential relations in soil microbiology.* Soil Science Society of America, Madison, Wisconsin, pp 97–117.

Tesarova M, Gloser J (1976) Total CO_2 output from alluvial soils with two types of grassland communities. Pedobiologia 16:364–372.

Toland DE, Zak DR (1994) Seasonal patterns of soil respiration in intact and clear-cut northern hardwood forests. Can J For Res 24:1711–1716.

Topp GC, Davis JL (1985) Measurement of soil water content using time-domain reflectometry (TDR): A field evaluation. Soil Sci Soc Am J 49:19–24.

Valentini R, Matteucci G, Dolman AJ, Schulze ED, Rebmann C, Moors EJ, Granier A, Gross P, Jensen NO, Pilegaard K, Lindroth A, Grelle A, Bernhofer C, Grunwald T, Aubinet M, Ceulemans R, Kowalski AS, Vesala T, Rannik U, Berbigier P, Loustau D, Guomundsson J, Thorgeirsson H, Ibrom A, Morgenstern K, Clement R, Moncrieff J, Montagnani L, Minerbi S, Jarvis PG (2000) Respiration as the main determinant of carbon balance in European forests. Nature 404:861–865.

Van Cleve K, Sprague D (1971) Respiration rates in the forest floor of birch and aspen stands in interior Alaska. Arctic Alpine Res 3:17–26.

Vertreg N, Penning de Vries FWT (1987) A rapid method for determining the efficiency of biosynthesis of plant biomass. J Theor Biol 128:109–119.

Virzo De Santo A, Alfani A, Sapio S (1976) Soil metabolism in beech forests of Monte Taburno (Campania Apennines). Oikos 27:144–152.

Weber MG (1990) Forest soil respiration after cutting and burning in immature aspen ecosystems. For Ecol Manage 31:1–14.

Wessman CA, Aber JD, Peterson DL, Melillo JM (1988) Foliar analysis using near infrared reflectance spectroscopy. Can J For Res 18:6–11.

Witkamp M (1966) Decomposition of leaf litter in relation to environment, microflora, and microbial respiration. Ecology 47:194–201.

Wilson KB, Hanson PJ, Baldocchi DD (2000) Factors controlling evaporation and energy partitioning beneath a deciduous forest over an annual cycle. Agric For Meteorol 102:83–103.

Wilson KB, Baldocchi DD, Hanson PJ (2001) Leaf age affects the seasonal pattern of photosynthetic capacity and net ecosystem exchange of carbon in a deciduous forest. Plant Cell Environ 24:571–583.

Woodwell GM, Botkin DB (1970) Metabolism of terrestrial ecosystems by gas exchange techniques: The Brookhaven approach. In Reichle DE (Ed) *Analysis of temperate forest ecosystems.* Springer-Verlag, New York, pp 70–85.

11. Soil Carbon Turnover

Julia B. Gaudinski and Susan E. Trumbore

The net carbon balance of forests is the small difference between very large fluxes of carbon uptake (via photosynthesis) and carbon loss (via respiration). If carbon uptake by forest vegetation exceeds C loss by vegetation and soils, the forest is a net sink of atmospheric carbon. Currently, regrowing temperate forest ecosystems of the eastern United States are storing C (Goulden et al. 1996; Greco and Baldocchi 1996; Hollinger et al. 1999; Curtis et al. 2002).

Estimates of net C balance can be made by eddy-covariance and biometric methods (direct measurements of carbon fluxes and/or stocks). Eddy covariance directly measures the net fluxes of C between atmosphere and forest at roughly half-hourly intervals. Those measurements are integrated to estimate an annual net C balance. The equipment can be automated and left in place to measure variations in annual C balance from year to year. However, eddy covariance gives little information on where the C is being stored in the ecosystem—in particular, whether storage occurs in vegetation or soils. Biometric methods calculate changes in C stocks from repeated surveys of vegetation and soils and, thus, can be used to determine where C is being stored. However, large spatial heterogeneity and errors associated with spatial scaling of field measurements prevent this method from detecting small changes in C stocks from year to year. Large discrepancies between eddy flux and biometric approaches have been found at two of five AmeriFlux sites in eastern deciduous forests of the United States (Curtis et al. 2002) and point to the need to resolve potential uncertainties in both approaches. One of the largest discrepancies is found at the Oak Ridge Reservation, the site of the Throughfall Displacement Experiment (TDE) (Chapter 22, this volume). Understanding where and how C is stored is important not only for resolving the present carbon balance of ecosystems, but also for predicting future ecosystem response to climate change.

A major uncertainty in biometric approaches for estimating ecosystem carbon budgets is quantifying current and future rates of soil C accumulation (or loss). Quantifying soil C stocks to within 20% is challenging, and in rocky soils, it can require hundreds of careful, labor-intensive measurements (Huntington et al. 1988; Fernandez et al. 1993). Quantifying C cycling rates through these stocks is even more difficult because soil C is made up of a variety of compounds that cycle on timescales ranging from days to millennia. We report C stocks and dynamics in soil organic matter at the Walker Branch Watershed in the Oak Ridge Reservation, Oak Ridge National Laboratory, Tennessee. Our approach to determining carbon dynamics in soils is to (1) separate soil organic matter (SOM) by depth, density, and size to isolate components that have different characteristic turnover times and (2) use radiocarbon to estimate the turnover time of each isolated fraction of soil organic matter. The ^{14}C approach takes advantage of radiocarbon produced by atmospheric thermonuclear weapons testing in the early 1960s (bomb ^{14}C; Fig. 11.1), which can be used as a tracer for the timing of photosynthetic uptake and C cycling rates in an ecosystem during the past several decades. Once we have estimates of C stocks and turnover times for isolated SOM fractions, we use a C–stock model to simulate changes in C inputs to the soil to assess the potential for future carbon storage.

Use of the bomb-^{14}C approach requires knowledge of the time record of ^{14}C in atmospheric CO_2 and in C inputs to soil organic matter since 1950. At Walker Branch, the local time history for ^{14}C as recorded in tree-ring cellulose agrees with the records for CO_2 in Northern Hemisphere air prior to 1990 (Fig. 11.1). After 1990, the Walker Branch record shows no decline in ^{14}C, although the record for Northern Hemisphere CO_2 shows a 4–8‰ per year decrease. This is largely because of local releases of ^{14}C that occurred (from an unkown source) on or around the study site after 1995 (Trumbore et al. 2002). Thus, the isotopic signatures for SOM fixed between 1990 and 1998 are not distinguishable from one another at this site, and we can only quantify SOM dynamics for C pools that cycle at rates greater than $\sim$10 years. However, this limitation does not affect the goals of this work, namely to quantify current and potential rates of soil C storage at Walker Branch, discuss these results in the context of eddy-covariance and biometric approaches also performed on Walker Branch, and put our results in the larger context of soil C accumulation potential for forests of the eastern United States.

Sites

The work discussed here was carried out at two sites on the Walker Branch Watershed. The first site is located on the upper slope of the ambient plot of the

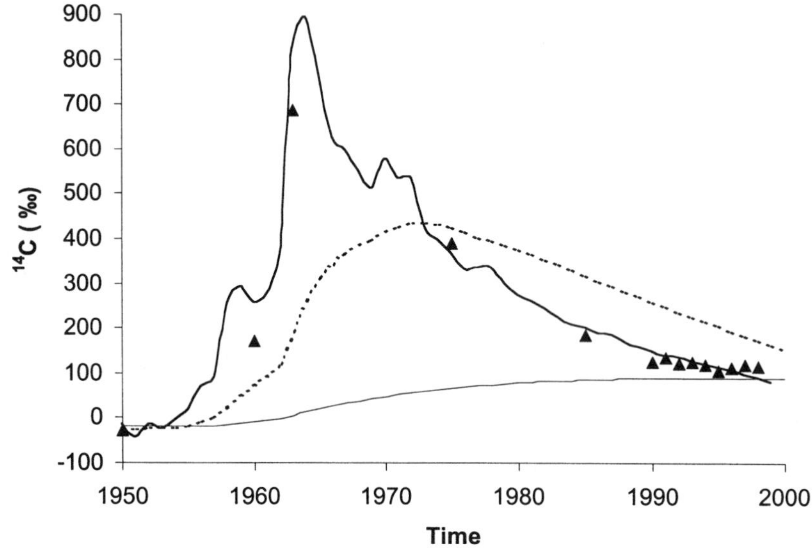

Figure 11.1. The global and local record of atmospheric ^{14}C of CO_2 since 1950. The thick solid line represents the time record of ^{14}C in the atmosphere (Northern Hemisphere) based on grapes grown in Russia (Burchuladze et al. 1989) for 1950 to 1976, direct atmospheric measurements for 1977 to 1996 (Levin and Kromer, 1997), and an assumed 4‰ per year decrease since 1996 (Levin and Hessheimer 2000). Filled triangles represent the local Walker Branch time record reconstructed with values for tree-ring cellulose from a *Quercus alba* tree taken on the Walker Branch Site (Trumbore et al. 2002). ^{14}C in cellulose from annual tree rings has been shown to be a good indicator of atmospheric ^{14}C of CO_2 (Hua et al., 1999). The accuracy of the ^{14}C measurement by accelerator mass spectrometry (AMS) is $\pm$ 6‰; thus, the error bars are smaller than the symbols shown. Note that none of the tree-ring cellulose values from 1990 to 1998 are significantly different from each other. Thin dashed and solid lines represent the amount of radiocarbon in a carbon pool with turnover times of 10 and 50 years, respectively (as modeled using our steady-state homogenous-carbon-pool model; see text for equations).

Throughfall Displacement Experiment (TDE) (Chapter 2, this volume). The second site (P5) is located about 1 km east in a very similar forest stand.

Methods

Soil Organic Matter Sampling

Samples for carbon and ^{14}C inventories were taken in 1998 from two soil pits (0.5 m $\times$ 0.5 m) dug with the quantitative pit methodology of Huntington et al. (1989) and Hamburg (1984) and modified as described in Gaudinski et al. (2000). This method involves sampling a large volume of soil to allow calculation of horizon-specific bulk densities. Samples which integrated over each soil horizon were collected from one of the pit faces for radiocarbon analysis and total C and N analyses. Carbon inventories are reported to a depth of about 80 cm.

Laboratory Analyses

Soil samples were separated into different fractions by a combination of hand sorting and density separations. Density separations are performed with a sodium polytungstate solution at $\rho = 2.1$ g cm^{-3}. Material that floats is defined as low density, whereas material that sinks is defined as high density and assumed to be associated with minerals. Organic horizons were considered to be 100% low-density material composed of litter that can be identified as coming from leaves (L_L) and roots (L_R) and humified (H) material that is comprised of pieces too small to be identified as plant tissues. Oi horizons were composed entirely of large pieces of leaf litter (L_L). Oe + Oa horizons were hand picked to separate L_L, L_R, and H fractions. Mineral horizons were comprised of low-density fractions (L_R and H) and high-density mineral-associated carbon (M). Size separations (sieving with an 80-μm sieve) and hand picking were performed on the low-density fractions from the A horizons to separate L_R and H fractions. These methods are described in further detail in Gaudinski et al. (2000).

The rate of incorporation of bomb-^{14}C in SOM fractions provides a measure of how fast C in a given fraction is exchanged with the atmosphere (on decadal timescales). This approach is particularly powerful when ^{14}C analysis can be performed on the same soil sampled at two different points in time (Trumbore 1993). The Plot 5 site was added to our study because we were able to obtain archived soils that were sampled in 1972. The archived P5 samples were dried, ground, and stored in a laboratory on Walker Branch. A pit description for the 1972 soil was not available; however, we discussed the 1972 sampling protocol with the original soil sampler (Todd, personal communication) and compared the archived and modern soil horizon colors to ensure that horizons sampled in 1998 were as similar as possible to those sampled in 1972. The 1972 pit remained open, and in 1998, we extended one of its walls for resampling. Density separations were performed on archived samples and the low-density fraction of the A horizon was separated into greater than and less than 80 μm fractions. Because the archived soil had been previously ground, hand picking to remove roots was not performed.

Radiocarbon Analyses

We converted all solid and gas samples to graphite by sealed-tube zinc reduction (Vogel 1992) and measured the ^{14}C content on an accelerator mass spectrometer (AMS) at the Center for Accelerator Mass Spectrometry at Lawrence Livermore National Laboratory. We express radiocarbon data as Δ^{14}C, the difference in parts per thousand (per mil or ‰) between the ^{14}C/^{12}C ratio in the sample compared to that of a universal standard (0.95 times the activity of oxalic acid I, decay corrected to 1950), which is defined as "Modern." All samples are corrected to a common δ^{13}C value of -25‰ to correct for the effects of mass-dependent isotopic fractionation on measured ^{14}C values. This correction accounts for the discrimination of atmospheric ^{14}C during fixation of CO_2 by plant photosynthesis (^{14}C is assumed to fractionate twice as much as ^{13}C). The reported ^{14}C values, therefore, reflect the atmospheric ^{14}C of CO_2 from which the C was originally fixed.

Interpretation of Radiocarbon Data

Radiocarbon is produced naturally in the upper atmosphere when high-energy cosmic rays collide with atomic nuclei to produce neutrons. The neutrons then collide with nitrogen to form ^{14}C via the following reaction:

$$^{14}_{7}N + ^{1}_{0}n \rightarrow ^{14}_{6}C + ^{1}_{1}H$$

^{14}C is a radioactive isotope that decays back to ^{14}N via beta decay (electron emission with a maximum energy of 0.155 Mev) with a half-life of 5730 years. Once produced, the ^{14}C quickly oxidizes to form $^{14}CO_2$ and follows the same physical and chemical pathways as $^{13}CO_2$ and $^{12}CO_2$. The estimated natural production rate of ^{14}C over the Earth's surface is 2.50 $\pm$ 0.50 atoms of ^{14}C cm^{-2} s^{-1}, and the steady-state natural $^{14}C/^{12}C$ ratio is $(1.2 \pm 0.2) \times 10^{-12}$ (Linick 1975). This $^{14}C/^{12}C$ ratio is referred to as "Modern" and, by definition, refers to the atmosphere of 1950.

Radiocarbon is also produced by aboveground thermonuclear explosions (bomb ^{14}C) when neutrons are released into the atmosphere. Atmospheric weapon testing in the late 1950s and early 1960s increased atmospheric ^{14}C by 10^{29} atoms (160 kg ^{14}C) or twice modern values (Linick 1975; Hesshaimer et al. 1994).

Globally, the $^{14}CO_2$ concentration has been decreasing since its peak values in the early to mid-1960s (Fig. 11.1). The decrease was rapid at first because of the dilution of the ^{14}C spike through exchange with carbon in the oceans and terrestrial biosphere. The rate of decline has slowed with time (Fig. 11.1), with the annual change of $\Delta^{14}C$ averaging ~ -8 ‰ y^{-1} in the 1980s and early 1990s. Between 1996 and 1999, the period of this study, rates of decrease slowed to 4–8‰ y^{-1} (Levin and Hesshaimer 2000). The accuracy of the ^{14}C measurement AMS with graphite and sealed-tube zinc reduction is $\pm$ 6‰ based on repeated analysis of intercalibration standards. In general, this level of accuracy allows us to resolve the time elapsed since organic matter was fixed from the atmosphere to within 1–2 years over the past five decades.

Turnover times for SOM fractions were estimated from their respective ^{14}C signatures and a steady-state, time-dependent model that tracks C and ^{14}C in each fractionated SOM pool. We assume that the C pools remain at steady state (inputs = losses) and that the C inputs added to each pool for each year have ^{14}C contents determined from the time record of atmospheric ^{14}C of CO_2 [see Trumbore et al. (1995) and Gaudinski et al. (2000) for further details]. The balance of ^{14}C in year t is given by

$$C_{(t)}\,R_{\mathrm{som}(t)} = IR_{\mathrm{atm}(t)} + C_{(t-1)} \times R_{\mathrm{som}(t-1)} - kC_{(t-1)}\,R_{\mathrm{som}(t-1)} - \lambda C_{(t-1)}\,R_{\mathrm{som}(t-1)}; \quad (1)$$

collecting terms, we have

$$R_{\text{som}(t)} = \frac{IR_{\text{atm}(t)} + (C_{(t-1)}\, R_{\text{som}(t-1)}\, (1 - k - \lambda))}{C_{(t)}}, \tag{2}$$

where

C = stock of carbon for the given C pool (in g C m^{-2})
I = inputs of C above and below ground (in g C m^{-2} y^{-1})
k = decomposition rate of SOM (in y^{-1})
$R = \left(\frac{\Delta^{14}C}{1000}\right) - 1$
R_{atm} = the ratio of ^{14}C in the atmosphere normalized to a standard
R_{som} = the ratio of ^{14}C in the given SOM pool: L, H, or M, normalized to a standard
λ = radioactive decay constant for ^{14}C = 1/8267 years
t = time (year) for which the calculation is being performed.

Note that the R_{som} at any time t depends not only on the $R_{atm(t)}$ but also on both C inventory and R_{atm} of previous years.

We varied turnover times (i.e., k^{-1}) until we matched the measured $\Delta^{14}C$ signature for the given SOM pool in the year of sampling (1998). This approach assumes that (1) the C pool being modeled is homogeneous and (2) there is no time lag between fixation of C from the atmosphere and the time when C is added to the SOM pool [i.e., C inputs each year have $^{14}C/^{12}C$ ratios equal to $R_{atm}(t)$]. Once a turnover time was estimated from ^{14}C data, the flux of C into and out of the pool were calculated as the stock of C in that pool divided by the turnover time.

Results

Soil C and ^{14}C Inventory

Soil organic matter inventories (not including roots) were 4610 g C m^{-2} for the soil pit dug on the ambient portion of the TDE (to a depth of 70 cm) and 5695 g C m^{-2} for the P5 site (to a depth of 80 cm; Table 11.1). Carbon concentrations decreased rapidly with depth with 46% C in the O horizons and 0.2 % C in the B horizons. Most of the SOM C stocks, however, resided in mineral soils, with an average of 615 g C m^{-2} in the O horizons (12% of the total) and 4540 g C m^{-2} in the mineral A, E and B horizons combined (88% of the total). Low-density (L_L + H) carbon made up 64% of the total soil carbon stocks. Humified organic material (H) makes up 90% of the total low-density C. Fine roots (< 2 mm in diameter) to a depth of 90 cm are assumed to contain 375 g C m^{-2} (live + dead roots with 47% carbon; Joslin and Wolfe 1999).

Radiocarbon contents of low-density fractions were consistently higher than those of dense fractions, though ^{14}C in both components decreased with soil depth (Fig. 11.2). Comparison of fractions separated from the bulk low-density material shows that it is made up of material that is higher in ^{14}C (fine roots) and lower in ^{14}C (humified fraction; H) relative to the bulk ^{14}C value (Fig. 11.2 and Table 11.2). Where low-density material was separated into components by sieving, the

Table 11.1. Soil organic matter inventory (not including roots)

					Low-density SOM			High-density SOM
Pit	Horizon	Bulk density[a] ($g\ cm^{-3}$)	Soil carbon ($g\ C\ kg^{-1}$ soil)	Bottom depth (cm)	Total C stock[b] ($g\ C\ m^{-2}$)	Leaf litter[c,d] ($g\ C\ m^{-2}$)	Humified[d] ($g\ C\ m^{-2}$)	Mineral associated[d] ($g\ C\ m^{-2}$)
TDE	Oi	—[e]	456	2	360	360	NA[f]	NA
	Oe + Oa	—	378	3	270	NA	270	NA
	A	0.44	50	8	615	NA	565	50
	E	1.01	7	50	2475	NA	885	1590
	Bt	1.18	3	70	890	NA	240	650
Total					4610			
P5	Oi	—	456	3	190	190	NA	NA
	Oe + Oa	—	378	5	410	NA	410	NA
	A	0.52	203	7	3095	NA	3025	70
	E	1.01	7	24	1245	NA	450	795
	B1/EB	1.13	2	41	430	NA	90	340
	Bt	0.92	2	63	325	NA	65	260
Total					5695			

[a] Gravel-free bulk density (i.e., less than 2 mm).
[b] Oi and Oe horizons calculated on a per area basis.
[c] On a dry weight basis.
[d] Low density means $\rho < 2.1\ g\ cm^{-3}$; high density means $\rho > 2.1\ g\ cm^{-3}$.
[e] Not available.
[f] NA = not applicable.

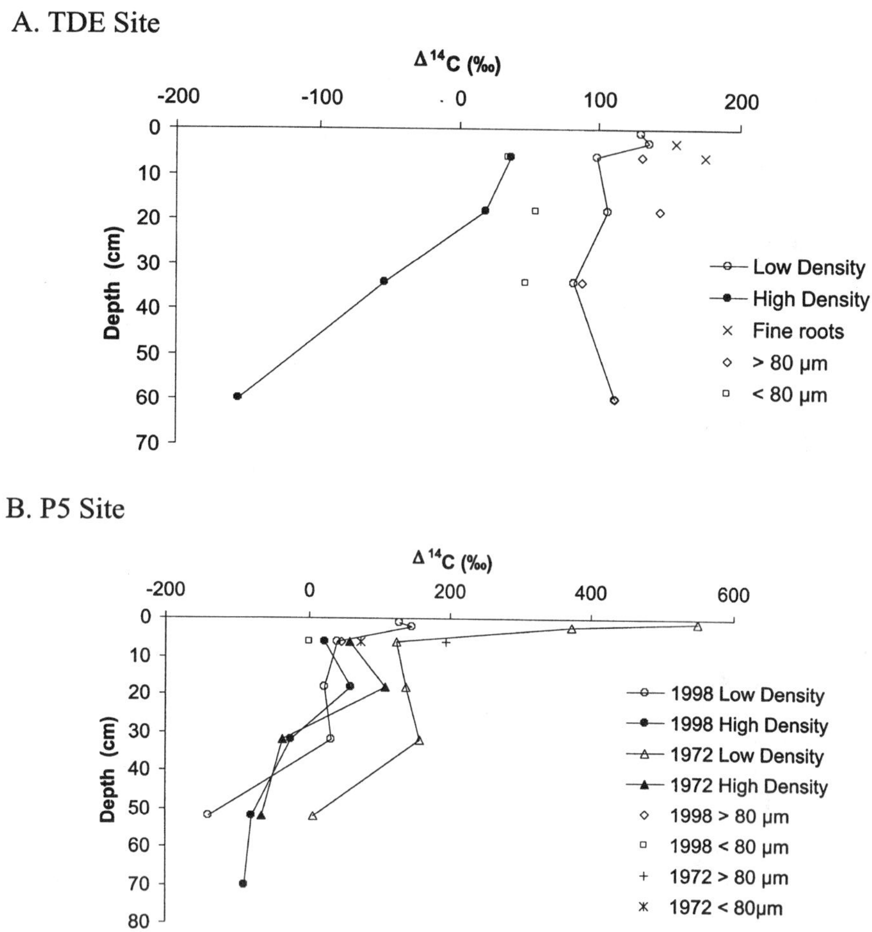

Figure 11.2. Radiocarbon profiles of low-density (open symbols) and high-density (closed symbols) soil organic matter and fine roots. The uppermost value in the profile represents the Oi horizon, which is bulk leaf litter that is still recognizable to the species level. The value directly below represents the humified component of the Oe + Oa horizon after roots and recognizable leaf parts have been removed. The remaining profile values are in the mineral soil and represent bulk low-and high-density soil organic matter samples or low-density samples sieved with an 80-μm sieve. When the bulk and sieved samples are shown together at the same depth, the bulk represents a mass-weighted composite of the greater than and less than 80 μm samples. At the P5 site, values are shown for archive (1972) and modern (1998) soils.

Table 11.2. ^{14}C values and associated turnover times (TTs) and corresponding fluxes

Horizon	Stock ($g\ C\ m^{-2}$)	$\Delta^{14}C$ of SOM (‰)	TT short (year)	TT long (year)	CO_2 flux ($g\ C\ m^{-2}\ y^{-1}$)	CO_2 flux horizon total ($g\ C\ m^{-2}\ y^{-1}$)	$\Delta^{14}C$ of SOM horizon total (‰)
TDE							
Oi (L_L)	358	130	**5**	68	NA	NA	NA
Oe + Oa (L_R)	—[a]	154	NA[b]		NA		
Oe + Oa (H)	269	136[c]	**6**	60	45		
A (LR)	—	175	NA		NA	6	97
A (H)	563	99[d]	91		6.2		
A (M)	46	37	189		0.2		
E (H)	883	107	83		10.6	14	84
E (M)	1582	19	435		3.6		
Bt (H)	240	113	80		3.0	3	81
Bt (M)	647	−156	1580		0.4		
P5							
Oi (L_L)	191	127	**5**	67	NA	NA	NA
Oe + Oa (L_R)	—	—	NA		NA		
Oe + Oa (H)	406	144[c]	**7–13**	55	58		
A (LR)	210		NA		NA	16	39
A (H)	3023	40	**195**		15.5		
A (M)	70	22	230		0.3		
E (H)	452	21	235		1.9	8	49
E (M)	796	59	142		5.6		
B1 (H)	91	31	210		0.4	1	−1
B1 (M)	342	−25	625		0.5		
Bt1 (H)	63	−139	1400		0.0	0.4	−87
Bt1 (M)	256	−80	830		0.3		

Note: Fluxes are derived by dividing inventory by the steady-state ^{14}C-derived TT. Where there is more than one possible TT shown, the bold values indicate the one used to calculate the flux. See text for further details.

[a] Not available.

[b] NA = Not applicable, see text for details.

[c] Represents the bulk $\Delta^{14}C$ value; however, picking of different components resulted in values from 130‰ to 136‰.

[d] Represents the $\Delta^{14}C$ value of a composite sample after fine roots have been picked out.

finer size fraction (< 80 μm) was lower in ^{14}C than the coarser size fraction (> 80 μm; Fig. 11.2).

Comparison of low-density organic matter sampled at the P5 site in 1972 and 1998 showed decreases in radiocarbon between the two sampling dates (Fig. 11.2B) ranging from 423‰ to 229‰ in the Oi and Oe + Oa horizons, respectively, to 296–84‰ in the mineral horizons. In contrast, the $\Delta^{14}C$ signature of high-density mineral-associated SOM samples are similar to samples from 1972 having $\Delta^{14}C$ values only 23–48‰ greater than those in 1998. The less than 80 μm fraction of the low-density material in the A horizon contains less radiocarbon than the greater than 80 μ fraction for both 1972 and 1998.

Radiocarbon values for the L_L, L_R, H, and M fractions and their associated ^{14}C-derived turnover times, and CO_2 fluxes (calculated by inventory divided by turnover time) are shown for each horizon below the Oi horizon in Table 11.2. Turnover times increase with depth from <10 to 1400 years for low-density material and from 200 to 1500 years in the high-density material. Calculated CO_2 fluxes associated with the decomposition of SOM decrease rapidly with depth from 45 to 60 g C m^{-2} y^{-1} in the Oe + Oa horizons to < 3 g C m^{-2} y^{-1} in the lower B horizons. Decomposition fluxes are not calculated for the Oi horizons because lagged radiocarbon inputs (from either stored photosynthate or components that reside on the plant for more than 1 year) make flux estimations too uncertain for material that cycles on less than 5-year timescales. As discussed earlier, the unusual ^{14}C history of this site further complicates the use of the atmospheric record of $^{14}CO_2$ for distinguishing among turnover times faster than 10 years at the Walker Branch site.

Discussion

Radiocarbon-Derived SOM Dynamics

Knowledge of C stock sizes and their average turnover times allows for evaluation of the potential for SOM to accumulate or lose C during the coming centuries under different scenarios of climate change or ecosystem productivity. The residence time of C in SOM pools is reflected in how much bomb ^{14}C they have incorporated since 1950 and in whether or not ^{14}C values have increased or decreased between 1972 and 1990 (Fig. 11.1). At Walker Branch, the greatest amount of bomb ^{14}C is in low-density fractions in the uppermost portion of the soil profile. The amount of bomb ^{14}C decreases with depth (Fig. 11.2). Most C with less than 10-year turnover times is in the uppermost soil horizons (to about a 5-cm depth) and represents a very small percentage (9%) of the total soil C stocks (excluding roots; Table 11.2). Carbon with turnover times of 10–100 years is found mostly between 5 and 40 cm in depth in both low- and high-density fractions and makes up 77% of the total SOM. Carbon cycling on timescales longer than 100 years predominates below a depth of 40 cm and constitutes ~14% of SOM stocks.

The model we use to estimate turnover times assumes that each fraction of SOM is homogeneous (with respect to turnover). Clearly, the low-density fraction of organic matter can be separated into different components with different cycling rates and, thus, $\Delta^{14}C$ values. This is especially true in surface horizons (Fig. 11.2A). For example, in the A horizon at the TDE site, roots make up 16% of the total sample mass and have ^{14}C signatures 130‰ higher than the > 80 μm fraction from the same horizon. In this case, removing roots increases the composite turnover time for the nonroot fraction from 65 to 90 years. The organic Oi and Oe + Oa horizons at Walker Branch, unlike the mineral horizons, have 1998 $\Delta^{14}C$ values above the $\Delta^{14}C$ of CO_2 for the current atmosphere. In such cases, two different k values predict the same $\Delta^{14}C$ values for the SOM fraction (Fig. 11.3). The $\Delta^{14}C$ values of organic matter for archived soils sampled in 1972 at the P5 site allow us to distinguish which of the two possible k values is the more reasonable. $\Delta^{14}C$ values in O horizons decreased between 1972 and 1998, indicating that faster turnover times are the best fit to the data (Fig. 11.3). Estimates of turnover time close to 10 years are limited in precision because of the lack of change in ^{14}C values for atmospheric CO_2 between 1990 and 1998 (Fig. 11.1). However, the radiocarbon data clearly show the Oi and Oe + Oa horizons have turnover times of < 10 years and 10 ± 5 years, respectively. For comparison, the Oe + Oa horizons at Howland, Maine and Harvard Forest, Massachusetts have turnover times of ~ 40 ±10 years, with the Howland site also constrained

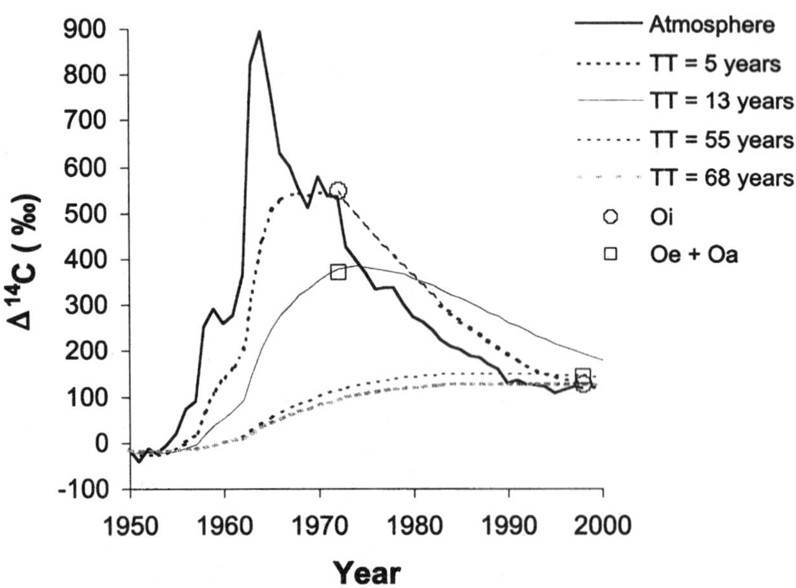

Figure 11.3. $\Delta^{14}C$ values for atmospheric CO_2 and organic horizon components for archived and modern soils and modeled (steady-state) turnover time curves. The Walker Branch input data are derived from tree-ring cellulose starting in 1990 (to account for the unusual atmospheric ^{14}C history at this site; see text for details).

by comparison of archived and contemporary soil samples (Gaudinski 2001). Our turnover-time estimates may, in fact, be too long by 1–5 years if there are any lagged ^{14}C inputs caused by carbohydrate storage prior to plant tissue (root or leaf) formation (Gaudinski et al. 2000; Trumbore 2000).

We estimate the flux of CO_2 derived from heterotrophic decomposition of each SOM fraction by dividing its stock by its ^{14}C-derived turnover time. We limited this analysis to the humified low-density and mineral-associated SOM fractions of the Oe + Oa and mineral horizons with turnover times of 10 years or more. Combined fluxes for low- and high-density fractions range from 45 to 58 g C m^{-2} y^{-1} in the Oe + Oa horizons, from 6 to 16 g C m^{-2} y^{-1} in the A horizons, and from 8 to 14 g C m^{-2} y^{-1} in the E horizons and are less than 3 g C m^{-2} y^{-1} in the lowermost B horizons (Table 11.2). The majority is from the Oe + Oa horizon (53%) and the A + E horizons (45%), whereas only 2% comes from the B horizons. Therefore, the total calculated decomposition flux from SOM decomposing on 10-year time scales or greater is 77 g C m^{-2} yr^{-1} (9% of the total soil respiration flux) with 6% from the Oe + Oa horizon and 3% from the mineral horizons. These fluxes represent an average over several years to decades and are not necessarily indicative of decomposition dynamics in any given year. We do not include the Oi horizon in this calculation because of our inability to determine turnover time accurately from radiocarbon for the reasons described.

Our results compare well to component-integration studies also performed on the Oak Ridge Reservation. Edwards et al. (1989), working in a *Quercus prinus* stand on Walker Branch, measured respiration rates for the whole soil profile, litter, and roots (autotrophic + heterotrophic sources) and estimated (by difference) that 32 g C m^{-2} y^{-1} came from decomposition of SOM in the mineral horizons (5% of the total soil respiration flux). Using a similar approach, Edwards and Harris (1977) found 23 g C m^{-2} y^{-1} came from the decomposition of SOM in the mineral horizons (2.2% of the total soil respiration flux) and 103 g C m^{-2} y^{-1} came from the O_2 horizon, which is comparable to the Oe + Oa horizon (10% of the total soil-respiration flux).

Potential for Soil C Accumulation

Using the soil C stock and turnover time information presented earlier, we can model the potential for the soils of Walker Branch to store carbon in response to climate change. To do this, we created a four-pool model that tracks C stocks:

$$\frac{dC}{dt} = \sum_{j=1}^{j=4} I_j - k_j C_j, \qquad (3)$$

where C_j represents the carbon stock in pool j, k_j equals (turnover time$_j$)$^{-1}$, and I_j equals the inputs to pool j. The four pools represent (1) leaf litter (Oi horizon), (2) the Oe + Oa horizon, (3) low-density SOM in the mineral horizons, and (4) high-density SOM in the mineral horizons. Inputs to each pool were calculated on the basis of initial steady-state conditions and the ^{14}C-derived k (turnover time).

The commonly cited process driving C sequestration in the coming century is an increase in forest net primary production (NPP) (caused by CO_2 or N fertilization) that will be larger than any decreases in overall turnover times (i.e., increased decomposition rate because of increases in mean annual temperature or precipitation changes). Thus, we modeled the response of Walker Branch soils to a 25% increase in NPP, which is similar to that observed in a young loblolly pine plantation after 2 years of CO_2 enrichment (at 565 ppm) (DeLucia et al. 1999). We ran the four-pool carbon model for average, high, and low cases. The high and low cases use the range of measured C stock and turnover-time values that respectively maximize or minimize C storage potential (Table 11.2 and Table 11.3). Our results indicate total C accumulation after 100 years of 505, 1140 and 235 g C m^{-2} for the average, high, and low cases, respectively (which are 39%, 63%, and 31% of the new steady-state value under the 25% NPP increase; Table 11.3 and Fig. 11.4A). The rate of C accumulation in the first year after the increase in NPP is between 30 and 120 g C m^{-2} y^{-1} (Fig. 11.4B). However, by the fifth year, the rates are between 7 and 22 g C m^{-2} y^{-1}, and after 100 years, they drop to between 1 and 6 g C m^{-2} y^{-1}. This rapidly decreasing rate of C storage is reflected in the *e*-folding times of 89, 38, and 151 years, respectively (Table 11.3). During the coming century, we estimate the average annual C accumulation rate in these soils to be between 2 and 11 g C m^{-2} y^{-1}(with 5 g C m^{-2} y^{-1} for the average case; calculated as the average for the initial 100–year period after a sustained 25% increase in NPP).

On timescales of human interest, C storage within the soil profile at Walker Branch will be predominantly within low-density humified organic components that have turnover times of a decade to several hundred years. Detrital C, such as leaf litter in the Oi horizon, will be a poor C sink because it turns over rapidly ($<$ 10 years), has small C stocks, and reaches steady state within years to decades (Fig. 11.5). The Oe + Oa and A + E horizons, however, have low-density C stocks with decadal turnover times and therefore, the potential to store hundreds of grams of carbon on human timescales (Fig. 11.5, Table 11.2). High-density, mineral-associated carbon stocks, which dominate total C stocks below the A + E horizon, have low inputs and long turnover times (200–2000 years) and can store large amounts of C ($>$ 100 g m^{-2}) only over several centuries (Fig. 11.5; Table 11.2).

Validity of the Steady-State Assumption and Current Rates of C Accumulation

The time-dependent ^{14}C model we used to determine turnover times and the flux estimates we have based on those turnover times both assume that Walker Branch soils are at steady state with respect to carbon. The ^{14}C and C-stock data available are too few to determine the validity of this assumption. However, there are several reasons why it might be reasonable to assume that C may be accumulating in soils at this site. First, a nearby (2 km away) eddy flux tower shows large amounts of annual C uptake at this site and there are published estimates that a

Table 11.3. Parameters used in soil carbon model and the resulting carbon increase after a 25% increase in net primary production

	Average			High			Low			Harvard Forest		
Pool	Stock ($g\ C\ m^{-2}$)	TT[a] (year)	Input[b] ($g\ C\ m^{-2}\ y^{-1}$)	Stock ($g\ C\ m^{-2}$)	TT (year)	Input[b] ($g\ C\ m^{-2}\ y^{-1}$)	Stock ($g\ C\ m^{-2}$)	TT (year)	Input[b] ($g\ C\ m^{-2}\ y^{-1}$)	Stock ($g\ C\ m^{-2}$)	TT (year)	Input[b] ($g\ C\ m^{-2}\ y^{-1}$)
Leaf litter	275	1	220	360	1	360	190	2	95	150	1	150
Oe + Oa	340	10	36	410	6	68	270	13	21	1800	40	45
LD	2660	168	16	4150	85	49	1170	250	5	1800	100	18
HD	1900	850	2	2300	200	12	1450	1500	1	4620	500	9
SS total[c]	5175			7220			3080			8370		
C increase at new SS[d]	1295			1805			770			2095		
C increase after 100 years[e]	505			1140			235			950		
e-folding time[f]	89			38			151			67		

Note: First three sets of columns refer to Walker Branch. The fourth set of columns refers to Harvard Forest, MA. See text for further details.

[a] TT = turnover time.

[b] Inferred steady-state inputs (i.e., stock/TT).

[c] Steady-state carbon inventory.

[d] Carbon stock increase due to 25% increase in net primary production after reaching new steady state.

[e] Carbon stock increase 100 years after net primary production increased by 25%.

[f] Time (years) for the stock to change to $1/e$ times the new steady-state value.

A.

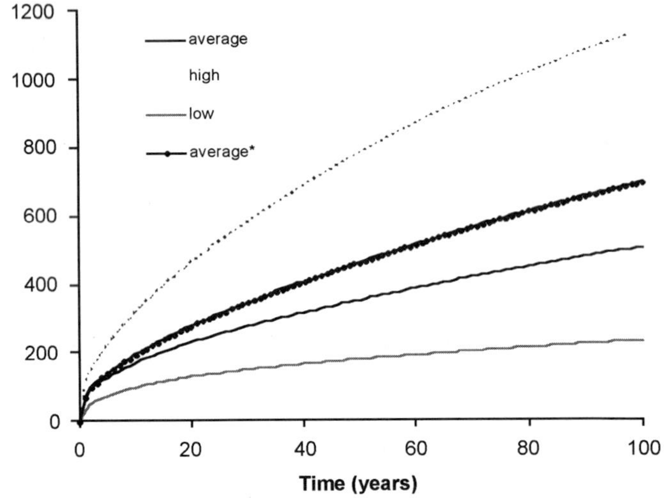

B.

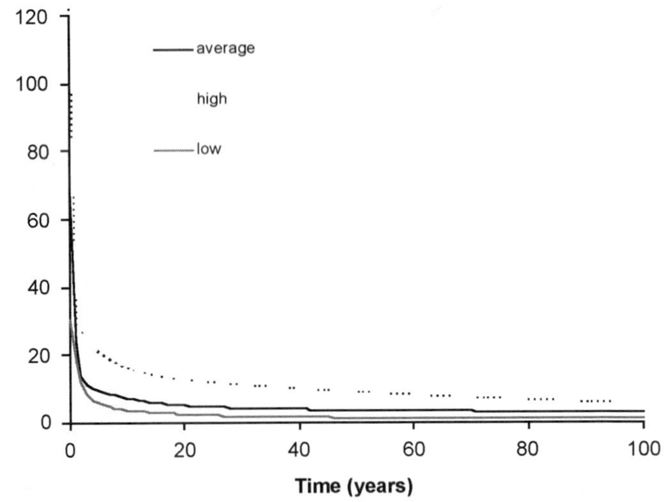

Figure 11.4. Soil carbon response after a 25% increase in NPP for the high, low, and average soil carbon stock and cycling rate scenarios (Table 11.3). (A) The three average, high, and low scenarios initially at steady state plus the average case initially accumulating 10 g C m^{-2} y^{-1}(average*). (B) Rate of change in soil C storage over time for the high, low and average case shown in (A).

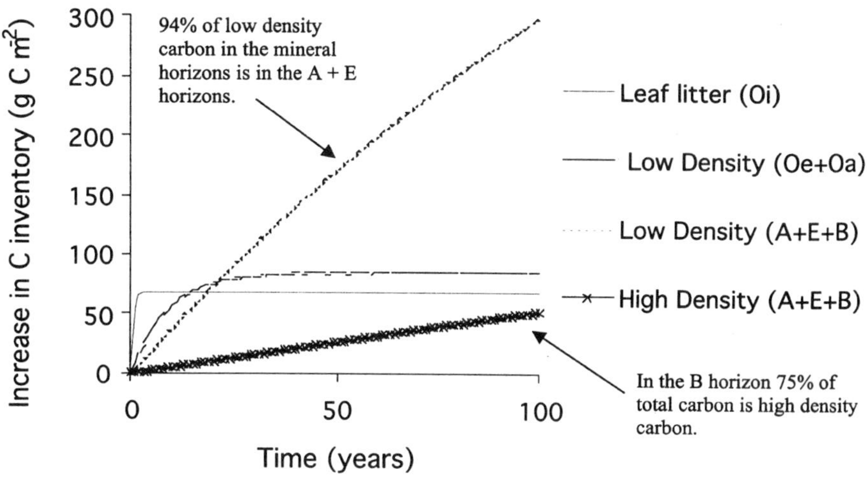

Figure 11.5. Carbon accumulation in response to a sustained 25% increase in NPP (average case) for different soil fractions.

large portion of that storage must be going into the soil (Malhi et al. 1999). Second, this site was selectively logged and grazed in the early 1900s, and the forest has regrown and been undisturbed since the 1940s. Finally, historic increases in atmospheric CO_2 concentrations and enhanced N deposition are thought to lead to increases in NPP that increase soil C storage.

To determine the magnitude of and any possible error associated with the steady-state assumption in an aggrading system, we used a non-steady-state accumulation model (Gaudinski et al. 2000) that matches the current C and ^{14}C stocks in 1998 with different values for inputs (I) and turnover times ($1/k$; where both I and $1/k$ are constants) and assumes that all C in the humified components of the Oe + Oa horizon has accumulated in the past 100 years. Inputs of 30–50 g C m^{-2} y^{-1} and turnover times of 6–7 years matched the C and ^{14}C data, respectively, and our model then predicts accumulation rates of $<< 1$ g C m^{-2} y^{-1} in 1998. Hence, C and ^{14}C stocks are consistent with a steady-state assumption for the Oe + Oa horizons. Turnover times of 6–7 years are on the low end of the measured range (6–13 years) (Table 11.2). However, turnover times of 10–15 years in an accumulating scenario produce ^{14}C values larger than any measured by 40–70‰ and thus provide further confirmation that this horizon is at steady state. Using this same approach for the low-density fractions of the A horizon (with inputs of 10–20 g C m^{-2} y^{-1} and turnover times between 250–350 years), a minimum of 200 years of accumulation is required to match the C and ^{14}C data. The model then results in 5–10 g C m^{-2} y^{-1} of accumulation in 1998.

Current accumulation rates in the E and B horizons will be even less than those in the O and A horizons because of their relatively small stocks of low-density SOM (Fig. 11.5, Table 11.1). Thus, our turnover time and flux estimations likely have little error from the steady-state assumption in the O and mineral E and B horizons. However, our steady-state assumption for the A horizon may not be entirely valid. If we initialize our average C stock model with accumulation in the low-density fraction of 10 g C m^{-2} y^{-1} (the A horizon contains 68% of the low-density material in the mineral horizons), the added C storage over 100 years (relative to a steady-state case) is less than 200 g C m^{-2} and still less than the range bracketed by our high scenario (Fig. 11.4A), thus having no effect on our estimate of potential C storage following a sustained 25% increase in NPP.

Our results of relatively low C-storage potential in Walker Branch soils are in accord with the results reported by Trettin et al. (1999), who measured changes in C stocks for eight long-term plots located in similar forest stands on Walker Branch. Their results showed C gains and losses between 1972 and 1993 in Oi and Oa horizons (~100–400 g C m^{-2} over the 11-year period), with most actually showing losses. The A horizons (0–15 cm) in their study tended to increase in % C (although not significantly) in five of the eight long-term plots. The B horizons (45–60 cm) actually show % C losses in five of their eight long-term plots between 1972 and 1993 (with significant losses in two plots). The net soil C balance shown by Trettin et al. (1999) for the whole soil profile (0–60 cm) between 1982 and 1993 is a trend toward C loss (significant C loss in only one plot). Johnson et al. (1988), however, show that soils (from 0–45 cm) on the Oak Ridge reservation adjacent to Walker Branch accumulated 1160 g C m^{-2} during a 15-year period (from 1980 to 1995; ~80 g C m^{-2} y^{-1} if the rate was constant). Time-series measurements of C stocks have inherent problems relating to large spatial variability; however, the data of Trettin et al. (1999) urge caution in any implicit assumption of C accumulation in soil C stocks, even in sites where accumulation seems very plausible.

Implications for Ecosystem Carbon Budgets of Walker Branch

Net forest C balance is the sum of C gains or losses in aboveground versus belowground biomass. At Walker Branch, eddy covariance techniques for a similar site 2 km away show net ecosystem production (NEP) of 580 g C m^{-2} y^{-1} (Wilson and Baldocchi 2001). Current estimates of aboveground biomass accumulation in wood on Walker Branch are 220 g C m^{-2} y^{-1} and for coarse woody roots are 45 g C m^{-2} y^{-1} (Chapter 22, this volume). Solving for the remaining belowground storage by difference would yield 315 g C m^{-2} y^{-1} of soil C storage [an increase that should be measurable with the methods of Trettin et al. (1999)]. Our results, however, show that the soils of Walker Branch are currently storing < 15 g C m^{-2} y^{-1}, and under changing climatic conditions, they likely have the potential to store no more than an additional 11 g C m^{-2} y^{-1}. Although some of the above-discussed assumptions introduce errors in our estimates, sustained storage rates >100 g C m^{-2} y^{-1} are not possible with reasonable scenarios of changing C inputs. This discrepancy will be discussed further in Chapter 22.

In forest ecosystems where eddy-covariance techniques show net uptake, an increase in soil C should not necessarily be assumed. As discussed earlier, Trettin et al. (1999) show a general trend (although significant in only one plot) of soil

C loss (for 0–60 cm) for eight long-term plots on Walker Branch between 1972 and 1993. An alternative biometric estimate of net ecosystem C storage at Walker Branch based on summing directly measured aboveground and belowground C fluxes (and assuming heterotrophic respiration is 50% of total soil respiration) yields an NEP of 250 g C m^{-2} y^{-1} (Curtis et al. 2002). Using the same estimates of C accumulation in wood and coarse roots discussed earlier in this chapter and in Chapter 22, soil C accumulation is then estimated to be close to zero and decreasing.

Beyond Walker Branch

Current soil C-storage rates at Walker Branch (< 15 g C m^{-2} y^{-1}) are similar to that of Harvard Forest, MA (10–30 g C m^{-2} y^{-1}) (Gaudinski et al. 2000), a more northerly mixed deciduous temperate forest in the eastern United States. We calculate that a 25% increase in NPP would cause C storage over the next century of 950 g C m^{-2} at Harvard Forest and 505 g C m^{-2} at Walker Branch using an average of measured C stocks and turnover times for both sites (Table 11.3) (Gaudinski et al. 2000). Total C-storage potential following this perturbation is larger at Harvard Forest primarily because of larger stocks and longer relative turnover times in Oe + Oa horizons (Table 11.3). Although such a gradient is not surprising (given the differences in climate), the increased annual C accumulation rate at both sites is still < 10 g C m^{-2} y^{-1} (calculated as the average over the initial 100-year period after a sustained 25% increase in NPP).

Our rates of current and potential C accumulation are in accord with research done in other natural ecosystems. Schlesinger and Lichter (2001) predict the potential for accumulation of < 60 g C m^{-2} y^{-1} in a young loblolly pine plantation as atmospheric CO_2 concentrations increase through time. In a literature review, Post and Kwon (2000) found that, after agricultural abandonment, average rates of soil C accumulation were 34 g C m^{-2} y^{-1} for reestablishing forests or grasslands. Agricultural soils in the United States are also estimated to be able to store $\sim$20 g C m^{-2} y^{-1} under certain management conditions that increase C inputs to soil (Lal et al. 1999). Rates of soil C storage in tropical forest soils are also on the order of tens of g C m^{-2} y^{-1} (Trumbore, unpublished data).

Conclusions

Using radiocarbon to constrain turnover rates of SOM, we have shown that bulk SOM consists of fractions that will respond at quite different rates to a change, such as an increase in NPP. Five percent of the carbon resides in leaf and root residues with turnover times of < 10 years. These pools will respond quickly to changes and have small inventories, so they are not expected to play a large role as C sinks or sources, although they may dominate year-to-year variation in soil heterotrophic respiration rates. Thirty-five percent of the C is stabilized by association with soil minerals and has turnover times > 100 years; although C stocks in this fraction are large, they are not expected to change dramatically during the next century of climate or even land-use change. Sixty percent resides in low density and humified organic matter in mineral soils with turnover times of 10–100 years; it is these fractions that will determine C storage potential of soils and that require further investigation as to their sources and what determines their turnover time in soils.

Our estimates of current and future C-storage capacity at Walker Branch are an order of magnitude smaller than those needed to reconcile eddy flux estimates of annual net carbon uptake and estimates of aboveground carbon storage at this site. This result suggests that methodological issues (perhaps those associated with an underestimation of nocturnal respiration of C by ecosystems) seriously affect eddy-covariance results from Walker Branch and should be investigated further.

References

Burchuladze AA, Chudy M, Eristavi IV, Pagava SV, Povinec P, Sivo A, Togonidze GI (1989) Anthropogenic ^{14}C variations in atmospheric CO_2 and wines. Radiocarbon 31:771–776.

Curtis PS, Hanson PJ, Bolstad P, Barford C, Randolph JC, Schmid HP, Wilson KB (2002). Biometric and eddy-covariance based estimates of ecosystem carbon storage in five eastern North American deciduous forests. Agric For Meteorol 113:3–19.

DeLucia EH, Hamilton JG, Naidu SL, Thomas RB, Andrews JA, Finzi A, Lavine M, Matamala R, Mohan JE, Hendrey GR, Schlesinger WH (1999) Net primary production of a forest ecosystem with experimental CO_2 enrichment. Science 284:1177–1179.

Edwards NT, Harris WF (1977) Carbon cycling in a mxed deciduous forest floor. Ecology 58:431–437.

Edwards NT, Johnson DW, McLaughlin SB, Harris WR (1989) Carbon dynamics and productivity. In Johnson DW, VanHook RI (Eds) *Analysis of biogeochemical cycling processes in Walker Branch Watershed.* Springer Verlag, New York, pp 197–232.

Fernandez IJ, Rustad LE, Lawrence GB (1993) Estimating total soil mass, nutrient content, and trace metals in soils under a low elevation spruce-fir forest. Can J Soil Sci 73:317–328.

Gaudinski JB (2001) *Belowground carbon cycling in three temperate forests of the eastern United States.* Doctoral Dissertation. University of California, Irvine, California.

Gaudinski JB, Trumbore SE, Erickson EA and Zheng S (2000) Soil carbon cycling in a temperate forest: Radiocarbon-based estimates of residence times, sequestration rates and partitioning of fluxes. Biogeochemistry 51:33–69.

Goulden ML, Munger JW., Fan SM, Daube BC, Wofsy SC (1996) Measurements of carbon sequestration by long-term eddy covariance—Methods and a critical evaluation of accuracy. Global Change Biol 2:169–182.

Greco S, Baldocchi DD (1996) Seasonal variations of CO_2 and water vapour exchange rates over a temperate deciduous forest. Global Change Biol 2:183–197.

Hamburg SP (1984) Effects of forest growth on soil nitrogen and organic matter pools following release from subsistence agriculture. In Stone EL (Ed) *Forest soils and treatment impacts.* The University of Tennessee, Knoxville, Tennessee, pp 145–158.

Hesshaimer V, Heimann M, Levin I (1994) Radiocarbon evidence for a smaller oceanic carbon dioxide sink than previously believed. J Exp Bot 43:1111–1119.

Hollinger DY, Goltz SM, Davidson EA, Lee JT, Tu K, Valentine HT (1999) Seasonal patterns and environmental control of carbon dioxide and water vapour exchange in an ecotonal boreal forest. Global Change Biol 5:891–902.

Hua Q, Barbetti M, Worbes M, Head J, Levchenko VA (1999) Review of radiocarbon data from atmospheric and tree ring samples for the period 1945–1997 AD. Iawa J 20:261–283.

Huntington TG, Johnson CE, Johnson AH, Siccama TG, Ryan DF (1989) Carbon, organic matter, and bulk density relationships in a forested spodosol. Soil Sci 148:380–386.

Johnson DW, Henderson GS, Todd DE (1988) Changes in nutrient distribution in forests and soils of Walker Branch Watershed, Tennessee, over an eleven-year period. Biogeochemistry 5:275–293.

Joslin JD, Wolfe MH (1999) Effects of minirhizotron installation disturbance upon root observation data. Soil Sci Soc Amer J 63:218–221.

Lal R, Follett RF, Kimble J, Cole CV (1999) Managing US cropland to sequester carbon in soil. J Soil Water Conserv 54:374–381.

Levin I, Hesshaimer V (2000) Radiocarbon—A unique tracer of the global carbon cycle dynamics. Radiocarbon 42:69–80.

Levin I, Kromer B (1997) Twenty years of atmospheric (CO_2)—C-14 observations at Schauinsland Station, Germany. Radiocarbon 39:205–218.

Linick TW (1975) *Uptake of bomb-produced carbon-14 by the Pacific Ocean.* Ph.D. Thesis. University of California San Diego, California.

Malhi Y, Baldocchi DD, Jarvis PG (1999) The carbon balance of tropical, temperate and boreal forests. Plant Cell Environ 22:715–740.

Post WM, Kwon KC (2000) Soil carbon sequestration and land-use change: Processes and potential. Global Change Biol 6:317–327.

Schlesinger WH, Lichter J (2001) Limited carbon storage in soil and litter of experimental forest plots under increased atmospheric CO_2. Nature 411:466–469.

Trettin CC, Johnson DW, Todd DE (1999) Forest nutrient and carbon pools at Walker Branch Watershed: Changes during a 21-year period. Soil Sci Soc Am J 63:1436–1448.

Trumbore SE (1993) Comparison of carbon dynamics in tropical and temperate soils using radiocarbon measurements. Global Biogeochem Cycles 7:275–290.

Trumbore S (2000) Age of soil organic matter and soil respiration: Radiocarbon constraints on belowground C dynamics. Ecol Appl 10:399–411.

Trumbore SE, Davidson EA, Decamargo PB, Nepstad DC, Martinelli LA (1995) Belowground cycling of carbon in forests and pastures of eastern Amazonia. Global Biogeochem Cycles 9:515–528.

Trumbore S, Gaudinski JB, Hanson PJ, Southon JR (2002) A whole-ecosystem carbon-14 label in a temperate forest. EOS 83:265,267–268.

Vogel JS (1992) A rapid method for preparation of biomedical targets for AMS. Radiocarbon 34:344–350.

Wilson KB, Baldocchi DD (2001) Comparing independent estimates of carbon dioxide exchange over 5 years at a deciduous forest in the southeastern United States. J Geophys Res 106(D24):34167–34178.

12. Rates of Coarse-Wood Decomposition

Donald E. Todd and Paul J. Hanson

Although coarse-wood (> 5 cm in diameter) decomposition rates were not evaluated as a part of the Throughfall Displacement Experiment (TDE) effort, a previous dataset from early Walker Branch studies (Todd et al. 1976) was available to address coarse-wood decomposition under ambient conditions. These data are included in this volume because woody decay is important to the ecosystem carbon budget of temperate deciduous forests (Chapter 22, this volume). A multiyear experiment to investigate site and species effects on decay rates and nutrient release or immobilization from decaying wood was established in March of 1972. Trees representing a range of wood characteristics (*Carya* sp., *Liriodendron tulipifera, Pinus echinata,* and *Quercus prinus*) were harvested during the winter of 1971/1972 and cut into sections having a constant length of 30 cm but variable diameters, ranging from 5 to 18 cm. Branch sections were placed on the Walker Branch watershed in valley bottom, midslope, and ridge-top sites. The branches were retrieved from the field twice annually through 1975, at which time the experiment was discontinued because the condition of the branch samples would not support further observations.

Mass loss began the first year after placement in the field, and branches for all species showed reduced mass loss over the measurement period (Fig. 12.1). Slope position did not alter branch-mass loss over time, but time of year and species had a significant effect on branch decomposition. *Liriodendron tulipifera* branches decayed at the fastest rates, losing 76% of their initial mass through June of 1975 for a mean annual mass-loss rate of 22.8% per year. The other species decayed

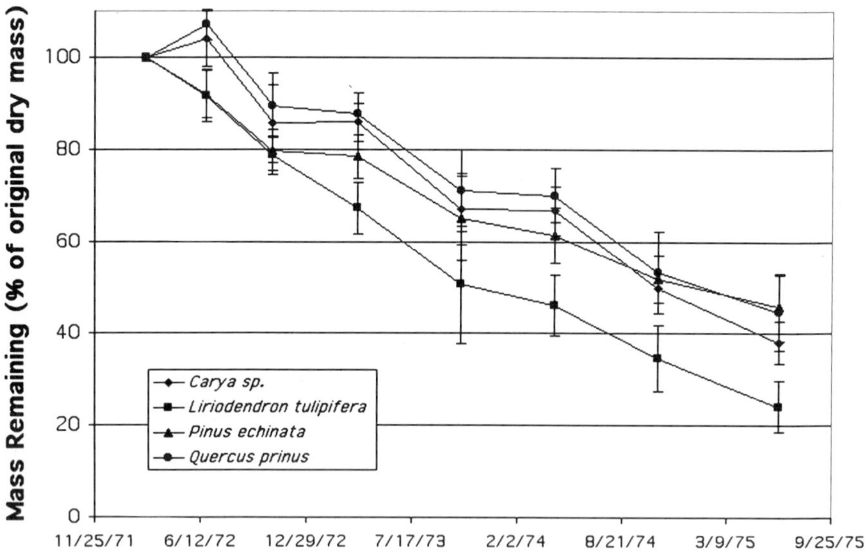

Figure 12.1. Relative decomposition of four species of wood during 4.5 years from November 1971 through September 1975 on ridge-top or side-slope positions of Walker Branch Watershed.

at slower rates, losing a mean of 57% of their initial mass by June of 1975 for a mean annual rate of 17% per year. The majority of the mass loss for *Carya, Pinus,* and *Quercus* was restricted to the warm summer months in the first 2 years with little change in mass during the winters of 1972/1973 and 1973/1974. Significant mass loss for *Carya, Pinus,* and *Quercus* was observed during the winter of 1974/1975. The pattern of *Liriodendron* mass loss was different, with more constant mass losses occurring during all years and seasons. The increased winter decomposition appears to be the primary reason yellow poplar had significantly less weight remaining at the end of the study than the other three species.

Ambient annual and growing-season (May through September) rainfall was above average during the period from 1972 through 1975, with mean annual and growing-season rainfall of 1648 and 589 mm, respectively. Mean annual temperature was near normal at 14.2°C. Conditions for the TDE during the period from 1993 through 2000 had a mean growing-season rainfall of 518 mm and mean annual temperatures of 14.5°C. The environmental differences between these two studies are not large, and we assume that coarse-wood decomposition on the TDE site would have followed similar temporal trends to the patterns observed by Todd et al. (1976).

To limit the effects of time on decomposition and to allow comparisons with other published studies, the decomposition constant k_d was calculated from the equation suggested by Olsen (1963):

$$M_t = M_o \, e^{(-k_d \, t)}, \tag{1}$$

Table 12.1. Species-specific coefficients for the exponential decomposition model [Eq. (1)]

Species	M_o	k_d (year^{-1})	R^2	n	Mean diameter (cm)
Carya sp.	1.13	0.296	0.76	33	10.1 ± 0.5
L. tulipifera	1.04	0.427	0.63	66	11.7 ± 0.3
P. echinata	0.96	0.223	0.54	60	10.6 ± 0.5
Q. prinus	1.15	0.282	0.54	101	10.2 ± 0.3

Note: The number of samples for each regression (n) and the mean diameter ± S.E. of the wood samples are also provided.

where M_t is the relative mass at sampling time t (in years), M_o is the original mass, and k_d is the decomposition rate constant (y^{-1}). Although M_o should theoretically be equal to 1, random error associated with the estimation of initial dry mass from the fresh mass/dry mass ratios allowed for a range of values. The values for k_d at Walker Branch ranged from 0.22 to 0.43 y^{-1} (Table 12.1). The highest values were found for *Liriodendron,* (0.43 y^{-1}) with the other species having a mean k_d of 0.27 y^{-1}. The k_d values derived from the data of Todd et al. (1976) represent an integrated measure of mass loss accounting for mineralization, fragmentation, and leaching losses (Lambert et al. 1980).

Table 12.2 summarizes bole-decay constants for deciduous and coniferous species of the eastern United States and other related work. The vast majority of these studies show decay constants below 0.1 y^{-1}, but the data are almost exclusively derived from measures of wood-density changes (i.e., evaluation of only a subset of the decaying material), which underestimate early mass loss associated with bark and outer sapwood decomposition. High k_d values for Walker Branch coarse-wood decomposition may have been the result of several factors: (1) The materials were observed for only the initial 3 years of decomposition; (2) the bole segments were placed directly on the surface of the litter (i.e., in good contact with a moisture source); and (3) the mass-loss data include losses of exterior bark and sapwood due to fragmentation. Some of the data from Mattson et al. (1987), which included estimates of mass loss from fragmentation, yielded k_d values approaching those of the Walker Branch observations. Exceptionally high values for bole-decomposition rate constants of 0.46 y^{-1} were reported by Lang and Knight (1979) for tropical forests, but more recent decay-constant estimates for tropical forest systems range from 0.11 to 0.19 y^{-1} (Chambers et al. 2000; Chambers et al. 2001).

Influence of Temperature

Chambers et al. (2000) also summarized published literature for 20 globally distributed studies and suggested that a respiration quotient (Q_{10}) of 2.4 was appropriate for the temperature sensitivity of wood decomposition. Applying the same analysis to the data for eastern North America in Table 12.2 yielded a Q_{10} of 2.1. In a laboratory-based analysis of the decomposition of woody roots, Chen et al. 2000 found that Q_{10} values were not constant. They varied with temperature with

Table 12.2. Exponential decay constant (k_d) for deciduous and coniferous species common to forests of the eastern United States and the associated mean annual temperature and precipitation for each study

Species	k_d (year^{-1})	Mean annual Temp. (°C)	Mean annual Precip. (mm)	Reference
Deciduous species				
Acer rubrum	0.08	12.6	1820	Mattson et al. 1987
A. saccharum	0.11	15.6	~1219	Onega and Eickmeier 1991
A. saccharum	0.09	~17	914	Chueng and Brown 1995
A. saccharum	0.086	15.6	~1219	Onega and Eickmeier 1991
Acer sp.	0.045	11.8	1180	MacMillan 1988
Carya sp.	0.17	12.6	1820	Mattson et al. 1987
Carya sp.	0.30	14.2	1352	This chapter
Carya sp.	0.035	11.8	1180	MacMillan 1988
Fagus grandifolia	0.019	11.8	1180	MacMillan 1988
L. tulipifera	0.43	14.2	1352	This chapter
L. tulipifera	0.11	12.6	1820	Mattson et al. 1987
Hardwoods	0.096	5.5	1400	Arthur et al. 1993
Nyssa sylvatica	0.126	12.6	1820	Mattson et al. 1987
Oxydendrum arboreum	0.033	12.6	1820	Mattson et al. 1987
Populus tremuloides	0.08	2	639	Alban and Pastor 1993
Q. alba	0.063	12.6	1820	Mattson et al. 1987
Quercus sp.	0.03	11.8	1180	MacMillan 1981
Quercus sp.	0.018	11.8	1180	MacMillan 1988
Q. coccinea	0.05	12.6	1820	Mattson et al. 1987
Q. prinus	0.28	14.2	1352	This chapter
Q. prinus	0.14	13	1820	Abbot and Crossley 1982
Q. prinus	0.059	13	1820	Abbot and Crossley 1982
Q. prinus	0.17	12.6	1820	Mattson et al. 1987
Conifer species				
Abies balsamea	0.03	3.4	~1400	Lambert et al. 1980
Picea glauca	0.071	2	639	Alban and Pastor 1993
Picea rubens	0.033	3.4	~1400	Foster and Lang 1982
Pinus resinosa	0.049	2	639	Alban and Pastor 1993
Pinus rigida	0.063	12.6	1820	Mattson et al. 1987
Pinus echinata	0.22	14.2	1352	This chapter
Tsuga canadensis	0.021	5.7	~762	Tyrell and Crow 1994

a value of 4 in the 5–10°C temperature range and a value of 1.4 at higher temperatures between 30°C and 40°C.

Influence of Moisture

The data in Table 12.2 do not suggest strong relationships between mean annual precipitation and k_d, but individual studies have provided data to show that moisture can play an important role in the rate of bole decay. Abbott and Crossley (1982) found that the decay constant was reduced by approximately half on xeric clear-cut sites as compared to uncut or mesic clear-cut locations (0.06 and 0.14 y^{-1}, respectively). Rate constants for decomposition from studies that evaluated both standing and fallen tree boles showed higher rate constants for the bole material with ground contact (Onega and Eickmeier 1991). Chambers et al. (2001) characterized the optimum and limiting moisture contents for wood respiration to be 2.5 and 0.5 g H_2O g^{-1} dry wood, respectively. At the opposite extreme of moisture Progar et al. (2000) found that excessively high moisture in the Pacific Northwest inhibited wood decomposition. Chueng and Brown (1995) studied mass loss of 5-cm-diameter, 0.75-m segments of *A. saccharinum* L. in a central Illinois bottomland forest and found a mass loss rate of 0.09 y^{-1}, which is lower than that observed on Walker Branch, but the bottomland conditions suggest wetter conditions that may reflect the wet inhibition also observed by Progar et al. (2000). Chen et al. (2000) demonstrated that dead coarse-root respiration also increased with moisture content.

Coarse-wood decomposition represents a potentially significant loss of carbon from eastern deciduous hardwood forests. Observations of bole decomposition from an early Walker Branch study (Todd et al. 1976) (Fig. 12.1) suggest mean rates of carbon losses to be applied to the coarse-dead-wood pool, but the influence of temperature and moisture have not been tested on Walker Branch. Published data suggest an exponential relationship between coarse-wood decomposition and temperature, but the correlation to wood moisture is unclear. Additional work will be required to identify appropriate expressions of the response of bole decomposition to site or wood water content or water potential.

References

Abbott DT and Crossley DA Jr. (1982) Woody litter decomposition following clear-cutting. Ecology 63:35–42.

Alban DH, Pastor J (1993) Decomposition of aspen, spruce, and pine boles on two sites in Minnesota. Can J For Res 23:1744–1749.

Arthur MA, Tritton LM, Fahey TJ (1993) Dead bole mass and nutrients remaining 23 years after clear-felling of a northern hardwood forest. Can J For Res 23:1298–1305.

Chambers JQ, Higuchi N, Schimel JP, Ferreira LV, Melack JM (2000) Decomposition and carbon cycling of dead trees in tropical forests of the central Amazon. Oecologia 122:380–388.

Chambers JQ, Schimel JP, Nobre AD (2001) Respiration from coarse wood litter in central Amazon forests. Biogeochem 52:115–131.

Chen H, Harmon ME, Griffiths RP, Hicks W (2000) Effects of temperature and moisture on carbon respired from decomposing woody roots. For Ecol Manage 138:51–64.

Chueng N, Brown S (1995) Decomposition of silver maple (*Acer saccharinum* L) woody debris in a central Illinois bottomland forest. Wetlands 15:232–241.

Foster JR, Lang GE (1982) Decomposition of red spruce and balsam fir boles in the White Mountains of New Hampshire. Can J For Res 12:617–626.

Lang GE, Knight DH (1979) Decay rates for the boles of tropical trees in Panama. Biotropica 11:316–317.

Lambert RL, Lang GE, Reiners WA (1980) Loss of mass and chemical change in decaying boles of subalpine balsam fir forest. Ecology 6:1460–1473.

MacMillan PC (1981) Log decomposition in Donaldson's Woods, Spring Mill State Park, Indiana. Am Midl Nat 106:335–344.

MacMillan PC (1988) Decomposition of coarse woody debris in an old-growth Indiana forest. Can J For Res 18: 1353–1362.

Matson KG, Swank WT, Waide JB (1987) Decomposition of woody debris in a regenerating , clear-cut forest in the Southern Appalachians. Can J For Res 17:712–721.

Olsen JS (1963) Energy storage and the balance of the producers and decomposers in ecological systems. Ecology 44:322–331.

Onega TL, Eickmeier WG (1991) Woody debris inputs and decomposition kinetics in a southern temperate deciduous forest. Bull Torrey Bot Club 118:52–57.

Progar RA, Schowalter TD, Freitag CM, Morrell JJ (2000) Respiration from coarse woody debris as affected by moisture and saprotroph functional diversity in Western Oregon. Oecologia 124:426–431.

Todd DE, Hendersen GS, Harris WF (1976) Nutrient release from decaying wood. Agronomy Abstracts. Agronomy Society of America. Madison, WI.

Tyrrell LE, Crow TR (1994) Dynamics of dead wood in old-growth hemlock-hardwood forests of northern Wisconsin and northern Michigan. Can J For Res 24:1672–1683.

5. Plant Growth and Mortality

13. Tree Seedling Recruitment in a Temperate Deciduous Forest: Interactive Effects of Soil Moisture, Light, and Slope Position

Jake F. Weltzin and Philip B. Allen

Predicted changes in global and regional precipitation regimes are likely to affect the distribution, structure, composition, diversity of plant communities [e.g., VEMAP Members (1995); Neilson and Drapek (1998); Bachelet and Neilson (2000)], and, in particular, forest ecosystems (Hanson and Weltzin 2000). Forests of the southeastern United States may be particularly vulnerable to changes in precipitation regimes and soil moisture contents because the increases in potential evapotranspiration predicted for this region may eventually exceed summer precipitation (National Assessment Synthesis Team 2000). These changes are predicted to occur at rates that may exceed the ability of forests to adapt through changes in species composition (Pastor and Post 1988; Davis 1989; Overpeck et al. 1991). Ultimately, the composition of forests under potential future climates will depend on interactive effects of regional and local abiotic conditions, biotic interactions, and species life-history traits. For example, coexistence mechanisms that govern the relative success of different tree species within a community will vary at each life-history stage of the tree (Nakashizuka 2001). Given adequate reproduction and seed dispersal, seedling recruitment and juvenile survivorship may dictate community composition (Harper 1977; Peet and Christensen 1987; Kobe 1996).

The objective of the current study was to investigate the interactive role of soil moisture, slope position, and light availability on recruitment of seedlings (i.e., the number of seedlings that survive one growing season relative to the number of seeds planted) for four tree species common to the Throughfall Displacement

Experiment (TDE) site and representative stands of upland oak–hickory forest: *Quercus prinus* L. (chestnut oak), *Cornus florida* L. (flowering dogwood), *Liriodendron tulipifera* L. (yellow poplar), and *Acer rubrum* L. (red maple). We hypothesized that if the availability of soil water constrains seedling recruitment (given sufficient seed dispersal), then rates of seedling recruitment would be greater on the wet than the dry plot. Because the low-slope positions receive runoff and subsurface flow from upslope positions, we predicted that soil moisture would be greater at low-slope positions with subsequent benefits to seedling recruitment. We further hypothesized that if low light constrains seedling recruitment, then recruitment rates would be reduced by provision of additional (artificial) shade. Our objective was to test these hypotheses with a manipulative experiment and a descriptive survey within the context of the overall experiment.

Methods

In autumn 1999, we established 24 (2 m $\times$ 1 m) plots on the TDE. On each of the wet and dry TDE plots, we established 12 plots, 6 at random locations each within the upper one-third (high slope) and lower one-third (low slope) of the slope. Each plot was covered with an exclosure (1.2 $\times$ 0.7 $\times$ 0.3-m tall) constructed of 1-cm wire mesh to exclude vertebrates. Exclosures were split at the middle and hinged at the corners to allow access to the plot for monitoring. Half of the plots at each TDE treatment/slope position combination were assigned at random to one of two shade treatments: unshaded (control) or shaded, which was produced with commercially available shade cloth (rated at 30% light reduction) secured to the top of the exclosure (thus, $n = 3$).

Seeds used in this study were collected on the site and near Knoxville, Tennessee, in autumn 1999 (or for *A. rubrum,* spring 2000). Seeds were visually examined for insects and pathological infection; particularly small or obviously empty seeds of *L. tulipifera* and *A. rubrum* were discarded. A subset of seeds for each species was tested for viability with 2,3,5-triphenyl tetrazolium chloride (Nyandiga and McPherson 1992). Viability of *Q. prinus, C. florida,* and *L. tulipifera* were 100%, 85%, and 15%, respectively. *A. rubrum* was not tested for viability prior to planting.

In December 1999, we planted 20 seeds of *Q. prinus* and *C. florida* at 12-cm spacing into each plot. In addition, we planted 60 seeds of *L. tulipifera,* 3 per hole, at 20 locations within each plot. In April 2000, we planted 20 *A. rubrum* seeds at 20 locations within each plot. All planted seeds were marked with galvanized nails, and leaf litter was redistributed after planting.

Starting in April 2000, we monitored seedling emergence and survival at weekly intervals for 12 weeks, and monthly thereafter until September 2000. On each monitoring date, dead seedlings were assigned a probable cause of mortality (e.g., dessication or defoliation). Maximum shoot elongation (as a measure of plant growth response) for each *Q. prinus* seedling was recorded after leaf fall; because only a few seedlings of any other species had emerged by or survived to that point, we did not measure their shoot elongation.

In September 2000, we sampled density (number/m^2) of naturally occurring *Q. prinus* in twelve 0.5-m × 1.0-m quadrats located at random within the upper- and lower-slope positions within the wet and dry treatments of the TDE. In addition, we subjectively chose four locations, or patches (~6 m in diameter), where the density of *Q. prinus* seedlings was apparently high, and determined density within a single 0.5-m × 1.0-m quadrat placed in the geographic center of the patch. These patches were located at the top of the ridge above but within 20 m of the TDE.

To assess the physiological response of seedlings to the different treatments over the course of the growing season, we analyzed leaf tissue for stable carbon-isotope composition ($\delta^{13}C$) (Brugnoli and Farquhar 2000). Δ integrates *ci/ca* over the active periods of photosynthesis and leaf formation and is frequently correlated to stomatal conductance and drought stress (Ehleringer 1990; Meinzer et al. 1992). In September 2000, we harvested one leaf from each of three *Q. prinus* seedlings selected at random from each plot. We did not similarly sample other species because they were rare or—as in the case of *C. florida*—had relatively little leaf tissue. Leaves were composited, oven-dried at 50°C for 48 h, ground to fine powder with a Wiley mill, and analyzed for $\delta^{13}C$ with a Finnigan MAT Delta Plus isotope ratio mass spectrometer.

Data for temperature of the surface of the soil for each plot were collected hourly between January and December 2000 with all-weather temperature dataloggers placed on the surface of the soil below the leaf-litter layer. Hourly temperature data were processed to obtain daily mean, minimum, and maximum temperatures for each plot. These daily data were then averaged by month for subsequent analysis. Soil-water content (%, v/v) in each plot was measured with a time-domain reflectometer (TDR) following the procedures of Topp and Davis (1985) for soils with high coarse fraction content. A pair of 35-cm TDR waveguides installed vertically into the center of each plot were monitored monthly between February and November of 2000. In May 2001, we measured attenuation of photosynthetically active radiation (PAR) at the top of the seedling canopy within each plot with an AccuPAR linear ceptometer.

Statistical Analyses

We used analysis of variance [ANOVA; SAS procedure GLM, SAS Institute (1989)] to evaluate main and interactive effects of water treatment, slope position, and shade treatment on proportional seedling emergence (emerged seedlings/seeds planted), survival (live seedlings/emerged seedlings), and recruitment (live seedlings/seeds planted) for each species through September 2000. Trends in seedling recruitment rates during the same time were investigated for main and interactive effects of water treatment, slope position, shade treatment, and monthly sample date with repeated-measures ANOVAs in a multivariate framework {MANOVAR; Pillai's Trace in SAS procedure GLM [SAS Institute (1989); von Ende (1993)]}.

We analyzed shoot elongation and carbon-isotope discrimination for *Q. prinus* with an ANOVA model. Monthly soil moisture contents and minimum, maximum, and mean soil surface temperatures were analyzed with a MANOVAR model. Prior to analysis, all data were tested for normality with the Shapiro-Wilk W-statistic (Shapiro and Wilk 1965). Data not normally distributed ($p < 0.05$) were transformed as appropriate following recommendations of Zar (1996). We used Fisher's protected least significant difference (LSD) (Fisher 1960) *a posteriori* mean separation tests for significant treatment effects ($p < 0.05$ unless otherwise indicated).

The water treatments and slope position within the TDE experimental design are unreplicated (Hurlbert 1984) primarily because of the cost involved in the large-scale implementation of the research project. However, Hanson et al. (1998) showed that individual measurements of soil moisture across an 8-m × 8-m grid of points within the TDE plots were not correlated with one another. Therefore, we assumed that individual plots were independent of one another with respect to soil moisture. Further, we confine our interpretation of results to the particular combinations of water treatment and slope position that comprise the experiment.

Results

Seedling Emergence, Survival, and Recruitment

No *A. rubrum* or *L. tulipifera* seeds emerged in any treatment combination. *Q. prinus* seedling emergence rates were unaffected by main or interactive effects of watering treatment, slope, or shade ($p > 0.16$). Mean seedling emergence rates (± SE) averaged 77% ± 2%, and emergence was complete by June 2000 (Fig. 13.1).

Quercus prinus survival and recruitment throughout and at the end of the growing season were unaffected by main or interactive effects of watering treatment, slope, or shade ($p > 0.08$). Recruitment rates increased through April and May 2000 as seedlings emerged but remained relatively constant into September (MANOVAR $p < 0.0001$) (Fig. 13.1). By September, seedling survival and recruitment averaged 84 ± 2% and 65 ± 2%, respectively.

Density of naturally occurring *Q. prinus* was not affected by main or interactive effects of water treatment or slope position ($p > 0.15$) and averaged 2.8 ± 0.5 plants/m^2. Density of *Q. prinus* seedlings in the subjectively chosen dense patches averaged 69 ± 0.5 plants/m^2.

Cornus florida emergence rates were about one-third those of *Q. prinus,* although most *C. florida* seedlings emerged about 1 month earlier than *Q. prinus* (Fig. 13.1). *C. florida* emergence tended to be greater in unshaded (31%) than in shaded (22%) plots ($p = 0.09$); seedling emergence was otherwise unaffected by water treatment or slope position.

Cornus florida survival and recruitment throughout the growing season did not differ between treatment combinations ($p > 0.10$). Seedling survival was relatively constant through June, but survival rates declined thereafter (MANOVAR $p < 0.0001$) (Fig. 13.1). At the end of the growing season, *C. florida* seedling survival and recruitment on high-slope positions averaged 29% and 8%, respec-

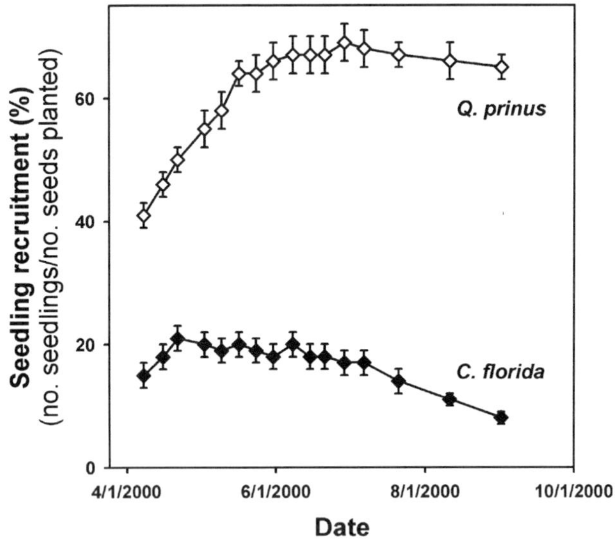

Figure 13.1. Recruitment (%) of *Q. prinus* and *C. florida* during 2000, calculated for each species as number of seedlings alive/number of seeds planted. Vertical bars represent 1 SE.

tively, and did not differ between dry and wet treatments ($p > 0.60$); in contrast, survival and recruitment at low-slope positions were four and seven times greater in wet (56% and 13%, respectively) than dry (8% and 3%, respectively) treatments (water × slope interactions, $p < 0.02$).

Quercus prinus Size and Carbon-Isotope Discrimination

Quercus prinus seedling shoot elongation was greater on high-slope (15 cm) than on low-slope (13 cm) positions on the dry treatment. On the wet treatment, the pattern was reversed; shoots were longer on low-slope (15 cm) than on high-slope (13 cm) positions (water × slope interaction, $p = 0.002$). *Q. prinus* shoot elongation was greater in unshaded (15 cm) than shaded (13 cm) plots only at the lower-slope position (light × slope interaction, $p = 0.02$). *Q. prinus* leaf carbon-isotope discrimination (Δ) at the end of the 2000 growing season was not affected by main or interactive effects of water treatment, slope position, or shade ($p = 0.40$) and averaged $21.04 \pm 0.12‰$.

Soil-Water Content and Soil Surface Temperature

Volumetric soil water content within the plots ranged from nearly 25% in early June to less than 10% in October and November (Fig. 13.2). Soil-water contents were also relatively low early in the growing season. However, water contents were dependent on the water treatment (water × date interaction, $p = 0.06$). In particular, soils were wetter in the wet treatment than the dry treatment on four of seven dates (February, July, September, and November; $p < 0.09$ for each

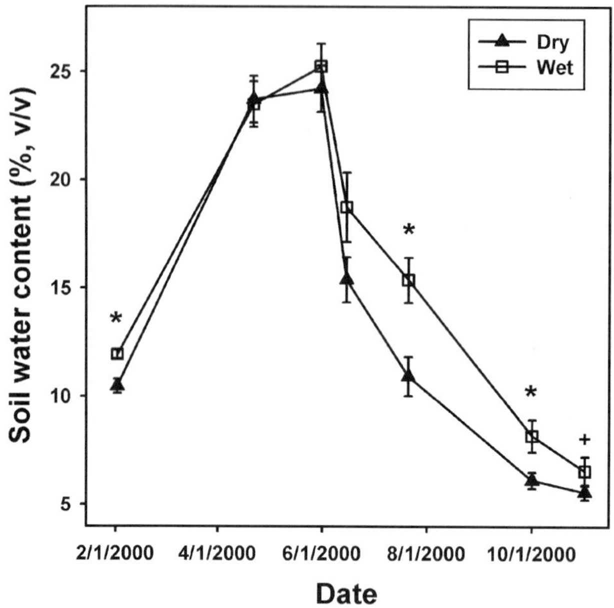

Figure 13.2. Soil-water content (%, v/v) in seedling recruitment plots on the wet and dry plots of the TDE during 2000. Vertical bars represent 1 SE. For means at each date: * $p < 0.003$; $^{+}$ $p < 0.09$.

date). Soil-water contents were greater in low-slope positions (16%) than in high-slope positions (14%). Soil-water contents were not affected by the shade treatment ($p = 0.92$) or by its interactions with other factors ($p > 0.14$).

Monthly mean soil surface temperatures ranged from 2.5°C in January to 22.4°C in August (data not shown). Mean temperatures were, on average, 0.5°C warmer at high-slope (16.5°C) than at low-slope (16.0°C) positions on the wet treatment but were not affected by slope position on the dry treatment (water × slope interaction, $p = 0.02$). Monthly minimum temperatures were about 0.4°C warmer at low-slope (14.0°C) than at high-slope (14.4°C) positions on the wet plot but were not affected by slope on the dry treatment (water × slope interaction, $p = 0.02$). Otherwise, mean and minimum temperatures were unaffected by the water or shade treatments ($p > 0.31$). Monthly maximum temperatures averaged 1.2°C warmer in unshaded than shaded plots at low-slope positions, but did not differ between unshaded and shaded plots at high-slope positions (light × slope interaction, $p = 0.03$). Interactive effects of month and the various treatments were either not significant or not interpretable.

Discussion

Seedling recruitment rates after one growing season were species-specific and ranged from 0% for *A. rubrum* and *L. tulipifera* to 65% for *Q. prinus*. Seedling recruitment reflected the combined effects of differential rates of seedling emergence, which ranged from 0% to 77%, and subsequent survival of emerged seedlings (Fig. 13.1).

For *Q. prinus,* results failed to support the hypothesis that seedling recruitment is constrained by soil-water content: Emergence, survival, and recruitment rates of seedlings from planted acorns did not differ between wet and dry plots, or between the relatively wet lower-slope position and relatively dry upper-slope position. This finding is supported by the density of naturally occurring *Q. prinus* seedlings, which did not differ across the TDE treatments or slope position. This conclusion is tempered by the lack of differences in soil-water content between the wet and dry TDE treatments during the period of seedling emergence (Fig. 13.2).

In fact, *Q. prinus* exhibited relatively high rates of emergence, which suggests that soil moisture was not particularly limiting to emergence and early establishment of this species on the TDE. This conclusion is supported by the lack of measurable differences in integrated water stress (i.e., carbon-isotope discrimination) and inconsistent responses in terms of shoot elongation between the wet and dry plots. This finding is also consistent with previous assessments of tree growth versus soil-water potentials, which indicated that of the five dominant tree species on the TDE, mature *Q. prinus* (and white oak, *Q. alba*) trees are least sensitive to low soil-water contents (Hanson et al. 2001). Extrapolations of the response of mature trees to seedlings are necessarily circumspect because the function or distribution of roots of mature trees are likely to differ from those of seedlings. Although the roots of seedlings are likely to occupy relatively shallow soil layers, there have been no species-specific investigations of seedling root distribution on the TDE.

Results for *Q. prinus* in this study are comparable to patterns observed in a demographic study of naturally occurring seedlings conducted on the TDE between 1993 and 1998 (Chapter 14, this volume). In 1998, densities of naturally occurring *Q. prinus* seedlings at high-slope positions on the TDE were quite comparable between the wet and dry treatment plots, and densities had increased ~two-fold since 1993 in both wet and dry treatments. However, in contrast with this study, where slope position and water treatment did not affect *Q. prinus* emergence or recruitment, densities of naturally occurring *Q. prinus* declined between 1993 and 1998 at mid-slope and low-slope positions, particularly in the dry plot (Chapter 14, this volume). Results for *Q. prinus* in this study can be also contrasted with a recent study of *Q. alba* seedlings on the TDE (Holmgren 1996). Mortality rates of *Q. alba,* 2 years after being transplanted as 1-year-old seedlings onto the TDE, ranged from 2% to 11% in wet and dry plots, respectively, and were greater at high- (10%) than low- (0%) slope positions (Holmgren 1996).

For *C. florida,* results tended to support the hypothesis that seedling recruitment is limited by soil-water content, although slope position also appeared to be important: seedling recruitment on low-slope positions was four times greater on the wet than the dry plot. This interpretation is consistent with Hanson et al. (2001), who observed that annual rates of mortality for saplings of *C. florida* were greater on the dry plot than on the wet plot of the TDE between 1995 and 1999.

Further, survivorship rates of naturally occurring *C. florida* seedlings on the TDE between 1993 and 1998 were greater in the low-slope position of the wet plot relative to other treatment-slope combinations (Chapter 14, this volume). *C. florida* typically has a shallow root system (Hinckley et al. 1979; Bahari et al. 1985); this could increase its susceptibility to low soil-water contents (i.e., drought stress) and contribute to the high rates of seedling mortality observed in all of these studies.

Lack of emergence by *A. rubrum* and *L. tulipifera* precludes adequate assessment of the relative importance of soil-water content to the demographics of these species. Lack of emergence suggests that (1) soil-water contents in both wet and dry TDE treatments, regardless of their differences, were insufficient for germination and emergence of these species or (2) other factors, such as low seed viability or possible seed predation, form the ultimate constraint on the emergence of these seedlings in the field. The lack of emergence by *A. rubrum* was somewhat surprising, because seedlings and saplings of this species are relatively well represented across the entire experimental site (personal observation; Hanson et al. 2001). In contrast, *L. tulipifera* seedlings (<15 cm tall) are uncommon across the site, particularly at upper-slope positions [where densities were 0.05 seedlings/m^2 in 1993; Holmgren (1996)]. Concordantly, *L. tulipifera* trees occur only on the lower third of the TDE slope (Hanson et al. 2001). The distribution of *L. tulipifera* seedlings and trees within this region is often attributed to shade intolerance or susceptibility to drought (Roberts et al. 1979; Wallace and Dunn 1980; Hanson et al. 2001). Similarly, results from previous research on the TDE indicate that *L. tulipifera* is the most sensitive to drought stress of the dominant overstory trees (Hanson et al. 2001).

As predicted, soil-water contents were greater at low-slope than at high-slope positions, perhaps because of surface run-on or subsurface flow from upslope positions. However, differences in soil-water content were not consistently reflected in integrated water stress, shoot growth, or recruitment of *Q. prinus*. However, for *C. florida,* higher soil-water contents at the low-slope position on the wet TDE plot may have favored seedling recruitment in unshaded plots.

Consistent with our original hypothesis that low light might constrain seedling recruitment, artificial shade tended to reduce emergence rates of *C. florida.* However, the mechanism for this is unclear because our shade treatments did not affect soil-water contents and had little or no effect on mean or minimum soil surface temperatures. Maximum temperatures were slightly higher in unshaded than in shaded plots, but only at low-slope positions. In contrast with the original hypothesis, provision of artificial shade did not affect recruitment rates for *Q. prinus.* The large seed size of this species may buffer its response to shade or other resource limitations (Saverimuttu and Westoby 1996; Walters and Reich 2000).

Smith and Huston (1989) hypothesized that the response of plants to dual and interactive gradients of light and water is characterized by a trade-off between drought tolerance and shade tolerance. The hypothesis is based on the observation that, under drier conditions, plants tend to allocate their biomass to belowground structures (i.e., to roots for water acquisition) rather than to aboveground structures (i.e., to shoots and leaf area). As such, plants would be sensitive to low water availability when growing under low light levels and sensitive to low light availability when growing under dry soil conditions. Although this seldom-tested hypothesis is supported by some empirical evidence (Kolb et al. 1990; Vance and Zaerr 1991), it is not supported by patterns of recruitment for *Q. prinus* or *C. florida* in this experiment. Similarly, in two specific tests of the hypothesis, one of which was conducted on the TDE, Holmgren (1996, 2000) found no trade-offs in the growth or survival of *L. tulipifera, Q. alba,* and *A. saccharum* along experimental gradients of light and water.

Results of this and other research indicate that plant species representative of typical upland deciduous forests in East Tennessee exhibit a wide range of sensitivity to experimentally imposed drought, ranging from little, if any, impact to significant increases in mortality (Hanson et al. 1995, 1998; Hanson et al. 2001). In particular, *C. florida* may be particularly sensitive to changes in precipitation regimes. Wetter climates in the future may contribute to enhanced establishment of seedlings and survival of saplings, whereas drier climates may reduce the likelihood of recruitment and survival of seedlings and saplings. Alternatively, wetter climates may promote pests and disease to which *C. florida* is prone, although this has yet to be tested empirically. In addition, *A. rubrum* saplings on the TDE are sensitive to differences in soil water content, although less so than *C. florida* (Hanson et al. 2001). In contrast, the survival of *Q. prinus* seedlings and the growth and survival of all dominant overstory tree species on the TDE (Hanson et al. 2001) were insensitive to differences in soil-water availability. Thus, species and life-history stage are both important determinants of the response of trees to gradients in resource availability [cf. Jones et al. (1994); Berkowitz et al. (1995); and Walters and Reich (1996, 2000)]. This differential response suggests that if future climates differ from current climates, the species composition of eastern deciduous forests may change from their current status (Chapter 25, this volume). Importantly, changes in climate may embody not only changes in mean conditions (e.g., annual or seasonal precipitation) but also increased incidence of extremes (e.g., drought), which may also affect community composition [e.g., Pederson (1998), Allen and Breshears (1998), Hanson and Weltzin (2000), Hanson et al. (2001)].

References

Allen CD, Breshears DD (1998) Drought-induced shift of a forest-woodland ecotone: Rapid landscape response to climate variation. Proc Natl Acad Sci 95:14839–14842.

Bachelet D, Neilson RP (2000) Biome redistribution under climate change. In Joyce LA, Birdsey R (Eds) *The impact of climate change on America's forests: A technical document supporting the USDA Forest Service RPA Assessment.* General Technical Report RMRS-GTR-59. USDA Forest Service, Rocky Mountain Research Station, Fort Collins, Colorado, pp 18–44.

Bahari ZA, Pallardy SG, Parker WC (1985) Photosynthesis, water relations, and drought adaptation in six woody species of oak–hickory forests in central Missouri. For Sci 31:557–569.

Berkowitz AR, Canham CD, Kelly VR (1995) Competition vs facilitation of tree seedling growth and survival in early successional communities. Ecology 76:1156–1168.

Brugnoli E, Farquhar GD (2000) Photosynthetic fractionation of carbon isotopes. In Leegood RC, Sharkey TD, von Caemmerer S (Eds) *Photosynthesis: Physiology and metabolism.* Kluwer Academic Publishers, Hingham, Massachusetts, pp 399–434.

Davis MB (1989) Lags in vegetation response to greenhouse warming. Clim Change 15:75–82.

Ehleringer JR (1990) Correlations between carbon isotope discrimination and leaf conductance to water vapor in common beans. Plant Physiol 93:1422–1425.

Fisher RA (1960) *The design of experiments.* Seventh edition. Hafner, New York.
Hanson PJ, Weltzin JF (2000) Drought disturbance from climate change: Response of United States forests. Sci Total Environ 262:205–220.
Hanson PJ, Todd DE , Edwards NT, Huston MA (1995) Field performance of the Walker Branch Throughfall Displacement Experiment. In Jenkins A, Ferrier RC, Kirby C (Eds) *Ecosystem manipulation experiments: Scientific approaches, experimental design, and relevant results:* Ecosystem Research Report Number 20. Commission of the European Communities, Brussels, Belgium, pp 307–313.
Hanson PJ, Todd DE, Huston MA, Joslin JD, Croker J, Augé RM (1998) *Description and field performance of the Walker Branch Throughfall Displacement Experiment: 1993–1996,* ORNL TM-13586. Oak Ridge National Laboratory, Oak Ridge, Tennessee.
Hanson PJ, Todd DE, Amthor JS (2001) A six-year study of sapling and large-tree growth and mortality responses to natural and induced variability in precipitation and throughfall. Tree Physiol 21:345–358.
Harper JL (1977) *Population biology of plants.* Academic Press, London.
Hinckley TM, Dougherty PM, Lassoie JP, Roberts JE, Teskey RO (1979) A severe drought: Impact on tree growth, phenology, net photosynthetic rate and water relations. Am Midl Nat 102:307–316.
Holmgren M (1996) *The interactive effects of shade and drought on seedling growth and survival.* Ph.D. dissertation. The University of Tennessee, Knoxville, Tennessee.
Holmgren M (2000) Combined effects of shade and drought on tulip poplar seedlings: Trade-off in tolerance or facilitation? Oikos 90:67–78.
Hurlbert SH (1984) Pseudoreplication and the design of ecological field experiments. Ecol Monogr 54:187–211.
Jones RH, Sharitz RL, Dixon PM, Segal DS, Scjmeoder RL (1994) Woody plant regeneration in four floodplain forests. Ecol Monogr 64:345–367.
Kobe RK (1996) Intraspecific variation in sapling mortality and growth predicts geographic variation in forest composition. Ecol Monogr 66:181–201.
Kolb TE, Steiner KC, McCormick LH, Bowersox TW (1990) Growth response of northern red-oak and yellow-poplar seedlings to light, soil moisture, and nutrients in relation to ecological strategy. For Ecol Manage 38: 65–78.
Meinzer FC, Saliendra NZ, Crisosto CH (1992) Carbon isotope discrimination and gas exchange in *Coffea arabica* during adjustment to different soil moisture regimes. Aust J Plant Physiol 19:171–84.
Nakashizuka T (2001) Species coexistence in temperate, mixed deciduous forests. Trends Ecol Evol 16:205–210.
National Assessment Synthesis Team (2000) *Climate change impacts on the United States: The potential consequences of climate variability and change.* US Global Change Research Program. Cambridge University Press, New York.
Neilson RP, Drapek RJ (1998) Potentially complex biosphere responses to transient global warming. Global Change Biol 4:505–521.
Nyandiga CO, McPherson GR (1992) Germination of 2 warm-temperate oaks, *Quercus-emoryi* and *Quercus-arizonica.* Can J Forest Res 22(9):1395–1401.
Overpeck JT, Bartlein PJ, Webb T III (1991) Potential magnitude of future vegetation change in eastern North America: Comparisons with the past. Science 254:692–695.
Pastor J, Post W M (1988) Response of northern forests to CO_2-induced climate change. Nature 334:55–58.
Pedersen, BS (1998) Modeling tree mortality in response to short- and long-term environmental stress. Ecological Modeling 105:347–351.
Peet RK, Christensen NL (1987) Competition and tree death. BioScience 37:586–595.
Roberts SW, Knoerr KR, Strain BR (1979) Comparative field water relations of four co-occurring forest tree species. Can J Bot 57:1876–1882.
SAS Institute (1989) *SAS/STAT user's guide, Version 6.* Fourth edition. SAS Institute, Cary, North Carolina.
Saverimuttu T, Westoby M (1996) Seedling longevity under deep shade in relation to seed size. J Ecol 84:681–689.
Shapiro SS, Wilk MB (1965) An analysis of variance test for normality (complete samples). Biometrika 52:591–611.
Smith TM, Huston MA (1989) A theory of the spatial and temporal dynamics of plant communities. Vegetatio 83: 49–69.
Topp GC, Davis JL (1985) Measurement of soil water content using time domain reflectometry (TDR): a field evaluation. Soil Sci Soc Amer J 49:19–24.
Vance NC, Zaerr JB (1991) Influence of drought stress and low irradiance on plant water relations and structural constituents in needles of *Pinus ponderosa* seedlings. Tree Physiol 8:175–184.
VEMAP Members (1995) Vegetation/ecosystem modeling and analysis project: Comparing biogeography and biogeochemistry models in a continental-scale study of terrestrial ecosystem responses to climate change and CO_2 doubling. Global Biogeochem Cycles 9:407–437.
von Ende CN (1993) Repeated measures analysis: Growth and other time-dependent measures. In Scheiner SM, Gurevitch J (Eds) *Design and analysis of ecological experiments.* Chapman and Hall, New York, pp 113–137.
Wallace LL, Dunn EL (1980) Comparative photosynthesis of three gap phase successional tree species. Oecologia 45:331–340.
Walters MB, Reich PB (1996) Are shade tolerance, survival, and growth linked? Low light and nitrogen effects on hardwood seedlings. Ecology 77:841–853.
Walters MB, Reich PB (2000) Seed size, nitrogen supply, and growth rate affect tree seedling survival in deep shade. Ecology 81:1887–1901.
Zar JH (1996) *Biostatistical analysis.* Prentice Hall, Upper Saddle River, New Jersey.

14. Response of Understory Tree Seedling Populations to Spatiotemporal Variation in Soil Moisture

Michael A. Huston, Milena Holmgren, and Michelle B. Kreh

Understory plants exist in an unproductive and stressful environment. Depending on the density of the forest canopy, they exist under moderate to deep shade conditions. Because they are small plants with relatively shallow root systems, they are subjected to water stress to a much greater degree than the more deeply rooted trees, which also have a greater drying effect on the soil through the transpiration of their larger canopies.

In the understory environment, with potential growth limitation by both light and water, plant growth and survival are expected to be extremely sensitive to variation in both of these limiting factors. Energetic and physiological trade-offs in the acquisition and use of both light and water are hypothesized to increase whole-plant light compensation levels under dry conditions, with the result that plants can be more shade tolerant under moist conditions than under dry conditions, and more drought tolerant with adequate light levels (Smith and Huston 1989). However, increased irradiance may also increase thermal and water stresses, particularly for small plants with reduced water storage and uptake capacity. Consequently, we can expect that under dry conditions, small plants may actually benefit from growing under the shade of large plants, which results in a net positive interaction (i.e., facilitation) between them (Holmgren et al. 1997). This shift from negative effects of shading on small plants under relatively moist conditions to positive effects of moderate shading under drier, stressed conditions is associated with nonlinear growth responses in relation to light intensity, which contradict the above-described trade-off model. Such positive interactions are more likely to be important under stressful, resource-poor conditions (Huston 1994, Holmgren et al. 1997; Callaway et al. 2002).

Not surprisingly, some field and experimental studies have found that shaded plants have lower growth or survival than sun plants at equivalent soil moisture levels (Jarvis et al. 1976, cited in Huston 1991; Kolb et al. 1990; Vance and Zaeer 1991; Fisher et al. 1991; Howe 1990; Abrams and Mostoller 1995; Withers 1979; Veenendaal et al. 1996; Coomes and Grubb 2000; Cole 2002), whereas other studies have found the opposite response, consistent with facilitation (Berkowitz et al. 1995; Holmgren 1996, 2000; Sack and Grubb 2002). Difficulties with experiments and field studies include the measurement and/or experimental control of both light and water availability, as well as the short duration of experiments using transplanted seedlings. Some of these difficulties also affect the interpretation of our results from the TDE, particularly the lack of control of the light environment.

The trade-off hypothesis predicts that a dry, shaded environment should be the least favorable combination for the growth and survival of most types of plant. Unfortunately, the preexisting natural pattern of light availability on the TDE site precludes testing this hypothesis, because of the absence of both dry environments with high light availability and wet environments with low light availability. Nonetheless, the 5-year duration of the study as well as the intensive spatiotemporal measurement of soil-water content provide a unique opportunity to assess the response of the seedlings of multiple tree species to variation in soil-water availability. Regardless of variations in the light environment, understory tree seedlings are expected to be much more responsive than larger size classes to the spatial and temporal variation in soil moisture that resulted from the interaction of the Throughfall Displacement Experiment (TDE) soil moisture manipulation with site topography and temporal variation in precipitation and evapotranspiration.

Here, we report on the spatial patterns of change in the diversity and abundance of understory tree seedlings over the first 5 years of experimental treatments.

Methods

Understory plants were surveyed in the summer of 1993, prior to the initiation of the TDE treatments, and again in the summer of 1998, after 5 years of treatment. Eight permanent plots, 3 × 3 m, were established at three topographic positions (upper slope, middle slope, and lower slope) in each of the throughfall displacement treatments (33% reduction, ambient, and 33% addition). A total of 72 plots were laid out with metal and plastic posts and plastic rope (Fig. 14.1). Data were recorded by species in all nine 1 × 1-m blocks within each plot.

For all stems < 200 cm in height, total height and two perpendicular measurements of canopy diameter were recorded (for vines, total length within the plot was recorded). In addition, stem diameter was measured with calipers near ground level for all stems > 30 cm in height. The number of leaves was recorded

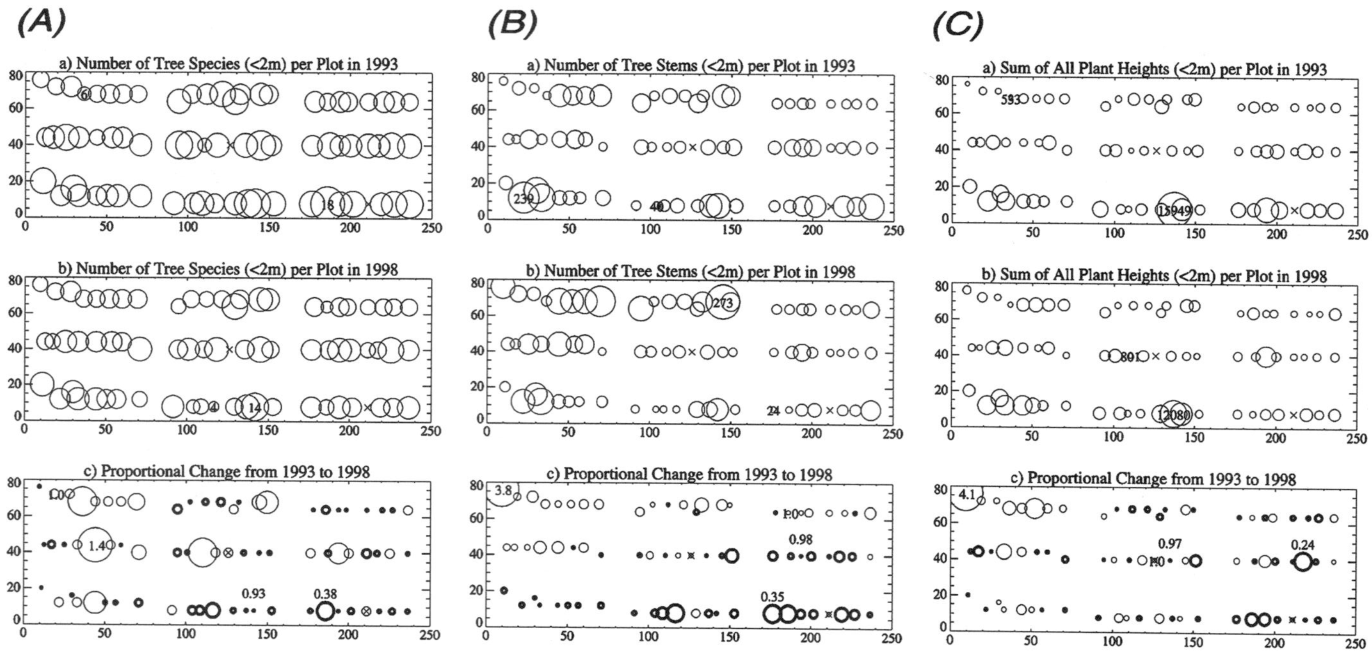

Figure 14.1. Plot locations and tree seedling patterns on the hillside of the TDE site for three community parameters. Each rectangle represents a map of the TDE site, showing plot locations in each of the nine slope–treatment combination. Upper slope is toward the top (north). (A) Number of tree species (< 2 m) per 3 × 3 m plot in the Throughfall Displacement Experiment in (a) 1993 and (b) 1998; (c) relative change between 1993 and 1998. Proportional increases in number of species are indicated with thin-line circles, and decreases with bold-line circles. The relative size of circles indicates relative level of variable, with maximum and minimum plots labeled with their respective values. X indicates the two plots where data were not collected in 1993 or 1998. (B) Number of tree stems (< 2 m) per 3 × 3 m plot in the TDE, (a), (b), and (c) as in (A). (C) Sum of heights of all plants < 2 m tall, including both woody and herbaceous; (a), (b), and (c) as in (A).

for all plants with fewer than 20 leaves. A visual estimate of percent cover, using Daubenmire cover classes (Daubenmire,1952) was made for each square meter. Voucher specimens were collected for plants that could not be identified in the field, and identifications were later determined in consultation with taxonomic experts.

Soil-water conditions were estimated for each of the 72 plots from measurements of volumetric percentage soil-water content (SWC) made using the time-domain reflectometry (TDR) method at each of the marked grid intersections (on 8-m centers) on the TDE site (see Chapter 2, this volume). A simple distance-weighting scheme with 25% increments was used to estimate SWC for each 3 × 3-m vegetation plot based on its position in relation to the nearest TDR measurement locations. For example, SWC for a plot within 1 m of a grid intersection was estimated as the TDR value for that location, whereas the SWC of a plot equidistant from four grid intersections was estimated as the mean of the four TDR values from those intersections.

The SWC values for each vegetation plot were summarized for two time periods (April–June and July–August) for each year (1993–1998). The mean, maximum, and minimum values were calculated for each time period in each year, and these values used to calculate means, maxima, and minima for the entire 6-year period. Five indices of SWC were calculated for the spring and fall of the 6-year period: average (mean of the annual means); minimum (minimum of the annual minima); average minimum (mean of the annual minima); maximum (maximum of the annual maxima); and average maximum (mean of the annual maxima).

The forest-floor light environment on the TDE site was characterized in 1994 by Holmgren (1996) using measurements of diffuse light transmittance made with an LAI-2000 canopy analyzer (Li-Cor, Inc., Lincoln, NE).

One plot in each survey could not be sampled due to recent treefall. Plot 37 (mid-slope ambient) was not sampled in 1998, and plot 69 (lower-slope dry) was not sampled in 1993. Both of these plots were left out of the following analysis.

Results

Clear topographic patterns in total seedling species richness and the distribution of many individual species were evident in 1993. By 1998, there had been significant changes in species diversity across the site, as well as changes in the abundance and distribution of particular species. There was great variation in the responses of individual species, with some increasing and some decreasing. The responses of some species were clearly related to the TDE treatments; others occurred relatively uniformly across the site, apparently in response to the interaction of topography with changes in local weather conditions [i.e., a shift to wetter conditions in comparison to the previous decade (Chapter 24, this volume)].

Light Environment

Light transmittance varied significantly among the nine combinations of treatment (dry, ambient, wet) and slope position (upper, middle, lower) on the TDE site. The lowest and least variable light transmittance (i.e., the darkest conditions) was in the three upper-slope areas (Table 14.1), which differed significantly from both the middle- and lower-slope positions ($p < 0.01$), which had higher mean light levels as well as higher variability. The dry treatment had lower light transmittance than either the ambient or wet treatments, but these differences were not statistically significant ($p \sim 0.15$). The somewhat surprising conclusion is that the driest areas also had the deepest shade. Because the light measurements were made after less than 1 year of experimental treatment, these differences most likely reflect the preexisting structure and species composition of the forest canopy, rather than any effects of the experimental treatments. Higher mortality of mature trees common in lower-slope positions, such as *Liriodendron tulipifera*, during the severe droughts of the late 1980s (see Chapter 24, this volume) undoubtedly contributed to the higher light availability in the mid- and lower-slope locations.

Changes in Seedling Community Structure

The strongest spatial pattern in tree seedling diversity was the topographic gradient of increasing diversity (number of tree species < 2 m height per 9 m^2) from the upper- to the lower-slope positions (Fig. 14.1A). In 1993, there were approximately twice as many tree species per plot at lower-slope positions as there were in upper-slope locations. After 5 years of treatment there was a consistent decline in species richness across the site, with a clear treatment effect. Seventy-eight percent of the plots in the dry treatment and 74% in the ambient declined in species richness, whereas only 42% of the plots declined in the wet treatment (Figs. 14.1A and 14.2A). The eight plots of the upper-slope wet treatment were the only slope–treatment combination with more plots gaining than losing species (Figs. 14.1Ac and 14.2Aa; note that two plots with eight species, and two with nine were unchanged, for a total of eight plots). Diagonally across the TDE site, the lower-slope dry treatment was the only slope–treatment combination in which all plots lost species (Figs. 14.1Ac and 14.2Ai).

The decline in species richness was related to changes in seedling density that were associated with hillslope position (Fig 14.1B). The greatest declines in seed-

Table 14.1. Light transmittance (% of full sunlight) in the nine slope position–treatment combinations of the TDE site

Treatment/ slope position	Wet	Ambient	Dry
Upper	20.4 ± 0.51	18.4 ± 0.73	16.9 ± 0.83
Middle	34.9 ± 1.92	31.4 ± 2.08	25.8 ± 2.59
Lower	31.1 ± 2.61	32.2 ± 3.99	23.2 ± 2.58

Note: Data are means and standard errors of measurements made directly above each planted seedling ($N \sim 40$ in each location) from Holmgren (1996).

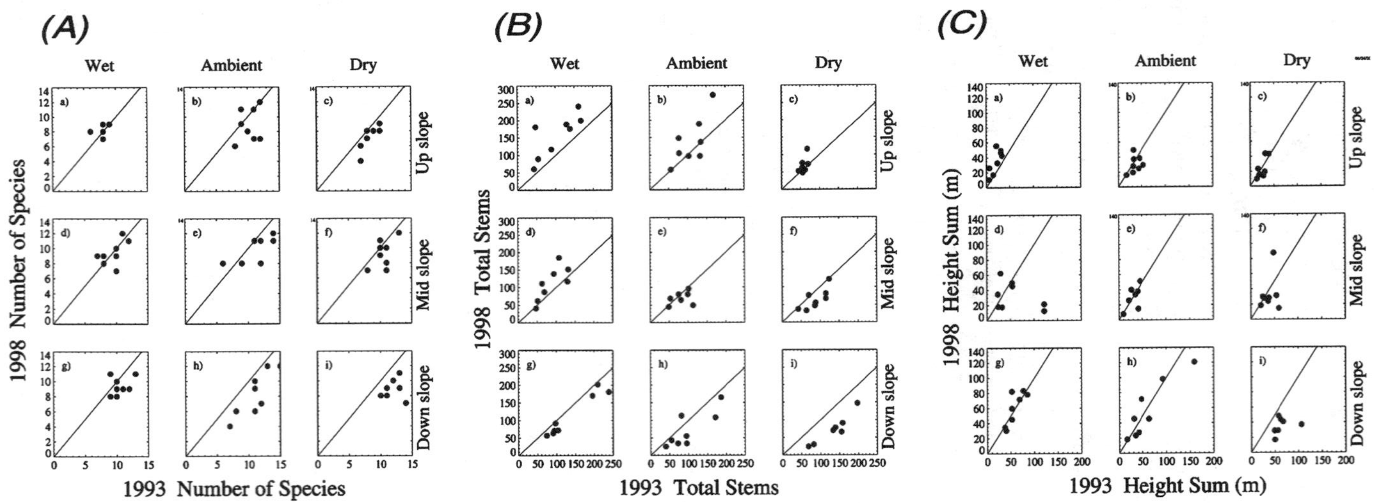

Figure 14.2. Changes between 1993 and 1998 in each of the 3 × 3 sample plots stratified by the nine combinations of slope position (down slope, mid-slope, and up slope) and treatment (wet, ambient, dry) on the TDE. The diagonal line represents a 1 : 1 relationship between 1993 and 1998 values (i.e., no change). (A) Change in number of species per plot. Note that the majority of plots declined in species number, with the exception of the upper-slope wet plot. (B) Change in number of stems < 2 m in height per plot. Note the diagonal symmetry in stem density change, with relatively little change in panels c, e, and g, primarily increases to the left and primarily decreases to the right. (C) Change in the sum of the heights of all plants (woody and herbaceous < 2 m in height per 3 × 3 sample plot. Note all plots in the upper-slope wet treatment increased in height sum, whereas all plots in the lower-slope dry treatment decreased.

ling density (number of stems < 2 m in height per 9 m^2) were in the lower-slope plots, where 96% of the plots declined in density. In contrast, 83% of the upper-slope plots increased, concentrated in the wet and ambient treatments. Mid-slope plots showed a monotonic trend of decreasing density along the treatment gradient of decreasing water availablity (i.e., greatest decline in the dry treatment, Fig. 14.2Bd–14.2Bf). As with species richness, the greatest contrast in responses was between the upper-slope wet treatment plots, all of which increased in stem density (Fig. 14.2Ba), and the lower-slope dry treatment plots, all of which decreased in stem density (Fig. 14.2Bi).

The height sum (sum of all plant heights including both woody and herbaceous plants less than 2 m in height per 9 m^2), was used as a general index of vegetation amount (mass or cover). Most slope–treatment combinations included plots with both increases and decreases in the height sum. As with species number and stem density, the strongest contrast was between the upper wet treatment and lower dry treatment plots. Again, all of the lower-slope plots in the dry treatment declined in height sum (Fig. 14.2Ci), whereas all of the upper-slope plots in the wet treatment increased (Fig. 14.2Ca).

The relationships of species richness with the density of stems and with the height sums were weak, particularly in 1993, when only the lower-slope ambient plots had a clear positive linear relationship between stem density and number of species (data not shown). In 1998, the lower-slope ambient plots had clear positive relationships between species richness and both stem density and height sum (Figs. 14.3Ah and 14.3Bh), whereas many other treatment–slope combinations also had patterns suggestive of positive trends, particularly if a single outlier was discarded (Figs. 14.3A and 14.3B). Changes in stem density were not conspicuously related to changes in species number (data not shown).

Community Structure Responses to Local Heterogeneity in Soil-Water Content

Comment on Interpretation of Correlations with Soil-Water Content

The results presented below are challenging to interpret, with some positive and some negative relationships, as well as many examples of no clear relationship. Spatial heterogeneity complicates these patterns, as it does most ecological analyses. With multiple factors varying spatially, the relationship between any two factors may be completely obscured by the effects of some other factor or factors. Such interactions confound the analysis and interpretation of experiments (Huston 1997, Huston et al. 2000; Huston and McBride 2002), but are likely to be even more important in the interpretation of field data (Huston 2002, O'Connor 2002).

Many factors in addition to soil water availability influence the distribution, survival, and growth of tree seedlings. These include the proximity and fecundity of mature trees as seed sources, as well as soil nutrients, light availability, herbivory, and trampling. Independently of variation in soil water content, any of these factors could cause a relative increase or decrease in the number and type of seedlings found in one location in relation to another location.

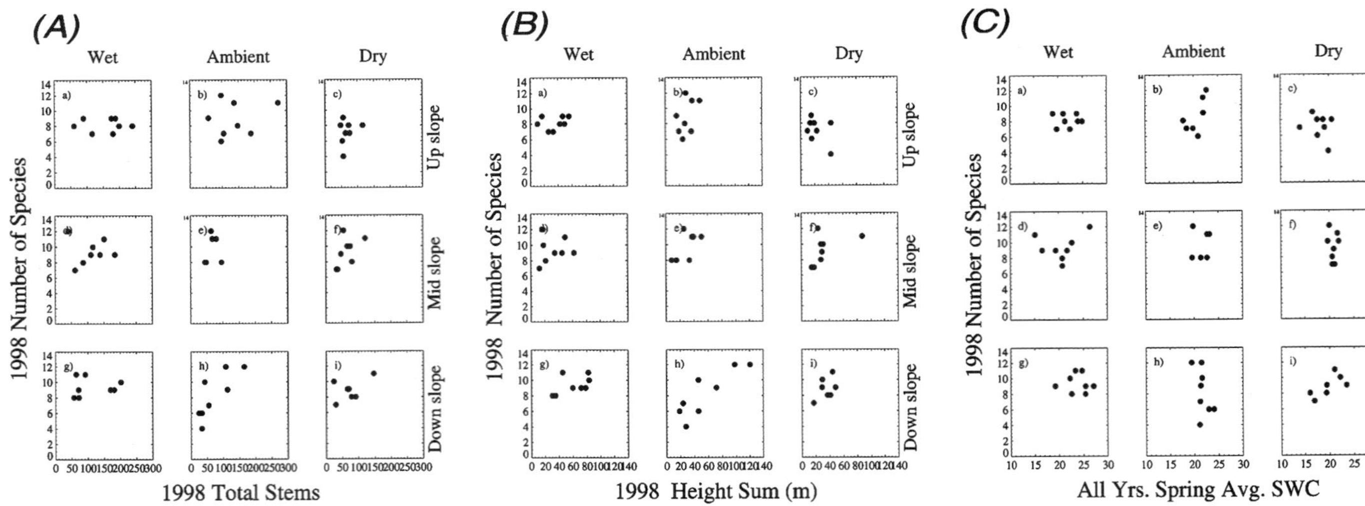

Figure 14.3. Total number of species present in each 3 × 3 m plot in 1998 in relation to (A) the total number of tree stems (< 2 m) in 1998 and (B) the sum of the heights of all woody and herbaceous plants present in each plot in 1998. Note the preponderance of positive slopes, suggesting that the effects of negative interactions (e.g., competition) on species diversity are relatively unimportant in most of the slope–treatment combinations. (C) Average spring SWC estimated for each 3 × 3 m plot (see Methods section). Note that some slope–treatment combinations have a very narrow range of average SWC (e.g., panel e), whereas other have a wider range (e.g., d,g).

Each of the five summary SWC variables (Average, Average Minimum, Absolute Minimum, Average Maximum, and Absolute Maximum) had differences between their patterns of variability within the slope-treatment areas, and also different relationships with the community structure variables (species richness, stem density, and height sum). In the results presented below, many linear patterns in response variables (species richness, stem density, and total vegetation (height sum) occur only in one of SWC summary variables, or are stronger one season than in another. Often, six or seven of the plots within a slope-treatment combination may have a linear relationship, with one or two outliers that do not fit the pattern. Given the large number of unknown and/or unmeasured variables that could influence these patterns, standard statistical approaches are useless, and approaches based on data stratification (e.g., into 9 slope-treatment combinations rather than three treatments) and quantile analyses (e.g., a focus on the maximum or minimum responses) offer the best opportunity to understand the processes underlying the patterns (Huston 2002).

Results

The mid-slope dry treatment consistently had the lowest variability, and the mid-slope wet treatment had the highest, for most SWC summary variables (e.g., Fig. 14.3C). There were few clear or consistent relationships between number of species and the various SWC summary variables. Only among the plots in the lower-slope ambient treatment was there a consistent linear pattern for species richness (and also for stem density and height sum), which was negative for most of the SWC summary variables, for both spring and summer values (e.g., Fig. 14.3Ch). Other slope–treatment combinations had linear patterns, often among six or seven of the plots with one or two outliers far off the line, which appeared only for one or two of the SWC summary variables (data not shown).

Although there was not a strong relationship of species richness, number of stems, or height sum with SWC variation within most of the slope–treatment combinations, these patterns were strongly influenced by physical and ecological conditions prior to the initiation of the TDE and are not likely to be a good indicator of treatment effects. A more sensitive indicator of vegetation response to changes in soil-water conditions between 1993 and 1998 is the change in the patterns over this time period. Both absolute and relative changes were evaluated and found to have similar patterns. Absolute changes (i.e., 1998 values − 1993 values) are presented here in relation to various SWC summary variables, to illustrate the consistency of some patterns and variability of others (Fig. 14.4). Changes in species number within the plots of the nine slope–treatment combinations showed consistent positive and consistent negative patterns, as well as no clear pattern in some of the areas (Fig. 14.4).

In several slope–treatment combinations, the change in species number was negatively related to various of the SWC variables; that is, more species were lost under wetter than drier conditions. These include two of the lower-slope combinations (wet and ambient), as well as the upper-slope dry treatment (Figs.

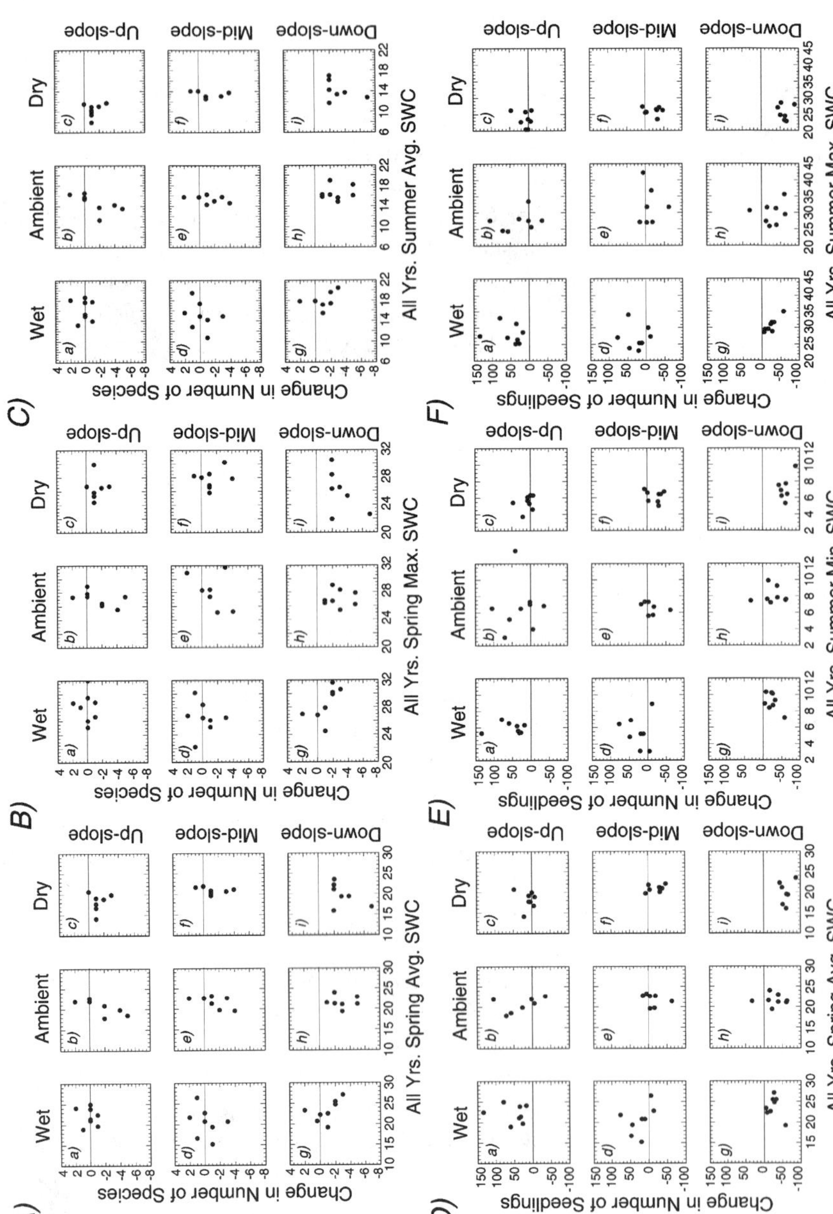
A)
B)
C)
D)
E)
F)
Wet
Ambient
Dry
Up-slope
Mid-slope
Down-slope
Change in Number of Species
Change in Number of Seedlings
All Yrs. Spring Avg. SWC
All Yrs. Spring Max. SWC
All Yrs. Summer Avg. SWC
All Yrs. Summer Min. SWC
All Yrs. Summer Max. SWC

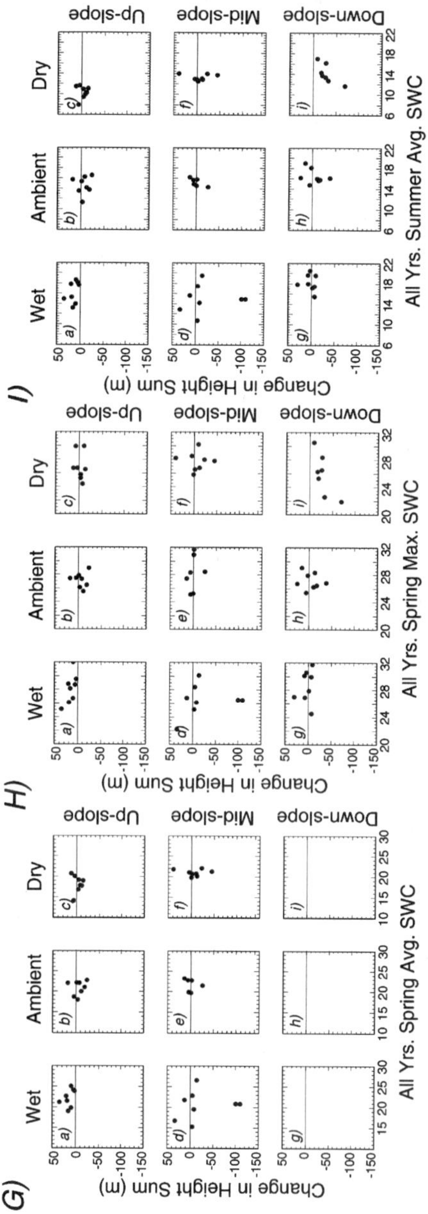

Figure 14.4. Changes in community structure variables from 1993 to 1998 in relation to various summary variables for SWC. (A–C) Change in the number of tree species present in each 3 × 3-m plot. (D–F) Change in the number of tree seedlings (stems < 2 m). (G–I) Change in the sum of the heights of all plants (< 2 m, woody and herbaceous). Note the occurrence of both positive and negative relationships and that many patterns change with changing SWC conditions.

14.4A–14.4Cc,g). Other plots had the expected positive relationship (assuming that increased water availability increased the probability of survival) between change in species number and at least one of the SWC summary variables, with fewer species being lost (or more gained) under wetter conditions, as in the upper-slope wet and ambient plots, the mid-slope wet plot, and the mid- and lower-slope dry plots (Figs. 14.4A–14.4Ca,b,d,f,i).

Changes in the number of seedlings (stems < 2 m) also show both positive and negative changes in relation to SWC variables. Two slope–treatment combinations, upslope ambient and lower-slope wet, have clear negative patterns (with one or no outliers) (Figs. 14.4Db and 14.4Fg), whereas others show indications of the expected positive patterns (particularly in the maximum response) for at least one SWC summary variable, (e.g., Figs. 14.4Ea,d, and 14.4Fc,f,i).

Height sum, an indicator of total plant mass, including both seedlings and herbaceous plants, also shows a mix of positive and negative responses across the slope–treatment combinations, generally in the same combinations that had patterns for species number, and number of stems. The lower-slope dry treatment shows a consistent positive response for most SWC variables (Figs. 14.4G–14.4Ii), whereas the three wet treatment slope positions tend to have negative responses (Figs. 14.4G–14.4Ia,d,g).

Patterns of Tree Species Distribution and Abundance

The above-described community-level responses are the net results of changes in the abundance of individual species, some of which may decline while other increase in response to the same change in soil-water conditions. Obviously, a positive response in some species could be canceled by a negative response in others, producing no net response in general properties such as species number or stem density. The responses of the individual species are a more direct indication of the mechanisms through which changes in soil-water conditions affect forest structure and diversity.

Twenty-nine species of tree seedlings were identified over the two surveys. Of these, 13 species had at least 72 individuals (an average of one per 3×3-m plot) in either 1993 or 1998 (Table 14.2). Six of these abundant species decreased in density between the surveys, whereas seven increased. Two species, sugar maple (*Acer saccharum*) and dogwood (*Cornus florida*) declined to less than 10% of their 1993 densities. Four species increased at least 50% over their 1993 densities: *Nyssa sylvatica*, *Quercus alba*, *Rhamnus caroliniana*, and *Sassafras albidum*.

The general pattern of density change within species was characterized by strong treatment and topographic effects. The two canopy tree species with the greatest absolute loss of individuals, *Acer rubrum* and *Lirodendron tulipifera*, had their greatest proportional reduction in the lower-slope dry treatment. *A. rubrum* decreased in density in all or most plots of the three dry treatment slope positions, with the reduction increasing from the upper portion to the lower portion of the slope (Fig. 14.5Ac,f,i). With the exception of the mid-slope ambient plot, *A. rubrum* had little change in stem density in the ambient and wet treatment slope

Table 14.2. Total number of tree seedlings (i.e., having stems < 2 m) on the TDE site surveyed in 1993 and 1998

Species	Total number 1993	Total number 1998	Difference 1998–1993	Fraction remaining 1998/1993
Declining species				
Acer rubrum	3798	3159	−639	0.83
Acer saccharum	82	3	−79	0.04
Carya sp.	109	70	−39	0.64
Carya tomentosa	88	83	−5	0.94
Cornus florida	241	17	−224	0.07
Liriodendron tulipifera	359	188	−171	0.52
Increasing species				
Nyssa sylvatica	305	507	202	1.66
Prunus serotina	187	221	34	1.18
Quercus alba	130	213	83	1.64
Quercus prinus	495	690	195	1.39
Quercus velutina	67	92	25	1.37
Rhamnus caroliniana	85	158	73	1.85
Sassafras albidum	727	1190	463	1.64
Minor species				
Albizia julibrissin	1	0	−1	
Amelanchier arborea	47	62	15	
Carpinus carolinina	37	24	−13	
Castanea dentata	2	3	1	
Cercis canadensis	43	45	2	
Diospyros virginiana	0	2	2	
Fagus grandifolia	14	12	−2	
Fraxinus americana	3	4	1	
Ilex montana	14	0	−14	
Juniperus virginiana	11	8	−3	
Liquidambar styraciflua	1	4	3	
Morus alba	2	2	0	
Morus rubra	9	8	−1	
Oxydendrum arboreum	11	10	10	
Pinus strobus	1	1	0	
Pinus spp.	2	1	−1	
Quercus spp.	17	5	−12	
Ulmus rubra	0	4	4	

Note: Abundant species are grouped according to those decreasing or increasing in numbers over the five year period, and uncommon species are shown together in one group.

positions. *L. tulipifera* increased in density in many plots in the middle- and upper-slope positions, particularly in the wet treatment, whereas it decreased in nearly all of the lower-slope plots, regardless of treatment (Fig. 14.5B).

Cornus florida, which declined across all treatment–slope combinations, had the smallest proportional reduction on the lower-slope wet treatment, although there was no net change in the mid-slope dry treatment. *N. sylvatica* increased in density across all treatment–slope combinations and had its greatest numeric increase in the three plots of the upper left corner of the site, upper and middle slope of the wet treatment and upper slope of the ambient treatment (Fig. 14.5C). *S. albidum* showed a similar pattern of maximum increase, but declined in the lower-slope wet and ambient treatments (Fig. 14.5D).

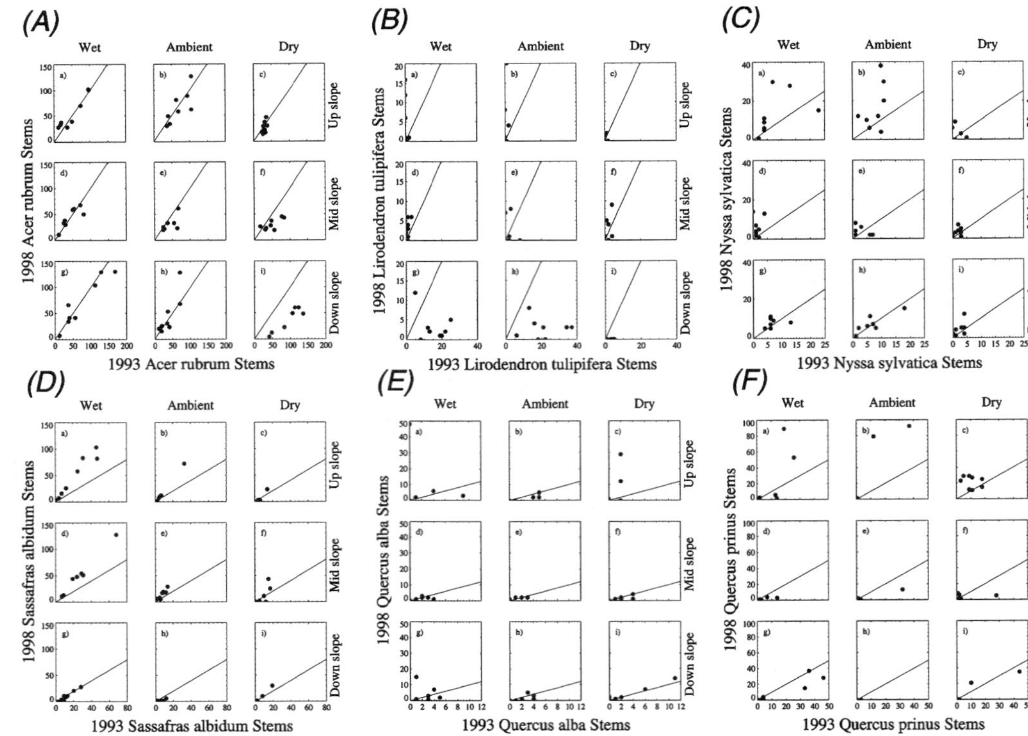

Figure 14.5. Change from 1993 to 1998 in the number of tree seedlings (stems < 2 m in height) per 3×3-m plot for various species. The diagonal line represents a 1 : 1 ratio between the 1993 values and the 1998 values (i.e., no change). Values to the left of the diagonal indicate in increase in that plot, and values to the right indicate a decrease. (A) *A. rubrum*; (B) *L. tulipifera*; (C) *N. sylvatica*; (D) *S. albidum*; (E) *Q. alba*; (F) *Q. prinus*.

Of the two most abundant oak species, *Q. alba* density remained constant or declined slightly in most plots, increasing significantly in only the upslope dry treatment and lower-slope wet treatment (Fig. 14.5Ec,g). *Q. prinus* increased substantially in a few plots in the upper-slope positions of all three treatments and declining in all of the other slope treatment combinations (Fig. 14.5F). *Carya tomentosa*, the most easily identified of the small hickories, remained relatively constant in all of the slope–treatment combinations.

Species Response to Local Heterogeneity in Soil Water Content

Changes in stem density over 5 years of treatment were related to relatively large differences in soil-water conditions associated with slope position and treatment for many species. At the scale of the entire experiment, most species tended to increase under wetter conditions and decrease where it was drier. A smaller range of variation in soil-water availability occurs within each slope–treatment combination as a consequence of local topography and perhaps other factors. In general, assuming an adequate seed supply and no other limiting factors, we would expect the survival and, thus, stem density of a species to be greater under wetter conditions and lower under dry conditions within a particular slope–treatment combination. One limiting factor that could be important under some conditions is competition from other species, that could reduce light or other resources. All other factors being equal, the density of competitors is likely to be highest under the most favorable conditions, which are probably the wet lower-slope areas of the TDE.

Acer rubrum is abundant across the site and had the largest numerical decrease of any species (Table 14.2), primarily in the dry treatment (Fig. 14.5A). Change in stem density within slope–treatment combinations was positively related to several SWC summary variables within the upper-slope wet treatment (Figs. 14.6Aa and 14.6Ba), and negatively related within the mid-slope dry treatment (Figs. 14.6Af and 14.6Bf). The counterintuitive negative relationship is potentially due to competition with other plants, as suggested by negative relationships between height sum and stem density in the mid- and lower-slope dry treatment positions (Fig. 14.6Cf,i).

Liriodendron tulipifera, which also decreased substantially (Table 14.2), had a slight tendency toward positive relationships between change in stem density and SWC variables (Fig. 14.6Da,b,e,h), as well as negative relationships with height sum (Fig. 14.6Fa,e).

Nyssa sylvatica density increased in all slope–treatment combinations, showing some negative relationships of stem density change with SWC (e.g., Figs. 14.6Gi and 14.6Hc,i) and at least one apparent positive relationship (Fig. 14.6Ha). The smaller increases in *N. sylvatica* density in the lower-slope positions may have been related to competition from other species, as suggested by the negative relationships between change in stem density and the change in height sum (Fig. 14.6Ig,h,i).

Sassafras albidum density increased substantially in some slope–treatment combinations (mid- and upper-slope wet treatment) but decreased in others

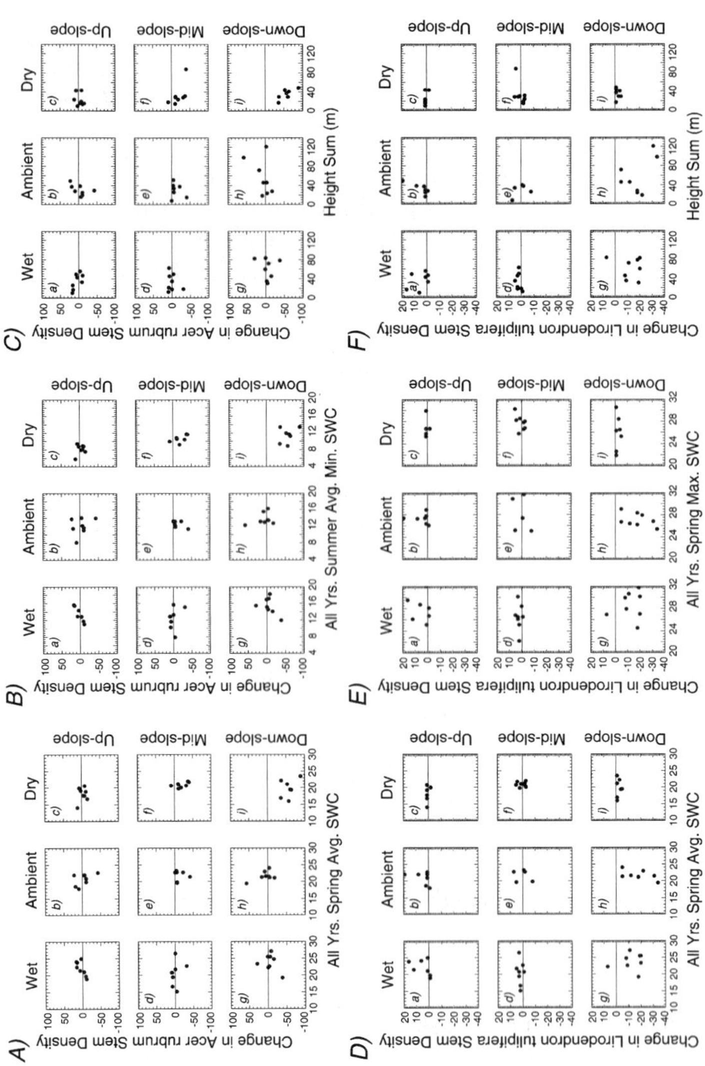
A)
B)
C)
D)
E)
F)
Change in Acer rubrum Stem Density
Change in Liriodendron tulipifera Stem Density
Wet
Ambient
Dry
Up-slope
Mid-slope
Down-slope
All Yrs. Spring Avg. SWC
All Yrs. Summer Avg. Min. SWC
Height Sum (m)
All Yrs. Spring Max. SWC

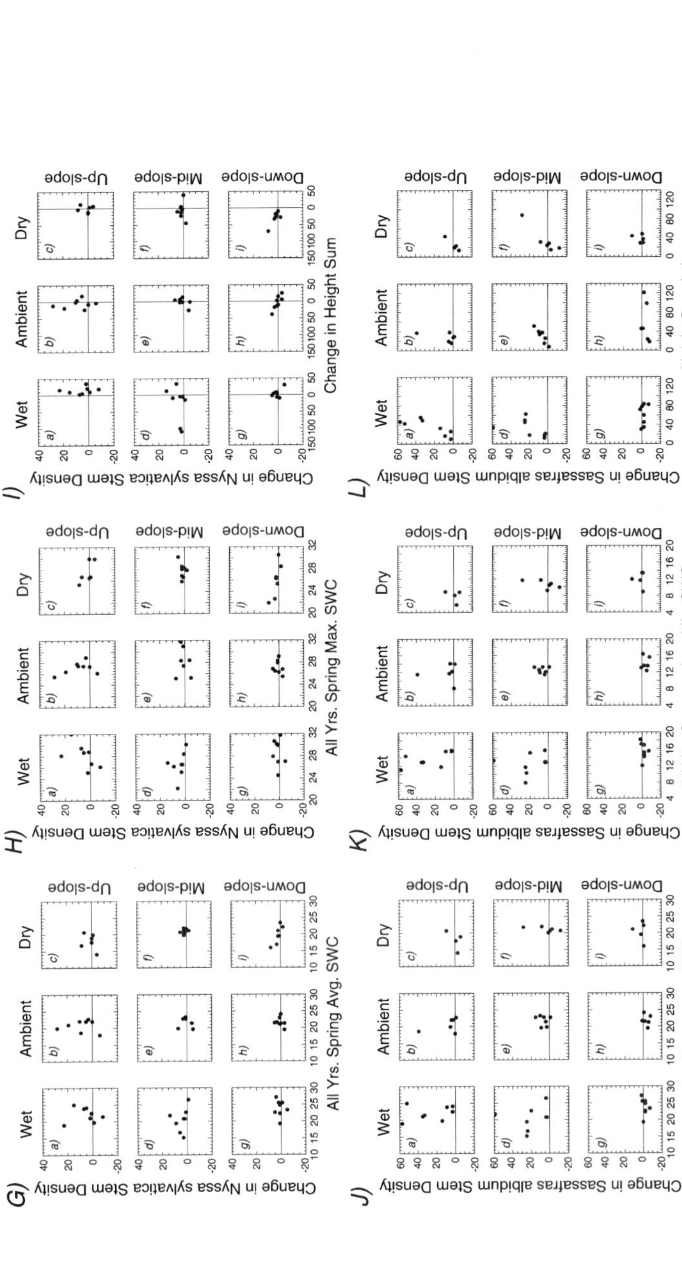

Figure 14.6. Change in seedling density (number of stems < 2 m per 3 × 3-m plot) between 1993 and 1998 for major tree species in relation to 2 (of the total of 10) summary variables for SWC and change in the sum of the heights of all woody and herbaceous plants < 2 m in height, an index of biomass and potential competitive interactions among the understory plants. Each panel of nine plots corresponds to the nine combinations of slope position (down-slope, mid-slope, and up-slope) and treatment (wet, ambient, dry) on the TDE. (A–C) *A. rubrum*; (D–F) *L. tulipifera*; (G–I) *N. sylvatica*; (J–L) *S. albidum*.

(lower-slope wet and ambient), showing no clear patterns of density change in relation to SWC (Figs. 14.6J and 14.6K). However, *S. albidum* showed numerous positive correlations with height sum (Fig. 14.6La,c,d,e,f,h), suggesting that it was not negatively affected by competition from other species and was itself contributing substantially to the height sum.

Cornus florida decreased precipitously across the site as saplings and adults (see Chapter 15, this volume), as well as plants <2 m in height, presumably as a consequence of the introduced dogwood anthracnose fungus. Seedling stem density declined in most plots, with a positive relationship between density change and SWC in the mid-slope wet treatment and a negative relationship in the upper-slope wet treatment (data not shown). The greatest declines in stem density were in the two plots with the largest height sum, in the lower-slope ambient treatment, suggesting a possible effect of competition (data not shown).

Quercus alba increased most in stem density in relatively dry plots within the upper-slope wet and dry areas. The greatest increase in density occurred in the driest plot of the upper-slope wet treatment, where there was also a negative relationship between change in stem density and the spring minimum SWC (data not shown). This plot also had a low height sum, suggesting that competition from other understory vegetation may have been weak in comparison with plots with higher height sum and slight or negative changes in stem density. In contrast, the change in *Q. alba* stem density increased with increasing SWC in the upper-slope dry treatment, where there was also a positive relationship with height sum (data not shown).

Changes in *Q. prinus* stem density were similar to those of *Q. alba*. *Q. prinus* increased most in stem density in relatively dry plots within the upper-slope wet and ambient areas (data not shown), but had no obvious trends in relation to soil water in the other slope–treatment combinations. Height sums in the plots with the highest *Q. prinus* stem densities were relatively low compared to some of the lower-slope plots, where *Q. prinus* densities remained stable or declined (data not shown).

Quercus velutina had both positive and negative relationships between change in stem density and SWC (data not shown). The positive relationships occurred primarily in the upper-slope and drier conditions, and the negative relationships in the lower-slope and wetter conditions. Relationships with height sum were negative in the upper- and lower-slope wet treatments, and positive in the upper-slope ambient and dry treatments.

Discussion

Understory tree seedling responses on the TDE exhibit all of the complexity one would expect of ecological phenomena (cf. Huston 2002). These include opposite correlations to a single variable under different conditions, such as the positive correlation in the wet treatment (Fig. 14.6Ba) and negative correlation in the dry treatment (Fig. 14.6Bf) of changes in *A. rubrum* stem density along a local soil moisture gradient, as well as many examples of no apparent response. Patterns of response and nonresponse of aggregate community variables, such as species richness and stem density, can be understood in the context of responses of the individual species. Only when most species respond similarly do clear patterns appear at the community level.

Responses to Variation in Soil-Water Content

Soil-water availability is not a simple phenomenon, regardless of how it is measured. Spatial patterns of soil-water availability on the TDE varied dramatically over time (see Chapter 2, this volume). Spatial heterogeneity in soil-water content is generally low under both extremely wet and extremely dry weather conditions and is maximum at some intermediate condition. In addition, topographic drainage alters water availability independently of weather and decouples the temporal dynamics of wetting and drying cycles between different positions on the landscape. Without continuous measurement of plant growth and physiological status in relation to soil-water availability (e.g., Chapter 6, this volume), it is difficult to determine the timing and value of the conditions most critical to the plants' survival or growth (which are not necessarily the same conditions). Consequently, determining the effects of soil-water content on the density and diversity of tree seedlings across the nine distinct slope–treatment combinations of the TDE, each with its own internal heterogeneity, is not a simple task.

Virtually none of the patterns presented here occur consistently for both the spring and summer values of all five SWC summary variables. This is not surprising, given the many other factors in addition to soil moisture that influence seedling survival and abundance. Two critical soil moisture properties varied among the nine slope–treatment combinations. First, the range of conditions (difference between maximum and minimum values) varied from low to high, depending on the local topography in each slope–treatment location. Second, the position of the values in the overall range of soil-water conditions over the entire site varied, from low soil-water availability in the driest slope–treatment combinations to higher soil-water availabilty in the wettest slope–treatment combinations. Given the high liklihood of nonlinear (and possibily unimodal) physiological and ecological responses to soil-water availability, we would expect contrasting patterns to appear in different slope–treatment combinations, as well as inconsistent patterns or no pattern where opposing processes were affecting the vegetation. Consequently, our interpretation is that if a clear linear pattern appears among at least six of the eight plots in a slope–treatment combination in at least 1 of the 10 SWC summary variables (5 summary variables for the spring and summer periods), this indicates a response to soil-water conditions that could be interpreted in terms of physiological or ecological processes.

Community-Level Responses

Two major patterns can be identified. First, there were clear changes across the entire TDE site that were independent of the treatments. Nearly all of the slope–

treatment combinations, including the ambient plots, declined in species number (Fig. 14.2A). The only location in which more plots gained species than lost species was the upper-slope wet treatment. In all other slope–treatment combinations, more plots lost than gained species. This general decline in species number must be the consequence of some change in conditions that affected all plots regardless of treatment. In addition, several individual species increased (e.g., *N. sylvatica*, *S. albidum*, *Q. prinus*, Figs. 14.5 and 14.6) or decreased [e.g., *C. florida* (data not shown) and *L. tulipifera* (Figs. 14.5 and 14.6)] independently of the treatments in subsets of the slope–treatment combinations.

One possible explanation for the sitewide responses is the weather during the 5 years between understory measurements. This period included wet years and dry years (see Chapter 2, this volume), but did not include any dry conditions as severe as the droughts of the late 1980s (see Chapter 24, this volume). Thus, soil-water conditions may have been somewhat wetter during the time the TDE was conducted than during the end of the 1980s, particularly in the wet and ambient treatments. In addition, some tree species may have been recovering from the 1980s droughts and increasing their reproductive output over the period of the TDE, as suggested below for *L. tulipifera* (see Chapters 14 and 15, this volume). In spite of any general factors that affected the entire site, it is also clear that very different processes were occurring in different slope–treatment combinations.

The second major pattern was the consistently opposite responses in two corners of the site. The greatest contrast was not between the driest (upper slope) dry plot and the wettest (lower slope) wet plot, but between the wettest (lower slope) dry treatment and the driest (upper slope) wet treatment. The upper slope wet treatment was the area with the fewest plots declining in species number (one out of eight), as well as the only slope–treatment combination in which all plots increased in both stem density and total height sum (Fig. 14.2). In contrast, the lower-slope dry treatment was the only slope–treatment combination in which all eight plots decreased in species number, in total stems, and in height sum (Fig. 14.2).

These two contrasting plots probably experienced the greatest relative change in soil-water conditions of all locations on the TDE site because of the interaction of treatment and topography. Prior to the initiation of the TDE, the lower-slope dry treatment was relatively wet, receiving runoff from the hillslope above it. However, with the initiation of the dry treatment, it experienced both a 33% reduction in direct throughfall and presumably a similar reduction in runoff (Chapter 23, this volume). In contrast, the upper-slope position of the dry treatment was dry prior to initiation of the TDE treatment, because it received little or no runoff. The imposition of the 33% reduction of throughfall in this location represented a smaller relative reduction in soil water than that experienced at the lower-slope positions.

A similar contrast occurred along the slope of the wet treatment. The upper-slope position was originally much drier than the lower-slope position, because it received little or no runoff from upslope. Consequently, the addition of 33% of each precipitation event to this dry site resulted in a much greater relative increase in soil water than that experienced at the bottom of the slope, which had always been a wetter environment.

These two contrasting locations are part of a pattern of change in stem density that was divided diagonally across the site, with the extreme responses in the lower-slope dry treatment, where all plots decreased in stem density (Fig. 14.2Bi) and the upper-slope wet treatment, where all plots increased in stem density (Fig. 14.2Ba). Similar to the upslope wet area, in both the adjacent upslope ambient area (Fig. 14.2Bb) and the adjacent mid-slope wet area (Fig. 14.2Bd), most of the eight plots increased in stem density. Diagonally across the TDE site, seven out of eight plots decreased in stem density in the lower-slope ambient area (Fig. 14.2Bh) and the mid-slope dry area (Fig. 14.2Bf), which are the two slope–treatment combinataions adjacent to the lower-slope dry area (Fig. 14.2Bi). Except for the extreme corner plots, this diagonal symmetry was not as evident for the changes in height sum (Fig. 14.2C).

This diagonally symmetrical pattern in the change in stem density indicates that the changes in species number resulted from different processes in different slope–treatment combinations. The lower-slope dry treatment, which was a relatively moist site prior to the initiation of the TDE, was occupied by plant species, both herbaceous and woody, that were adapted to these more mesic conditions and had presumably survived previous droughts because of the wetter conditions at lower slope positions. In addition, it is likely that the individual plants that were present, particularly the abundant *A. rubrum* (Fig. 14.5Ah) had developed carbon allocation patterns (e.g., root : shoot ratios) adapted to moist, rather than dry, environments. The 33% throughfall reduction, plus the reduction in runoff from upslope, was the apparent cause of the mortality of at least half of the *A. rubrum* and all of the *L. tulipifera* stems (Figs. 14.5A and 14.5B). Most of the *A. rubrum* decline for the entire TDE site occur in this single slope–treatment combination. Other tree species in this location changed relatively little in stem density or increased slightly (e.g., *N. sylvatica*, *S. albidum*, *Q. alba*, *Q. prinus*, and *Q. velutina* (Figs. 14.5 and 14.6). This topographic pattern of mortality (i.e., highest mortality in normally wetter landscape positions) is similar to the mortality pattern in adult trees reported by Elliott and Swank (1994), following the severe late 1980s drought at the Coweeta Hydrological Laboratory in North Carolina.

In contrast, the upper-slope wet treatment area was a dry, ridge-top environment prior to the initiation of the treatments. In spite of relatively high seedling densities (Fig. 14.2B), total plant cover was low in comparision to mid- and lower-slope positions, as indicated by the low height sums for the plots (Fig. 14.2C). The dry soils, in combination with the relatively deep shade (Table 14.1), created conditions unfavorable for the growth of most tree species and eliminated the dense herbaceous growth found in the mid- and lower-slope positions. With the 33% augmentation of throughfall input of the TDE treatment, this area became much more favorable for the growth and survival of tree seedlings, leading to a substantial increase in both stem density and height sum in all eight of the plots (Figs. 14.2B and 14.2C). All of the major tree species increased dramatically in stem density in at least some of the plots (Fig. 14.5), including *A. rubrum*, which approximately doubled its density in three of the plots, and changed relatively little in the others (Fig. 14.5A).

Understanding the variation in understory changes across the TDE site probably requires a more complex hydrologic analysis than has been applied at the site. In particular, the surface and subsurface drainage pathways could potentially extend the dry treatment effects into the mid- and lower-slope ambient areas because of the diagonal slope across the site, roughly perpendicular to the diagonal symmetry discussed earlier. Particularly in the dry treatment, the downslope flow direction tends toward the ambient plots (see Chapter 2, Fig. 2.5, this volume), which could have resulted in drying of the mid- and lower-slope ambient plots if their soil water regime was supplemented by runoff from the adjacent area that became the dry treatment. The flow direction in the wet treatment also tends to angle toward the ambient plot, although the impact appears to be primarily along the western edge of the ambient treatment (see Chapter 2, Fig. 2.5, this volume). Although altered hydrologic drainage might explain the reduction in stem density in the lower-slope ambient area, stem density also decreased in the lower-slope wet treatment, which would seem to be a sufficient distance from the dry treatment to be unaffected.

In the context of the possible drainage-augmented treatment effects, the preexisting light environment on the site may have contributed to the observed responses in a manner consistent with the light-water trade-off hypothesis discussed earlier. The unexpected pattern of light availability, with the upper-slope positions being significantly darker than the mid- and lower-slope and the dry treatment mid- and lower-slope areas being significantly darker than the same slope positions in the ambient and wet treatments (Table 14.1), has the consequence that the two slope–treatment combinations with the consistently opposite responses both had low light availability. The light-water trade-off hypothesis predicts that plant survival under these conditions should be extremely sensitive to soil-water content. Indeed, the shift from moist lower-slope conditions to treatment-caused dry conditions produced a decrease in tree species richness, tree stem density, and total vegetation height sum in all eight plots in this slope–treatment combination (Figs. 14.2Ai, 14.2Bi, and 14.2Ci). In contrast, in the low light environment of the formerly dry upper-slope wet treatment, the wetter conditions produced an increase in stem density and height sum in all 8 plots, while species richness remained unchanged (Figure 14.2Aa,Ba,Ca).

Whatever the causes, there was a clearly a strong slope position–treatment interaction. The lower-slope positions all declined in stem density, and the consistency and amount of decline increased from wet to dry treatments (Fig. 14.2Bg,h,i). The upper-slope positions all increased in stem density, and the consistency and strength of the increase increased from dry to wet (Fig. 14.2Ba,b,c). Along the mid-slope positions, stem density primarily increased in the wet treatment, decreased slightly in the ambient, and decreased more strongly in the dry treatment.

Responses of Individual Species

The strongest patterns discussed above are the consequence of the facts that most species (particularly *A. rubrum*) declined in density in the lower-slope dry treatment and that most species (except *A. rubrum*) increased in density in the upper-slope wet treatment.

The most surprising response may be that of *A. rubrum*, a species that is widely noted as increasing in abundance in eastern forests (Abrams 1998) and that was the most abundant seedling species on the TDE site (Table 14.2). In contrast to the increased basal-area growth of *A. rubrum* saplings on the wet treatment (see Chapter 15, Fig. 15.1, this volume), seedling densities changed little in the wet treatment, but decreased dramatically in the two lower-slope positions of the dry treatment (Fig. 14.5A). As discussed earlier, these two locations probably experienced the greatest decrease in soil-water availability of any slope–treatment combination on the TDE site, so preexisting root : shoot ratios and other facultative adaptions to a moist environment may have made seedlings in this area more susceptible to drought stress than seedlings that developed under drier conditions. The favorable effects of increased soil water on *A. rubrum* survival are clearly seen in the positive response to local variation in SWC in the upper-slope wet treatment (Figs. 14.6Aa and 14.6Ba). These mortality responses are consistent with the sensitivity of sapling transpiration to soil water potential observed by Wullschleger et al. (1998).

Surprisingly, the change in *A. rubrum* density in response to local variation in SWC in the mid- and lower-slope dry treatment tends to be negative (Figs. 14.6Af,i and 14.6Bf,i). Although this might result from an unfavorable carbon allocation pattern, as discussed earlier, it might also be caused by the negative effects of competition from other understory plants. The change in *A. rubrum* seedling density in these two lower-slope dry treatment plots was negatively related to the height sum, which is an index of total understory mass or volume and potentially related to the intensity of compeition (Fig. 14.6Cf,i). Changes in height sum were positively correlated with SWC in the lower-slope position, suggesting that most species were responding favorably to local variation in SWC (Fig. 14.4i).

The density change in two species clearly was clearly independent of the TDE treatment. *C. florida*, which experienced extensive mortality in the sapling and mature stages (Hanson et al. 2001; Chapter 15, this volume), also declined by more than 90% in seeding abundance (Table 14.2). This decline was associated with the spread of dogwood anthracose, caused by an introduced fungus. The fungal outbreak was regional and likely exacerbated by the severe droughts of the 1980s (see Chapter 24, this volume; Clinton et al. 1993). In addition to the increased seedling mortality, the recruitment of new seedlings was reduced by the mortality of most of the seed-producing mature dogwoods. There was 100% mortality of *C. florida* seedlings in 61 of the 70 plots censused in both years. However, of the three plots in which *C. florida* numbers increased slightly, two were in the lower-slope wet treatment, consistent with higher sapling growth in the wet treatment (see Chapter 15, Fig. 15.1, this volume), whereas one was in the mid-slope dry treatment, which was not consistent with the lower-sapling growth in the dry treatment (see Chapter 15, Fig. 15.1, this volume).

The response of seedlings of *L. tulipifera*, which, along with *Q. alba* and *Q. prinus*, is one of the most abundant tree species on Walker Branch Watershed, was also affected by lingering effects of the severe droughts of the late 1980s. The diameter growth of mature *L. tulipifera* decreased dramatically in 1986, the second year of severe drought (Barlar, 2000; see Chapter 24, this volume), concurrent with a major outbreak of the specialist yellow-poplar weevil (*Odontopus calceatus*) (Huston, personal observation). The combination of drought and defoliation by weevils resulted in extreme crown thinning in most of the mature *L. tulipifera* on the watershed, many of which have subsequently died (Huston and Barlar, personal observation). This severe impact on *L. tulipifera* had two major effects on the TDE site. First, the reduction in crown size and leaf density increased the light availability for understory plants, particularly in the lower-slope positions where this species is most abundant (Table 14.1). Second, the reduced vigor and mortality of the mature *L. tulipifera* greatly reduced the recruitment of seedlings, particularly in the lower-slope areas where the adults (Hanson et al. 1998, Fig. 3) and seedlings had previously been abundant. *Odontopus* abundance remained high until the mid-1990s, and the diameter growth of most mature *L. tulipifera* on the watershed remained far below the predrought rates, in spite of a return to high rainfall, and dropped again to extremely low levels during 1994 (Barlar, 2000, see Chapter 24, this volume), the wettest year between 1993 and 2000 (Table 2.3, Chapter 2, this volume). Following 1994, the growth of most mature *L. tulipifera* began to recover (Barlar 2000), and as indicated by the trees on the TDE site, recovery has continued through 2000 (Fig. 15.2, Chapter 15, this volume).

The dramatic growth decline and gradual recovery of mature *L. tulipifera* provide the context for understanding the dynamics of the seedling population. Not surprisingly, seedling densities dropped to zero in the lower-slope dry treatment, where *L. tulipifera* seedling density was low even prior to the initiation of the TDE treatments. However, densities also declined significantly in all but 1 of the 16 plots in the lower-slope ambient and lower-slope wet treatment areas (Fig. 14.5Bg,h,i), where *L. tulipifera* seedling densities were initially much higher than anywhere else on the TDE site. This decline in these lower-slope positions, where seedlings and adults were formerly abundant and vigorous, is surprising, particularly given the positive treatment response of the adult *L. tulipifera* on the TDE site, most of which occur on the lower-slope position (Fig. 15.2, Chapter 15, this volume). Furthermore, in a study of transplanted seedlings on the TDE site, Holmgren (1996) found the expected moisture-related slope and treatment mortality responses over 2 years (1994 and 1995), although, surprisingly, mortality was nearly as high in the lower-slope wet treatment (63%) as it was in the lower-slope dry treatment (75%). The absence of any relationship between density change and height sum provides no evidence for the role of competition in these areas

(Fig. 14.6Fg,h). The decline in seedling densities in these two areas may be more a consequence of reduced seed rain than of increased mortality, which averaged 20% per year in the mid- and lower-slope ambient areas in Holmgren's (1996) seedling transplant study.

The substantial increase in *L. tulipifera* seedling densities in all of the other slope–treatment combinations, except the upper-slope dry treatment, may be less surprising than it first appears. The gradual recovery of the growth rates, and presumably reproductive output, of mature individuals during the course of the TDE suggests that the seed availability for this species may have increased substantially between the pretreatment period and the later years of the experiment. Coupled with the presence of a mature *L. tulipifera* near the border of the upper-slope wet and ambient areas, as well as individuals in the mid-slope ambient and dry treatment areas, the relatively normal precipitation plus treatment-augmented soil-water conditions of 1996 and 1997 may have allowed the seedling densities to recover somewhat from the extremely low levels found in the mid- and upper-slope areas in 1993.

Of the remaining common species, most followed the basic pattern of total stem density described previously, with the greatest increases in the northwest corner of the site (upper-slope wet) and smaller increases toward the dry treatment and lower-slope positions, with little change in density in the lower-slope dry treatment (except for *N. sylvatica*, which increased slightly in density). The most consistent increases in density were in two bird-dispersed early successional species, *S. albidum* and *N. sylvatica* (Figs. 14.5C, 14.5D, 14.6G–14.6L). *N. sylvatica* is commonly a mid-canopy and suppressed understory species, whereas *S. albidum* seems to require more light to survive. Neither species increased substantially in the lower-slope positions, where competition may have been a factor. *N. sylvatica* showed a clear negative relationship between the change in its density and the change in total height sum, an index of total vegetation amount that is likely to be correlated with the intensity of competition (Fig. 14.6Ig,h,i). *S. albidum* also showed a negative trend between change in stem density and the change in height sum (data not shown).

The two abundant oaks, *Q. alba* and *Q. prinus*, increased significantly in density only in the upper-slope positions, regardless of treatment (Figs. 14.5E and 14.5F), with only a few of the individual plots experiencing a major increase in density. In addition to being the driest slope position in all of the treatments, the upper-slope positions had lower light levels than mid- and lower-slope position, suggesting that these oaks are surprisingly shade tolerant, even under relatively dry conditions. Nonetheless, both species showed clear responses to local variation in SWC, and the responses appeared to reverse between the wet and dry treatments. Change in stem density was negatively correlated with SWC in the upper-slope wet treatment and positively correlated in the upper-slope dry treatment (data not shown). *Q. velutina* was much less abundant over the TDE site, but increased slightly in some plots within all but one of the nine slope–treatment combinations. *Q. velutina* tended to have positive relationships between stem density change and SWC in the upslope plots and negative relationships in the lower-slope plots (data not shown). Some slope–treatment combinations had negative relationships with height sum (data not shown), suggesting negative effects of competition.

Conclusion

Seedling responses to the throughfall displacement treatments were strongly species-specific, with some species decreasing and others increasing. Most species decreased in density in the dry treatment and also in the lower-slope portions of the ambient and wet treatment. Those species that increased in density tended to do so in the upper-slope areas of the wet and ambient treatments, as well as in the mid-slope area of the wet treatment. The strongest positive and negative density responses were in the two plots that experienced the greatest relative change in soil moisture conditions as a result of the experimental treatments. These plots also had relatively low light availability, which the light–water trade-off hypothesis predicts should make plant survival very sensitive to variation in soil-water availability. The upper-slope wet treatment plot likely experienced the greatest increase in soil moisture availability and the lower-slope dry treatment probably had the greatest relative decrease in soil moisture availability. In numerous cases, strong responses in the upper-slope portion of a treatment (including the ambient) were the opposite of the responses in the lower-slope portion, producing no net effect for the treatment as a whole. This indicates that some of the treatment responses cannot be detected and understood unless they are subdivided into slope position–treatment combinations (cf. Huston 2002).

Responses of the individual species indicate that climatic changes analogous to the TDE treatments are likely to have effects on tree regeneration that are topographically restricted, rather than uniformly distributed across the landscape. Specifically, an increase in precipitation is likely to extend species distributions to formerly drier, upslope portions of the landscape while potentially eliminating some species from the lower-slope areas they formerly occupied. A decrease in precipitation may have the strongest effects on formerly wetter, lower-slope portions of the landscape, reducing or eliminating some of the species adapted to those conditions and having relatively little effects on the abundance and diversity of species occupying drier portions of the landscape.

Over the 5-year period covered by this study, there were some changes that appeared to be independent of the treatments, such as the decline in the number of species in most plots across the site. Some of the changes in the ambient areas were similar to those in the adjacent treatment areas. Possible causes of these general responses include directional climate changes over the course of the experiment, delayed physiological and ecological recovery from the effects of previous climatic conditions, specifically the severe drought of the late 1980s, and unanticipated interactions between treatment and ambient plots caused by alteration of downslope drainage.

References

Abrams MD (1998) The red maple paradox. BioScience 48:355–364.

Abrams MD, Mostoller SA (1995) Gas exchange, leaf structure, and nitrogen in contrasting successional tree species in open and understory sites during a drought. Tree Physiol 15:361–370.

Barlar DG (2000) *Temporal and spatial variation in annual growth of yellow-poplar (Liriodendron tulipifera, L.) at Walker Branch Watershed, Tennessee.* Masters Thesis, School of Forestry, The University of Tennessee, Knoxville, Tennessee.

Berkowitz AR, Canham CD, Kelly VR (1995) Competition vs. facilitation in early successional communities. Ecology 76:1156–1168.

Callaway RM, Brooker RW, Choler P, Kikvidze Z, Lortie CJ, et al. (2002) Positive interactions among alpine plants increase with stress. Nature 417:844–847.

Clinton BD, Boring LR, Swank WT (1993) Canopy gap characteristics and drought influences in oak forests of the Coweeta Basin. Ecology 74:1551–1558.

Cole P (2002) *Factors affecting the distribution of an exotic grass species, Microstegium vimineum.* Ph.D. Dissertation, The University of Tennessee, Knoxville, Tennessee.

Coomes DA, Grubb PJ (2000) Impacts of root competition in forests and woodlands: A theoretical framework and review of experiments. Ecol Mono 70:171–207.

Daubenmire RF (1952) Forest vegetation of northern Idaho and adjacent Washington, and its bearing on concepts of vegetation classification. Ecol Mono 22:301–330.

Elliott KJ, Swank WT (1994) Impacts of drought on tree mortality and growth in a mixed hardwood forest. J Veg Sci 5:229–236.

Fisher BL, Howe HF, Wright SJ (1991) Survival and growth of *Virola surinamensis* yearlings: Water augmentation in gap and understory. Oecologia 86:292–297.

Kolb TE, Steiner KC, McCormick LH, Bowersox TW (1990) Growth response of northern red-oak and yellow-poplar seedlings to light, soil moisture, and nutrients in relation to ecological strategy. For Ecol Manag 38:65–78.

Hanson PJ, Todd DE, Huston MA, Joslin JD, Croker J and Augé RM (1998) *Description and field performance of the Walker Branch Throughfall Displacement Experiment: 1993–1996.* ORNL/TM-13586, Oak Ridge National Laboratory, Oak Ridge, Tennessee.

Hanson PJ, Todd DE, Amthor JS (2001) A six-year study of sapling and large-tree growth and mortality responses to natural and induced variability in precipitation and throughfall. Tree Physiol 21:345–358.

Holmgren M (1996) *The interactive effects of shade and drought on seedling growth and survival.* Ph.D. Dissertation. University of Tennessee, Knoxville, Tennessee

Holmgren M (2000) Combined effects of shade and drought on tulip poplar seedlings: trade-off in tolerance or facilitation? Oikos 90:67–78.

Holmgren M., Scheffer M, Huston MA (1997) The interplay of facilitation and competition in plant communities. Ecology 78:1966–1975.

Howe HF (1990) Survival and growth of juvenile *Virola surinamensis* in Panama: Effects of herbivory and canopy closure. J Trop Ecol 6:259–280.

Huston MA (1991) Use of individual-based forest succession models to link physiological whole-tree models to landscape-scale ecosystem models. Tree Physiol 9:293–306.

Huston MA (1994) *Biological Diversity: The Coexistence of Species on Changing Landscapes.* Cambridge University Press, Cambridge, England.

Huston MA (1997) Hidden Treatments in Ecological Experiments: Re-evaluating the Ecosystem Function of Biodiversity. Oecologia 110:449–460.

Huston MA (2002) Introductory Essay: Critical issues for improving predictions. In Scott JM, Heglund PJ, Morrison ML, et al. (Eds) *Predicting Species Occurrences,: Issues of Scale and Accuracy,* Island Press, Washington, District of Columbia, pp 7–21.

Huston MA, McBride AC (2002) Evaluating the relative strengths of biotic versus abiotic controls on ecosystem processes. In Loreau M, Naeem S, Inchausti P (Eds) *Approaches to Understanding Biodiversity and Ecosystem Function,* Oxford University Press, Oxford, England, pp 47–60.

Huston MA, Aarssen LW, Austin MP, Cade BS, Fridley JD, Garnier E, Grime JP, Hodgson J, Lauenroth WK, Thompson K, Wardle DA (2000) No consistent effect of plant diversity on productivity. Technical comment on "Plant diversity and productivity experiments in European grasslands," A. Hector et al., Science, Nov. 5, 1999. *Science* 289:1255a (www.sciencemag.org/cgi/content/full/289/5483/1255a).

Jarvis, P.G., G.B. Grace, and J.J. Landsberg. 1976. Coniferous forest. In Monteith JL (Ed) *Vegetation and the Atmosphere, Volume 2,* Academic Press, London, England, pp 246–272.

O'Connor RJ (2002) The conceptual basis of species distribution modeling: Time for a paradigm shift? In Scott JM, Heglund PJ, Morrison ML, et al. (Eds) *Predicting Species Occurrences,: Issues of Scale and Accuracy.* Island Press, Washington, District of Columbia, pp 25–33.

Sack L, Grubb PJ (2002) The combined impacts of deep shade and drought on the growth and biomass allocation of shade-tolerant woody seedlings. Oecologia 131:175–185.

Smith TM, Huston MA (1989) A theory of the spatial and temporal dynamics of plant communities. Vegetatio 83: 49–69.

Vance NC, Zaeer JB (1991) The influence of drought stress and low irradiance on plant water relations and structural constituents in needs of *Pinus ponderosa* seedlings. Tree Physiol 8:175–184.

Veenendaal EM, Swaine MD, Agyeman VK, Blay D, Abebrese IK, Mullins CE (1996) Differences in plant and soil water relations in and around a forest gap in West Africa during the dry season may influence seedling establishment and survival. J Ecol 84:83–90.

Withers JR (1979) Studies on the status of unburnt *Eucalyptus* woodland at Ocean Grove, Victoria. V. The interactive effects of droughting and shading on seedlings under competition. Aust J Bot 27:285–300.

Wullschleger SD, Hanson PJ, Tschaplinski TJ (1998) Whole-plant water flux in understory red maple exposed to altered precipitation regimes. Tree Physiol 18:71–79.

15. Tree and Sapling Growth and Mortality

Paul J. Hanson, Donald E. Todd, Darrell C. West,
Nelson T. Edwards, M. Lynn Tharp, and William A. Simpson Jr.

Plant-dry-matter accumulation ultimately depends on the yield of carbon building blocks (i.e., nonstructural carbohydrates) from the difference between carbon assimilation (Chapter 3, this volume) and autotrophic respiration (Chapter 4, this volume). Accumulated sugar and starch reserves in roots and shoots (Chapter 5, this volume) represent the primary compounds, along with stored elements, that must be present to support mass accumulation in the stems of saplings and trees. Waring and Pitman (1985) proposed a hierarchy of photosynthate allocation priorities for trees that considers stem growth to be a relatively low allocation priority, suggesting that changes in stem-growth rates would be a sensitive indicator of water stress response. Sustained low stem-growth rates have also been linked to mortality (Kohyama and Hara 1989; Pedersen 1998; Swaine et al. 1987; Tainter et al. 1984). Because growth and mortality are the integrated result of of physiological responses to environmental stress, they represent key end points for investigations of forest response to changing precipitation regimes. This chapter describes the annual growth and mortality of saplings and large trees during 7 full years of throughfall manipulation (1994–2000) and provides a quantitative description of the response of tree growth to soil-water deficits. The Throughfall Displacement Experiment (TDE) observations are also contrasted with published data for other deciduous hardwood forests, and their application for predicting growth responses to soil-water deficits throughout the eastern deciduous hardwood forest is discussed.

Methods

Sapling Growth

In February and March of 1994, 10 transects were established for observations of sapling growth and mortality across the three TDE plots from lower- to upper-slope positions. Although other species were considered for these measurements, only *Acer rubrum* L. and *Cornus florida* L. were distributed across the TDE in sufficient numbers for inclusion. Saplings ranged from 10- to 60-mm diameter at breast height (dbh), with the majority from 10 to 40 mm. Height measurements were not included because the crowns were broad without predominant main shoots and because height growth was minimal in the low light understory environment of our closed canopy stand. Starting at spring leafout, biweekly measurements of stem diameter at a permanently marked location on each sapling's main stem (typically between 1 and 1.5 m above the ground) were conducted until sapling growth had ceased for that year. Each stem diameter value was the mean of three replicate diameter measurements made with a digital caliper (0.01-mm resolution) from three different angles around the marked point of measurement. The mean of replicate measures from different angles was required to minimize the impact of noncircular stem cross sections. Sapling stem diameters were converted to basal area to express mean daily sapling growth rates per plant in $mm^2\ d^{-1}$ or integrated annual sapling growth in $mm^2\ y^{-1}$.

Tree Growth

Measurements of all individuals greater than 0.1 m dbh and a record of their presence or mortality were recorded annually. *Quercus alba* L., *Q. prinus* L., *A. rubrum, Liriodendron tulipifera* L., and *Nyssa sylvatica* Marsh. trees greater than 0.2 m dbh (170 trees) were fitted with dendrometer bands for biweekly measurements of stem circumference during each growing season. These five species made up almost 80% of the basal area of the experimental area (Hanson et al. 2001a, 2001b). A single dendrometer measurement consisted of duplicate digital caliper measurements (0.01-mm resolution) of the distance between two reference holes in stainless-steel dendrometer bands (25.4 mm wide $\times$ 0.2 mm thick) installed around the circumference of each tree (McLaughlin and Downing 1996). Measured changes in the circumference of each tree were combined with information on its initial stem diameter to obtain the change in stem basal area over time ($mm^2\ d^{-1}$ or $mm^2\ y^{-1}$). Dendrometer bands were installed on *Q. alba, Q. prinus,* and *A. rubrum* trees prior to the 1993 growing season, and bands for *L. tulipifera* and *N. sylvatica* were added in February of 1994. All dendrometer bands were installed during the dormant season, ahead of the initial growth measurements, to eliminate potential first-year bias in the dendrometer band measurements (Keeland and Sharitz 1993).

Winter stem shrinkage (Winget and Kozlowski 1964, 1965) or stem shrinkage coincident with leafout was observed to varying degrees for all five species measured in this study. Therefore, cumulative growth and/or biweekly growth rate integrals were calculated only after recovery from shrinkage. Because basal area change over the 2-week measurement periods is large when compared to diurnal shrink/swell patterns (Haasis 1934; Kozlowski 1971; Kramer and Kozlowski 1979; Kozlowski 1982), it was not necessary to correct measured circumference growth for diurnal patterns of stem shrinkage.

Statistical Analyses of Growth and Mortality

Individual trees and saplings were considered to be the experimental unit for analyses of the effects of treatment and year-to-year climate impacts. To account for bias in growth rates caused by pretreatment plant size, the initial basal area of individual saplings was used as a covariate in analysis-of-variance tests of treatment effects. Similarly, for the large trees, measured pretreatment growth rates (*Acer, Quercus*) and/or their initial basal areas (*Liriodendron, Nyssa*) were used as covariates. All covariates were highly significant and, therefore, appropriate for their intended purpose. Chi-square analyses were used to test for treatment effects on sapling mortality rates. Nonlinear regression was used to evaluate relationships between growth rate, phenology, and soil-water potential. All statistical analyses were conducted with SPSS 6.1 for Macintosh (SPSS, Inc.).

Variation in Wood Density Under Water Stress

To evaluate the hypothesis that mature tree mass was changing independently of volume, we studied annual ring volume growth and wood density. The primary objective was to determine if quantifying and including annual density together with the basal-area increment would make a significant difference in estimated dry biomass production.

Tree increment cores were collected from all canopy-dominant trees used for dendrometer growth observations (*Acer, Nyssa, Liriodendron,* and *Quercus*) and analyzed with X-ray densitometry. This scheme resulted in samples from 232 trees across all treatments. Two 5-mm-diameter cores were extracted from opposite sides of each tree to obtain an average ring width estimate. Each core was extracted to a depth between 150 and 200 mm and included tissue for the period from 1970 to 1999 and often more years. All tree cores were processed for X-ray tomography (CT) examination similar to that described by Tuskan et al. (1999). Each core was dried under vacuum at room temperature until no weight change could be detected. Each core, containing 30–50 annual rings, was scanned with a 420-kV, 3-mA collimated, 0.474-mm-thick X-ray beam with a Scientific Measurement Systems Model B201 CAT scanning device (Elliott et al. 1994). The X-ray density data from the reconstructed two-dimensional digital image for each core were stored as an array of numbers representing local X-ray attenuation. The middle scan of this three-dimensional array (a 0.474-mm-wide slice through the longitudinal axis of the tree core) was extracted for two-dimensional analysis of X-ray density at 158-μm segments along each core. Raw X-ray density units per segment were converted to mass with species-specific multipliers derived from measured sapwood densities and the known volume of wood sampled by CT scans: 0.00198 g unit^{-1} for *A. rubrum,* 0.00178 g unit^{-1} for *N. sylvatica,* 0.0019 g unit^{-1} for *L. tulipifera,* 0.00184 g unit^{-1} for *Q. alba,* and 0.00188 g unit^{-1} for *Q. prinus.* Multipliers differed because organic composition, anatomical features, cell structure, and mineral content of cell compounds differ from species to species. Increments along each CT scan were partitioned to years of growth by (1) initial identification of X-ray density minima with the distance between minima representing one annual growth cycle and (2) visual verification of these annual estimates against the actual tree cores to identify false minima.

Results

Sapling Growth

Annual sapling growth varied by species, TDE treatment, and year (Fig. 15.1). Mean annual basal-area growth rates ranged from 18 to 33 mm^2 y^{-1}, with the highest and lowest rates observed for the wet and dry treatment saplings, respectively. Over the 7-year period, the mean annual sapling basal area increment of *A. rubrum* was significantly higher ($p < 0.01$) in the wet plot compared to the ambient and dry plots (Fig. 15.1). The mean annual basal-area growth of *C. florida* saplings was not only significantly higher in the wet plot but also significantly lower in the dry plot compared to the ambient plot ($p = 0.01$). The highest annual growth for *A. rubrum* was observed in 1996 and 1997, but the highest growth rates for *C. florida* were measured in 1994, the year of lowest *A. rubrum* growth. The lowest growth rates for *C. florida* were measured in 1997, one of the "wet"

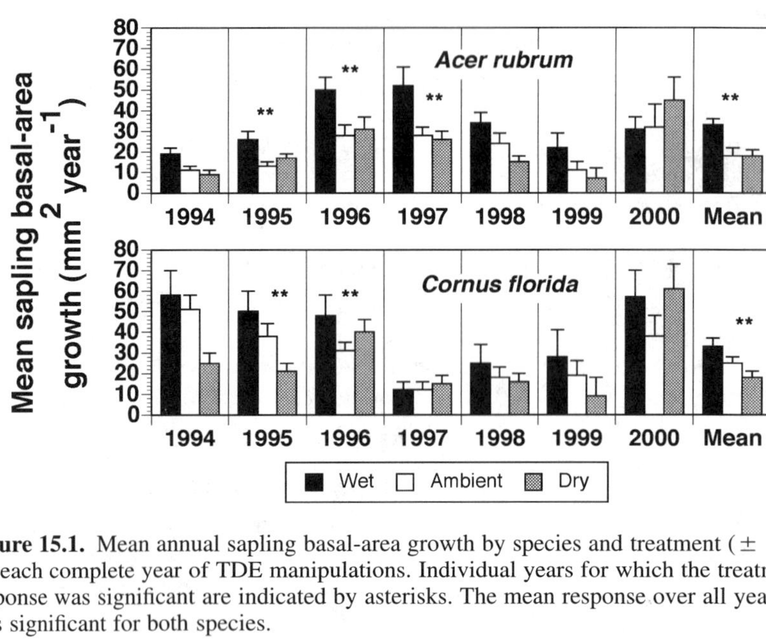

Figure 15.1. Mean annual sapling basal-area growth by species and treatment (± SE) for each complete year of TDE manipulations. Individual years for which the treatment response was significant are indicated by asterisks. The mean response over all years was significant for both species.

growing seasons. Sapling stem growth of *A. rubrum* and *C. florida* is largely limited to the first half of the growing season in 1994, 1995, and 1996 (Hanson et al. 2001a); thus, most sapling stem diameter growth occurs before the development of late-season water deficits (Chapter 2, this volume). During the 2000 growing season, the mean growth of understory saplings increased by a factor of 2–3 over that observed in 1998 and 1999, and treatment differences were not present (Fig. 15.1). Optimum growth conditions (i.e., minimal water stress) early in 2000 limited the impact of the TDE treatments in 2000.

Tree Growth

Unlike the sensitivity of sapling growth, no significant treatment responses of mean tree diameter growth were observed with dendrometer bands through the 2000 growing season (Fig. 15.2). Because of a concern that very large-tree (dbh > 0.35 m) growth responses might be masking a treatment response of smaller canopy trees (dbh between 0.1 and 0.35 m) when averaged in with the other trees, we further evaluated treatment responses with respect to tree size for all trees in the annual dbh survey. This test confirmed no significant treatment effects on tree growth (data not shown). Two species, *L. tulipifera* and *N. sylvatica,* have begun to show trends toward reduced and increased growth on the dry and wet plots, respectively (Fig. 15.2, year 2000). Although not yet significant, it is anticipated that this species-specific response would become significant if the chronic TDE manipulations are sustained for a long time period.

Notwithstanding the lack of treatment responses for large-tree basal-area growth, substantial differences were observed between species and years (Fig. 15.2). *Q. prinus* consistently had the highest annual growth rate of the five species monitored, but *Q. alba* and *A. rubrum* were nearly as productive. Annual growth of *N. sylvatica* was less than half that of *Quercus* and *Acer* trees. For all species except *Liriodendron,* maximum annual growth occurred during 1994 or 1996, and minimum annual growth occurred during 1993 or 1995. In the drought years of 1993 and 1995, annual basal-area growth was reduced when compared to growth in the wet growing seasons of 1994 and 1996, but severe late-season droughts in 1998 and 1999 had little impact on basal-area growth in those years. *L. tulipifera* exhibited a unique pattern of growth over time, showing increasing growth rates from 1994 through 2000. Hanson et al. (2001a) hypothesized that slow growth of *L. tulipifera* during the first 3 years of the study resulted from intense herbivory by the yellow-poplar weevil (*Odontopus calceatus;* Burns and Gibson 1968). In more recent years (1999 and 2000), basal area growth of *L. tulipifera* has reached rates as high as or higher than those of *Acer* and *Quercus.*

Although it is commonly assumed that large trees have many deep roots that provide protection against periodic droughts (Kozlowski 1982; Abrams 1990), large-tree growth reductions associated with drought conditions in 1993, 1995, 1998, and 1999 were apparently driven only by the reduced soil-water potentials of the upper 0.35 m of the soil horizon (Fig. 2.11, Chapter 2, this volume). Soil-

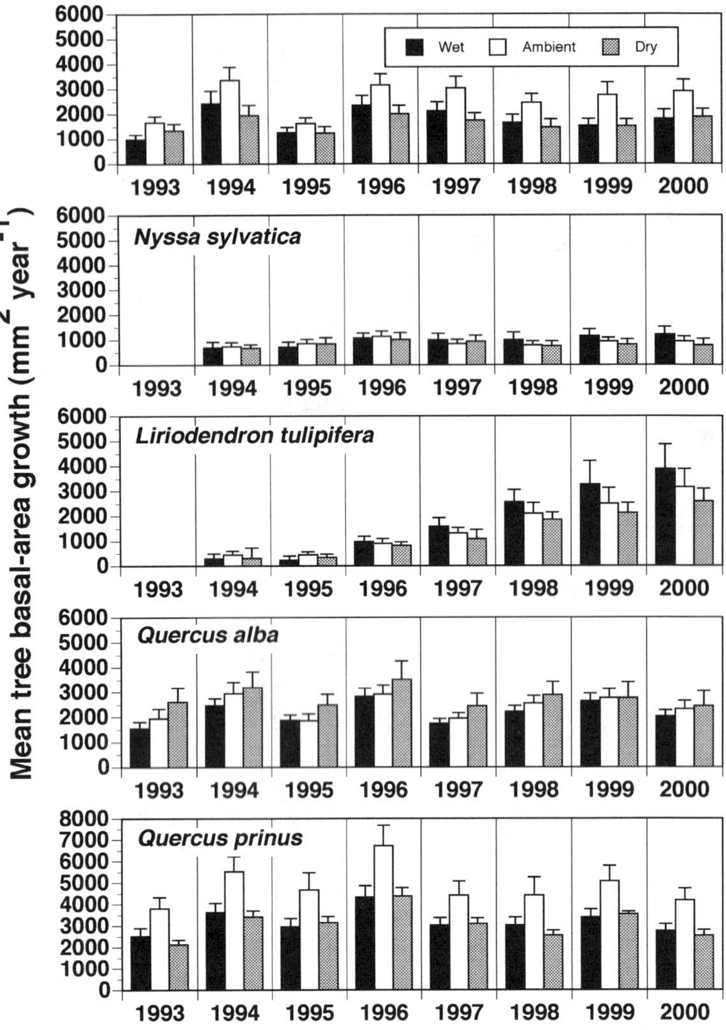

Figure 15.2. Mean annual tree (stems > 0.1 m dbh) basal-area growth by species and treatment for the pretreatment year (1993) and all other years receiving sustained TDE manipulations (1994–2000). Analyses of covariance using pretreatment growth (*A. rubrum, Q. alba,* and *Q. prinus*) or initial basal area (*N. sylvatica* and *L. tulipifera*) as covariates have not shown any significant treatment effect through 2000.

water potentials in deeper soil horizons (0.35–0.7 m) never fell below −1 MPa, suggesting that ample water resources were available with depth. Because the growth reductions in dry years were associated with only shallow-soil drying, we concluded that the number and/or conductivity of deep roots of eastern deciduous forests are insufficient to sustain unencumbered growth during drought. However, deep roots may still provide water sufficient to sustain low levels of leaf function (Chapters 3 and 7, this volume) and water use (Chapter 6, this volume).

Timing of Bole and Branch Growth

Measurements of branch elongation and of diameter change of terminal branches of overstory trees were conducted simultaneously with bole diameter measurements to determine if the timing of bole diameter growth was a good indicator of growth taking place throughout the tree. Terminal branches (28 branches 10–20 mm in diameter) from 8 trees of 2 species (*Acer, Quercus*) accessible from canopy access towers were marked for elongation and diameter measurements before budbreak and were subsequently measured along with bole circumference change throughout the 1997 growing season. The rates of leaf elongation and acorn development (*Quercus*) on these branches were also observed, but they were not expected to correlate with either bole- or branch-growth rates because they are accomplished early in the spring and late in the summer, respectively. The relative growth for leaf length, branch elongation, and diameter growth of branches and boles is plotted in Fig. 15.3. As expected, leaf expansion and branch elongation occurred early in the growing season and preceded the development of branch and bole diameter. Species had little impact on timing of bole and branch development, and the bole and branch diameter increases were essentially synchronous in time. These data suggest that tree mass from allometric equations can be proportionately allocated to branch and bole growth throughout the growing season. Measurable acorn diameter growth was initiated after stem growth was 50% complete, but completed expansion at the same time as did the tree boles and branches.

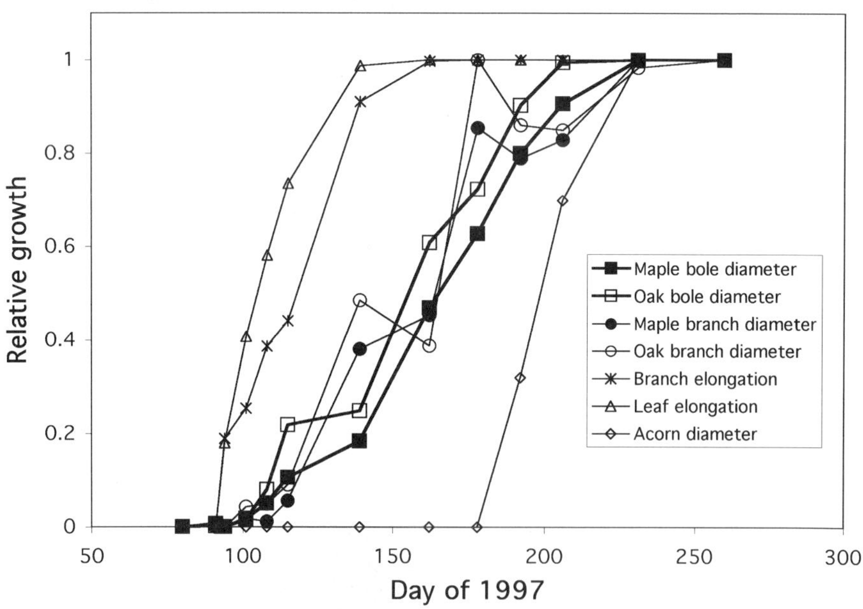

Figure 15.3. Growth relative to the maximum for each tissue for leaf elongation, branch elongation, bole diameter change at dbh, branch diameter change, and the diameter change of acorns throughout the 1997 growing season.

Mortality

Through the 2000 growing season, the mortality of *C. florida* saplings was more than double the mortality of *A. rubrum* saplings, but both species showed a similar long-term trend (Fig. 15.4). The greatest annual mortality of both *A. rubrum* and *C. florida* was observed in 1998. Significant treatment effects for annual mortality of *A. rubrum* were found only in 1998 and 1999, but *C. florida* mortality showed

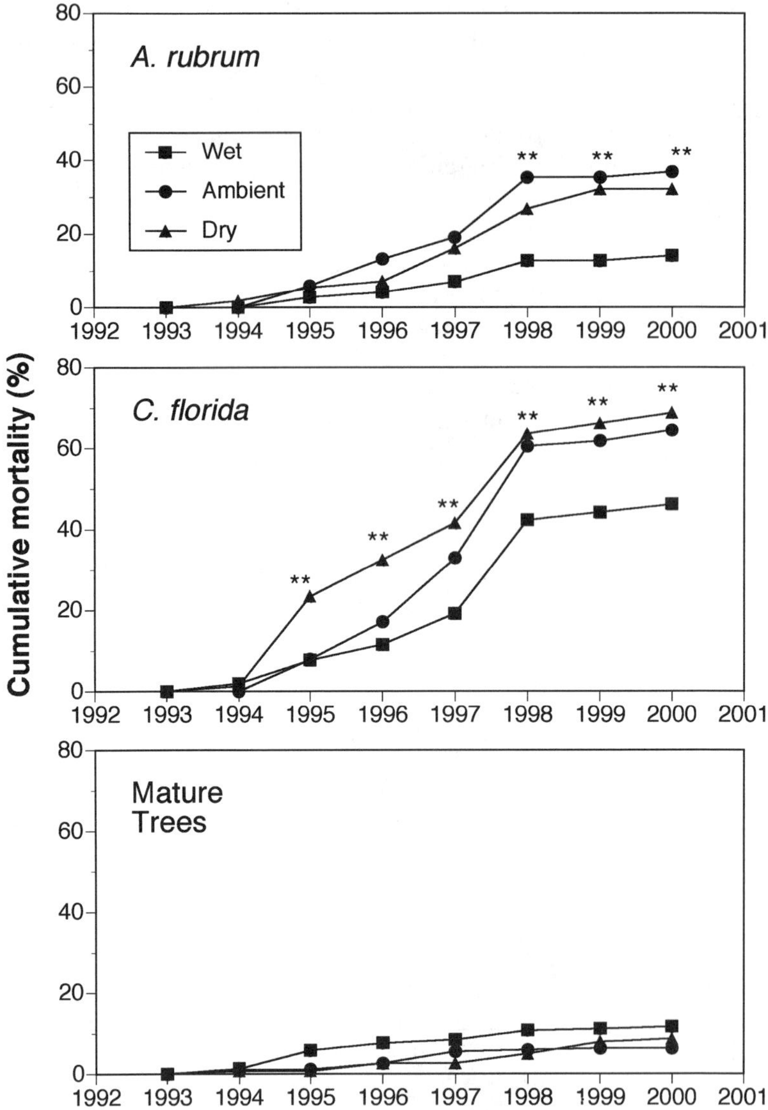

Figure 15.4. Cumulative mortality for two sapling species (*A. rubrum* and *C. florida*) and the mean large-tree response by treatment from 1993 through 2000. Significant differences between treatments are indicated by asterisks.

significant treatment effects in all years except 1994. At the end of 2000, cumulative mortality on the wet treatment plot was significantly lower than that on the ambient and dry plots for both species. In the first 3 years of the study, the pattern of mortality with respect to treatments was different. Following the drought of 1995, mortality of *C. florida* was significantly enhanced by the dry treatment, whereas *A. rubrum* showed little mortality and no significant treatment response (Fig. 15.4). The change in the overall response pattern over time emphasizes the need for long-term observations of chronic-change scenarios.

Although *A. rubrum* and *C. florida* saplings had comparable aboveground stem and branch structure and foliar physiology (Gebre et al. 1998; Tschaplinski et al. 1998), *C. florida* had a higher mortality rate than *A. rubrum* in most years (Fig. 15.4). Rooting depth differences are one possible cause for this difference. *C. florida* has a shallow root system (Hinckley et al. 1979, 1981; Bahari et al. 1985) and, therefore, may be sensitive to drought conditions leading to high mortality rates. Alternatively, because *Cornus* anthracnose symptoms (Hiers and Evans 1997) were ubiquitous across the plots (data not shown), it is possible that *Cornus* mortality resulted from the fungal pathogen *Discula,* with possible interactions with drought. Reduced mortality of *C. florida* and *A. rubrum* saplings on the wet plot indicates that a wetter future would sustain establishment and survival of trees in the understory but a drier future might reduce understory success of susceptible species (see also Chapter 14, this volume).

From 1993 through 2000, 66 out of 763 trees with stems greater than 0.1 m dbh died for an annual average rate of ~1% of the individuals per year. Sixteen trees died from known causes unrelated to the imposed treatments (i.e., wind throw or lightning). For the remaining trees, annual mortality rates showed little change by TDE treatment or year, reaching cumulative mortalities of 11.6%, 6.3%, and 8.6% in the wet, ambient, and dry plots, respectively, at the end of 2000. Differences among treatments were not significant for trees. Pedersen (1998) hypothesized that severe drought years may represent "inciting stresses" that would lead to the eventual mortality of large trees, but analysis of mortality of Midwestern overstory oaks located in Arkansas, Illinois, Indiana, and southern Ohio did not show increased mortality of overstory oaks in the first year following an inciting stress (Pedersen 1999). Large-tree mortality on the TDE was rather conservative at ~1% per year, with little evidence of accelerated mortality following the severe 1995 and 1998 droughts. It is possible that inciting stresses may not operate within a few years in large trees, but rather may take longer periods to be expressed. Further discussion of the impact of climatic patterns on the long-term patterns of growth and mortality of the Walker Branch forest can be found in Chapter 24 (this volume).

Variation in Wood Density Under Water Stress

No substantial changes in wood density were observed from 1980 to 2000 or as a function of annual diameter increment (Fig. 15.5). However, small significant differences were observed over time for *N. sylvatica* and *Quercus* sp., and with increasing annual diameter increment for *N. sylvatica, L. tulipifera,* and *Quercus*

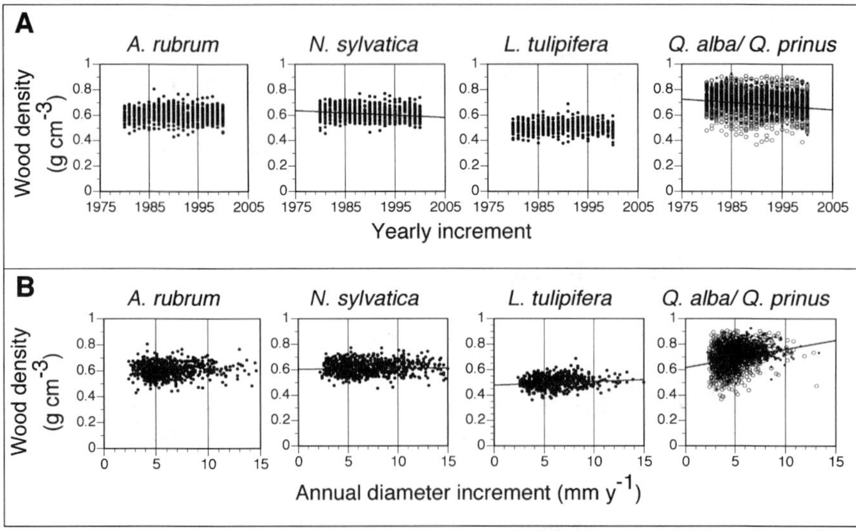

Figure 15.5. Relationship between wood density and (A) year of growth or (B) the annual diameter increment of a given year's tree ring by species. Lines are added to those graphs that show a significant slope from linear regression.

sp. Although statistically significant, these changes would have minimal impacts on conclusions derived solely from other measures of growth. Analysis of covariance contrasting pretreatment (1989–1993) and posttreatment (1994–2000) wood density was conducted to test for the impacts of TDE treatments on wood density. That analysis showed no significant response for *A. rubrum, N. Sylvatica, Q. alba,* and *Q. prinus,* but a slightly higher wood density of dry plot *L. tulipifera* wood was found (0.51 g cm^{-3} versus 0.49 g cm^{-3}; $p = 0.005$). Similar to the above-listed small trends, this change is unlikely to impact growth assumptions based on standard tree diameter measurements. Furthermore, the same test for posttreatment wood density for 1998–2000 wood was not significant for any species.

All of these results confirm the application of diameter measures of tree growth and subsequent extrapolations to mass from allometric relationships as being appropriate for drawing conclusions about changing tree mass over time or in response to interannual or imposed precipitation regimes.

Discussion

Stem Growth and Soil-Water Potential

Chapters 3, 6, and 7 (this volume) clearly show that TDE treatments and seasonal drought can have a strong effect on physiological processes, yet there was no simple relationship between variation in growing-season precipitation and annual basal-area growth of large trees. Can this be explained? Hanson and Weltzin (2000) and Hanson et al. (2001a) pointed out that there is commonly a disconnect between active growth periods and late-season droughts in the eastern deciduous hardwood forests. In years without soil-water deficits (i.e., 1994, 1996, and 1997), the basal-area growth of saplings and trees showed initiation of stem growth following canopy-leaf development, maximum rates of stem growth in late May and early June, and the completion of stem basal-area growth by early August in all years (Figs. 15.6 and 15.7) (Hanson et al. 2001a). Growth reductions in early August always took place even though canopy foliage remained physiologically active throughout late-summer periods (Wilson et al. 2000a, 2000b, 2001; Chapter 3, this volume). Thus, basal-area growth (a spring and early summer phenomenon) was nearly complete before the onset of late summer droughts. Severe spring droughts (if they occurred) might represent a significant water stress event with direct impact on current-year basal-area growth.

To fully evaluate the relationship between soil-water deficits and the phenology of growth throughout the year, we combined basal-area growth and soil-water potential data for all dendrometer trees from 1993 through 1999 to fit an empirical model of seasonal growth and its response to soil water potential. Cumulative tree basal-area growth in the absence of stress ($\mathrm{CBA}_{\mathrm{optimum}}$) was described by a logistic function of the following form (Hoerl 1954):

$$\mathrm{CBA}_{\mathrm{optimum}} = a - a\, e^{\mathrm{b}(\mathrm{DOY}-S)^{\wedge}d}, \tag{1}$$

where DOY is the current calendar day, *S* is the day of the year when growth begins, *a* is the cumulative basal-area growth attained for the time period under consideration in (mm^2) (i.e., typically one growing season), and *b* and *d* are constants that allow flexibility in the shape of the growth curve. The first derivative of this function represents the mean daily basal-area growth rate ($\mathrm{DBA}_{\mathrm{optimum}}$) (in $\mathrm{mm}^2\ \mathrm{d}^{-1}$) and takes the following form:

$$\mathrm{DBA}_{\mathrm{optimum}} = -a\ b\ d\ (\mathrm{DOY} - S)^{d-1}\ e^{\mathrm{b}(\mathrm{DOY}-S)^{\wedge}d}, \tag{2}$$

To estimate observed daily basal-area growth rate under all conditions ($\mathrm{DBA}_{\mathrm{observed}}$), Eq. (2) was made a function of soil-water potential as follows:

$$\mathrm{DBA}_{\mathrm{observed}} = \mathrm{DBA}_{\mathrm{optimum}} \left(\frac{\mathrm{WP}_{min} - \mathrm{SWP}}{WP_{min}}\right), \tag{3}$$

where SWP is the soil-water potential of the soil containing active roots and $\mathrm{WP}_{\mathrm{min}}$ is the soil-water potential at which growth ceases. Hanson et al. (2001a) used an alternate equation for describing intra-annual tree basal-area growth rates, but we substitute Eq. (3) because it also provides an estimate of cumulative annual basal-area growth (*a*). Substituting the value of 1 for *a* in Eq. (1) or (2) yields relativized values for cumulative growth or daily growth rates. Nonlinear regression was used to fit observed daily basal-area growth ($\mathrm{mm}^2\ \mathrm{d}^{-1}$) for individual trees or saplings to Eq. (3). The SWP values used for the tree growth equation were for the 0–0.35-m soil depth because it contains 74% of all fine roots for the TDE (Joslin and Wolfe 1998; Chapter 16, this volume). The SWP values used for the sapling growth equation were taken from estimates for the 0.15-m depth that was expected to better reflect the effective sapling rooting depth.

Figure 15.6 contrasts the fitted response surface of Eq. (3) with observed data for *A. rubrum.* Although there was substantial variation in the magnitude of growth rates for each species, a pattern similar to that seen in Fig. 15.6 was found for each of the other trees and the saplings (Hanson et al. 2001a). Maximum growth rates occurred before the end of June in each year (i.e., before day 180), and all basal area growth was completed before late September (i.e., day 270) in both wet and dry years. Although annual basal area growth of large trees is insensitive to late-season drought because of the temporal disconnect mentioned earlier, the response surface (Fig. 15.6) clearly shows that early-season daily growth rates were related to soil-water potential. In addition, visual inspection of the observed biweekly growth data (Fig. 15.7, upper graph) and predicted daily growth rates for *A. rubrum* (Fig. 15.7, lower graph) further suggest that Eq. (3) has captured key features of tree basal-area growth dynamics. Declining soil-water potentials resulted in a linear decline to zero growth at a critical limiting soil-water potential. The lower graph of Fig. 15.7 further illustrates this point. Dashed lines represent the potential growth phenology for each year, and the heavy solid lines are the predicted growth rate attained in each year. Growth reductions were 38%, 5%, 26%, 5%, 4%, 10%, 14%, and 6% for each year from 1993 to 2000. The largest growth reductions were predicted to occur in 1993 and 1995, when drought conditions overlapped a significant portion of the potential growth phenology in each year. Clearly, spring or early-summer drought conditions have the greatest potential to impact current-year growth rates.

The linear decline in growth with decreasing soil-water potential [Fig. 15.6 and Eq. (3)] is consistent with the response of *Pinus strobus* L. annual basal-area growth to mean growing-season soil-water potential (Vose and Swank 1994; McNulty and Swank 1995). Similar patterns were reported by Buell et al. (1961) for eastern hardwoods and by Bréda and Granier (1996), who concluded that early wood development in *Q. petraea* was independent of soil-water deficits. Phipps (1961) concluded that soil-water was never low enough in Ohio deciduous forests to limit the development of early sapwood. In support of this conclusion, Robbins (1921) reported that levels of March–June precipitation were positively correlated with *Quercus* growth in Missouri. Tree-ring chronologies for Walker Branch *Q. alba* and *Q. prinus* trees from 1970 through 2000 show 1980 as a minimum growth year (Fig. 24.6; Chapter 24, this volume). Evaluation of rainfall patterns for 1980 do, in fact, show that spring drought was a likely contributor to the low growth in that year (data not shown).

Model estimates of the soil-water potential leading to total basal-area growth inhibition for trees (Table 15.1) indicate that *L. tulipifera* is the most sensitive to soil-water stress with a WP_{min} value of -1.1 MPa. The data for *N. sylvatica* indicate that it might be the least sensitive to soil drying (WP_{min} of -2.1 MPa). WP_{min} values for *A. rubrum* and the two *Quercus* species were all centered on

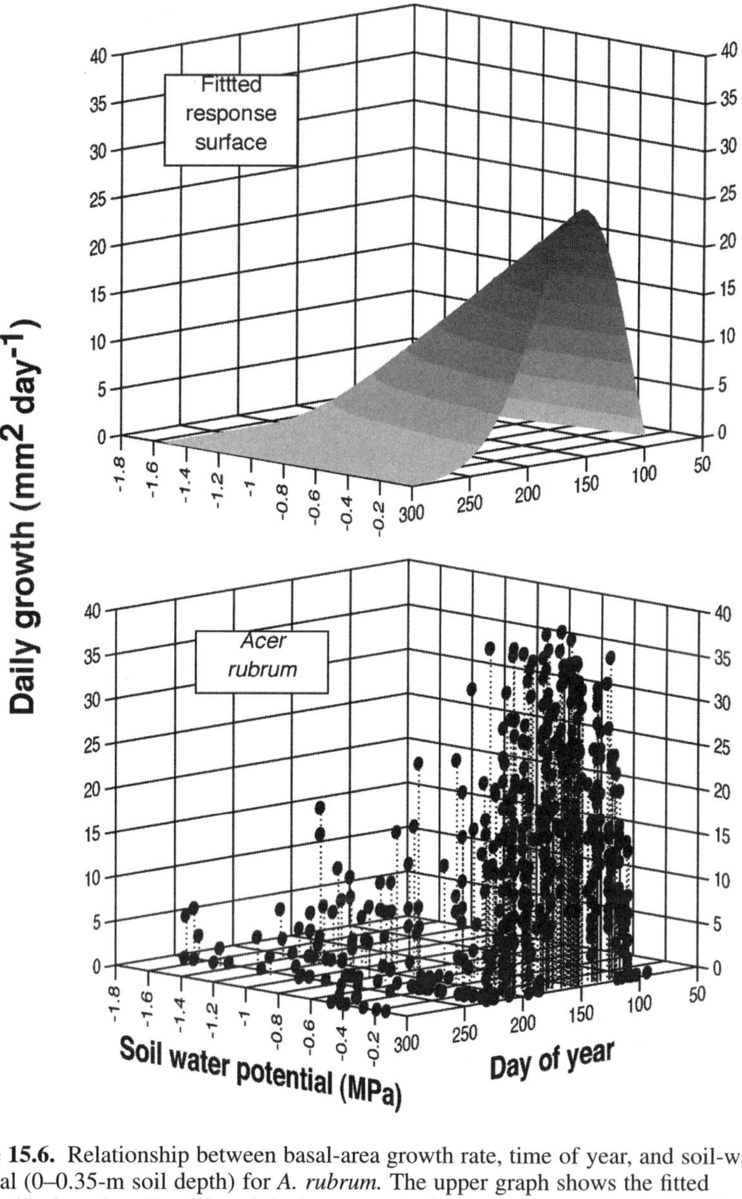

Figure 15.6. Relationship between basal-area growth rate, time of year, and soil-water potential (0–0.35-m soil depth) for *A. rubrum.* The upper graph shows the fitted relationship based on Eq. (3) and the lower graph shows the measurements for all years.

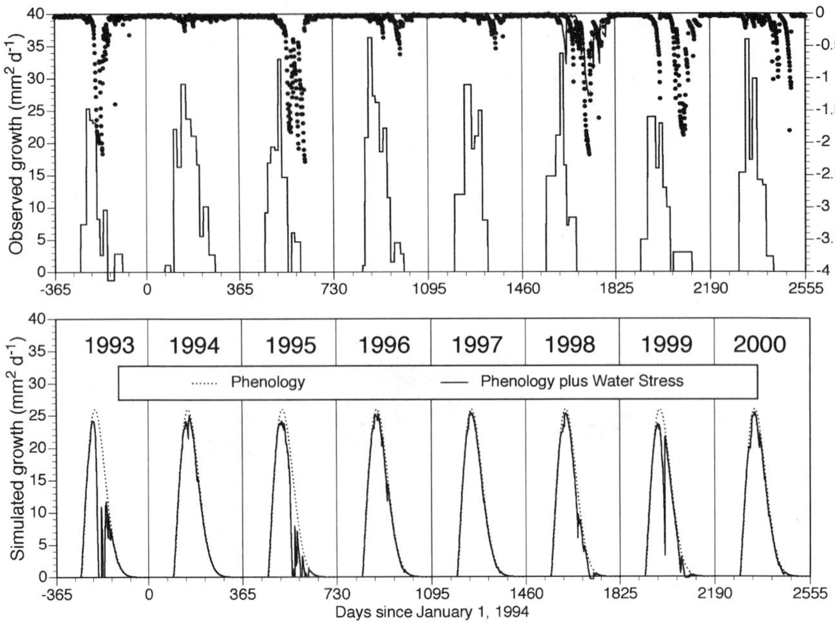

Figure 15.7. Multiyear pattern of observed growth and soil-water potential (upper graph) and simulated patterns of stem growth in the absence and presence of soil-water deficits (lower graph). Water stress was predicted to reduce growth 38%, 5%, 26%, 5%, 4%, 10%, 14%, and 6% respectively, from 1993 to 2000. Data and simulations are for *A. rubrum.*

Table 15.1. Large-tree species-specific regression coefficients (± SE) for the relationship between daily (or cumulative) basal-area tree growth and the day of the year and soil-water potential in the top 0.35 m of the soil profile [Eq. (3)] for five tree species and the mean

	Regression coefficients					
Species	*a* (mm^2)	*b*	*d*	*S* (DOY)	WP_{min} (MPa)	r^2
A. rubrum	2251 ± 81	−2.6E-4 ± 2E-4	1.93 ± 0.17	103 ± 5	−1.6 ± 0.2	0.46
L. tulipifera	2064 ± 132	−1.4E-5 ± 3E-5	2.52 ± 0.51	90 ± 13	−1.1 ± 0.2	0.32
N. sylvatica	1161 ± 57	−5.5E-5 ± 7E-5	2.23 ± 0.27	93 ± 7	−2.1 ± 0.5	0.32
Q. alba	3394 ± 220	−1.2E-3 ± 3E-4	1.48 ± 0.07	~90	−1.7 ± 0.3	0.27
Q. prinus	5218 ± 256	−9E-4 ± 2E-4	1.55 ± 0.06	~90	−1.6 ± 0.2	0.36
Mean	2851 ± 77	−1.6E-4 ± 1.3E-4	1.94 ± 0.16	86 ± 6	−1.6 ± 0.1	0.23

Note: Equation (3) has five constants: One controls the magnitude of the growth response (*a*), two control the shape of the curve (*b, d*), and the other two are the day of growth initiation (*S*) and the soil-water potential at which growth ceases (WP_{min}). The nonlinear solutions for individual oak species would not iterate to reasonable finite solutions if S was allowed to vary, so it was fixed for those regressions.

the traditional permanent wilting point for vegetation of − 1.5 MPa (Brady 1974). The relative sensitivity of *L. tulipifera* growth to soil-water potential is consistent with interspecies observations of foliar conductance sensitivity to soil-water potential (Hinckley et al. 1979, 1981; Bahari et al. 1985). It also agrees with interspecies rankings of drought sensitivity based on saturated foliar osmotic potential (Tschaplinski et al. 1998). For saplings, *A. rubrum* and *C. florida* had similar WP_{min} below − 3 MPa. The large differences in the predicted WP_{min} values for saplings and trees stem from the choice of the soil depth at which SWP measurements were obtained. Different reference depths must be used to reflect differences in rooting depths for saplings and trees. Work with stable-isotopic tracers is being conducted on the TDE to verify rooting depths of seedlings, saplings, and mature trees (Weltzin, 2001 personal communication).

Typically, relationships between daily stem growth and soil-water potential are not published in the literature. However, the relationships between soil-water status and other components of plant growth or physiology, such as root growth (Teskey and Hinckley 1981; Kuhns et al. 1985) and foliar gas exchange (Hinckley et al. 1981; Ellsworth and Reich 1992; Bréda et al. 1993; Epron and Dreyer 1993), are commonly reported. The coefficients in Tables 15.1 and 15.2 represent a good summary of tree and sapling basal-area growth for a range of soil-water conditions. These coefficients should have direct application to growth models, but such applications must be done with caution. The soil-water potential values used in the tree regressions in Table 15.1 were for the upper 0.35 m of the soil horizon, but those for the sapling regressions in Table 15.2 were referenced to the SWP at 0–0.15 m. If other studies use soil-water measurements at other depths to represent a different rooting zone, the relationships would change. Relationships might also change with soil nutrient availability and stand age.

Diameter Growth and Mass Accumulation

Measurements of tree growth are usually confined to observations of changing diameter (basal area) and/or height. As a relative index of tree growth, such measurements are easy to conduct, repeatable, and cost-effective. Aboveground mass accumulation, however, is the component of tree growth needed to address questions about the response of stand net primary production and carbon accu-

Table 15.2. Sapling regression coefficients (± SE) for the relationship between daily (or cumulative) basal-area growth and the day of the year and soil-water potential in the top 0–0.15 m of the soil profile [Eq. (3)] for two species and the mean

	Regression coefficients					
Species	a (mm^2)	b	d	S (DOY)	WP_{min} (MPa)	r^2
A. rubrum	35.5 ± 2.9	−1.3E-4 ± 2.5E-4	2.18 ± 0.41	104 ± 8	−3.8 ± 1.6	0.33
C. florida	56.5 ± 3.6	−5.6E-6 ± 2E-5	2.83 ± 0.74	94 ± 17	−3.1 ± 0.8	0.46
Mean	45.6 ± 2.4	−7.2E-6 ± 2E-5	2.78 ± 0.59	94 ± 13	−3.5 ± 0.8	0.38

Note: The equation has five constants: One controls the magnitude of the growth response (a), two control the shape of the curve (b, d), and the other two are the day of growth initiation (S) and the soil-water potential at which growth ceases (WP_{min}).

mulation (Chapter 22, this volume). Therefore, the relationship between spatial measures of tree growth (e.g., diameter change) and mass accumulation must be known. For this purpose, we use site-specific allometry, relating tree dbh to the mass of boles, branches, and leaves of the trees (Harris et al. 1973). However, changing wood density in response to the TDE treatments or to water stress driven by interannual differences in precipitation could be an alternative carbon allocation strategy for mature trees under water stress. Altered mass accumulation associated with wood synthesis following measurable diameter growth (i.e., cell expansion) is a hypothetical mechanism for such change (Edwards and Hanson 1996; Sprugel and Benecke 1991). Such a change might not be observed using traditional growth measurements based on changes in tree dimensions (e.g., dendrometer circumference or dbh-tape diameter observations) and would invalidate their application. Furthermore, if drought altered the timing of diameter growth near the ground versus growth within the branches of the canopy, inappropriate conclusions regarding mass accumulation might also be made.

Other Studies of Tree Growth and Drought

Previous work on tree-growth responses to soil-water deficits relied on naturally occurring interannual differences in rainfall patterns. Hursh and Haasis (1931) reported large-tree defoliation and mortality in the southern Appalachians during the 1925 drought when May–August precipitation was 32% below the long-term average and soil-water contents (~0.3-m depth) declined to 8–11% (v/v). They also reported that foliar browning (i.e., wilting) was severe for *C. florida, O. arboreum,* and *Q. prinus,* whereas species of *Fraxinus, Robinia,* and *Carya* showed less injury. Small (1961) reported foliar wilting and subsequent mortality for a number of hardwoods growing on shallow soils of the New Jersey Piedmont in 1957, when summer rainfall (May–August) was only 52% of the reference years. Midseason wilting, foliar browning, and early senescence were observed on understory species of the TDE during the drought of 1998, and overstory senescence was accelerated (Chapter 17, this volume). A late-1998 survey of the Oak Ridge reservation and the surrounding region also showed evidence for early canopy senescence for trees on shallow soils and south-facing slopes.

Orwig and Abrams (1997) analyzed annual radial growth responses from tree rings of *Q. alba, Q. velutina, L. tulipifera, C. glabra,* and *N. sylvatica* from 1900 through 1993 to identify how each species responded to severe droughts recorded in 1930, 1942, 1954, and 1966. Those years had growing-season precipitation levels between 54% and 82% of the 60-year average (557 mm) for their study site in Virginia, which is similar to the manipulated rates of precipitation in the TDE study. With the exception of *Q. alba,* Orwig and Abrams (1997) found no significant difference in the growth of these species for any of the targeted drought years, but they concluded that growth in years following severe drought episodes was reduced relative to the long-term trend. Abrams et al. (1998) evaluated tree-ring variation between 1984 and 1995 for 10 hardwood species in central Pennsylvania to determine if drought impacts varied by growth location (i.e., xeric ridges, dry mesic barrens, mesic valleys, and wet mesic riparian sites). They found that trees growing on xeric ridges were most affected by the occurrence of drought. Unfortunately, they did not characterize drought in terms of soil-water potential, making direct comparisons with this study difficult. Buell et al. (1961) contrasted tree growth of a number of tree species growing on shallow soils in a normal (1956) versus a dry year (1957 with 48% less rainfall) and measured dramatic midsummer stem shrinkage during drought. However, the shrinkage was reversible and total growth was similar in the wet versus dry years because growth occurred before late-season droughts.

Tree-ring studies of hardwoods that correlate annual growth increment with climate also support the idea that spring drought conditions will produce larger growth effects than late-season droughts of similar magnitude (Fritts 1960; Buell et al. 1961; Phipps 1961; Fritts 1962; Penninckx et al. 1999). Borchert (1998) described a similar phenomenon for tropical rainforests. Observations of large-tree growth in the TDE study and the articles summarized here show that differences in seasonal timing of rainfall will have a greater impact on basal-area growth of eastern deciduous forests than percentage changes in rainfall quantity.

In contrast to the results presented in this chapter, Jenkins and Pallardy (1995), using tree-ring analysis, found a significant correlation between annual basal-area increment of *Q. coccinea* and *Q. velutina* in the Missouri Ozarks and measures of drought severity. However, their observations were made in a region of substantially lower precipitation, and it is reasonable to anticipate stronger relationships between growth and soil-water deficits in areas where rainfall is more limiting. During an unusually severe drought in central Missouri in 1976 (34% of normal rainfall), when average soil water potential in the upper 0.45 m of soil declined to −2.6 MPa, Hinckley et al. (1979) reported 19% reductions in stem circumference growth of dominant white oaks. Hanson and Weltzin (2000) point out that forests near the prairie–forest border (i.e., Missouri) might be at a higher risk for drought impacts than those farther east, where rainfall is more abundant.

Conclusions

Eight years of sapling and mature tree growth measurements on the upland, *Quercus*-dominated, temperate deciduous forests of eastern Tennessee have failed to demonstrate simple relationships between the occurrence of atmospheric drought (i.e., below-average precipitation inputs) and plant-growth reductions. However, detailed analysis of deviations from seasonal patterns of optimum growth driven by soil-water deficits do show a consistent and predictable pattern that should have broad application for tree and ecosystem plant-growth models. Differences in seasonal timing of rainfall are likely to have a greater impact on basal-area growth of eastern deciduous forests than mean annual reductions in rainfall, and spring droughts that overlap periods of maxium stem and branch growth would be the most damaging to forest productivity.

References

Abrams MD (1990) Adaptations and responses to drought in *Quercus* species of North America. Tree Physiol 7: 227–238.

Abrams MD, Ruffner CM, Morgan TA (1998) Tree-ring responses to drought across species and contrasting sites in the ridge and valley of central Pennsylvania. For Sci 44:550–558.

Bahari ZA, Pallardy SG, Parker WC (1985) Photosynthesis, water relations, and drought adaptation in six woody species of oak-hickory forests in central Missouri. For Sci 31:557–569.

Brady NC (1974) The Nature and Property of Soils (8th Edition). MacMillan Publishing Company, New York, pp. 190–192.

Borchert R (1998) Responses of tropical trees to rainfall seasonality and its long term changes. Clim Change 39: 381–393.

Bréda N, Cochard H, Dreyer E, Granier A (1993) Water transfer in a mature oak stand (*Quercus petraea*): Seasonal evolution and effects of a severe drought. Can J For Res 23:1136–1143.

Bréda N, Granier A (1996) Intra-and interannual variations of transpiration, leaf area index and radial growth of a sessile oak stand (*Quercus petraea*). Ann Sci For 53:521–536.

Buell MF, Buell HF, Small JA, Monk CD (1961) Drought effect on radial growth of trees in the William L. Hutcheson Memorial Forest. Bull Torrey Bot Club 88:176–180.

Burns DP, Gibson LP (1968) The leaf-mining weevil of yellow-poplar. Can Entomol 100:421–429.

Edwards NT, Hanson PJ (1996) Stem respiration in a closed-canopy upland oak forest. Tree Physiol 16:433–439.

Elliott J, Anderson P, Davis GR, Wong FSL, Dover SD (1994) Computed tomography Part II: The practical use of a single-source and detector. JOMJ Min Met Mat Soc 46:11–19.

Ellsworth DS, Reich PB (1992) Water relations and gas exchange of *Acer saccharum* seedlings in contrasting natural light and water regimes. Tree Physiol 10:1–20.

Epron D, Dreyer E (1993) Long-term effects of drought on photosynthesis of adult oak trees [*Quercus petraea* (Matt.) Liebl. and *Quercus robur* L.] in a natural stand. New Phytol 125:381–389.

Fritts HC (1960) Multiple regression analysis of radial growth in individual trees. For Sci 6:334–349.

Fritts HC (1962) The relation of growth ring widths in American beech and white oak to variations in climate. Tree-Ring Bull 25:2–10.

Gebre GM, Tschaplinski TJ, Shirshac TL (1998) Response in water relations of several hardwood species to throughfall manipulation in an upland oak forest during a wet year. Tree Physiol 18:299–305.

Haasis RW (1934) Diametrical changes in tree trunks. Publication No. 450. Carnegie Institution of Washington, Washington, District of Columbia.

Hanson PJ, Weltzin JF (2000) Drought disturbance from climate change: Response of United States forests. Science Total Environ 262:205–220.

Hanson PJ, Todd DE, Amthor JS (2001a) A six year study of sapling and large-tree growth and mortality responses to natural and induced variability in precipitation and throughfall. Tree Physiol 21:345–358.

Hanson PJ, Todd DE, Riggs JS, Wolfe ME, O'Neill EG (2001b) *Walker Branch Throughfall Displacement Experiment data report: Site characterization, system performance, weather, species composition, and growth.* ORNL/CDIAC-134, NDP-078A. Carbon Dioxide Information Analysis Center, US Department of Energy, Oak Ridge National Laboratory, Oak Ridge, Tennessee.

Harris WF, Goldstein RA, Henderson GS (1973) Analysis of forest biomass pools, annual primary production and turnover of biomass for a mixed deciduous forest watershed. In Young HE (Ed) *Proceedings IUFRO Symposium Working Party on Forest Biomass.* University of Maine Press, Orono, Maine, pp 43–64.

Hiers JK, Evans JP (1997) Effects of anthracnose on dogwood mortality and forest composition of the Cumberland Plateau (USA). Conser Biol 6:1430–1435.

Hinckley TM, Dougherty PM, Lassoie JP, Roberts JE, Teskey RO (1979) A severe drought: Impact on tree growth, phenology, net photosynthetic rate and water relations. Am Midl Nat 102:307–316.

Hinckley TM, Teskey RO, Duhme F, Richter H (1981) Temperate Hardwood Forests. In Kozlowski TT (Ed) *Water deficits and plant growth, Vol VI.* Academic Press, New York, pp 153–208.

Hoerl AE Jr. (1954) Fitting curves to data. In Perry JH (Ed) *Chemical business handbook,* McGraw-Hill, New York, pp 20–55 through 20–77.

Hursh CR, Haasis FW (1931) Effects of 1925 summer drought on southern appalachian hardwoods. Ecology 12: 380–386.

Jenkins MA, Pallardy SG (1995) The influence of drought on red oak group species growth and mortality in the Missouri Ozarks. Can J For Res 25:1119–1127.

Joslin JD, Wolfe MH (1998) Impacts of water input manipulations on fine root production and mortality in a mature hardwood forest. Plant Soil 204:165–174.

Keeland BD, Sharitz RR (1993) Accuracy of tree growth measurements using dendrometer bands. Can J For Res 23:2454–2457.

Kohyama T, Hara T (1989) Frequency distribution of tree growth rate in natural forest stands. Ann Bot 64:47–57.

Kramer PJ, Kozlowski TT (1979) *Physiology of woody plants.* Academic Press, New York.

Kozlowski TT (1982) *Water supply and tree growth: Part 1 Water deficits.* For Abst 43:57–95.

Kozlowski TT (1971) Measurement of cambial growth. In Kozlowski TT (Ed) *Growth and development of trees.* Academic Press, New York, pp 168–195.

Kuhns MR, Garrett HE, Teskey RO, Hinckley TM (1985) Root growth of black walnut trees related to soil temperature, soil water potential, and leaf water potential. For Sci 31:617–629.

McLaughlin SB, Downing DJ (1996) Interactive effects of ambient ozone and climate measured on growth of mature loblolly pine trees. Can J For Res 26:670–681.

McNulty SG, Swank WT (1995) Wood $\delta^{13}C$ as a measure of annual basal area growth and soil water stress in a *Pinus strobus* forest. Ecology 76:1581–1586.

Orwig DA, Abrams MD (1997) Variation in radial growth responses to drought among species, site, and canopy strata. Trees 11:474–484.

Pedersen BS (1998) The role of stress in the mortality of midwestern oaks as indicated by growth prior to death. Ecology 79:79–93.

Pedersen BS (1999) The mortality of midwestern overstory oaks as a bioindicator of environmental stress. Ecol Appl 9:1017–1027.

Penninckx V, Meerts P, Herbauts J, Gruber W (1999) Ring width and element concentrations in beech (*Fagus sylvatica* L.) from a periurban forest in central Belgium. For Ecol Manage 113:23–33.

Phipps RL (1961) *Analysis of five years dendrometer data obtained within three deciduous forest communities of* Neotoma. Research Circular No. 105. Ohio Agricultural Experiment Station, Wooster, Ohio.

Robbins WJ (1921) *Precipitation and the growth of oaks at Columbia, Missouri.* Agriculture Experiment Station Research Bulletin 44. University of Missouri, College of Agriculture, Columbia, Missouri.

Small JA (1961) Drought response in William L. Hutcheson Memorial Forest, 1957. Bull Torrey Bot Club 88:180–183.

Sprugel DG, Benecke U (1991) Measuring woody–tissue respiration and photosynthesis. In Lassoie JP, Hinckley TM (Eds) *Techniques and approaches in forest tree ecophysiology.* CRC Press, Boca Raton, Florida, pp 329–355.

Swaine MD, Lieberman D, Putz FE (1987) The dynamics of tree populations in tropical forest—A review. J Trop Ecol 3:359–366.

Tainter FH, Fraedrich SW, Benson DM (1984) The effect of climate on growth, decline, and death of northern red oaks in the western North Carolina Nantahala Mountains. Castanea 49:127–137.

Teskey RO, Hinckley TM (1981) Influence of temperature and water potential on root growth of white oak. Physiol Plant 52:363–369.

Tschaplinski TJ, Gebre GM, Shirshac TL (1998) Osmotic potential of several hardwood species as affected by manipulation of throughfall precipitation in an upland oak forest during a dry year. Tree physiol 18:291–298.

Tuskan G, West D, Bradshaw HD, Neale D, Sewell M, Wheeler N, Megraw B, Jech K, Wiselogel A, Evans R, Elam C, Davis M, Dinus R (1999) Two high-throughput techniques for determining wood properties as part of a molecular genetics analysis of hybrid poplar and loblolly pine. Appl Biochem Biotech 77–79:55–65.

Vose JM, Swank WT (1994) Effects of long-term drought on the hydrology and growth of a white pine plantation in the southern Appalachians. For Ecol Manage 64:25–39.

Waring RH, Pitman GB (1985) Modifying lodgepole pine stands to change susceptibility to mountain pine beetle attack. Ecology 66:889–897.

Wilson KB, Baldocchi DD, Hanson PJ (2000a) Quantifying stomatal and non-stomatal limitations to carbon assimilation resulting from leaf aging and drought in mature deciduous tree species. Tree Physiol 20:787–797.

Wilson KB, Baldocchi DD, Hanson PJ (2000b) Spatial and seasonal variability of photosynthetic parameters and their relationship to leaf nitrogen in a deciduous forest. Tree Physiol 20:565–578.

Wilson KB, Baldocchi DD, Hanson PJ (2001) Leaf age affects the seasonal pattern of photosynthetic capacity and net ecosystem exchange of carbon in a deciduous forest. Plant Cell Environ 24:571–583.

Winget CH, Kozlowski TT (1964) Winter shrinkage of forest trees. J For 62:335–337.

Winget CH, Kozlowski TT (1965) Seasonal basal growth area as an expression of competition in northern hardwoods. Ecology 46:786–793.

16. Fine-Root Growth Response

J. Devereux Joslin and Mark H. Wolfe

Long-term changes in soil-water regimes resulting from climate change could have profound effects, not only on aboveground ecosystem processes and productivity but also on belowground processes. Such belowground effects would have corresponding impacts upon nutrient and carbon cycling in forest ecosystems (McClaugherty et al. 1982; Nadelhoffer et al. 1985; Joslin and Henderson 1987; Hendrick and Pregitzer 1996a; Joslin et al. 2000). The ability of forest ecosystems to survive and adapt to changes in climate will depend in part on how plants balance resources among photosynthetic organs, stems and branches, reproductive organs, and root systems in a complex process that plant physiologists have long been struggling to understand (Bradford and Hsiao 1982; Kramer 1983).

Some evidence exists that tree species adapted to dry climatic regimes generally have higher root : shoot ratios (R : S) and deeper root systems than species that are more suited to mesic climatic regimes [e.g., van Buijtenen et al. (1976); Caldwell and Richards (1986); Commeau and Kimmons (1989)]. Such evidence has probably contributed to the general concept that trees frequently respond, following exposure to periods of dry conditions, by increasing R : S ratios and/or by increasing rooting depth (Kozlowski et al. 1991). Aboveground production in Douglas fir (*Pseudotsuga menziesii* Franco) was lower in relation to fine-root production on more droughty, nutrient-poor soils in Keyes and Grier (1981). The development of a more extensive root system at moderately low soil-water potentials was suggested by Teskey and Hinckley (1981) to be a contributing factor to drought tolerance by *Quercus alba* L. Hendrick and Pregitzer (1996b) observed "episodic" root production at depths greater than 50 cm that appeared to be associated with periods of high water demand.

The following hypotheses regarding belowground forest responses to long-term decreases in water availability are consistent with the above-cited. The inverse of each response is hypothesized in response to increases in water availability:

1. Increases in fine-root biomass (FRB) and root length density
2. Increases in R : S, especially the FRB-to-foliar biomass ratio
3. Increased depth of rooting and/or increased density of fine roots at greater depths.

The primary objective of belowground research as part of the Throughfall Displacement Experiment (TDE) (Hanson et al. 2001) was to address the three hypotheses by comparing fine-root responses across the three water treatments. A fourth potential impact, treatment effects on fine-root turnover, is also being examined. Data from minirhizotrons, pretreatment and posttreatment root coring, soil pits, and root ingrowth cores have been collected over 6.5 years. Net changes in absolute values for FRB, root length density, R : S, and root distribution have been examined. In addition, 7 years of minirhizotron observations on root elongation across treatments have been used as indicators of the treatment effects on fine-root production, ratios of fine-root production to foliar production, and fine-root turnover rates. Treatment effects on both gross and net fine-root production and mortality patterns have been examined, both temporally and with depth.

Although the major thrust of this research is to examine the belowground response to increasingly dry or wet moisture regimes, the data collected in this experiment also allow the pursuit of a secondary objective: the examination of factors acting for short time intervals to control the timing and magnitude of tree root elongation. The debate about the relative importance of environmental variables, particularly soil temperature and soil-water potential, and endogenous plant factors has been a subject of research for decades (Reed and McDougal 1937; Rogers 1939; Rogers and Head 1969; Lyr and Hoffman 1967; Deans 1979; Teskey and Hinckley 1981; Hendrick and Pregitzer 1996b; Vogt et al. 1996). Although it is clear from these studies that both external climatic variables and endogenous factors interact to control the timing of root elongation, the predominant pulse in root elongation in mature deciduous trees of the temperate zone generally occurs in late spring and early summer, with the most pronounced peak immediately following shoot growth [e.g., Lyr and Hoffman (1967); Head (1973); Kaushal et al. (1989)]. Using our long-term dataset on root elongation observed in minirhizotrons, we were able to examine the timing of root growth across a wide range of soil temperatures and soil-water potentials. Through the examination of these data, we attempted (1) to determine relationships between root elongation and both soil-water potential and soil temperature, (2) to quantify the impact of phenology on root elongation, (3) to develop a multivariate regression equation to predict root elongation using the three variables of soil temperature, soil-water potential, and phenology, and (4) to delineate soil-water potentials that were optimum for root growth.

Methods

Pretreatment and Posttreatment FRB Sampling

Pretreatment sampling of standing FRB on the experimental site was initiated in January 1993 and completed by late March 1993. Samples were collected from soil depths of 0–30 cm, 30–60 cm and, 60–90 cm with a gasoline-powered auger with a 10-cm (diameter) by 30-cm (length) saw-tipped sample tube. Sampling was carried out across all treatment areas and across three slope positions (upper, middle, and lower) within the treatment areas. A stratified sampling scheme provided 270 samples (90 samples per treatment; 30 per slope position within each treatment) at 0–30 cm, 207 samples ($\sim$69 per treatment) at 30–60 cm, and 54 samples ($\sim$18 samples per treatment) at 60–90 cm. Samples were carried to the laboratory, where they were refrigerated until they could be processed to determine the standing pretreatment FRB by treatment, depth, slope, and root-size class ($<$1 mm in diameter, 1–2 mm, 2–5 mm, and $>$5 mm). [See Joslin and Wolfe (1999) for more details on methods.] For purposes of simplifying data presentation, FRB is defined in this chapter as the oven-dry weight of all root material less than 2 mm in diameter.

Posttreatment sampling of standing FRB on the experimental site was initiated in November 1999 and completed by mid-March 2000. Samples were collected in an identical manner as the pretreatment samples from 0 to 30 cm and from 30 to 60 cm, except the focus was placed on the upper- and lower-slope positions. The stratified sampling scheme provided 360 core samples at 0–30 cm: 240 samples (80 per treatment $\times$ slope position) on the upper slope and 120 samples (40 per treatment $\times$ slope position) on the lower slope. At 30–60 cm, it provided 120 samples: 60 at each of the upper- and lower-slope positions and 40 at each of three treatments (20 per treatment $\times$ slope position). To determine FRB, these cores were processed by procedures identical to those used in 1993 (Joslin and Wolfe 1999).

To avoid rocks during coring and to account for rock volume more accurately at the depth of 60–90 cm, we excavated soil pits posttreatment on the upper slope. Nine pits were excavated during February–March 2000 and an additional 6 pits in November–December 2000, thus providing 5 pits from each of the 3 treatments (15 pits total). One hundred percent of the soil, roots, and rocks were removed from each pit, which consisted of a 0.147-m^3 rectangular prism [70 cm by 70 cm by 30 cm (60–90-cm depth interval)]. Samples were carried to the laboratory, where they were refrigerated until they could be processed to determine the standing posttreatment FRB by treatment, depth, slope and root-size class.

Minirhizotrons

Minirhizotron tubes were installed during February and March 1994 in each of the three TDE treatments. Altogether, 30 tubes where installed: 15 along the upper slope (5 in each treatment) and 15 along the lower slope. The tubes where installed at a 45° angle to the slope of the ground with a guide and a 5.0-cm-diameter bucket auger (Taylor et al. 1990). Tubes were inserted to a vertical depth of 90 cm except where precluded by the rocks. The maximum depth achieved for most tubes fell between 60 and 90 cm. The minirhizotron tubes, 5.7 cm in diameter and 130 cm long, were etched with a single set of 130 equally spaced grids, highlighted with fluorescent orange paint, and capped on the soil end (after Hendrick and Pregitzer 1992).

Video imaging of root growth was conducted using a Circon color agricultural camera during seven growing seasons: 1994 through 2000. In each year, imaging was begun in early to mid-April, prior to budbreak, and continued biweekly through the end of October. From 13 to 15 observation periods were recorded during each of the 7 years, with the exception of 1995, when equipment failure prevented observations from July through September. Video images were digitized with PC-based MSU-ROOTS software (Michigan State University Instructional Media Center; Enslin et al. 1994) to measure root length production and to track the changes in the condition of fine roots during the study. All roots were categorized at any point in time as (1) new with respect to the prior sampling date, (2) previously existing and still alive, (3) dead, or (4) missing. If a root became "missing" prior to being categorized as dead and did not reappear at a future photographic session, it was considered dead from the time of disappearance. Fine-root production and mortality measurements were made as described by Hendrick and Pregitzer (1993). Further details on minirhizotron installation and data collection can be found in Joslin and Wolfe (1998, 1999).

Ingrowth Cores

The ingrowth-core method was used as an independent method of comparing the relative quantity of FRB production across treatments (Persson 1980; Joslin and Henderson 1987; Persson and Ahlstrom 1991). In February and March 1993, subsequent to pretreatment FRB sampling, root-free ingrowth cores were placed into the auger holes created in the soil profile by the pretreatment sampling coring itself.

Ingrowth cores were constructed from soil taken from a pit located adjacent to the TDE experimental treatment area. Soil was removed from the pit in 30-cm increments. In the greenhouse, roots and rocks were separated from the soil by dry sieving through a 5-mm sieve, and the soil was thoroughly mixed and air-dried. The rock-free bulk density (BD_s) of each depth interval was estimated based on v_p, the volume of the pit, v_r, the volume of rocks and roots determined by displacement in water, fw_s, the fresh weight of soil minus rocks and roots, and m_s, the moisture content of the remaining soil:

$$\mathrm{BD}_s = \frac{\mathrm{fw}_s - m_s}{v_p - v_r}.$$

Ingrowth cores were reconstructed for each depth interval to the calculated BD_s by manually repacking soil into 10-cm (diameter) by 30-cm PVC pipe segments. To facilitate removal of soil cores, each tube segment had to be bisected lengthwise and rejoined with packing tape prior to soil packing. Cores were stored frozen until field placement.

Between February 1 and March 14, 1993, the frozen ingrowth cores were removed from their split PVC pipe segments in the field, placed in 5-mm mesh plastic bags, and inserted in precored holes to the appropriate depth. At 0–30 cm, 81 cores were implanted, 27 per treatment (26–29 per treatment, 8–11 per treatment–slope position combination); 146 cores were implanted at 30–60 cm (47–50 per treatment, 14–18 per treatment–slope position combination); 43 cores were implanted at 60–90 cm (14–20 per treatment, 4–8 per treatment–slope position combination).

In January 1996, ingrowth cores were harvested after three growing seasons (1993–1995) by careful hand excavation. Because TDE treatments were not actually begun until July 1993, plants experienced less than three full growing seasons of the treatments during this period. Mesh plastic bags identified the original core boundary during the excavation process. Roots growing into ingrowth cores were cut at the mesh-bag interface. Excavated cores were stored frozen until the roots were washed from the soil, dried, and sorted according to the methods described earlier for pretreatment soil cores.

Data Analysis

Production, Mortality, and Turnover

For the purposes of minirhizotron data presentation and statistical analysis, the primary experimental unit used was the summed response for each individual minirhizotron tube across a given depth interval and time interval. Data are presented separately from upper- and lower-slope positions and from all three soil-depth intervals (0–30 cm, 30–60 cm, and 60–90 cm). However, because soil-water treatment differences were considerably more distinct on the upper slope throughout the study, the major focus of the analyses is on the upper slope. Furthermore, the upper slope provided the most consistent tree species composition (*Quercus prinus*–*Q. alba* overstory) across all locations. Similarly, the greatest FRB and root activity occurred in the surface 30 cm of the soil profile. Previous FRB-pool-size measurements revealed that 74% of the fine roots (<2-mm in diameter) were positioned in the surface 30 cm of the soil profile (Joslin and Wolfe 1999). Therefore, the focus of the analyses is on this portion of the profile.

The summed data from each of these tubes on root production, root mortality, and net root production were analyzed. Thus, whether annual or 7-year comparisons between treatments are being presented, they are based on summed data from five minirhizotron tubes per treatment. Statistical comparisons are restricted to annual and 7-year treatment differences. Minirhizotron data were collected almost entirely during the growing season (April to October), as mentioned earlier. During the summer of 1996, the video equipment failed in late June. This resulted in a "data gap" for the remainder of that growing season. Because most of this period of missing data corresponded with the most severe drought of the research period and because our data indicate minimal root growth occurring below -0.6 MPa of water potential, we believe that very little root growth occurred in any of the treatments during most of this time. The 7-year totals for fine-root production and mortality do not include data from this period nor from any winter period. Checks for the occurrence of winter root elongation, from January 1998 and March 1998 minirhizotron measurements, revealed that winter accounted for less that 5% of the annual total.

Data collected during the first year (1994) immediately following minirhizotron installation strongly suggest that excessive root proliferation occurred during this year as a result of root-pruning effects and soil-disturbance impacts on nutrient availability (Joslin and Wolfe 1999). Although the magnitude of minirhizotron measures of root growth appears to have been amplified during this period, there is no evidence for any treatment-related biases. The data from this first year appear to correctly reflect, across the three treatments, the phenology of fine-root production and mortality as well as the relative quantity (if not the absolute amount) of organic matter involved in these processes. Hence, raw data from this first year are presented and analyzed along with data from the subsequent 4 years.

Statistical tests for the main effects of treatment on an annual basis were conducted for each year with an analysis of estimated coefficients based on response curves (Gurevitch and Chester 1986; Meredith and Stehman 1991). Polynomial response functions were fitted to 13 repeated measures of each experimental unit with orthogonal contrasts. The estimated coefficients of these functions were then treated as primary data and analyzed with univariate analysis of variance (ANOVA), with the GLM procedure of SAS (SAS Institute 1989). Statistical tests for differences between treatments for the entire 7 years were conducted with univariate ANOVA on summed data across the appropriate tubes within a treatment and the appropriate depth interval. Covariance analysis of variance was applied to annual stand foliar production values, with the 1993 data as the covariate. The main effects of year and treatment were evaluated with significance determined from p-values below 0.10.

Fine-Root Turnover Index

Fine-root turnover refers to the process by which roots grow, die, and decompose. In come cases, it has been defined as the ratio of annual fine-root production to the either the maximum or minimum belowground standing crop of fine roots (Dahlman and Kucera 1965; Gill and Jackson 2000). Whereas this ratio provides an index of the rate at which new roots are being produced to replace those that are dying, we prefer a definition that acknowledges that the rate of turnover is really the average rate for all three processes involved: production, mortality, and decomposition (Joslin and Henderson 1982, 1987; Santantonio and Hermann 1985). Although it is commonly assumed that ecosystems are near equilibrium with regard to the rates of these three processes, so that there is no net long-term change in the biomass of roots over time, this is clearly not the case in some ecosystems (particularly aggrading ones) or for certain time intervals. Hence, although the ratio of one of these processes, such as root production, to the standing crop is sometimes used as an index of turnover rate, truly the ideal measure would be one that employs the average of all three rates.

Here, we use the average of indices of two of these processes; production and mortality. It should be noted that we have no data for this site on root decomposition rate. To provide this index of turnover rate, the total sum of root elongation (an index of production) plus the length of roots exhibiting mortality was divided by the initial FRB. Note that because it is difficult to convert minirhizotron elongation and mortality measurements into FRB production estimates (in units of kg ha^{-1} y^{-1}, for example), these measurements cannot provide absolute turnover estimates. Instead, these measures are used as an index of fine-root turnover for comparison purposes across treatments. A method for converting minirhizotron data to FRB production estimates is described below.

Factors Controlling Root Elongation

In order to examine the effects of soil-water potentials, temperature, and time of year on root elongation, only data from the surface 30 cm of the soil profile on the upper slope were used. As mentioned earlier, this zone provided the widest range of soil-water potentials and the most uniform species composition. Data from 31, 2-week intervals were used for statistical analyses across 3 years of data collection during the growing season (early April to late October, 1995 to 1997). Data from 1994 were not used in the analyses because previous comparisons have demonstrated that data from the first year after minirhizotron installation commonly were biased by disturbance effects from installation (Joslin and Wolfe 1999). Data for July through October 1995 were missing because of equipment failure.

Upper-slope data were stratified by the three treatment plots (wet, ambient, and dry), and composited sets of data on (1) root-elongation index, (2) soil-water potential, and (3) soil temperature were used in the analyses. Methods used to measure these variables and to conduct this analysis have been described in detail in Joslin et al. (2001). Our measure of root-elongation intensity (REI) was the mean value from five minirhizotrons in a given treatment plot for the sum of all recorded increases in root length in the surface 30 cm during each 2-week interval. Soil-water potential (Ψ) equaled the mean value of all Ψ measurements taken in the top 35 cm of soil for the upper slope of a given treatment plot during a given 2-week time interval. Soil temperature (ST) was defined as the mean value of all 10-cm ST measurements for the upper slope of a given treatment and time interval.

A fourth variable, termed the "phenology index," was created to characterize the season of the year in a semiquantitative manner. Based on previous data on *Quercus–Carya* forests (Teskey and Hinckley 1981; Joslin, 1983; Joslin and Henderson 1987), which indicated that REI peaks approximately 1 month after completion of leaf expansion, we assumed that the influence of phenology would be maximal on this date. Based upon 6 years of leaf-expansion data, June 10 was chosen as the "zero" phenology-index date, being 30 days after the average date of 95% completion of leaf expansion for years 1992 through 1997 (May 11; range: May 5 to May 18). Each 2-week interval was then assigned a number equal to the number of days that its midpoint fell before or after June 10. Seasonal patterns of leaf expansion were determined from an index of relative canopy coverage (see Chapter 17, this volume).

Simple correlation was used to determine the strength of the relationship between REI and each of the three independent variables alone (soil-water potential, soil temperature, and phenology). Then, the stepwise regression procedure (SAS Institute 1989) was used to develop the best predictive model for REI from these same three independent variables, using linear, nonlinear, and log-linear procedures, including the use of log plots and log-log plots. To use the log plot with Ψ, it was necessary to reverse its sign because the log of a negative number is nonsensical.

Converting Minirhizotron Data to FRB Production Estimates

Estimates of the production of FRB were determined from a regression developed between (1) the length of new roots present on minirhizotron faces at one point in time and (2) the FRB of "new" roots per unit of soil volume collected in 17 cores (10 cm in diameter $\times$ 30 cm long) at the same point in time. Data collected from this single point in time of moderately high growth during 1 year (late June 1998) were extrapolated to the entire study period. Cores were collected from the top 30 cm and washed through a 0.25-mm sieve. From the fine roots collected from each core, "new" roots were separated based on physical resemblance to roots observed on minirhizotron faces in the surface 30 cm for the same time interval. Key characteristics were color (white to pale brown), degree of firmness (softness and lack of "woody" firmness that results following suberization), and morphology.

The developed relationship between length of "new" roots (from minirhizotrons) and the oven-dry mass of fine "new" roots per unit soil volume obtained from the cores at the same point in time was generalized to the entire soil profile examined in the study (the top 90 cm). This extrapolation was based on the following assumptions: (1) The one point in time used to establish this relationship is representative of the relationship between periodic root elongation observed on minirhizotron faces and periodic biomass production measured in soil cores; (2) the fine roots in the surface 30 cm (which account for 74% of FRB for the surface 90 cm) are representative of the remainder of the profile; (3) there are no slope position differences in this relationship; and (4) few roots exist below 90 cm, so their contribution to FRB biomass production is minimal at best.

To obtain root-biomass-production estimates for the roots 2–5 mm in diameter, extrapolation from the <2-mm-diameter roots was performed, based on the relative amounts of standing root biomass in these two size classes throughout all depth intervals and slope positions. Because minirhizotrons and ingrowth cores virtually never encounter roots >2 mm, we have no data on the rate of growth in this size class. In this case, root thickening is the primary form of root growth. Estimates from other research on root production in *Q. alba* [Joslin 1983; Joslin and Henderson 1987, using an entirely different measurement approach, the "compartment flow model" or "budget approach" (Joslin and Henderson 1982; Santantonio and Herman, 1985)] found that production per unit standing root biomass was 30–35% higher in the smaller (<2 mm) size class compared to that of the 2–5-mm size class. Hence, the extrapolated estimate of root production based on the standing-root-biomass relationship was reduced by this percentage.

Results

Pretreatment FRB Distribution

Fine-root biomass in the top 90 cm of the soil profile was distributed over the entire site so that ~74% of the fine roots were in the surface 30 cm of soil, 16% in the 30–60-cm layer, and 9% in the 60–90-cm layer (Fig. 16.1) (Joslin and Wolfe 1999). These distribution percentages were quite consistent across all treatments and slope positions, with the greatest variability at the deepest depth interval. Fine roots were rare below 90 cm, although no systematic measurements were made below this depth. These data correspond well with minirhizotron observations of root elongation 2 years after installation, wherein 66–79% of root elongation occurred in the surface 30 cm, 14–33% in the 30–60-cm layer, and 0–7% in the 60–90-cm layer (Fig. 16.2) (Joslin and Wolfe 1999).

In general, FRB tended to be slightly higher (8–32%) in the wet plot prior to treatments than in the ambient or the dry plots, on both the upper- and lower-slope positions (Fig. 16.1). Significant ($p < 0.05$) pretreatment differences between wet and dry plots existed only on the upper slope at both the 0–30-cm (14% less in the dry plot) and the 60–90-cm (55% lower) depth intervals. On the lower slope, the wet and the dry plots never differed significantly prior to treatments; however, on the lower slope, the wet plot was significantly higher than the ambient plot at both the 0–30-cm (19% less in the ambient) and the 60–90-cm intervals (68% lower in ambient, not shown).

Minirhizotron Results

The minirhizotron data exhibited large differences in the amount of root elongation across different slope positions during the first growing season (Fig. 16.2) (Joslin and Wolfe 1999). These differences were not seen in the later minirhizo-

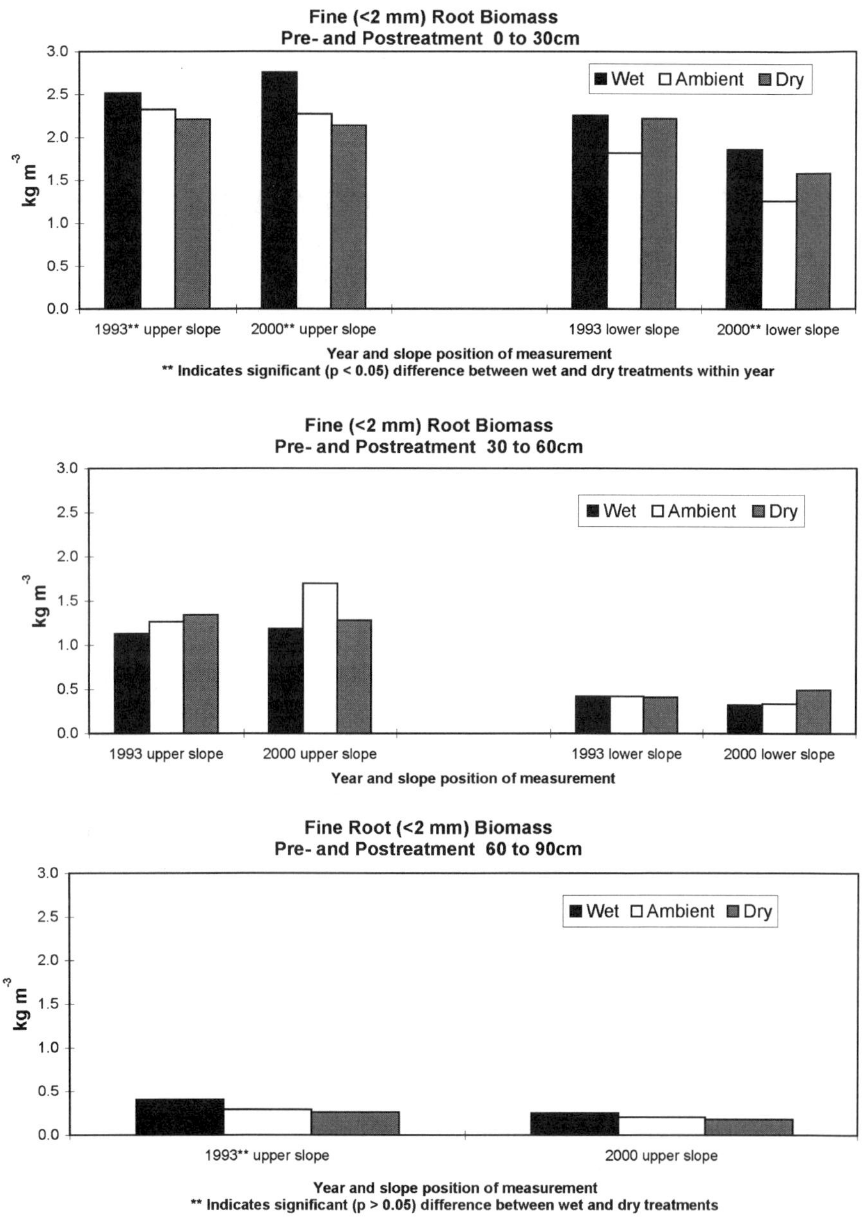

Figure 16.1. Fine-root (<2 mm diameter) Biomass changes from 1993 to 2000 by depth increment (0–30 cm, 30–60 cm, and 60–90 cm), by treatment (wet, ambient, and dry), and by slope position (upper slope, lower slope). All data are expressed as kilograms of oven-dried root material per cubic meter of rock-free soil. No data collected for lower slope at 60–90 cm in 2000. Data for upper slope at 60–90 cm in 2000 were collected by quantitative soil pits; all other data were collected by soil coring.

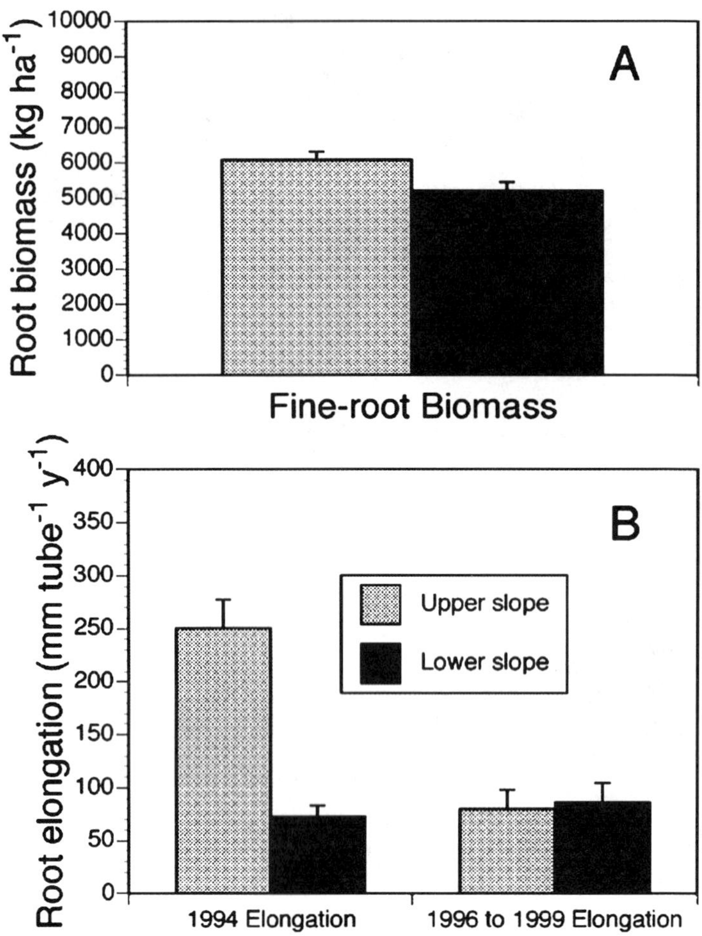

Figure 16.2. Comparison of upper- and lower-slope positions with respect to (1) standing pretreatment fine-root (<2 mm) biomass, (2) 1994 total root elongation observed in minirhizotrons (millimeters of elongation per tube), and (3) mean annual total root elongation observed in minirhizotrons over a 4-year period (1996 through 1999). All data depicted are from the surface 30 cm of the soil profile. In contrast to Fig. 16.1, all biomass data are expressed as kilograms of oven-dried root material per hectare of surface area. In the 1994 data (B), elongation from the two slope positions was significantly different ($p < 0.0001$).

tron data, nor in the preinstallation FRB data. The first year's (1994) minirhizotron data indicate that, in the surface 30 cm, root elongation was about four times greater on the upper slope than on the lower slope, and root-elongation rates were about three times greater than those measured from either slope position during the same seasonal period 2 years later (1996). In pointed contrast, 2 years later, the minirhizotron data indicate that slope positions are not significantly different with regard to root elongation and that the 1996 root elongation data are consistent with the absence of any significant slope position differences in FRB.

The above evidence suggests that the disturbances of both roots and soil during minirhizotron installation result in certain biases in the data collected in the months immediately following installation. The major bias results from excessive root proliferation, especially in the surface portion of the soil profile. Root pruning and growth-stimulating nutrient release resulting from soil disturbance are likely reasons for the observed root proliferation near the newly installed minirhizotron.

Factors Controlling Elongation

Minirhizotron measurements of REI from the second, third, and fourth years of study were used to evaluate the relative importance of two major environmental factors and of phenology in determining the seasonal timing of REI for the upper-slope portion of the study area (Joslin et al. 2001). Soil-water potential (Ψ) and soil temperature (ST) were the primary environmental variables examined with respect to their control over REI, and the data were used to delineate Ψ values that were optimum for growth. Phenology (PHEN), evaluated as a factor independent of Ψ and ST, was considered to be a largely genetically determined variable contributing to the timing of REI. A multivariate regression equation that attempted to employ all three of these variables, ST, Ψ, and PHEN, was developed to predict REI. Minirhizotron measurements of REI collected over three growing seasons in the surface 30 cm of the soil profile were composited and analyzed via multiple regression techniques, with corresponding sets of data on Ψ, ST, and PHEN.

Neither simple linear-regression analysis nor stepwise multiple-regression analysis resulted in a statistically significant correlation between REI and ST (nor the $\log_{10}$ of ST) during the April through October growing season (Joslin et al. 2001). The simple linear relationship between Ψ and REI was, however, statistically significant ($p < 0.02$) but not powerful ($r = +0.25$) (Joslin et al. 2001). At near-zero Ψ levels (i.e., between 0 and -0.2 MPa,), a very wide range of REI existed, with the highest REI rates occurring in the same Ψ range as relatively low rates of growth. In contrast, at low (moderately negative) Ψ levels, REI became consistently reduced; that is, below a Ψ of -0.4 MPa, REI never reached 15% of the maximum rate. A highly significant ($p < 0.0001$) relationship between Ψ and REI was obtained when the relationship was examined as log linear, with $\log_{10}$ $(-\Psi)$ versus $\log_{10}$ (REI). Across the range of observed Ψ (-0.025 to -1.1 MPa), REI was never observed to totally cease. Although not used in developing this regression, 1998 data indicate that below a Ψ of -1.5 MPa (see Fig. 16.3, September 8, 1998), root elongation appeared to cease.

The simple linear relationship between the phenology index (PHEN) and REI was highly significant ($p < 0.0001$), with $r = -0.40$. During the months of May and June (PHEN < 30), there is a very wide range of REI, but this period contains the three largest REI rates. The 24 highest REI intensity levels all occurred between late April and very early August (PHEN < 60). In contrast, after the first

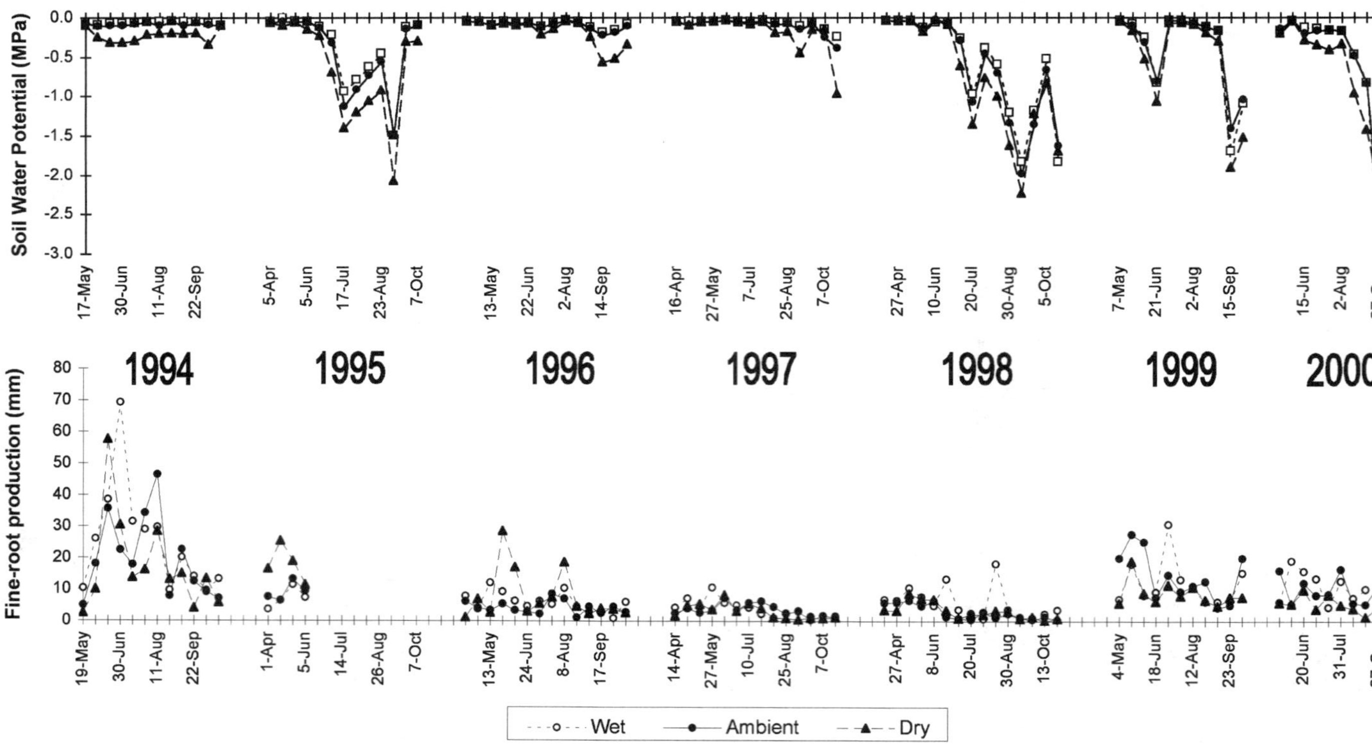

Figure 16.3. The bottom half of the figure shows periodic REI (mean net elongation per tube per 14-day interval) observed at minirhizotron faces in the surface 30 cm, upper slope, over the six growing seasons by treatment. The upper half of the figure juxtaposes the soil-water potential for the upper 35 cm by treatment for the corresponding period.

week of August (August through October), the REI level was never as high as this spring–early summer period, despite the fact that from mid-August to mid-October, soil moisture conditions were commonly within the "optimum range" for growth ($\Psi > -0.3$ and ST between 15°C and 21°C), with log(PHEN) versus log(REI) yielding the best description of the relationship between PHEN and REI ($p < 0.0001$), with $r = -0.49$ (Joslin et al. 2001).

In order to develop an equation to attempt to predict the timing of REI, we used the SAS Stepwise nonlinear regression procedure and the relationship of REI to the three dependent variables of PHEN, ST, and Ψ. As mentioned earlier, ST did not contribute significantly and so was dropped from the equation. The $\log_{10}$(PHEN) turned out to be the most important variable ($p < 0.0001$), with $\log_{10}(-\Psi)$ also contributing significantly ($p = 0.0014$).

$$\log_{10}(\text{REI}) = 1.33 - 0.31 \log(-\Psi) - 0.23 \log(\text{PHEN}).$$

The R^2 for the regression model was 0.32.

Seasonal and Annual Patterns of Root Elongation

One of the most notable results following 7 years of minirhizotron observations is the near absence of statistically significant differences between treatments in observed REI rates on either the upper- or lower-slope positions. Although there were some interesting trends and patterns, differences were not large nor did they reach significance at the 0.05 level.

Primarily, we concentrated our minirhizotron analyses on the surface 30 cm of the upper slope, for the several reasons described in the Methods section: (1) Soil-water treatment differences were greater on the upper slope; (2) tree species composition varied the least across treatments on the upper slope; (3) the greatest FRB occurred in the surface 30 cm of the soil profile; and (4) minirhizotron observations of root activity were highest and least variable in the surface 30 cm. Data from the lower slope are also summarized in this subsection.

In most years and treatments, the major portion of fine-root elongation on the upper slope, as observed in minirhizotrons, was completed by early July (Fig. 16.3), regardless of soil moisture levels during the latter half of the growing season. Of course, no data were available after June in the second year of observation (1995). Significant root growth in the first year (1994) did continue until early August, but data from that year appear to be abnormally high for reasons discussed earlier. One exception to this trend was a slight growth spurt in early August in the wet and the ambient treatments in 1996, a relatively wet year.

During the final 3 years of the study, marked dry periods began in July and lasted through the end of the growing season (Fig. 16.3). One interesting trend appeared to occur during 1998 and 1999: Roots in the wet plot, and sometimes the ambient plot, appeared to respond with short-lived root growth spurts during brief periods of wetting in the midst of these dry spells (Fig. 16.3: see August 17, 1998, July 15, 1999, and October 7, 1999). Such growth spurts in the latter

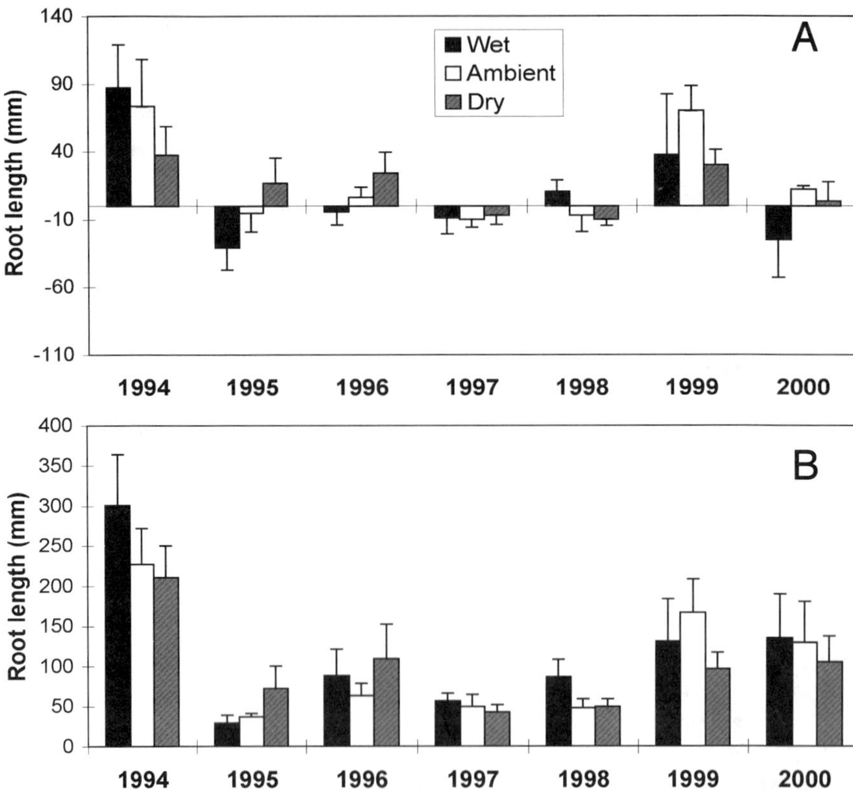

Figure 16.4. Annual quantities of net (A) and gross (B) root growth (i.e., elongation expressed in millimeters per minirhizotrons tube) by treatment for 1994–1999 for the surface 30 cm of the upper slope. Net growth = gross growth minus mortality; gross growth = mean of sums of new root-elongation measurements over a particular growing season for all five minirhizotron tubes within a treatment, expressed in millimeters per tube. Error bars indicate standard errors of the mean, based on annual cumulative net growth for each minirhizotron. Treatment differences for net growth were significant ($p < 0.1$) only for 1996, and gross growth treatment differences were not significant.

half of the growing season were never observed in the dry plot during these 3 years.

During 1995 and 1996, both gross and net root elongation were greatest in the dry plot (Fig. 16.4A and 16.4B). The higher level of gross fine-root production in the dry plot compared to the other two treatments was marginally significant ($p < 0.10$) only during 1996 (Joslin and Wolfe 1998). In 1997, gross and net production were comparable across all treatments. By 1998, the first of three consecutive dry years, the initial trend appears to have reversed itself. Both gross and net root growth was highest in the wet plot in 1998, whereas both gross and net root growth were higher in the wet and ambient plots than in the dry plots in 1999. Again, differences were neither large nor statistically significant.

On the lower slope, seasonal and year-to-year patterns were even less well defined than on the upper slopes. Major growth spurts were observed in every month of the growing season, rather than primarily occurring during May to July as on the upper slope (Fig. 16.5). The only apparent factor limiting growth was soil-water potential, because growth spurts were not observed when the soil-water potential dropped below -10 MPa.

Treatment Effects on Root Elongation and Mortality: Summing the First 7 Years

Figure 16.6 depicts the summation, across the first 6 years, of gross fine-root production, of fine-root mortality, and of net fine-root production (the difference between the first two values) for the surface 30 cm of the upper slope and the lower slope. Treatment differences were not statistically significant in gross root production, in mortality, or in net root growth at either slope position. In fact, net root growth is surprisingly similar (40–100 mm per tube) across all treatments and slope positions.

On the upper slope, the wet plot had the highest rates of both gross root production (18% and 15% higher than that in ambient and dry plots, respectively) and mortality (23% and 20% higher than the dry and ambient plots). In sharp contrast, on the lower slope, the dry plot had the highest rates of both gross root production and mortality, and the ambient plot the lowest. This combination of highest rates of both mortality and production rates in the wet plot on the upper slope resulted in comparable amounts of net root production across the three treatments. In fact, on both slopes, long-term gross production rates correlated highly with long-term mortality rates, resulting in comparable net production across all treatments and slope positions.

When the sum of root production and root mortality divided by initial FRB was used as an index of root turnover, treatment indices on both slope positions were very similar across treatments and slope positions (Fig. 16.7). The higher root production and mortality in the wet plot on the upper slope, would appear to be best explained by higher pretreatment FRB (9–14% higher than the ambient and dry plots) and higher final FRB (21–29%) higher in the wet plot. Similarly, on the lower slope, the low rates of root production and mortality in the ambient plot are consistent with lower pretreatment FRB (20–22% lower) and lower final FRB (20–30% lower) in the ambient plot. In both slope positions, the net result is virtually no difference between treatments in fine-root-turnover indices.

Net Changes in FRB Between 1993 and 2000

On the upper slope during the 7 years, no significant changes in the total FRB were observed within any of the treatments. Apparent reductions between 1993 and 2000 in the upper-slope deep-interval (60–90 cm) FRB are most likely attributable to the different techniques of root sampling used at this depth in the 2 years. Pit sampling (year 2000) produced rock-free soil concentrations of FRB that are ~70% the magnitude of those obtained by core sampling (1993), because core sampling is restricted to relatively rock-free areas.

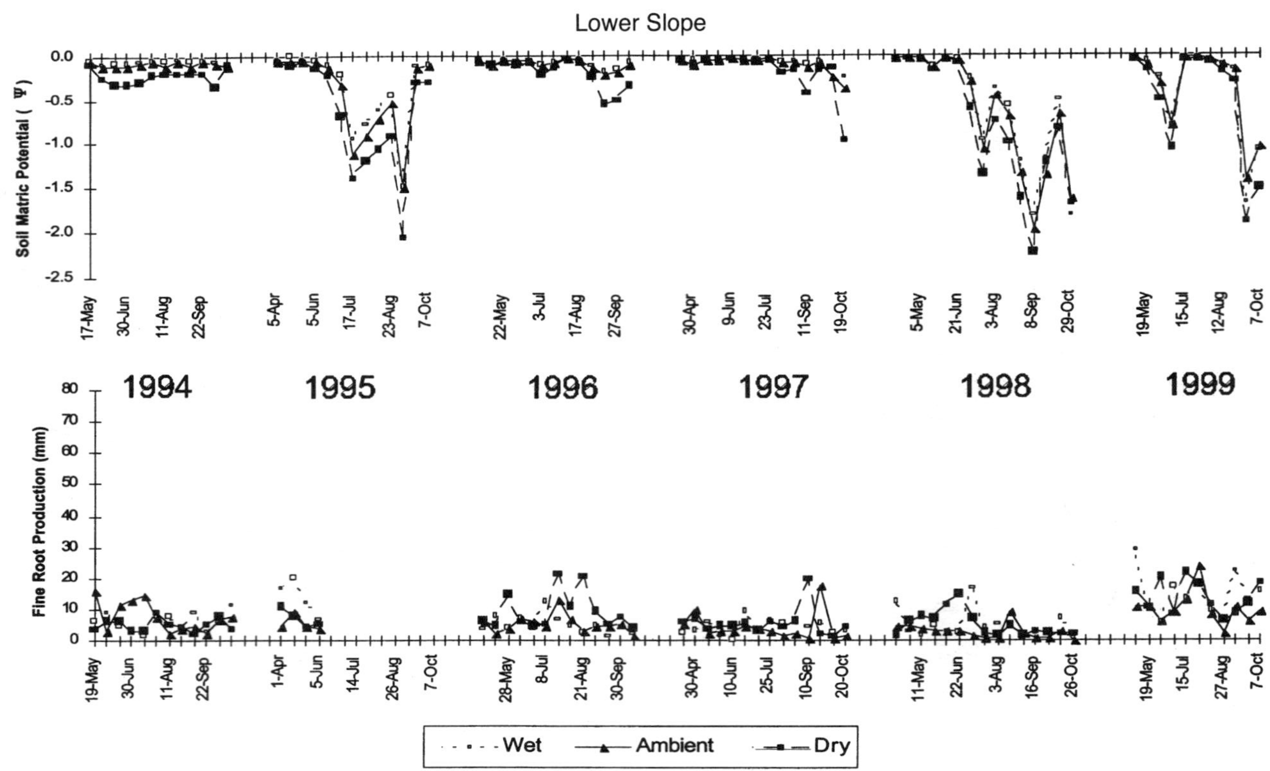

Figure 16.5. The bottom half of the figure shows periodic REI (mean net elongation per tube per 14-day interval) observed at minirhizotron faces in the surface 30 cm, lower slope, over the six growing seasons by treatment. The upper half of the figure juxtaposes the soil-water potential for the upper 35 cm by treatment for the corresponding period.

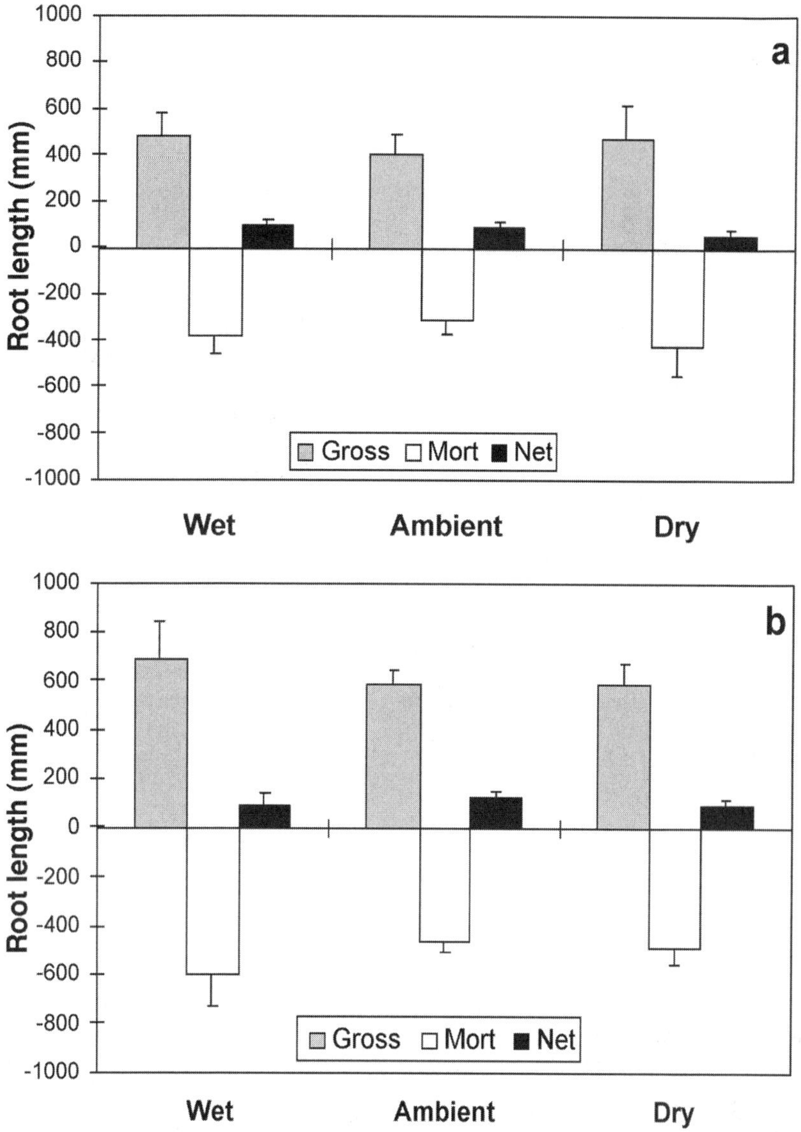

Figure 16.6. Summation of 6 years (1995 through 2000) of gross root growth, root mortality, and net root growth by treatment and slope position [(a) lower slope, (b) upper slope] in the surface 30 cm. Each value represents the mean of the summed root length data for five minirhizotron tubes, expressed in millimeters per tube. Error bars indicate the standard errors of the mean. No treatment differences are statistically significant.

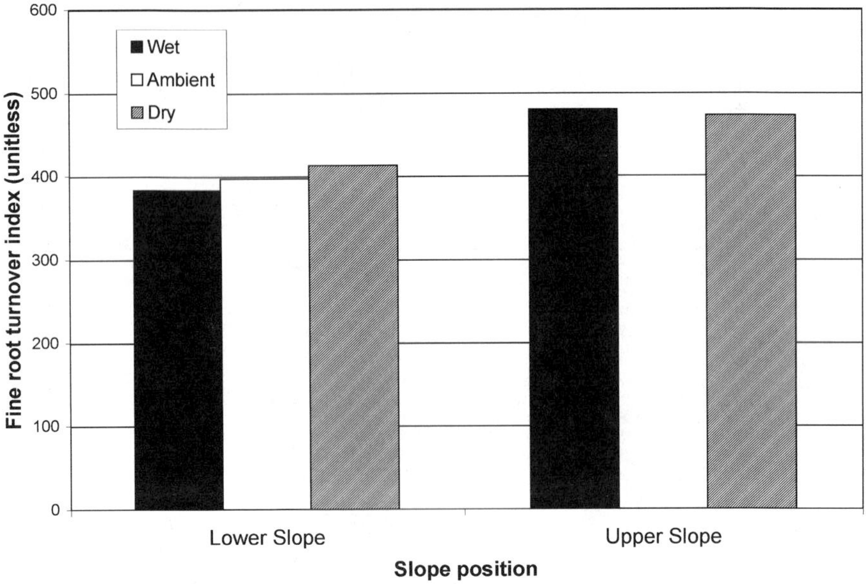

Figure 16.7. Indices of fine-root turnover for the surface 30 cm from 1995 to 2000 by treatment and slope position. The unitless indices, for each treatment and slope position, are based on the mean total length of root elongation per minirhizotron tube for the entire period plus the mean total length of root mortality per tube divided by the initial mean fine-root biomass for that treatment and slope position. All data used are from the surface 30 cm of the soil profile.

Across the lower slope, FRB declined in all three treatments in the surface 30 cm. This "slope effect," irrespective of treatment, may reflect the higher overall sensitivity of lower-slope species [e.g., *Liriodendron tulipifera* L. (yellow-poplar) and *Acer saccharum* L. (sugar maple) to drought in general, as the years 1998–2000 included extended drought periods during the latter 3 months of the growing season (Figs. 16.4A and 16.5). In upper-slope positions, only 1% of the overstory consists of *L. tulipifera,* but they represent 15% of the lower slope overstory (Hanson et al. 2001). Across the lower slope, 95% of the overstory *L. tulipifera* trees are greater 25 m in height, and *L. tulipifera* forms a "supercanopy." Because this species is noted for its "free" or "indeterminate" aboveground growth pattern, it is not surprising that fine-root growth appears to occur fairly randomly throughout the growing season on the lower slope (Fig. 16.5), in contrast to the upper slope, where most root growth is completed by early August (Fig. 16.4A). Root growth is quite common on the lower slope in September and October. The 1998, 1999, and 2000 droughts all occurred late in the growing season, thus perhaps greatly reducing both *L. tulipifera* carbon fixation and fine-root production.

After 7 years of treatments, an examination of changes that have occurred in the FRB with respect to treatment and to treatment × depth × slope position reveal a couple of interesting trends. In some cases, the trends are confirmed by statistically significant changes or interactions. For both the upper and lower slopes, the distribution of FRB shifted so that, relative to the wet plot, less FRB occurred near the surface (upper 30 cm) in the ambient and dry plots. Quite the opposite occurred at deeper depths in these two treatments, where relatively more FRB was measured in the dry and ambient plots. The above-described pattern is illustrated by the fact that in the surface soil layer (0–30 cm), the FRB in the wet plot, which in pretreatment had been greater than the other two treatments, became increasingly greater during the 7 years. This occurred on both the upper and lower slopes (Fig. 16.1). On the lower slope, the wet plot FRB became significantly ($p < 0.05$) higher than for both of the other two treatments, whereas prior to treatment, it had not been significantly different from the dry plot. On the upper slope, the wet plot FRB also became significantly ($p < 0.05$) higher than both the other two treatments, whereas prior to treatment, it had not been significantly different from the ambient plot. In the upper-slope surface interval, the ambient and dry plot FRB went from pretreatment levels that were 9% and 14% lower than that of the wet plot, respectively, to posttreatment levels that were 21% and 29% lower. It is worth noting, however, that this latter change was not confirmed by 7 years of minirhizotron observations, which indicated virtually no differences among treatments in net root elongation in the upper-slope surface 30 cm (Fig. 16.6a).

The above effect in the surface 30 cm was generally "counteracted," so to speak, by the fact that, at deeper depths, the pretreatment higher levels of the wet plot FRB tended to disappear or become reversed. By year 2000, wet plot FRB, on the both upper and lower slopes, was no longer significantly higher than either that of the ambient or dry plot, at any of the deeper depth intervals. On the upper slope, this change was particularly striking for the dry versus the wet plot at 60–90 cm and for the ambient plot at 30–60 cm (Fig. 16.1). On the upper slope at 60–90 cm, wet plot FRB had been significantly ($p < 0.05$) higher (55%) than dry plot FRB prior to treatments, whereas after 7 years, this difference was no longer significant (28%).

On the lower slope, a shift was especially marked for the dry versus the wet plot at 30–60 cm (Fig. 16.1). At this depth on the lower slope, the dry plot FRB increased by 22%, whereas the wet plot FRB declined by 23%. As a result, the dry plot FRB became 66% ($p < 0.05$) higher than the wet plot FRB, in contrast to pretreatment levels where the dry plot FRB had actually been lower than the wet plot FRB.

Estimates of Root Biomass Production from Minirhizotron Measurements

Comparisons of new root production in soil cores in the surface 30 cm of the upper slope with new root elongation observed on minirhizotron tubes in 1998 produced a relationship in which 1 mm of observed root elongation per minirhizotron tube was, on average, equivalent to 2.16 g m^{-2} of FRB production. Combining all treatments, the 5-year annual mean amount of root elongation per tube in the surface 30 cm for the period 1995–1999 was 72 mm y^{-1} for the upper slope (Figs. 16.4B and 16.6a) and 74 mm y^{-1} for the lower slope (Fig. 16.6b). Extrapolating these data to the surface 90 cm of soil on the basis of root biomass distribution by depth, the estimated rate of FRB production for this 5-year period was 224 g m^{-2} y^{-1}. Extrapolation to the diameter root size class of 2–5 mm resulted in an estimate of an additional 96 g m^{-2} y^{-1} for this size class, producing a total estimated root biomass production rate for <5-mm roots of 320 g m^{-2} y^{-1}.

Foliar Production and Fine Root : Foliage Ratio

Annual stand-level foliar biomass production (as indexed by litterfall) was not significantly affected by either the throughfall displacement treatments or by interannual differences in weather during the first 4 years. Annual production of aboveground litterfall at the stand level was nearly constant at ~500 g m^{-2} y^{-1} (Chapter 17, this volume). However, during 1998, 1999, and 2000, litter production was actually significantly higher in the dry plot compared to the wet plot (Chapter 17, this volume). These foliar results, coupled with the absence of statistically significant treatment effects on the long-term net increment in FRB, as indexed by net-root-elongation measures in minirhizotrons and by direct measurements by coring techniques, lead us to conclude that there is no support in this study for the hypothesis that exposure to dry conditions induces an increase in fine root : foliar biomass ratio. In fact, surprisingly, during the latter 3 years it appears that dry conditions are inducing a decrease in fine root : foliar biomass ratio.

Ingrowth Core Results

After three growing seasons, the mean FRB in the extracted ingrowth cores was very similar to those in the pretreatment cores. Cores were installed in February and March 1993, prior to treatment initiation in July 1993, halfway through the first growing season. Covariate analysis of variance, using the FRB of the matching pretreatment root cores as the covariate, revealed no significant ($p < 0.05$) treatment differences at any depth interval or slope position. All interactions between treatment and slope position and treatment and depth were also statistically insignificant. In the ingrowth cores, a significantly higher percentage of the FRB was very fine roots (<1 mm in diameter) than in the pretreatment cores. In general, the very fine-root biomass was 20–25 % higher in the ingrowth cores compared to the pretreatment cores, and the biomass of roots 1–2 mm in diameter was, on average, ~50% less in the ingrowth cores. This result appears to reflect simply the time lag that is required for the thickening of newly produced very fine roots into 1–2-mm-diameter roots.

There was a shift in FRB in ingrowth cores at 60–90 cm that is consistent with the pattern observed at that depth interval for changes in net FRB on the upper slope. In pretreatment cores, the total FRB at this depth interval had been significantly higher in the wet plot compared to the other two treatments (Fig. 16.1). In contrast, after three growing seasons, FRB in ingrowth cores from this depth was highest in the dry plot, with no significant differences among treatments.

Discussion

Fine-Root Biomass and Root : Shoot Ratio

Belowground research on TDE was designed to examine three hypothesized responses to long-term decreases in water availability: increases in (1) FRB and root length density, (2) R : S, especially FRB to foliar biomass ratio, and (3) depth of rooting. Minirhizotron observations and net changes in total FRB measured with soil cores and pits both point toward rejection of both hypotheses (1) and (2). One of the most notable findings from these 7 years of minirhizotron observation of root elongation is the near absence of treatment effects on the net amount of root elongation (the gross amount of elongation minus mortality). This finding, corroborated by the absence of treatment effects on FRB changes after 7 years, appears to be the result of the resilience of this forest ecosystem in maintaining a relatively constant fine-root mass over the long term.

In spite of short-term negative net root production during certain periods, which was most evident in the dry plot, the stands were able to respond opportunistically with spurts of root growth during periods of favorable water potential, especially early in the growing season. In this ecosystem, the major pulse of root elongation appears to be genetically programmed to occur in late spring and early summer (Joslin et al. 2001), a pattern similar to that in many deciduous forests in much of the temperate zone (Joslin 1983; Hendrick and Pregitzer 1996b). There was, however, a trend during 1998 and 1999 for the roots in the ambient plot and especially the wet plot to respond, at times, to short periods of soil wetting during the latter half of the growing season with small spurts of root growth. Meanwhile, during these years, roots in the dry plot showed no such significant root-growth response. Such pulses were partially responsible for higher overall gross root production in the wet plot on the upper slope; however, this higher rate of fine-root production was countered by comparably higher root mortality. Although these treatment differences could be interpreted as resulting from a higher allocation of carbon belowground as soil moisture increases, they were not statistically significant and their magnitude was consistent with higher pretreatment FRB in the wet plot on the upper slope.

In seedling studies, absolute increases in FRB with decreasing water availability have only been found in 1 case out of 10 experiments across 5 studies (Torreano and Morris 1998; Osunubi and Davies 1981; Bongarten and Teskey 1987; Tomlinson and Anderson 1998; Pallardy and Rhoads 1993). In stand-level irrigation studies, although FRB did increase as water availability declined in four of seven cases, in only one case was this increase statistically significant (DeVisser et al. 1994; Bredemeier et al. 1998; Devakumar et al. 1999; Gower et al. 1992; Dickman et al. 1996).

A second, more remarkable result is the absence in our study of higher R : S ratios in the dry plot compared to other treatments. Even more surprising is the apparent shift toward a decrease in R : S ratio in the dry plot compared to the wet plot. In sharp contrast, in three out of the five seedling studies where R : S was examined, statistically significant shifts toward greater R : S ratios in response to reduced irrigation were found (Joslin et al. 2000). Furthermore, three long-term irrigation studies on plantations or forests found higher R : S ratio responses to lower rates of water application (Gower et al. 1992; DeVisser et al. 1994; Devakumar et al. 1999). The results of Devakumar et al. (1999) revealed a large decrease in R : S ratio with irrigation over a 10-year period. However, the results of certain experiments indicate that such R : S responses may be highly species-specific (Osunubi and Davies 1981; Pallardy and Rhoads 1993).

Rooting Depth

Our results regarding net changes in FRB after 7 years of treatment give some support to the hypothesis that exposure to increasingly dry conditions may cause increased rooting at greater depths, as well as reduced root production near the surface. Some support is also available for the inverse—that exposure to increasingly wet conditions may reduce rooting at greater depths. Comparisons between the wet plot and the dry plot after 7 years on both the upper slope at 60–90 cm and on the lower slope at 30–60 cm provide limited support to this conclusion. Similarly, comparisons between the wet plot and the ambient plot on the upper slope at 30–60 cm provide support. Such trends toward deeper rooting in the dry and ambient plots compared to the wet plot are also consistent with changes in surface rooting patterns. For both the upper and lower slopes, the distribution of FRB shifted so that, relative to the wet plot, less FRB occurred near the surface (the upper 30 cm) in the ambient and dry plots.

In the literature, the rooting depth hypothesis is supported by the greater rooting depth that was observed in two of the three cases where it was examined (Osunubi and Davies 1981; Torreano and Morris 1998). Osunubi and Davies (1981) found deeper rooting in seedlings of a deep-rooted tree species *Quercus robur* L. (English oak) but not in a shallow-rooted one *Betula pendula* L. (silver birch). In stand-irrigation studies, rooting depth changes were only examined in two cases, one resulting in a significant increase as the result of irrigation (Dickman et al. 1996) and the other resulting in a significant decline with irrigation (Devakumar et al. 1999).

In our study, the possibility that root systems are tapping supplies of water deep in the soil profile during dry periods has not been eliminated. Although a very high bulk density (~ 1.75 g cm^{-3}) in the zone of maximum clay accumulation (B2t) of the argillic horizon appears to restrict most root growth to the upper 90 cm, it appears that some roots have penetrated this barrier. Such penetration below 90 cm was observed in all 15 pits excavated to determine FRB. Although the density of fine roots is generally meager at this depth, the occurrence of fine-root growth indicates that such penetration does occur and could lead to significant water uptake, especially during dry periods.

Effect of Soil-Water Potential on Root Elongation

In our 1995–1997 analysis of minirhizitron data for the upper slope, root elongation never appeared to cease entirely nor did a high soil-water content ever appear to limit REI (Joslin et al. 2001). Teskey and Hinckley (1981) found that the relationship between log($-\Psi$) and log(root intensity) appeared linear at moderately high soil-water contents (i.e., when -3.0 MPa $< \Psi < -0.5$ MPa). On the other hand, our data do indicate that, under the driest conditions during this period (i.e., Ψ below -0.35 MPa), REI never reached 15% of the maximum rate observed at less negative Ψ values (Joslin et al. 2001). In contrast, in 64% of the growth intervals at more positive Ψ's, REI was greater than the maximum REI of the seven intervals when Ψ was below -0.35 MPa. During all 7 of these "dry" intervals, REI was much lower (35–95% lower) than the mean REI for the remaining 72 data points when Ψ was greater than -0.35. Other field research on *Q. alba, Q. rubrum* L. (northern red oak), and *Juglans nigra* L. (black walnut) shows root elongation declining markedly at Ψ's between -0.4 and -0.8 MPa and virtually ceasing at about -1.2 MPa (Larson and Whitmore 1970; Teskey and Hinckley 1981; Kuhns et al. 1985). Our results with this mix of upland hardwood species generally agree with this conclusion (Joslin et al. 2001).

Effect of Soil Temperature on Root Elongation

Our results indicating no statistically significant relationship between root elongation and soil temperature during the growing season (Joslin et al. 2001) are not in agreement with substantial literature that indicates temperature having an important influence on REI (Lyr and Hoffman 1967; Ovington and Murray 1968; Hermann 1971; Larson 1974; Teskey and Hinckley 1981; Kuhns et al. 1985; Burke and Raynal 1995). The optimum ST for *Quercus* root growth in Larson (1974) was 24°C, whereas Teskey and Hinckley (1981), with *Q. alba*, Kuhns et al. (1985), with black walnut, and Burke and Raynal (1995), with northern hardwoods, discovered that ST was the primary factor affecting REI at STs below 17–18°C. Admittedly, our data were collected over a somewhat narrow ST range of 10–21°C. However, even when we examined across the ST range of 10–15°C, where other studies found temperature a major factor, or when we eliminated periods when Ψ would have most likely limited REI ($\Psi < -0.2$), we still found no statistically significant relationship between ST and REI.

Effect of Phenology on Root Elongation

In our study covering 1995 through 1997, the significance of the role of phenology on the upper slope is perhaps best illustrated by the fact that all the highest (top 24 out of 81) REI measurements occurred during late April to early August (Joslin et al. 2001). This pattern appeared in spite of the fact that during late summer and early autumn, ST was continually in a range considered optimum for *Quercus* trees (Larson 1974; Teskey and Hinckley 1981) and the soil Ψ was greater than -0.3 and should not have been limiting in almost half these intervals (10 out of 24). Reich et al. (1980) found a similar pattern with mature *Q. alba* and invoked the concept of genetically controlled phenological programming. This theory is in conflict with the concept of the physiological optimum, which posits that root growth is highly plastic, opportunistic, and responsive to ideal soil conditions (e.g., Teskey and Hinckley 1981). Although environmental variables, especially soil-water potential (Ψ), are highly important factors controlling root elongation in our study, our data are consistent with some degree of phenological programming. In addition to our results and those of Reich et al. (1980), root growth in a number of studies of temperate-zone upland hardwood species seems to be reduced in late summer and early autumn, relative to late spring and early summer, regardless of moisture conditions (Teskey and Hinckley 1981; McClaugherty et al. 1982; Kuhns et al. 1985; Joslin and Henderson 1987; Hendrick and Pregitzer 1996b). In fact, in a meta-analysis of 17 published cases involving north-temperate-zone deciduous tree species, by far the most active time for root elongation was June and July (Joslin et al. 2001). In trees growing in a climate with a typical mid-summer drought, this pattern may be an adaptation to avoid the typical summer drought period.

Minirhizotron data from the lower slope of our own study seems to support this limitation on the role of phenology in "controlling" the timing of root growth. On the lower slope, fine-root elongation during the 6 years was more evenly spread out over the months of the growing season than on the upper slope, especially when soil-water potential is taken into consideration. The different pattern on the lower slope may be related to the fact that *L. tulipifera* is a canopy dominant only on the lower slope, and the importance of *Quercus* species is greatly reduced. Unlike the *Quercus* species, *L. tulipifera* exhibits a "free-growth" pattern with regard to the production of shoots and foliage and may do the same with regard to root production.

Factors Influencing Patterns of Mortality

There was no apparent seasonal pattern exhibited by fine-root mortality, which seemed to remain relatively steady throughout the growing season across all 7 years of our study. Edwards and Norby (1999) reported a similar absence of seasonal pattern in root mortality in *A. saccharum* and *A. rubrum*. One might expect higher mortality rates during the warmer months because various studies have indicated higher mortality rates in response to artificial temperature treatments (Edwards and Norby 1999) and shorter root life spans across increasing temperature gradients (Hendrick and Pregitzer 1993; Pregitzer et al. 1997, 2000). Higher soil temperatures have been assoicated with higher root respiration rates, nutrient mineralization rates, and pathogen and herbivore rates (Gill and Jackson 2000). Other studies indicate higher root mortality during leaf senescence in northern hardwood forests (McClaugherty et al. 1982; Hendrick and Pregitzer 1992, 1996b). One confounding factor that might account for the absence of higher recorded mortality in mid- and late-summer and autumn at our site is the frequent occurrence of low soil moisture during these periods. Because mortality is difficult to detect until decomposition begins and because low soil moisture greatly inhibits decomposition, poor detection of mortality during these periods is possible.

Fine-Root Response Effects on Belowground Carbon Allocation and Stand Carbon Budgets

After 7 years of minirhizotron data collection and an examination of net changes in FRB after 6.5 years of treatment, almost no evidence of treatment effects on the allocation of carbon belowground has emerged. The minirhizotron data give some slight indication of higher allocation of carbohydrates belowground in the wet—the opposite of that hypothesized; however, this pattern was not statistically significant, and the small magnitude of this difference is proportional to the initially higher FRB in the wet plot. These data also indicate slightly higher rates of mortality in the wet plot, also nonsignificant and proportional to the higher initial FRB.

The combination of slightly higher production and higher mortality, based on minirhizotron data, resulted in estimates of no net changes in FRB. Direct measurements of FRB between 1993 and 2000 indicate that slight net increases in FRB may have occurred in the wet plot, relative to the other two treatments, but only in the surface 30 cm of the lower slope. These changes, however, are quite small and marginally significant statistically. On the other hand, slight net decreases in FRB in the wet plot may have occurred at 60–90 cm on the upper slope, relative to the other two treatments. In the context of the belowground carbon budget, these changes are small indeed.

Although these data seem to indicate no effect of treatments on belowground carbon allocation over the length of the study to date, increased proliferation of roots lower in the soil profile, if real, could result in a small shift in carbon distribution in soil. Perhaps greater carbon sequestration under drier conditions could eventually result in increased deep carbon allocation, as roots begin to decay because this carbon should be afforded greater protection by the heavy clay content of the lower horizons (Jardine et al. 1990) and cooler soil temperatures. Such a shift toward increased rooting with depth under drier conditions deserves further investigation.

Fine-Root Response Effects on Belowground Nutrient Cycling

In light of the fact that treatment effects on belowground carbon allocation were minimal, if any, during the first 6 years of this study, it is unlikely that there would be large impacts via fine roots on the nutrient cyling process during this time period. Perhaps the most likely scenario would involve the possible shift in root density under drier conditions toward fewer roots near the surface and more roots at deeper intervals. This shift conceivably could result in a lower efficiency of nutrient uptake in the surface, where litter decomposition is paramount. On the other hand, it may simply reflect a slight shift toward higher nutrient availability in the B horizon relative to the A horizon in the dry plot (Johnson et al. 2002; Chapter 23, this volume). Such a shift could result in better exploitation of nutrients from the deeper horizons and/or greater movement of nutrients into these zones resulting from a greater quantity of FRB turnover there.

Conversely, the shift toward slower litter decomposition and accumulation of litter with increasing dryness, observed by Johnson et al. (2002), could be a possible factor in the slighter higher rates of root turnover and higher net accumulation of fine roots in the surface 30 cm of the wet plot. Related to this effect, of course, is the lower nutrient availability observed in soil solutions in the dry plot (Johnson et al., 2002; Chapter 23, this volume). Higher leaf decomposition rates in the wet plot should result in higher mineralization rates for most nutrients. As mentioned earlier, root proliferation may occur in such zones of nutrient accumulation (Drew and Saker 1978; Eissenstat and Caldwell 1988; Pregitzer et al. 1993). Also, higher nitrogen concentrations in fine roots (Eissenstat et al. 2000) and higher soil nitrogen availability (Nadelhoffer et al. 1985; Nadelhoffer 2000) have both been observed to reduce fine-root longevity (increasing turnover rates) and to increase root respiration rates (Reich et al. 1998; Pregitzer et al. 1998).

Fine-Root Response Effects on Water Uptake

We have presented some evidence that the root systems, especially of mature overstory trees, may be flexible enough to shift their rooting to deeper zones of water availability. Although the evidence is by no means convincing, if found to be true, it would not be particularly surprising, in light of the extant literature. We did, however, expect the dense argillic horizon (bulk density averages 1.75 g cm^{-3} at ~70 cm) to be a more significant barrier to downward expansion of roots. This appears to be the case with saplings and certain shallow-rooted understory species [e.g., dogwood (*Cornus florida* L.)], which, in fact, are the species that have suffered significantly higher mortality in the dry plot (Chapter 15, this volume).

On the other hand, the mature trees have broken through the densest zone of soil, although the "breakthroughs" are widely dispersed spatially. This exploration of deeper soil layers may be critical to their survival under dry conditions and may partially explain the rather small treatment differences in growth and foliar production produced by the mature trees. Future examination should look more closely at treatment differences not only in rooting between 60 and 90 cm, but also at rooting below 90 cm.

References

Bongarten BC, Teskey RO (1987) Dry weight partitioning and its relationship to productivity in loblolly pine seedlings from seven sources. For Sci 33:255–267.

Bradford KJ, Hsiao TC (1982) Physiological responses to moderate water stress. In Lange OL, Nobel PS, Osmond CB, Ziegler H (Eds) *Encyclopedia of plant physiology, Vol 12B*. Springer-Verlag, Berlin, pp 264–324.

Bredemeier M, Blanck K, Dohrenbusch A, Lamersdorf N, Meyer AC, Murach D, Parth A, Xu YJ (1998) The Solling roof project—site characteristics, experiments, and results. For Ecol Manage 101:281–293.

Burke MK, Raynal DJ (1995) Fine root growth phenology, production, and turnover in a northern hardwood forest ecosystem. Plant Soil 162:135–146.

Caldwell MM, Richards JH (1986) Competing root systems: morphology and models of absorption. In Givnish TJ (Ed) *On the economy of plant form and function*. Cambridge University Press, Cambridge, England, pp 251–273.

Comeau PG, Kimmins JP (1989) Above-and below-ground biomass and production of lodgepole pine on sites with differing soil moisture regimes. Can J For Res 19:447–454.

Dahlman RC, Kucera CL (1965) Root productivity and turnover in native prairie. Ecology 46:84–89.

Deans JD (1979) Fluctuations of the soil environment and fine root growth in a young Sitka spruce plantation. Plant Soil 52:45–51.

Devakumar AS, Prakash PG, Sathik MBM, Jacob J (1999) Drought alters the canopy architecture and micro-climate of *Hevea brasiliensis* trees. Trees 13:161–167.

DeVisser PH, Beier BC, Rasmussen L, Kreutzer K, Steinber N, Bredemeier M, Blanck K, Farrell EP, Cummins T (1994) Biological response of five forest ecosystems in the EXMAN project to input changes of water, nutrients and atmospheric loads. For Ecol Manage 68:15–29.

Dickman DI, Nguyen PV, Pregitzer KS (1996) Effects of irrigation and coppicing on above-ground growth, physiology, and fine-root dynamics of two field-grown hybrid poplar clones. For Ecol Manage 80:163–174.

Drew MC, Saker LR (1978) Nutrient supply and the growth of the seminal root system in barley. III. Compensatory increases in growth of lateral roots, and in rates of phosphate uptake, in response to localized supply of phosphate. J Exp Bot 29:435–451.

Edwards NT, Norby RJ (1999) Belowground respiratory responses of sugar maple and red maple saplings to atmospheric CO_2 enrichment and elevated air temperature. Plant Soil 206:85–97.

Eissenstat DM, Caldwell CC (1988) Seasonal timing of root growth in favorable microsites. Ecology 69:870–873.

Eissenstat DM, Wells CE, Yanai RD, Whitbeck JL (2000) Building fine roots in a changing environment: Implications for root longevity. New Phytol 147:33–42.

Enslin WR, Pregitzer KS, Hendrick RL (1994) *MSU ROOTS: A PC-based program to quantify plant roots.* Center for Remote Sensing, Michigan State University, East Lansing, Michigan.

Gill RA, Jackson RB (2000) Global patterns of root turnover for terrestrial ecosystems. New Phytol 147:13–31.

Gower ST, Vogt KA, Grier CC (1992) Carbon dynamics of Rocky-Mountain Douglas-fir—Influence of water and nutrient availability. Ecol Monogr 62:43–65.

Gurevitch J, Chester ST Jr. (1986) Analysis of repeated measures experiments. Ecology 67: 251–255.

Hanson PJ, Todd DE, Amthor JS (2001) A six year study of sapling and large-tree growth and mortality responses to natural and induced variability in precipitation and throughfall. Tree Physiol 21:345–358.

Head GC (1973) Shedding of Roots. In Kozlowski TT (Ed) *Shedding of Plant parts.* Academic Press, New York, pp. 273–293.

Hendrick RL, Pregitzer KS (1992) The demography of fine roots in a northern hardwood forest. Ecology 73:1094–1104.

Hendrick RL, Pregitzer KS (1993) The dynamics of fine root length, biomass, and nitrogen content in two northern hardwood ecosystems. Can J For Res 23:2507–2520.

Hendrick RL, Pregitzer KS (1996a) Applications of minirhizotrons to understand root function in forests and other natural ecosystems. Plant Soil 185:293–304.

Hendrick RL, Pregitzer KS (1996b) Temporal and depth-related patterns of fine root dynamics in northern hardwood forests. J Ecol 84:167–176.

Hermann RK (1971) Growth and production of tree roots: A review. In Marshall KK (Ed) *The belowground ecosystem: A Synthesis of plant-associated processes.* Range Science Department Science Series No 26. Colorado State University, Fort Collins, Colorado, pp 7–28.

Jardine PM, Wilson GV, McCarthy JF, Luxmoore RJ, Taylor DL, Zelazny LW (1990) Hydrogeochemical processes controlling transport of dissolved organic carbon through a forested hillslope. J Contam Hydrol 6:3–19.

Johnson DW, Hanson PJ, and Todd DE Jr. (2002) The effects of throughfall manipulation on soil leaching in a deciduous forest. J Environ Qual 31:204–216.

Joslin JD (1983) *The quantification of fine root turnover in a white oak stand.* PhD thesis. University of Missouri, Columbia, Missouri.

Joslin JD, Henderson GS (1982) A test of the budget method: A revised approach to the measurement of fine root turnover. In Muller RN (Ed) *Central Hardwood Forest Conference, Proceedings of the 4th Meeting.* University of Kentucky, Lexington, Kentucky, pp 220–228.

Joslin JD, Henderson GS (1987) Organic matter and nutrients associated with fine root turnover in a white oak stand. For Sci 33:330–346.

Joslin JD, Wolfe MH (1998) Impacts of water input manipulations on fine root production and mortality in a mature hardwood forest. Plant Soil 204:165–174.

Joslin JD, Wolfe MH (1999) Disturbances during minirhizotron installation can affect root observation data. Soil Sci Soc Am J 63:218–221.

Joslin JD, Wolfe MH, Hanson, PJ (2000) Effects of altered water regimes on forest root systems. New Phytol 147: 117–129.

Joslin JD, Wolfe MH, Hanson, PJ (2001) Factors controlling the timing of root elongation intensity in a mature upland oak stand. Plant Soil 228:201–211.

Kaushal P, Guehl JM, Ausenac G (1989) Differential growth response to atmospheric carbon dioxide enrichment in seedlings of Cedrus atlantica and Pinus nigra ssp. Laaricio var. Corsicana. Can J For Res 19:1351–1358.

Keyes MR, Grier CC (1981) Above-and below-ground net production in 40 year-old Douglas-fir stands on low and high productivity sites. Can J For Res 11:599–605.

Kozlowski TT, Kramer PJ, Pallardy SG (1991) The physiological ecology of woody plants. 2nd edition. Academic Press, San Diego, California.

Kramer PJ (1983) Water relations of plants. Academic Press, New York.

Kuhns MR, Garrett HE, Teskey RO, Hinckley TM (1985) Root growth of black walnut trees related to soil temperature, soil water potential, and leaf water potential. For Sci 31:617–629.

Larson MM (1974) Effects of soil moisture on early growth of oak seedlings. For Res Rev (Ohio Agric Res Dev Ctr) 74:10–12.

Larson MM, Whitmore FW (1970) Moisture stress affects root regeneration and early growth of red oak seedlings: Influence of soil temperature. For Sci 16:495–498.

Lyr H, Hoffman G (1967) Growth rates and growth periodicity of tree roots. Int Rev For Res 2:181–236.

McClaugherty CA, Aber JD, Melillo JM (1982) The role of fine roots in the organic matter and nitrogen budgets of two forested ecosystems. Ecology 63:1481–1490.

Meredith MP, Stehman SV (1991) Repeated measures experiments in forestry: Focus on analysis of response curves. Can J For Res 21:957–965.

Nadelhoffer KJ (2000) The potential effects of nitrogen deposition on fine root production. New Phytol 147:131–139.

Nadelhoffer KJ, Aber JD, Melillo JM (1985) Leaf litter production and soil organic matter dynamics along a nitrogen availability gradient in southern Wisconsin (USA). Ecology 66:1377–1390.

Osunubi O, Davies WJ (1981) Root growth and water relations of oak and birch seedlings. Oecologia 51:343–350.

Ovington JD, Murray G (1968) Seasonal periodicity of root growth of birch trees. In Ghilarov M S, Kovda VA, Novichkova-Ivanova LN, Rodin LE, Sveshnikova VM (Eds) *Methods of productivity studies in root systems and rhizosphere organisms.* SSR Acad of Sciences, Nauka Publishing, Leningrad, Russia, pp 146–154.

Pallardy SG, Rhoads JL (1993) Morphological adaptations to drought in seedlings of deciduous angiosperms. Can J For Res 23:1766–1774.

Persson H (1980) Fine root dynamics in a Scots pine stand with and without near-optimum nutrient and water regimes. Acta Phytogeogr Suec 68:101–110.

Persson H, Ahlstrom K (1991) The effects of forest liming and fertilization on fine root growth. Water Air Soil Pollut 54:365–375.

Pregitzer KS, Hendrick RL, Fogel R (1993) The demography of fine roots in response to patches of water and nitrogen. New Phytol 125:575–580.

Pregitzer KS, Kubiske ME, Yu CK, Hendrick RL (1997) Relationships among root branch order, carbon, and nitrogen in four temperate species. Oecologia 111:302–308.

Pregitzer KS, Laskowski MJ, Burton AJ, Lessard VC, Zak DR (1998) Variation in sugar maple root respiration with root diameter and soil depth. Tree Physiol 18:665–670.

Pregitzer KS, King JS, Burton AJ, Brown SE (2000) Responses of fine roots to temperature. New Phytol 147:105–115.

Reed HS, MacDougal DT (1937) Periodicity in the growth of the orange tree. Growth 1:371–373.

Reich PB, Teskey RO, Johnson PS, Hinckley TM (1980) Periodic root and shoot growth in oak. For Sci 26:590–598.

Reich PB, Walters MB, Tjoelker MG, Vanderklein D, Buschena C (1998) Photosynthesis and respiration rates depend on leaf and root morphology and nitrogen concentration in nine boreal tree species differing in relative growth rate. Funct Ecol 12:395–405.

Rogers WS, Head GC (1969) Factors affecting the distribution and growth of roots of perennial woody species. Proc Easter Sch Agr Sci, University of Nottingham 15:280–295.

Rogers WS (1939) Root studies. VIII. Apple root growth in relation to root-stock, soil seasonal, and climatic factors. J Pomol Hortic Sci 17:99–130.

Santantonio D, Hermann RK (1985) Standing crop, production, and turnover of fine roots on dry moderate, and wet sites of mature Douglas-fir in western Oregon. Ann Sci For 42:113–142.

SAS Institute (1989) *SAS/STAT1 Users Guide. Version 6.* 4th edition, Vol 1 and 2. SAS Institute, Cary, North Carolina.

Taylor HM, Upchurch DR, McMichael BL (1990) Applications and limitations of minirhizotrons for root studies. Plant Soil 129:29–35.

Teskey RO, Hinckley TM (1981) Influence of temperature and water potential on root growth of white oak. Physiol Plant 52:363–369.

Tomlinson PT, Anderson PD (1998) Ontogeny affects response of northern red oak seedlings to elevated CO_2 and water stress. II. Recent photosynthate distribution and growth. New Phytol 140:493–504.

Torreano SJ, Morris LA (1998) Loblolly pine root growth and distribution under water stress. Soil Sci Soc Am J 62:818–827.

van Buijtenen JP, Bilan V, Zimmerman RH (1976) Morpho-physiological characteristics related to drought resistance in Pinus taeda. In Cannell MGR, Last FT (Eds) *Tree physiology and yield improvement.* Academic Press, London, pp 348–359.

Vogt KA, Vogt DJ, Palmiotto PA, Boon P, O'Hara J, Asbjornsen H (1996) Review of root dynamics in forest ecosystems grouped by climate, climatic forest type and species. Plant Soil 187:159–219.

17. Canopy Production

Paul J. Hanson, Donald E. Todd, and J. Devereux Joslin

Annual production of canopy foliage and its subsequent senescence in autumn, coupled with flower, seed, and terminal-branch production and senescence are among the quintessential processes defining deciduous forests. The annual renewal of canopy foliage and the production of seed crops are energy-intensive processes requiring the formation of stem, leaf, and flower primordia in previous seasons, the storage of carbohydrate supplies, and the translocation and storage of nutrients (Kozlowski and Pallardy 1997). Litter collection data provide direct information about annual carbon inputs to the forest floor. They also are a primary set of measurements from which the maximum annual leaf area index can be derived. Prior to the initiation of this study, changing precipitation inputs were hypothesized to have various impacts on the production of foliage, flowers, seeds, and terminal branches based on potential changes in carbohydrate supply or available nutrients (Gholz et al. 1990). Three specific hypotheses, however, were developed to include potential effects of both wetting and drying on canopy production:

1. Changes in stored carbohydrates and/or changes in current photosynthate available at the time of leaf production were predicted to increase canopy production as water inputs were amplified and to decrease production as they were reduced.
2. Drier than normal conditions were predicted to lead to immobilization of nutrients in soil organic matter and, as a result, to limit canopy production. This

limitation of productivity could occur even though the total pool of essential elements on the dry plot might increase.

3. Increases in water throughput were predicted to cause the loss of base cations (e.g., K, Ca, and Mg) from the forest ecosystem and thereby to result in reduced canopy production.

Methods

To evaluate canopy leaf production and litter inputs to the forest stand, fallen leaves, and other materials (i.e., twigs, seeds, etc.) were collected within each of the three treatment blocks of the Throughfall Displacement Experiment (TDE) from 1992 (pretreatment) through 2000. Fallen leaves and other forms of litter were collected from 147 rectangular baskets (53 × 38 cm) located near the central 49 (7 rows × 7 columns) grid intersections of each of the 3 treatment blocks of the TDE. Litter collections were made periodically throughout the year (typically mid-May, late August, and biweekly from October 1 through December 1 of each year).

Once collected, the litter samples were dried for 24–48 hours at 70°C. The individual litter samples were sorted into leaves, woody material, seeds, and flowers in all years. In 1994, 1995, and 1999, the foliage component of litter was sorted into the following species categories *Acer* sp., *Cornus florida, Liriodendron tulipifera, Nyssa* plus *Oxydendrum, Pinus* sp., *Quercus alba, Quercus prinus,* other *Quercus/Carya* sp., and miscellaneous species.

Values of the leaf area index for the treatment plots were derived from area–leaf-litter-mass relationships for the litter collected. The total mass of foliar litter was divided by the basket collection area (0.2014 m^2) to yield annual foliar litter production on a ground-area basis. This value was divided by the mean, species-weighted leaf-litter mass per unit area (LMA; g m^{-2}) of the litter for each year to obtain estimates of the annual leaf area index (LAI) across the experimental area.

Seasonal patterns of stand leaf area development were determined from the ratio of understory to overstory light penetration determined from the daily sum of hourly observations collected above the forest canopy (mean of one or two locations) and at 1.5 m above the ground below the forest canopy (mean of three to six locations). This canopy–light ratio (CLR) [Eq. (1)] was calculated as 1 minus the ratio of the daily sum of understory photosynthetically active radiation (PAR_u; PAR at 1.5 m) to the daily sum of overstory incident PAR (PAR_o)

$$\mathrm{CLR} = 1 - \frac{\mathrm{PAR}_u}{\mathrm{PAR}_o}. \tag{1}$$

Because of the presence of tree boles and branches, preleafout and postsenescence baselines for the CLR were not zero. Therefore, to express the CLR on a 0 to 1 scale, it was necessary to adjust the ratio for the light penetration during leafless periods as shown in Eq. (2).

$$\text{RelLAI} = \frac{\text{Observed CLR} - \text{Baseline}}{\text{Maximum CLR} - \text{Baseline}}, \tag{2}$$

where RelLAI is the relative LAI on the 0 to1 scale for a given date. However, because the baseline ratio resulting from the presence of boles and branches changed with solar elevation throughout a seasonal cycle, we found it necessary to use more than one baseline for the calculation of the seasonal pattern of RelLAI. Although the baseline CLR in the absence of leaves would vary continuously with solar elevation, we found that use of the preleafout baseline for days 80 through 180 and the postsenescence baseline for days 181 through 350 yielded an acceptable RelLAI pattern. It is important to note that the RelLAI values are representative of the development of leaf area not leaf mass (i.e., they may overpredict the rate of annual leaf mass accumulation). Approximations of leaf mass development or a direct estimate of LAI could also be made from the same data with a light-extinction approach (Hutchinson and Baldocchi 1989; Vose et al. 1995).

Results

Leaf Emergence and Canopy Development

With the exception of the 1996 growing season, canopy leaf emergence was initiated between April 3 and April 9 [day of year (DOY) 93 to 99] in all years (Table 17.1 and Fig. 17.1). In 1996, a cold spring extended the dormant season by nearly 2 weeks, with leaf emergence not occurring until April 21 (DOY = 111). Notwithstanding differences in the leaf out date each year, canopy development approached maximum values by day 140 in each year. The TDE treatments have had no discernable impact on canopy development.

The overall mean soil temperature on the initial date of leaf emergence (leafout date; LOD) was 12.4°C, but ranged from 9.8°C in 1993 to 15.2°C in 1999. Treatment effects on LOD were not observed.

Table 17.1. Fitted coefficients for annual canopy leaf development for the equation $\text{RelLAI} = 1 - 1\,e^{b(\text{DOY-LOD})^d}$

Variable	1993	1994	1995	1996	1997	1998	1999	2000
LOD	96	99	94	111	93	94	96	97
b ($\times 10^{-2}$)	−0.6	−0.6	−1.36	−4.75	−6.84	−2.84	−2.15	−1.53
d	2.49	2.11	1.69	1.35	1.02	1.39	1.44	1.55
r^2	0.98	0.99	0.99	0.99	0.99	0.98	0.97	0.99

Note: In the equation, RelLAI is the relative level of canopy development on a 0 to 1 scale, b and d are constants controlling the shape of the curve, DOY is the calendar day of the year, and LOD is the date on which visible leaf/stem elongation begins. LOD is derived from a curve-fitting procedure but is nominally the date on which visible bud expansion is evident from the ground.

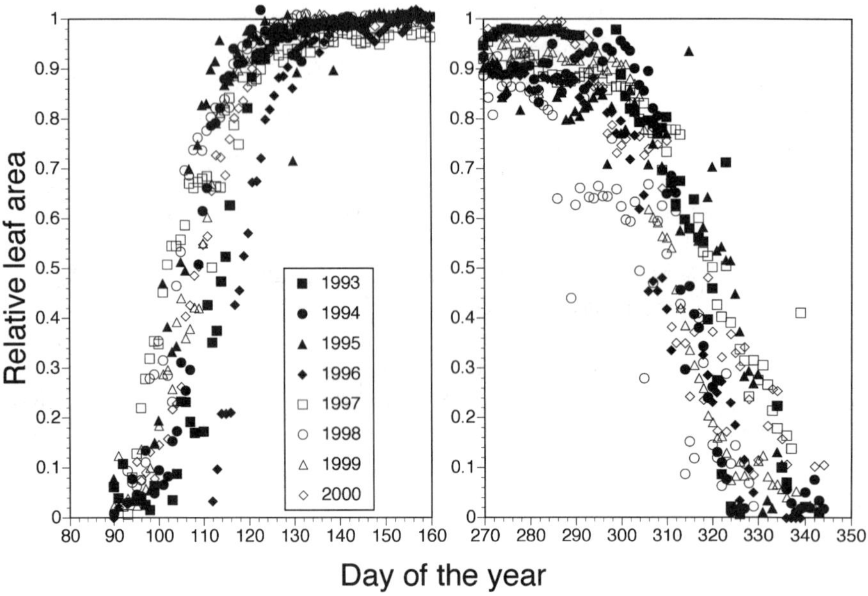

Figure 17.1. Pattern of relative leaf area from 1993 through 2000. Fitted equations for annual leaf development (days 80 to 150) are provided in Table 17.1 and maximum LAI values to associate with these relative patterns can be found in Table 17.2.

Leaf Senescence

Unlike the specificity of the year-to-year variation observed for the initiation of leafout in the spring, canopy senescence was harder to resolve from the light-penetration-based index of canopy LAI (Fig. 17.1). Declines in RelLAI were apparent in all years, starting around day 230, but canopy senescence was earlier in 1998, with observed early litter fall from *Acer, Liriodendron, Nyssa,* and *Oxydendrum.* Significant autumn leaf drop was initiated consistently between days 290 and 300 in all years, but the resolution of RelLAI data make the determination of a specific date or species-specific dates impossible.

Mass of leaf-litter collections in the litter baskets by mid-September of each year was also used to test for treatment differences in the rate of leaf senescence. We conducted covariate analysis of variance for 1994 through 2000 litterfall using the early 1993 litterfall as the covariate to determine if the TDE treatments affected the rate of canopy leaf senescence. No significant treatment impacts were observed for 1995, 1996, 1997, or 2000. Leaf-litter accumulation by mid-September on the wet plot was reduced relative to ambient and dry plot litterfall in both 1994 ($p = 0.00$) and 1999 ($p = 0.01$), suggesting potential inhibition of senescence by precipitation increases in some years. Accelerated leaf senescence of the dry plot trees was only observed in 1998 ($p = 0.001$), when leaf-litter fall in the dry plot was 50% greater than in either the wet or ambient plots.

Leaf Area Index

Because the TDE site is composed of 22 different species, the estimation of LAI from foliar litter production and species-weighted LMA is at best an approximation. LMA characteristics vary by species and, within a species, by height in the canopy, making simple extrapolations to the stand very difficult. For example, the mean LMA for the 2000 leaf litter was 61 g m^{-2} for *A. rubrum,* 74 g m^{-2} for *L. tulipifera,* and 68 g m^{-2} for *N. sylvatica* and ranged from 79 to 104 g m^{-2} for the *Quercus* species on the site. We used the mass fraction of each species' contribution to total leaf-litter mass and their mean LMA to derive annual estimates of mean LMA. The mean LMA weighted by species contributions did not vary greatly among years. We observed LMA values of 80, 82.9, 87.5, 82.6, 88.8, 80.8, 79.5, and 77.9 g m^{-2} in 1993 through 2000, respectively. We did not find significant differences in LMA caused by the TDE treatments. Mean leaf-litter mass multiplied by the species-weighted LMA for each year gave an 8-year mean LAI for the TDE site of 6.2 m^2 m^{-2} (Table 17.2). Herbivory leaf losses of 8–10% per year (Shure et al. 1998) were not reflected in estimates of leaf-litter mass from the litter baskets, suggesting that the LAI estimates from the product of leaf-litter mass and species-weighted LMA might be somewhat low. We have not adjusted the values in Table 17.2 because we do not have herbivory estimates for every year.

Annual Foliar Production

From 1994 (the first full year of TDE treatments) through 1997, no significant differences were observed between treatments, and annual leaf-litter production was very stable with a mean value of 509 g m^{-2} y^{-1} (Table 17.2). Following the two wet seasons of 1996 and 1997, mean litter production in 1998 dropped 4% and remained lower than the 1994–1997 mean in 1999. The mean litter production in 2000 increased to a level 8% higher than the established 1994–1997 mean. Analysis of the change in mean leaf-litter production in 1998 revealed a significant ($p = 0.01$) treatment × time interaction: Dry plot foliar production remained constant, whereas the ambient and wet plot leaf-litter production fell. This treatment effect remained significant in 1999 ($p = 0.00$) and became greater in 2000 ($p = 0.01$). Leaf production never decreased in years following severe late-season droughts (1995, 1998, and 1999), but reduction in ambient and wet plot leaf litter production were observed following the 1996 and 1997 wet years.

Figure 17.2 contrasts the percent contribution of various species to total foliar production across the TDE site in 1999 with the basal-area contribution of the same species. Ring-porous species (*Quercus* and *Carya*) represent more than half (58%) of the annual foliar production. *Acer* sp.contribute 26% of the annual leaf production and the common midcanopy species (*Nyssa* and *Oxydendrum*) make up 18% of the total. The overall distribution of leaf-litter production among species reflects their basal-area contribution. Treatment differences in foliar litter mass production described above were not attributed to shifts in species composition (data not shown). Early increases in dogwood (*C. florida*) mortality on

Table 17.2. Annual canopy litter production (mean ± SE) by year and treatment, broken down into leaf and nonfoliar components

Plot	1993	1994	1995	1996	1997	1998	1999	2000
Leaf litter (g dry mass m^{-2})								
Wet	—	513 ± 11	493 ± 13	508 ± 13	522 ± 11	474 ± 10a[a]	471 ± 11a	497 ± 12a
Ambient	—	483 ± 16	504 ± 14	514 ± 14	497 ± 13	455 ± 13a	483 ± 11a	537 ± 18b
Dry	—	511 ± 13	523 ± 10	546 ± 15	533 ± 12	534 ± 19b	536 ± 14b	610 ± 17c
Mean	496 ± 10	503 ± 8	507 ± 7	523 ± 8	518 ± 12	488 ± 9	497 ± 9	548 ± 10
Covariate *p*-Values	na[b]	0.25	0.10	0.12	0.89	0.01	0.00	0.00
Total nonfoliar litter (g dry mass m^{-2})								
Wet	—	147 ± 16a	255 ± 32	97 ± 15	127 ± 19	97 ± 15	278 ± 31	64 ± 5
Ambient	—	111 ± 12b	236 ± 19	98 ± 15	89 ± 14	72 ± 8	146 ± 28	63 ± 8
Dry	—	97 ± 8b	229 ± 19	84 ± 10	142 ± 20	89 ± 11	188 ± 24	80 ± 8
Mean	86 ± 7	118 ± 7	240 ± 14	93 ± 8	119 ± 10	86 ± 6	204 ± 17	69 ± 4
p-Values	0.29	0.01	0.73	0.72	0.10	0.30	0.00	0.20
Acorn production (g dry mass m^{-2})								
Mean	na	na	60 ± 8	9 ± 2	32 ± 4	1 ± 0.3	87 ± 10	2 ± 0.7
LAI (m^2 m^{-2})								
Mean	6.2	6.0	5.7	6.3	5.8	6.0	6.25	7.0

Note: The respective *p*-values for treatment main effects are from analysis of covariance for leaves which uses the 1993 leaf-litter collections as the covariate and analysis of variance for the nonfoliar litter. Also shown are mean (± SE) annual acorn production for 1995 through 2000 and LAI estimates for the site.

[a] Different letters following the means indicate siginificant differences among treatments.

[b] na = not applicable.

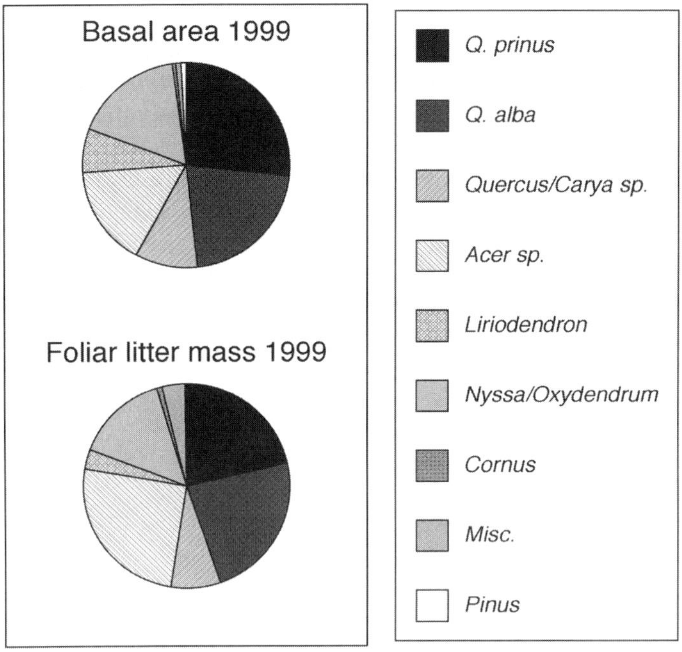

Figure 17.2. Percent contribution by species or species group to 1999 stand basal area (m^2 ha^{-1}) and 1999 foliar litter production (g m^{-2}).

the dry plot and cumulative losses of *A. rubrum* and *C. florida* from the dry and ambient plots (Chapter 15, this volume) certainly reduced canopy leaf production on the dry and ambient plots versus the wet plot. However, the overall contribution of sapling foliage to the stand total leaf-litter production is less than 1% and is not responsible for the opposite trend seen in the foliar-litter-production numbers for the stand.

Nonfoliar Litter Production

Nonfoliar litter production (flowers, seeds, twigs, and bark) showed considerably higher year-to-year variation than leaf-litter production, ranging from 69 to 240 g m^{-2} y^{-1} (Table 17.2). The maximum nonfoliar litter production observed in 1995 and 1999 was associated with large acorn crops, but wind damage in May 1995 and ice-storm damage in late 1998 also contributed to larger than normal nonfoliar litter collections in those years.

Seed Production

Mast production from *Quercus* and *Carya* species was minimal in most years ($<$ 10 g m^{-2}), but significant mast production was observed in 1995 and 1999 (Table 17.2). The TDE throughfall manipulations did not affect mast production (data not shown), and years of high mast production were not correlated with environmental conditions, stem growth (Chapter 15, this volume), or dormant-season carbohydrate levels (Chapter 5, this volume). Mechanisms for mast production have not been elucidated (Cecich and Sullivan 1999; Olson and Boyce 1971) and remain a largely stochastic phenomenon, but evidence does exist for genetic control over acorn production rates within individual trees (Healy et al. 1999; Greenberg 2000). Although *A. rubrum* flowers profusely in all years, early-spring frosts often reduced optimum seed development, and *A. rubrum* seed production was sporadic from year to year. *A. saccharum* seed production, which occurs in the summer as opposed to early spring, was only abundant in 1994. In the heart of the range of *A. saccharum* (i.e., northern states), abundant seed years have been reported every second or third year (Garrett and Graber 1995). *L. tulipifera* seed production is common in all years (data not shown). Seed production by *Cornus, Nyssa, Oxydendrum, Prunus, Pinus,* and other less common species occurs every year, but is not an obvious stand-level carbon sink in any of the years observed.

Discussion

Growing-Season Duration

The active growing season for our deciduous upland *Quercus* forest is defined by the period during which the canopy supports leaves. At Walker Branch, the total growing-season length varied by as much as 18 days, ranging from 187 to 203 days from 1993 to 2000. Longer growing seasons were associated with warmer spring periods, but simple and robust predictive relationships between date of leafout and air, soil, or cumulative temperature indices were not identified in this study.

Foliar Production and Influence of Nutrient Cycling

The pattern of foliar production did not agree with the initial hypothesis. There was no evidence to suggest that throughfall reductions (dry plot) would lead to reduced canopy leaf production. The wet and ambient plots' leaf-litter production was significantly ($p < 0.01$) different from the dry plot beginning in 1998, with both the ambient and the wet plots declining, whereas the amount of dry plot litter production remained constant (Table 17.2). This separation of treatments held relatively steady—and statistically significant—through 1999 and 2000, when foliar production increased in all plots. Previous-year soil-water conditions were expected to be a good predictor of foliar production because most of the principal species in this *Quercus*-dominated forest exhibit fixed and determinant foliar growth (Kramer and Kozlowski 1979). For example, leaf buds are set in these *Quercus* species by September of the previous year, and virtually all foliar production the following spring occurs in April and May from these set buds. However, previous-year soil-water conditions, as well as dormant-season carbohydrate reserves (Chapter 5, this volume), did not correlate well with leaf-litter production. The greatest decline in the wet and ambient plots occurred in 1998 following two consecutive years of high soil-water content. In contrast, all plots increased foliar litter production in 2000, following 2 consecutive years of late-season drought. Foliar litter production in 1996 was also unresponsive to the drought year of 1995.

The determinate growth pattern of the *Quercus*-dominated TDE forest may partially explain the lack of a response of foliar production to drought years. Canopy development in this forest is normally complete by the middle of May in all years, well ahead of the development of soil-water deficits. In addition, because leaf development in *Quercus* and many other species is at least 50% dependent on current rather than stored carbohydrate reserves (Dickson et al. 2000), overwintering stored reserves would need to fall to very low levels before impacts on canopy production were observed. McLaughlin (1980) showed that a large white *Quercus* tree could have carbohydrate reserves of 54 kg of sugar, sufficient to support the replacement of the leaf canopy three times.

A hypothesis that could explain both the 1998 short-term decline in litter production in the wet and ambient plots is a connection to base cation losses from the site, which may have resulted in reduced soil availability of these cations during the growing season. Johnson et al. (2002; Chapter 23, this volume) have calculated the fluxes of nutrients through the soil profile by treatment and found that throughout the study, Ca and Mg fluxes have been the highest in the wet treatment and the lowest in the dry. During most years, the fluxes of these two nutrients are on the order of three times higher in the wet treatment. In other words, the increased quantity of water moving through the soil profiles of the wet and ambient treatments appears to be removing more base cations from these plots than from the dry plot. Consistent with this finding is the observation that Ca and Mg concentrations in B horizon soil solutions of the dry treatment were nearly twice as high as for the wet plot during the growing season from 1996 through 2000, indicating a higher availability of these nutrients for tree uptake and incorporation in the foliage.

This hypothesis is also consistent with the temporal pattern of foliar litter production. The droughts of 1998, 1999, and 2000 reduced the fluxes of Ca and Mg during the sampling period, because of reductions in soil-water flux. Foliar litter production increased in all treatments during 1999 and again in 2000. This argument is weakened somewhat, however, by the absence of increased concentrations of Ca and Mg in B horizon soil solutions during the 1998–2000 drought period compared to previous wet years.

The availability of soil base cations—principally Ca and Mg—for plant uptake has been demonstrated to be reduced in many different forest types by a combination of soil leaching and tree uptake and storage in biomass (or removal via harvesting) (e.g., Johnson and Todd 1990). Such reductions in soil base-cation levels can be accelerated by the deposition of acidic anions in precipitation (Gilliam et al. 1996; Mitchell et al. 1996), by conversion of pines to hardwoods (Johnson et al. 1995), by harvest removal (especially whole-tree harvesting), or by increased rainfall inputs (as demonstrated in this study). Such reductions in soil base cations has led to speculation and concern about the development of cation deficiencies in both coniferous forests and deciduous forests in many regions of North America and Europe (e.g., Bernier and Brazeau 1988; Schulze 1989; Joslin et al. 1992; Cook et al. 1997; Friedland and Miller 1999). Specific concerns about the potential for base-cation nutrient deficiencies in hardwood forests have been raised, but foliar deficiencies have only rarely been demonstrated (Cote et al. 1995; Long et al. 1997; Timmer and Teng 1999).

Interannual Variability in Canopy Production

We were surprised by the stability of interannual quantity of leaf production because one of our initial hypotheses was that canopy leaf production would be an early indication of precipitation change. From 1993 through 1997 (including a severe drought in 1995), mean leaf-litter production varied by only 26 g m^{-2} (range 496–523 g m^{-2}), and this variability increased to only 60 g m^{-2} during the period from 1998 to 2000, when we have significant treatment effects.

A few studies that we describe briefly have published interannual patterns of litter production that go beyond 1 or 2 years of observations. Bray and Gorham (1964) summarize observations for an *Acer* forest near Toronto, Canada, for the period from 1957 through 1961. Over that 5-year period (1959 was not measured) foliar, bud scale, and fruit production averaged 310 g m^{-2} y^{-1} with additional inputs of 130 g m^{-2} y^{-1} from branch and bark materials. As in our observations, they concluded that annual foliar production was relatively stable from year to year (range from 280 to 320 g m^{-2} y^{-1}), but stem and bark contributions to litterfall were much more variable (50–320 g m^{-2} y^{-1}). Burton et al. (1991) showed stable foliar litter production from 1988 to 1990 for five northern hardwood sites in Michigan ranging from 344 to 369 g m^{-2} y^{-1}. Reproductive-litter (seeds and flowers) production showed greater variability and ranged from 29 to 108 g m^{-2} y^{-1}. Sykes and Bunce (1970) followed litter production in a mixed deciduous forest (*Quercus/Fraxinus*) from 1966 through 1968 in England and found mean annual foliar, bract/flower/fruit, and twig inputs of 323, 54, and 103 g m^{-2} y^{-1}, respectively. They also observed that interannual differences in foliar litter production were much lower than for twigs and other woody materials and attributed variable woody material inputs to wind events. In general, if foliar litter production is separated from total litter production, limited interannual variability is found for hardwood forests.

We found published data from only two studies that used similar collection methods to the TDE study, and that had a longer record of litter production. Magill et al. (2000) reports an 8-year record (1989–1996) of litter production (components not defined) for a *Quercus*-dominated hardwood stand in central Massachusetts exposed to various levels of nitrogen and sulfur inputs. Although the experimental fertilizations had no impact on litter production, interannual patterns showed some increase with time from mean values around 270 g m^{-2} y^{-1} in 1989 to 340 g m^{-2} y^{-1} between 1994 and 1996. Large changes in litter production from 1990 to 1992 suggested sensitivity to certain environmental variables, but causes were not discussed. This Harvard Forest dataset continues to be collected, and although the quantity of leaf-litter production is 32% less than observed at Walker Branch, the interannual pattern from 1993 through 1999 is very similar (Magill, personal communication). The longest published record of litter production for a North American hardwood forest (Knutson 1997) is a 17-year record of total litterfall (leaves, bud scales, catkins, buds, fruits, and twigs and small branches) for an *Acer/Tilia/Ostrya* forest in northeastern Iowa for 1978 through 1994. Knutson (1997) found that total litterfall ranged from a low of 504 to a high of 785 g m^{-2} y^{-1} during that period. Knutson (1997) found no relationship between annual rates of litterfall and either annual or growing-season precipitation. Unfortunately, foliar data were not presented separately from total litterfall in their paper.

Multiyear Patterns of Litterfall for Walker Branch Watershed

Leaf-litter-production estimates for Walker Branch Watershed are available from a number of earlier studies, including observations in 1969 (Grizzard et al. 1976), 1973 (Comiskey 1978), 1980 (Mann and West 1981), and 1989 (Garten et al. 1994). Grizzard et al. (1976) reported mean values of foliar, reproductive components, and twigs of 378, 51, and 41 g m^{-2} y^{-1} respectively, across several vegetation types occurring on Walker Branch Watershed in 1969. For the 1973 litterfall season, Comiskey (1978) found comparable means of 372 and 26 g m^{-2} yr^{-1} for foliar and reproductive/twig components, respectively, and an additional 10 g m^{-2} y^{-1} of frass. Comiskey (1978), assuming that frass fall equaled 86% of insect consumption of leaf material, estimated that insect consumption of leaf material was 3.2% of total leaf production. In 1980, Mann and West (1981), studying five hardwood plots on the Oak Ridge Reservation similar to those on Walker Branch, found average foliar and nonfoliar litterfall rates of 385 and 29 g m^{-2} y^{-1}, respectively, in 1980. Garten et al. (1994) also collected leaf litterfall across Walker Branch Watershed ridges and slopes for 1989 and found a mean foliar litterfall rate of 402 g m^{-2} y^{-1}. Although the data are not from the same locations as the litter collections from the TDE study, a plot of all these combined with the TDE plot observations (Fig. 17.3) indicates that leaf-litter production may have increased from 350 to 400 g m^{-2} y^{-1} in 1969 to values near 500 g m^{-2} y^{-1} today. However, because most of the increase takes place since 1990, we may instead just be looking at site-to-site differences and long-term stability in leaf-litter production.

Permanent-plot forest-survey data for Walker Branch from 1967 through 1983 (Johnson et al. 1987) document a clear reduction in live biomass (and basal area) associated with tree mortality in the first few years of the 1980s, and there was very likely a temporary reduction in foliar litterfall during this time. However, we have no direct records of litterfall for that period. The data from Fig. 17.3 suggest that any reductions in litter production following the insect-induced mortality in the early 1980s was short-lived.

Significant changes in foliar litter production reported starting in 1998 as a part of the TDE are equal to or greater in magnitude than the maximum interannual

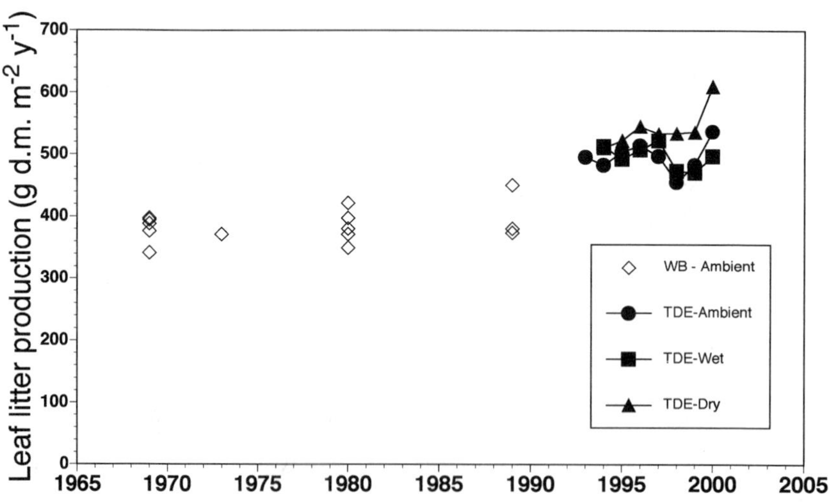

Figure 17.3. Multiyear pattern of leaf-litter production for Walker Branch Watershed, including data from ridge, valley, and side slope locations. Data from the TDE study are plotted for 1993–2000. Data for other similar sites on the Walker Branch Watershed were obtained from the following years of field observations: 1969 (Grizzard et al. 1976), 1973 (Comiskey 1978), 1980 (Mann and West 1981), and 1989 (Garten et al. 1994).

variation in foliar litter production, suggesting that cumulative impacts of long-term precipitation change may very well lead to altered canopy leaf production and understory light and temperature environments. Changes of this type have a very real potential to affect interspecific competition among regenerating understory species and lead to long-term shifts in species composition and associated biodiversity.

References

Bernier B, Brazeau M (1988) Foliar nutrient status in relation to sugar maple dieback and decline in the Quebec Appalachians. Can J For Res 18:754–761

Bray JR, Gorham E (1964) Litter production in forests of the world. Adv Ecol Res 2:101–157.

Burton AJ, Pregitzer KS, Reed DD (1991) Leaf area and foliar biomass relationships in northern hardwood forests located along an 800 km acid deposition gradient. For Sci 37:1041–1059.

Cecich RA, Sullivan NH (1999) Influence of weather at time of pollination on acorn production of *Quercus alba* and *Quercus velutina.* Can J For Res 29:1817–1823.

Comiskey CE (1978) *Aspects of the organic carbon cycle on Walker Branch Watershed: A study of land/water interaction.* Ph.D. thesis. The University of Tennessee, Knoxville, Tennessee.

Cook RB, Joslin JD, McLaughlin SB, Eagar C, Knoepp JD, Trettin C (1997) Terrestrial ecosystems effects report. In *1997 National Acid Precipitation Assessment Program's interim report to the U.S. Congress.* National Acid Precipitation Assessment Program, Washington, District of Columbia.

Cote B, O'Halloran I, Hendershot WH, Spankie H (1995) Possible interference of fertilization in natural recovery of a declining sugar maple stand in southern Auebed. Plant Soil 168–169:471–480.

Dickson RE, Tomlinson PT, Isebrands JG (2000) Allocation of current photosynthate and changes in tissue dry weight within northern red oak seedlings: Individual leaf and flush carbon contribution during episodic growth. Can J For Res 30:1296–1307.

Friedland AJ, Miller EK (1999) Major-element cycling in a high-elevation Adirondack forest: Patterns and changes, 1986–1996. Ecol Appl 9:958–967.

Garrett PW, Graber RE (1995) Sugar maple seed production in northern New Hampshire. Research Paper No. 697. USDA Forest Service, Northeastern Forest Experiment Station, Radnor, Pennsylvania, pp 1–6.

Garten CT Jr., Huston MA, Thomas CA (1994) Topographic variation of soil nitrogen dynamics at Walker Branch Watershed, Tennessee. For Sci 40:497–511.

Gholz HL, Ewel KC, Teskey RO (1990) Water and forest productivity. For Ecol Manage 30:1–18.

Gilliam FS, Adams MB, Yurish BM (1996) Ecosystem nutrient responses to chronic nitrogen inputs at Fernow Experimental Forest, West Virginia. Can J For Res 26:196–205.

Greenberg CH (2000) Individual variation in acorn production by five species of southern Appalachian oaks. For Ecol Manage 132:199–210.

Grizzard T., Henderson GS, Clebsch EEC, Reichle DE (1976) *Seasonal nutrient dynamics of foliage and litterfall on Walker Branch Watershed, a deciduous forest ecosystem.* ONRL/TM–5254. Oak Ridge National Laboratory, Oak Ridge, Tennessee.

Healy WM, Lewis AM, Boose EF (1999) Variation of red oak acorn production. For Ecol Manage 116:1–11.

Hutchinson BA, Baldocchi DD (1989) Forest Meteorology In Johnson DW, VanHook RI (Eds) *Analysis of biogeochemical cycling processes in Walker Branch Watershed.* Springer-Verlag, New York, pp 21–95.

Johnson DW, Todd DE (1990) Nutrient cycling in forests of Walker Branch watershed, Tennessee: Roles of uptake and leaching in causing soil changes. J Environ Qual 19:97–104.

Johnson DW, Henderson GS, Harris WF (1987) Changes in aboveground biomass and nutrient content on Walker Branch Watershed from 1967 to 1983. In Hay RL, Woods FW, DeSelm H (Eds) *Proceedings of the Central Hardwood Forest Conference.* The University of Tennessee, Knoxville, Tennessee, pp 487–496.

Johnson DW, Binkley D, Conklin P (1995) Simulated effects of atmospheric deposition, harvesting, and species change on nutrient cycling in a loblolly pine forest. For Ecol Manage 76:29–45.

Johnson DW, Hanson PJ, Todd DE Jr. (2002) The effects of throughfall manipulation on soil leaching in a deciduous forest. J Environ Qual 31:204–216.

Joslin JD, Kelly JM, Van Miegroet H (1992) Soil chemistry and nutrition of North American spruce-fir stands: Evidence for recent change. J Environ Qual 21:12–30.

Knutson RM (1997) An 18-year study of litterfall and litter decomposition in a northeastern Iowa deciduous forest. Am Midl Nat 138:77–83.

Kozlowski TT, Pallardy SG (1997) *Growth control in woody plants.* Academic Press, San Diego, California.

Kramer PJ, Kozlowski TT (1979) *Physiology of woody plants.* Academic Press, New York.

Long RP, Horsley SB, Lijia PR (1997) Leaf nutrition and photosynthetic performance of sugar maple (*Acer saccharum*) in stands with contrasting health conditions. Tree Physiol 17:169–178.

Magill AH, Aber JD, Berntson GM, McDowell WH, Nadelhoffer KJ, Melillo JM, Steudler P (2000) Long-term nitrogen additions and nitrogen saturation in two temperate forests. Ecosystems 3:238–253.

Mann LK, West DC (1981) Whole-tree harvesting: First year progress report—Impacts on productivity and nutrient change. ONRL/TM–7873. Oak Ridge National Laboratory, Oak Ridge, Tennessee.

McLaughlin SB, McConathy RK, Barnes RL, Edwards NT (1980) Seasonal changes in allocation by White Oak (*Quercus alba* L.). Can J For Res 10:379–388.

Mitchell MJ, Raynal DJ, Driscoll CT (1996) Biogeochemistry of a forested watershed in the central Adirondack Mountains: Temporal changes and mass balances. Water Air Soil Pollut 88:355–369.

Olson DF Jr., Boyce SG (1971) Factors affecting acorn production and germination and early growth of seedlings and sprouts. In *Oak symposium proceedings,* Northeastern Forest Experiment Station, USDA Forest Service, Upper Darby, Pennsylvania, pp 44–48.

Schulze E-D (1989) Air pollution and forest decline in a spruce (*Picea abies*) forest. Science 244:776–777.

Shure DS, Mooreside PD, Ogle SM (1998) Rainfall effects on plant herbivore processes in an upland oak forest. Ecology 79:604–617.

Sykes JM, Bunce RGH (1970) Fluctuations in litter-fall in a mixed deciduous woodland over a three-year period 1966–68. Oikos 21:326–329.

Timmer VR, Teng Y (1999) Foliar nutrient analysis of sugar maple decline: Retrospective vector diagnosis. In Horsley SB, Long RP (Eds) *Sugar maple ecology and health: Proceedings of an international symposium;* June 2–4, 1998, Warren, Pennsylvania. General Technical Report NE-261. USDA Forest Service, Northeast Research Station, Radnor, Pennsylvania.

Vose JM, Sullivan NH, Clinton BD, Bolstad PV (1995) Vertical leaf area distribution, light transmittance, and application of the Beer-Lambert Law in four mature hardwood stands in the southern Appalachians. Can J For Res 25:1036–1043.

6. Response of Other Organisms

18. Foliar Chemistry and Herbivory

Donald J. Shure, Peter D. Mooreside, Rebekah E. Chapman, and Allan D. Wilson

The atmospheric accumulation of greenhouse gases during the next 100 years is expected to produce significant but complex alterations in regional precipitation patterns (Schneider 1993). The physiological stresses associated with changes in precipitation may increase the susceptibility of tree species to damage by insect herbivores (Mattson and Haack 1987a; Ayres 1993). Trees may adjust their foliar chemistry in meeting such moisture-related stresses. However, progressive damage by herbivores may limit the extent of any physiological or biochemical adjustments to soil moisture deficits. Further examination of these reciprocal interactions between plants and herbivores seems especially relevant considering the growing evidence supporting the important role of insect herbivores as possible regulators of forest ecosystem processes (Schowalter et al. 1986; Ayres 1993; Schowalter 2000).

Considerable progress has been made toward understanding the complex interactions between plant foliar chemistry and herbivore populations. Evidence has been accumulating that links insect herbivory with the levels of plant phenolics (i.e., tannins) in hardwood species (Schultz and Baldwin 1982; Coley et al. 1985; Schultz 1988; Tuomi et al. 1988; Coley and Aide 1991; Herms and Mattson 1992; Tuomi 1992; Dudt and Shure 1994; but see Bernays et al. 1989). For example, inverse correlations exist between plant phenolics and insect damage along resource gradients of light or nutrients (Mole et al. 1988; Landsberg and Ohmart 1989; Shure and Wilson 1993; Dudt and Shure 1994). Changes in plant carbon/nutrient balance offer a plausible explanation for the phenotypic expression of plant phenolics across these resource gradients (Herms and Mattson 1992). This hypothesis suggests that plants will preferentially allocate carbon for growth when conditions are favorable, whereas they may accumulate and use carbon for differentiation products like phenolic defenses when high light and/or low nutrients promote photosynthesis in excess of growth (Bryant et al. 1983; Chapin et al. 1987; Mole et al. 1988; Fajer et al. 1992; Herms and Mattson 1992; Hunter and Schultz 1995; Hunter and Forkner 1999; Agrell et al. 1999). Nevertheless, the lack of consistent correlations between plant defensive chemistry and insect herbivory imply that insect herbivores may be limited by other aspects of foliar chemistry, including nitrogen (Mattson 1980; White 1984), toughness (Coley 1983), or water content (Scriber 1984). Climatic conditions, such as drought, may operate indirectly on insect herbivores through their effects on plant foliar chemistry.

The initial stages of drought may lead to greater foliar nutrient and carbohydrate concentrations through plant osmotic adjustments (Hsiao et al. 1976; White 1984; Mattson and Haack 1987a, 1987b). Insect performance may also increase during initial drought periods if foliar nutrients and carbohydrates accumulate (Stowe et al. 1994). However, the potential for plant roots to take up nutrients will gradually decline as drought intensifies, and differentiation products, such as phenolics, may accumulate as moderate drought retards growth more than it does photosynthesis (Koslowski 1982; Wardlaw 1990). Severe drought stress should ultimately limit photosynthesis as well as nutrient and water uptake and lead to decreased production of carbon-based phenolic defenses (Ehleringer and Cook 1984; Horner 1990; Chapin, 1991). Nevertheless, limited nutrient and water availability during severe drought should render plant foliage largely unsuitable for insect herbivores, despite the decline in production of plant phenolics (White 1984; Mattson and Haack 1987b; Louda and Collinge 1992; Shure et al. 1998). Direct insect mortality would be expected during severe drought. More studies are needed to fully test these nonlinear responses, especially because drought effects can be strongly insect-feeding-guild-specific (Larsson 1989; Louda and Collinge 1992; Shure et al. 1998; Schowalter 2000).

Our project at Walker Branch was designed to examine the effect of rainfall modification on plant–consumer processes. We addressed the question of how southeastern forest tree species adjust their foliar chemistry in meeting the stresses imposed by rainfall alteration during global climate changes. Different life-cycle stages of several important tree species were used to determine whether rainfall-induced changes in foliar chemistry were similar throughout the life cycle. Our approach also enabled us to assess the impact of adjustments in plant foliar nutrients or defense allocation on the extent of herbivore damage by different feeding guilds of insect herbivores. This approach permitted an evaluation of the potential role of different consumers in mediating the response of producer populations to precipitation changes. Our study thus offered an opportunity to provide a mechanistic understanding of the role of climate changes in influencing the critical interactions between hardwood trees and their coadapted herbivores.

Methods

Study Area

Our study was initiated in 1992 during the pretreatment phase of the Throughfall Displacement Experiment (TDE). The specific study area offered a uniform slope, consistent soil type, and a relatively homogenous vegetation distribution within an upland oak forest. The soils at Walker Branch are primarily typic Paleudults, quite acid, cherty, infertile, and highly permeable (Garten and Taylor 1992; Garten 1993). Detailed information on stand composition, hydrology, soils, and biogeochemistry at the Walker Branch Watershed and the TDE are presented elsewhere (Johnson and Van Hook 1989; Hanson et al. 1995; Chapter 2, this volume).

Field Sampling

We selected white oak (*Quercus alba*), chestnut oak (*Q. prinus*), and red maple (*Acer rubrum*) trees for sampling purposes. These species are the dominant overstory species on the TDE site. Nine trees of each of three diameter size classes [$<$ 2.5 cm diameter at breast height (dbh), $\geq$ 2.5 cm $<$ 10 cm dbh, and $\geq$ 10 cm dbh] were selected for each species within each of the 80-m $\times$ 80-m plots of the TDE. These size classes approximate seedling, sapling, and mature life stages. Three individuals of each size class were randomly selected from the bottom, middle, and top sections of each plot. A total of 243 trees (81 per species) were permanently marked in June 1992 to monitor changes in foliar chemistry and insect herbivory throughout the study. Tree sampling was initiated in late August 1992 and continued in early July and late August of 1993 through 1997, although seedlings were not sampled in July because of the limited leaves available from these smaller trees.

Sampling entailed removing 10 leaves from each tree on each of the 2 sample dates per growing season. We used extension pole clippers to remove two small branches (up to a height of 14 m) from larger trees. Five leaves were randomly selected per branch. Hand sampling of leaves was possible from seedlings and small saplings. The 10 leaves per tree were bagged, marked, placed in styrofoam containers, and returned to the laboratory for processing. These procedures, as well as the subsequent herbivory and plant chemical analyses, have been used successfully in past studies (Shure and Wilson 1993; Dudt and Shure 1994; Shure et al. 1998).

Sample Analysis

All leaves obtained in late August of each year were photocopied prior to chemical analysis. Each photocopy indicates the cumulative leaf area damaged or removed by different insect feeding guilds (strip feeders, leaf skeletonizers, pit feeders, leaf miners, leaf rollers, and leaf galls) throughout the growing season (Coulson and Witter 1984; Hargrove et al. 1984). Characteristic damage patterns were discernable for all components of each feeding guild (Shure et al. 1998). We used a magnetized digitizer to determine total leaf area and percentage of leaf area damaged by each feeding guild.

Plant chemical analyses were conducted to determine foliar nitrogen content, leaf water content, total phenols, and hydrolyzable and condensed tannin concentration on all 11 sample dates. The detailed procedures and their potential limitations have been presented elsewhere (Shure and Wilson 1993; Shure et al. 1998). The 10 leaves from each tree were composited and oven-dried at 35°C for 48 h, ground to powder with a Wiley Mill (1.27-mm mesh size), and frozen until analyzed. Subsamples of the powdered leaf material were analyzed for foliar nitrogen content through the Plant and Soil Analysis Laboratory at the University of Georgia. A 50% methanol extraction procedure was used as a precursor for plant phenolic analyses (Bate-Smith 1981). The Folin–Denis technique was used for total phenolics (Swain and Hillis 1959). Hydrolyzable tannins (ellagitannins only) were measured using an acetic acid–sodium nitrate procedure (Bate-Smith 1972). Condensed tannins were estimated as proanthocyanidins with a butanol–HC1 technique (Bate-Smith 1981).

Data Analysis

We used a three-way (3 × 3 × 3) factorial design for analysis of variance (ANOVA) in our data sets. The ANOVAs included species (*Q. alba, Q. prinus,* and *A. rubrum*), treatment (wet, ambient, and dry), and tree diameter size class (seedling, sapling, and mature) as main effects. All ANOVAs were nested to account for replicate samples (trees). Restricted least significant difference tests were used to locate differences in pairs when ANOVAs were significant (Carmer and Swanson 1973). Statements about main effects have been tempered when interaction terms were significant.

Results

Rainfall

Rainfall at the TDE varied widely throughout the study. A fairly severe mid-summer drought developed during the first year of the TDE in 1993. Total rainfall in June and July 1993 (9.2 cm) was less than that in any year since 1982. In contrast, precipitation exceeded the long-term average during June and July 1994. A second drought developed during the 1995 growing season (11.7 cm; June–August) following a relatively wet spring (17.5 cm in May). Considerable rainfall occurred from May through July 1996 (61.4 cm) until dry weather occurred during August (6.8 cm). A second wet growing season followed in 1997, as monthly precipitation from May through August (56.1 cm total) exceeded the long-term average (44.5 cm total; 1982–1997).

Foliar Chemistry

Foliar nitrogen concentrations exhibited definite patterns at the TDE (Table 18.1). Nitrogen levels in *Quercus* leaves averaged about 2.0% (dry weight) early in each growing season; *Q. prinus* leaves consistently had higher nitrogen content than

Table 18.1. Foliar nitrogen levels (% dry mass) in tree species sampled on treatment plots at the TDE from 1994 to 1997

	Q. alba			*Q. prinus*			*A. rubrum*			
Year	Wet	Ambient	Dry	Wet	Ambient	Dry	Wet	Ambient	Dry	*F*
1994 J	1.84	1.98	1.93	2.08	2.15	2.13	1.65	1.56	1.57	0.64
A	1.90	1.87	1.87	1.99	2.07	2.13	1.46	1.49	1.45	0.48
1995 J	1.92	1.99	1.91	2.02	2.06	2.09	1.72	1.65	1.59	0.38
A	1.83[a]	1.81[a]	1.76[a]	1.90[a]	1.93[a]	1.93[a]	1.58[a]	1.48[b]	1.40[c]	4.84**
1996 J	1.99	2.00	1.94	2.05	2.15	2.08	1.71	1.67	1.59	0.46
A	1.81	1.79	1.78	1.84	1.86	1.79	1.57	1.44	1.49	2.23
1997 J	2.16[a]	2.13[ab]	1.98[b]	2.28[a]	2.31[a]	1.99[b]	1.88[a]	1.72[ab]	1.58[b]	11.70**
A	2.12[a]	2.14[a]	2.01[a]	2.09[b]	2.27[a]	1.95[c]	1.82[a]	1.67[b]	1.56[b]	11.23**

Note: Results are presented for early July (J) and late August (A) of each year. The *F*-values in the final column reflect the level of treatment effects in the 3 × 3 × 3 factorial ANOVA, with different letters (superscripts) indicating means that are significantly different.
*$p < 0.05$.
**$p < 0.01$.

Q. alba leaves. In contrast, nitrogen levels in *Acer* foliage were always lower (about 1.6%) at that time. All three species declined seasonally in leaf nitrogen, but the extent of this decline appeared to be weather related. Large seasonal declines were evident during the droughts of 1993 (Shure et al. 1998) and 1995 and at the end of the dry period in late August 1996. Much less seasonal decline was apparent in the wet summer of 1994. Foliar nitrogen levels in all three species remained especially high throughout the wet 1997 growing season. These basic patterns of foliar N concentrations have remained similar for all tree size classes and at different elevations within the study area (Shure et al. 1998; current ANOVAs for 1994–1997).

Throughfall displacement had only limited effects on foliar nitrogen concentrations (Table 18.1). No treatment effects developed for leaf nitrogen levels in *Quercus* from 1994 to 1996. However, leaf nitrogen was consistently lower in dry than ambient plots in 1997; the differences were significant for *Q. prinus*. In contrast, nitrogen levels in *Acer* leaves were usually higher in wet than ambient plots and lower in dry than ambient plots from 1995 through 1997. These trends were significant in August of 1995 and 1997 (Table 18.1). Treatment differences in *Acer* were greater in 1997, when foliar nitrogen levels were especially high.

Leaf phenolic concentrations in August 1994 (Fig. 18.1) reflect the general patterns observed during the pretreatment and early treatment phases of our study (1992–1994). Total phenolics in *Q. alba* consistently exceeded levels in *Q. prinus* and *A. rubrum*. *Q. alba* trees often had significantly higher total phenolics in mature canopy-dwelling trees than in understory conspecifics ($p < 0.05$ on seven of the eight sample dates). An inverse correlation existed between carbon allocation to hydrolyzable or condensed tannins in *Quercus* versus *Acer* trees (Fig. 18.1). *Q. alba* leaves contained significantly higher ($p < 0.0001$ on all dates) levels of hydrolyzable tannins, whereas *Acer* trees produced higher levels of condensed tannins; *Q. prinus* leaves were intermediate in both cases. Hydrolyzable-tannin levels in each species were generally similar for all tree size classes (Shure et al. 1998). In contrast, condensed tannins were often higher ($p < 0.05$ on six dates) in mature trees than in understory trees.

The effect of throughfall displacement on plant phenolics was related to weather conditions. Throughfall displacement had no clear effects on plant phenolics during the wet 1994 growing season (Fig. 18.1). However, treatment effects were evident during the dry 1995 summer (Table 18.2). Condensed tannin concentrations in all three tree species were significantly lower on the dry plot relative to the wet plot in early July 1995. In contrast, hydrolyzable tannins in both *Quercus* species were significantly elevated above ambient levels in wet and dry plots. These early-season treatment differences were offset as severe soil-water deficits developed on all three plots by late August 1995 (Table 18.2). Condensed tannin levels in *Acer* leaves were unexpectedly low in all plots throughout the drought in 1995. Treatment effects were again absent during the wet 1996 growing season (Table 18.3).

Different patterns of leaf phenolics emerged during the second consecutive wet growing season in 1997 (Table 18.3). Phenolic concentrations in *Quercus* were consistently elevated in dry treatment plots, although the differences between

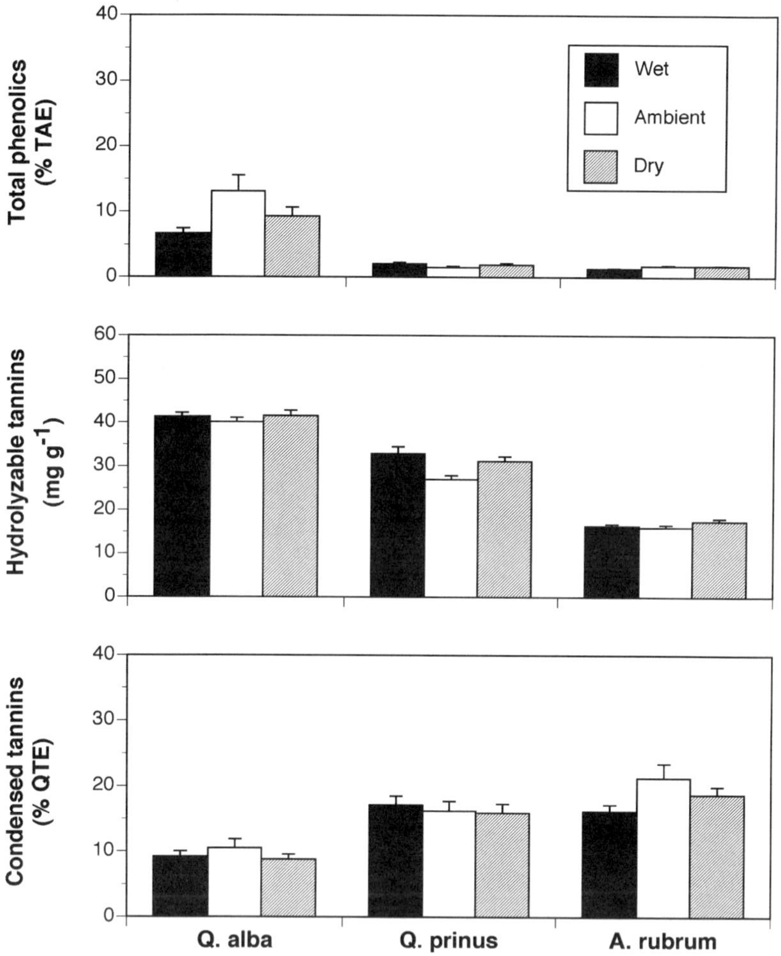

Figure 18.1. Levels of foliar phenolics (means ± 1 SE) in tree species sampled on treatment plots at the TDE in August 1994. Total phenols are expressed as percentage of tannic acid equivalents (% TAE), hydrolyzable tannins as milligrams of hexahydroxydiphenoylglucose equivalents per gram of leaf tissue, and condensed tannins as percentage of Quebracho tannin equivalents (% QTE).

ambient and dry plots were only significant for hydrolyzable tannins. *Quercus* foliage thus tended to be lower in nitrogen and water content (Shure, unpublished data) and higher in hydrolyzable tannins in dry than in ambient plots in 1997. Condensed tannin levels were unusually high in *Q. prinus* and *Acer* foliage throughout 1997.

Insect Herbivory

Insect herbivory varied among plant species and in relation to rainfall fluctuations (Table 18.4). Total leaf area damaged was significantly higher ($p < 0.0001$) on

Table 18.2. Levels of foliar phenolics in tree species sampled on treatment plots at the TDE in July and August 1995

		Q. alba			*Q. prinus*			*A. rubrum*			
Date	Measure	Wet	Ambient	Dry	Wet	Ambient	Dry	Wet	Ambient	Dry	*F*
July 1995	TP	16.8	15.6	12.3	2.0	1.9	1.5	1.8	1.7	1.5	0.88
	HT	52.2[a]	46.3[b]	51.5[a]	36.3[a]	31.9[b]	34.3[a]	15.9[a]	16.1[a]	16.7[a]	5.95**
	CT	9.0[a]	8.7[ab]	6.1[b]	20.1[a]	15.7[b]	14.5[b]	8.9[a]	7.3[ab]	5.7[b]	9.00**
August 1995	TP	5.9	7.4	8.4	2.0	1.2	2.4	1.6	1.6	2.1	1.98
	HT	46.3	46.5	43.6	33.8	29.2	36.1	17.6	17.9	17.1	0.40
	CT	9.7	13.2	8.5	16.7	14.7	11.7	9.1	10.3	7.4	0.72

Note: Total phenols (TP) are expressed as percentages of tannic acid equivalents, hydrolyzable tannins (HT) as milligrams of hexahydroxydiphenoylglucose equivalents per gram of leaf tissue, and condensed tannins (CT) as percentage of Quebracho tannin equivalents. The *F*-values in the final column reflect the level of treatment effects in the 3 × 3 × 3 factorial ANOVA, with different letters (superscripts) indicating means that are significantly different.

$^{**}p < 0.01$.

Table 18.3. Levels of foliar phenolics in tree species sampled on treatment plots at the TDE in August 1996 and July and August 1997

		Q. alba			*Q. prinus*			*A. rubrum*			
Date	Measure	Wet	Ambient	Dry	Wet	Ambient	Dry	Wet	Ambient	Dry	*F*
August 1996	TP	8.7	9.6	5.6	2.0	1.7	1.7	1.6	2.1	1.8	1.98
	HT	43.0	45.2	40.7	33.0	30.4	31.8	17.6	17.1	18.3	0.40
	CT	10.9	13.3	10.2	19.6	20.1	20.8	11.5	13.4	10.9	0.72
July 1997	TP	13.9	23.0	30.4	1.9	2.9	4.0	1.5	2.1	1.7	2.33
	HT	35.7[b]	39.0[b]	43.2[a]	31.4[b]	29.4[b]	36.7[a]	14.9[b]	17.2[ab]	18.9[a]	12.47**
	CT	7.3[a]	8.2[a]	9.0[a]	22.1[b]	34.2[a]	41.2[a]	20.0[a]	29.6[a]	19.8[a]	3.67*
August 1997	TP	20.9	26.4	27.7	2.7	2.2	3.2	1.4	1.8	1.9	0.64
	HT	38.3[b]	39.0[b]	43.7[a]	35.0[a]	32.0[b]	37.1[a]	14.0[a]	15.0[a]	14.0[a]	7.04**
	CT	12.1	15.7	17.7	30.7	33.4	38.4	23.9	33.7	29.9	2.89†

Note: Total phenols (TP) are expressed as percentages of tannic acid equivalents, hydrolyzable tannins (HT) as milligrams of hexahydroxydiphenoylglucose equivalents per gram of leaf tissue, and condensed tannins (CT) as percentage of Quebracho tannin equivalents. The *F*-values in the final column reflect the level of treatment effects in the 3 × 3 × 3 factorial ANOVA, with different letters (superscripts) indicating means that are significantly different.

*$p < 0.05$.

**$p < 0.001$.

†$p = 0.058$.

Table 18.4. Annual leaf area damaged (%) by major feeding guilds of insect herbivores on tree species sampled on treatment plots at the TDE in 1994–1997

		Q. alba			*Q. prinus*			*A. rubrum*			
Date	Guild	Wet	Ambient	Dry	Wet	Ambient	Dry	Wet	Ambient	Dry	*F*
1994	Strip feeders	7.21	10.17	7.35	5.53	5.71	5.24	8.28	8.02	6.11	1.95
	Skeletonizers	5.55^{a}	3.59^{b}	3.15^{b}	5.37^{a}	4.93^{a}	4.07^{a}	0.22^{a}	0.12^{a}	0.31^{a}	3.02*
	Total	13.3	14.3	10.9	11.1	10.8	9.6	8.8	8.5	6.7	2.64
1995	Strip feeders	11.16	12.80	12.67	11.09	10.18	10.94	5.81	5.31	4.14	0.02
	Skeletonizers	3.69	3.03	3.56	3.02	1.81	2.22	0.46	0.22	0.26	2.47
	Total	15.3	16.5	16.7	14.4	12.2	13.5	6.8	6.2	4.9	0.18
1996	Strip feeders	5.81	5.64	7.34	5.59	7.28	4.40	4.44	5.22	4.56	0.75
	Skeletonizers	3.32	2.86	2.89	4.02	2.29	3.67	0.45	0.34	0.63	2.13
	Total	9.4	8.8	10.5	9.8	9.8	8.2	5.4	6.0	5.8	0.00
1997	Strip feeders	18.67^{a}	15.29^{ab}	14.28^{b}	17.99^{a}	13.86^{a}	15.61^{a}	8.41^{a}	7.11^{a}	5.24^{a}	3.53*
	Skeletonizers	3.35^{a}	1.55^{b}	1.87^{b}	1.59^{a}	1.78^{a}	1.75^{a}	0.22^{a}	0.32^{a}	0.25^{a}	3.51*
	Total	22.4^{a}	17.5^{b}	17.1^{b}	19.8^{a}	16.3^{a}	17.5^{a}	8.9^{a}	7.7^{a}	5.7^{a}	3.93*

Note: Totals reflect all six feeding guilds. The *F*-values in the final column reflect the level of treatment effects in the $3 \times 3 \times 3$ factorial ANOVA, with different letters (superscripts) indicating means that are significantly different.
$^{*}p < 0.05$.

Quercus than on *Acer* trees throughout the study; damage was always highest on *Q. alba* trees. Strip feeders (caterpillars) were the most important insect feeding guild for all three tree species. Leaf skeletonizers were also important herbivores on *Quercus,* but not on *Acer* trees. These two functional feeding guilds generally accounted for more than 95% of the leaf area damaged on each tree species. Leaf miners, pit feeders, leaf galls, and leaf rollers caused much less damage. Leaf mining was greater on *Quercus* trees than on *Acer* trees, whereas pit feeding was greater on *Acer* trees. Leaf galls and leaf rollers caused <0.1% of the leaf area damaged on each tree species. These patterns have remained similar for different size classes of each tree species over most of our study (Shure et al. 1998). However, herbivore damage on both *Quercus* species increased significantly from seedling to mature size classes in 1996 ($F = 7.57$, $p < 0.01$) and 1997 ($F = 7.69$, $p < 0.01$).

Leaf area damaged at the TDE was only partially correlated with interannual rainfall totals (Table 18.4). Herbivory on all species was relatively high during the wet 1994 summer. Strip feeding in 1994 was significantly greater than during the drought of 1993 (Shure et al. 1998). Skeletonizing remained similar during both of these years. Total herbivory on both *Quercus* species actually increased during the drought of 1995 as a result of greater consumption by strip feeders. In contrast, *Acer* had lower insect damage in 1995 than in 1994 because of less strip feeding. Total herbivory on *Quercus* dropped significantly during the wet 1996 summer when strip feeding was much lower; damage on *Acer* remained similar in 1995 and 1996. Herbivore damage was especially high during the second consecutive wet growing season in 1997. Total herbivore damage averaged 19.0%, 17.9%, and 7.4% in *Q. alba, Q, prinus,* and *Acer* trees, respectively. Strip feeding was very high on all three species in 1997, when leaf skeletonizing on both *Quercus* species was at its lowest.

Throughfall displacement had only limited effect on herbivore damage levels at the TDE (Table 18.4). Herbivore damage was occasionally elevated above ambient levels on wet plots and below ambient levels on dry plots; the reverse response never developed. This trend was observed for *Acer* trees in all years but 1996 and for *Q. alba* and *Q. prinus* trees in 1997 and 1994, respectively. However, the differences were only significant for elevated damage levels on *Q. alba* trees in wet plots in 1997. Nevertheless, the occurrence of this trend was interesting, given the conflicting patterns of foliar damage in response to interannual rainfall fluctuations at the TDE.

Discussion

Overall Patterns of Foliar Chemistry

Several patterns have emerged from our studies of foliar chemistry and rainfall fluctuations at the TDE (Shure et al. 1998). The tree species selected for study are slow-growing, canopy-dominant species occupying a rather nutrient-poor upland oak forest (Garten 1993). These species are thus expected to employ relatively high levels of foliar phenolics in limiting herbivore damage and subsequent tissue replacement costs (Coley et al. 1985). In fact, all three species generally contained greater foliar phenolic concentrations and somewhat lower nitrogen content than do slow-growing species in other temperate deciduous forests (Mattson 1980; Coley and Aide 1991; Shure and Wilson 1993). Our findings have also indicated only minimal variations in foliar nitrogen and, to some extent, phenolic concentrations in different size classes of each species (Shure et al. 1998). However, condensed tannin and total phenolic concentrations were often significantly higher in mature, canopy-dwelling trees than in smaller, understory conspecifics (Shure et al. 1998). The buildup of more structurally based condensed tannins in well-insolated canopy trees fits the carbon/nutrient balance hypothesis and supports the growing trend of strong positive correlations between light availability and plant phenolic production (Larsson et al. 1986; Bryant et al. 1987; Herms and Mattson 1992; Shure and Wilson 1993; Dudt and Shure 1994; Hunter and Forkner 1999; Agrell et al. 1999). Clear size-class differences were usually lacking for hydrolyzable tannins, which may reflect the ease of maintenance of similar constitutive levels of this more mobile category of tannins (Shure et al. 1998). These overall patterns were generally present from 1992 through 1997, despite significant changes in moisture relationships.

Leaf Nitrogen

The tree species studied at the TDE all exhibited similar foliar nitrogen responses to rainfall fluctuations over the 6-year study. The relative ratios of foliar nitrogen levels among these three species remained similar each year and indicated the especially low leaf nitrogen levels in *Acer* trees from this nitrogen-limited habitat (Garten 1993). Foliar nitrogen levels in each species were always relatively high in early July following the annual overwinter recharge of soil moisture on the Walker Branch Watershed (Hanson et al. 1995; Chapter 2, this volume). Nitrogen uptake was thus enhanced during spring, when foliage development was in progress. Rainfall patterns during each growing season determined the magnitude of seasonal declines in foliar nitrogen concentrations. Foliar nitrogen levels in all species dropped more sharply in dry than wet summers, suggesting the difficulty of nitrogen uptake during drought. Any drought-induced shift to water uptake by roots at deeper, more nutrient-impoverished soil strata would further reduce foliar nitrogen concentrations. In addition, nutrients in the forest floor litter at the TDE undergo greater immobilization rates during drought (Joslin et al. 2000). In contrast, enhanced decomposition and more rapid nitrogen mineralization rates during wetter periods should facilitate nitrogen uptake, especially if root production and root growth increase during short-term favorable periods (Joslin et al. 2000). The unusually high foliar nitrogen levels in *Quercus* and *Acer* throughout 1997 suggests optimal soil moisture conditions (see Chapter 2, this volume), especially on wet and ambient plots. The consistent increase in leaf nitrogen content of *Acer* trees from ambient to wet plots and decrease from ambient to dry plots in 1995 and 1997 indicates the subtle impact of moisture reductions on this nitrogen-limited species.

Foliar Phenolics

Plant phenolics exhibited complex responses to rainfall fluctuations throughout the study. Our initial findings from 1992 to 1994 suggested that phenolic concentrations were directly correlated with rainfall changes (Shure et al. 1998). Most measures of phenolics dropped during the 1993 drought and rebounded to varying degrees over the much wetter 1994 growing season. The magnitude of the drought in 1993 was apparently sufficient to limit photosynthesis or phenolic biosynthesis (Gershenzon 1984) and, thus, the level of tannins in plant foliage. Decreases in foliar nitrogen during the 1993 drought failed to produce a corresponding increase in phenolics even though plant growth rates were low in 1993 (Shure et al. 1998). Foliar phenolics responded differently during the wet summer of 1994. The failure of total phenols and hydrolyzable tannins to rebound following the 1993 drought may have reflected a preferential utilization of carbon for growth and not defense (Bryant et al. 1983; Herms and Mattson 1992; Tuomi 1992); plant growth rates were high in 1994. Carbon may have been unavailable for widespread phenolic production during the wet growing season (Tuomi et al. 1988; Marquis and Whelan 1994; Shure et al. 1998). Instead, the increased production of condensed tannins in all species in 1994 could represent an adaptive trade-off. Condensed tannins have been hypothesized to be especially effective against microbial pathogens (Zucker 1983). Fungal damage was significantly higher on leaf material during the wet 1994 summer (Shure et al. 1998).

The differential response of each category of phenolics continued from 1995 to 1997. Hydrolyzable-tannin concentrations in *Quercus* foliage rebounded in 1995 when the growth of *Quercus* trees was slowed during drought (Hanson et al. 2001). This buildup of hydrolyzable tannins should occur early in the growing season (Shure and Wilson 1993; Hunter and Forkner 1999; this chapter) when leaf nitrogen levels peak and insect herbivory is usually highest (Reichle et al. 1973; Schowalter 2000). This more mobile category of tannins should prove especially effective against insect herbivores early in the growing season (Feeny 1970; Zucker 1983). In contrast, condensed tannin levels in *Q. alba* and especially *Acer* leaves were reduced throughout the severe drought of 1995. Overwinter carbon storage may have been allocated to hydrolyzable-tannin production before the drought had developed in 1995. Condensed-tannin levels rebounded somewhat in *Acer* foliage in 1996 and then reached especially high levels in all species during the second wet growing season in 1997. The buildup of condensed tannins during 1994, 1996, and 1997 should have provided defense against the fungal pathogens that caused increased foliar damage during these wet growing seasons (Shure et al. 1998; unpublished data D. Shure). Condensed-tannin levels were usually highest during late August when fungal damage is expected to peak at the end of the growing season (Shure and Wilson 1993; Hunter and Forkner 1999;

Schowalter 2000). Interestingly, the relatively low growth rates of *Quercus* and *Acer* trees in 1997 (Hanson et al. 2000) may have enabled greater carbon allocation for foliar phenolics during this second consecutive wet growing season. Further research is obviously needed to verify these carbon-allocation scenarios. However, many of these suggested shifts in production of different categories of tannins would fit the carbon/nutrient-balance hypothesis and lend possible support to the purported use of hydrolyzable and condensed tannins as major defenses against insect herbivores and microbial pathogens, respectively (Zucker 1983). Nevertheless, the reverse directionality in phenolic responses across the treatment plots in 1995 (dry) versus 1997 (wet) underscores the complexity of nonlinear responses that can be expected during rainfall changes in the future (Horner 1990).

Patterns of Insect Herbivory

Insect herbivory at Walker Branch was relatively high throughout the 6-year study. Annual damage on *Q. alba* (10 to 19%, overall average of 13.4%) and *Q. prinus* (7.7–17.5%, overall average of 11.2%) trees at the TDE has consistently exceeded average levels reported for broadleaf tree species occupying temperate deciduous forests (Schowalter et al. 1986; Landsberg and Ohmart 1989; Coley and Aide 1991; Schowalter 2000). Past estimates generally range from 3% to 17% with overall averages of 7.5% (Coley and Aide 1991) and 8.8% (Landsberg and Ohmart 1989). Reichle et al. (1973) reported a 3-year average of 7.7% of the leaf area removed from canopy trees in a *Liriodendron* forest at the Walker Branch Watershed. Recent estimates of leaf area damaged on mature *Q. rubra* trees at the Coweeta Long-Term Ecological Research (LTER) site in North Carolina have ranged from 8% to 19% (Reynolds and Crossley 1997; Hunter and Forkner 1999). Damage levels in North Carolina were higher at lower elevations (Reynolds and Crossley 1997). However, our damage estimates were comparable to or less than those reported for *Quercus* seedlings and saplings in Missouri forests (Sork et al. 1993; Marquis and Whelan 1994; Wold and Marquis 1997). In contrast, annual insect damage on *A. rubrum* foliage at the TDE (5.7–8.0%) was similar to levels reported from North Carolina (Reynolds and Crossley 1997; Hunter and Forkner 1999). Schowalter (2000) has provided a comprehensive review of the limitations associated with quantitative estimates of insect consumption (see Reichle et al. 1973) and an assessment of the additional costs of sap-sucking insects, which have been excluded from most studies.

Use of a guild approach afforded an especially useful means of assessing the possible reasons for variations in overall herbivore damage levels (Larsson 1989; Hunter and Schultz 1995; Schowalter 2000). Leaf strippers and leaf skeletonizers were the major insect feeding guilds at the TDE. These guilds exhibited high levels of damage on *Quercus* trees despite the relatively low levels of foliar nitrogen and high levels of phenolics. Coley and Aide (1991) reported an average foliar nitrogen content of 2.25% for tree species in temperate deciduous forests. The especially low levels of foliar nitrogen in *Acer* at the TDE (1.4–1.88%) may prevent skeletonizers from using these trees as a major food source (Shure et al.

1998). Foliar nitrogen concentrations can significantly influence insect feeding (Mattson 1980; Scriber and Slansky 1981; White 1984; Mattson and Scriber 1987; Schowalter 2000) because of the metabolic constraints imposed by much lower nitrogen concentrations in leaf than insect tissues (Mattson 1980; Ayres 1993). As a result, insect herbivory on leaves containing less that 2.0% nitrogen is quite limited (Fox and McCauley 1977). These findings suggest that insect feeding at the TDE may be finely tuned to weather-induced fluctuations in foliar nitrogen concentrations.

Considerable difficulty has occurred in predicting the response of insect herbivores to moisture fluctuations (Gershenzon 1984; Mattson and Haack 1987a; Larsson 1989; Ayres 1993; Stowe et al. 1994; Schowalter et al. 1999; Schowalter 2000). This difficulty has often stemmed from the varied responses of specific herbivores to changing ratios of foliar nutrients and plant phenolics as water stress increases (Bazzaz et al. 1987; Louda and Collinge 1992). Insect damage at the TDE was directly correlated with rainfall totals from 1992 to 1994 (Shure et al. 1998). However, our results over the full 6 years suggest a more complicated set of causal relationships. The two major insect feeding guilds at the TDE exhibited contrasting responses to foliar quality during wet versus dry growing seasons. Leaf skeletonizing on *Quercus* remained directly correlated with rainfall totals until heavy consumption by strip feeders occurred in 1997. Leaf strippers were unaffected by drought throughout the study, but often increased in importance during wet growing seasons. These contrasting guild responses produced the complex pattern of foliar damage observed at the TDE site, which suggests either subtle effects of rainfall fluctuations on plant foliar chemistry or possible interguild interactions.

The high levels of insect herbivore damage at the TDE may reflect insect population buildups or enhanced feeding during favorable growing seasons. Heavy rainfall in 1994, 1997, and early 1996 facilitated water and nitrogen uptake and enhanced the overall palatability of leaf tissue. Carbon was apparently allocated preferentially for growth during 1994 and 1996 (Hanson et al. 2001). The limited phenolic production in these wet years may have promoted an adaptive shift to condensed versus hydrolyzable tannins. Foliar damage by leaf strippers and leaf skeletonizers increased accordingly as foliar quality improved during favorable periods. The second consecutive wet summer in 1997 produced very high foliage quality and elevated levels of insect damage by leaf strippers. Foliar damage in 1997 was especially high on the wet treatment plot, where leaf nitrogen and water content were usually higher and tannin concentrations were often somewhat lower than on the dry plot. Nevertheless, the failure of foliar damage to decline as expected during the severe drought of 1995 suggests a possible carryover effect on insect performance (Rossiter 1994). Greater numbers or quality of caterpillar eggs produced at the end of the favorable wet growing season in 1994 may have resulted in increased foliar damage by strip feeders during 1995. In addition, foliar nitrogen limitations at the end of the 1995 drought may have impacted egg production or egg quality and contributed to fewer leaf strippers during 1996. Maternally governed carryover effects of this type have been doc-

Table 18.5. Annual carbon losses (kg ha^{-1}) through insect herbivore damage of leaf tissue from species sampled on treatment plots at the TDE in 1996 and 1997

	1996			1997		
	Wet	Ambient	Dry	Wet	Ambient	Dry
Q. alba	82.8	46.8	28.4	202.5	90.0	45.0
Q. prinus	35.1	31.1	62.6	72.9	50.0	130.5
A. rubrum	30.2	38.7	36.9	50.9	48.2	35.6
Total	148.1	116.6	127.9	326.3	188.2	210.6

Note: Data were obtained by multiplying unit-area litterfall estimates (kg ha^{-1}, dry mass) for each species at each treatment plot (Chapter 17, this volume) by percent leaf area damaged each year at that location (this chapter) and assuming carbon represents 45% of leaf mass for these species (Coleman 1973).

umented (Rossiter 1994; Mousseau and Fox 1998) and may help clarify the complex interannual patterns of insect damage observed at the TDE.

Rainfall fluctuations at the TDE influence insect herbivores, principally through moisture-induced changes in foliar nitrogen concentrations. Foliar nitrogen changes may enhance food quality (Mattson 1980; White 1984) or lead to carry-over effects on egg production or egg quality. Plant phenolics appear to play more of a secondary role at the nitrogen-limited TDE site because insect damage and tannin levels were less well correlated (Shure et al. 1998). In response, the relatively high levels of insect herbivory at the TDE should exert important influences on ecosystem processes (Schowalter 2000). For example, overall carbon losses (Table 18.5) through insect damage on *Quercus* and *Acer* foliage averaged 130.9 k_g ha^{-1} across the TDE site in 1996 (low herbivory) and 241.7 k_g ha^{-1} in 1997 (highest herbivory). A significant fraction of this carbon will be transferred directly to the forest floor as insect frass [86% of insect consumption at Walker Branch; Comiskey (1978)] throughout the growing season. Herbivory can promote premature leaf abscission and greenfall (Risley and Crossley 1993), increase the accessibility of damaged leaf surfaces to entry by microbial pathogens, and facilitate additional light penetration through missing tissue to understory seedlings and saplings (Schowalter 2000). Herbivore removal of leaf tissue also represents a potential loss of overwinter carbon storage, depending on the degree of compensatory recovery of photosynthetic leaf surface (Schowalter 2000) by tree species. Climate-induced modifications of these herbivore-mediated pathways may ultimately contribute to ecosystem disruption and forest destabilization in the future.

References

Agrell J, McDonald EP, Lindroth RL (1999) Responses to defoliation in deciduous trees: Effects of CO_2 and light. In Hofgaard A, Ball JP, Danell K, Callaghan TV, (Eds) *Animal responses to global change in the north.* Ecological Bulletins 47. MUNKSGAARD International Publishers, Malden, Massachusetts, pp 84–92.

Ayres MP (1993) Plant defense, herbivory, and climate change. In Kareiva PM, Kingsolver JG, Huey RB (Eds) *Biotic interactions and global change.* Sinauer Associates, Sunderland, Massachusetts, pp 75–94.

Bate-Smith EC (1972) Detection and determination of ellagitannins. Phytochemistry 11:1153–1156.

Bate-Smith EC (1981) Astringent tannins of the leaves of *Geranium* species. Phytochemistry 20:211–216.

Bazzaz FA, Chiarello NR, Coley PD, Pitelka LF (1987) Allocating resources to reproduction and defense. BioScience 37:58–68.

Bernays EA, Cooper-Driver G, Billener M (1989) Herbivores and plant tannins. Adv Ecol Res 19:263–302.

Bryant JP, Chapin FS III, Klein DR (1983) Carbon/nutrient balance of boreal plants in relation to vertebrate herbivory. Oikos 40:357–368.

Bryant JP, Chapin FS III, Reichart PB, Clausen TP (1987) Response of winter chemical defense in Alaska paper birch and green alder to manipulation of plant carbon/nutrient balance. Oecologia 72:510–514.

Carmer SG, Swanson MR (1973) An evaluation of ten pairwise multiple comparison procedures by Monte Carlo methods. J Am Stat Assoc 68:66–74.

Chapin FS III (1991) Integrated responses of plants to stress. BioScience 41:29–36.

Chapin FS III, Bloom AJ, Field CB, Waring RH (1987) Plant responses to multiple environmental factors. BioScience 37:49–57.

Coleman DC (1973) Soil carbon balance in a successional grassland. Oikos 24:195–199.

Coley PD (1983) Herbivory and defensive characteristics of tree species in a lowland tropical forest. Ecol Monogr 53:209–233.

Coley PD, Aide TM (1991) Comparison of herbivory and plant defenses in temperate and tropical broad-leaved forests. In Price PW, Lewinsohn TM, Fernandes GW, Benson WW (Eds) *Plant-animal interactions: Evolutionary ecology in tropical and temperate regions.* John Wiley & Sons, New York, pp 25–49.

Coley PD, Bryant JP, Chapin FS III (1985) Resource availability and plant anti-herbivore defense. Science 230: 895–899.

Comiskey CE (1978) *Aspects of the organic carbon cycle on Walker Branch Watershed: A study of land-water interaction.* Ph.D. thesis. The University of Tennessee, Knoxville, Tennessee.

Coulson RN, Witter JA (1984) *Forest entomology: Ecology and management.* John Wiley & Sons, New York.

Dudt JF, Shure DJ (1994) The influence of light and nutrients on foliar phenolics and insect herbivory. Ecology 75: 86–98.

Ehleringer JR, Cook CS (1984) Photosynthesis in *Encelia farinosa* Gray in response to decreasing water potential. Plant Physiol 75:688–693.

Fajer ED, Bowers MD, Bazzaz FA (1992) The effect of nutrients and enriched CO_2 environments on production of carbon-based allelochemicals in *Plantago:* A test of the carbon/nutrient balance hypothesis. Amer Nat 140:707–723.

Feeny PP (1970) Seasonal changes in oak leaf tannins and nutrients as a cause of spring feeding by winter moth caterpillars. Ecology 51:565–581.

Fox LR, Macauley BJ (1977) Insect grazing on *Eucalyptus* in response to variation in leaf tannins and nitrogen. Oecologia 29:145–162.

Garten CT Jr. (1993) Variation in foliar ^{15}N abundance and the availability of soil nitrogen on Walker Branch Watershed. Ecology 74:2098–2113.

Garten CT Jr., Taylor GE Jr. (1992) Foliar $\delta^{13}C$ within a temperate deciduous forest: Spatial, temporal, and species sources of variation. Oecologia 90:1–7.

Gershenzon J (1984) Changes in the level of plant secondary metabolites under water and nutrient stress. In Timmermann BN, Steelink C, Loewus FA (Eds) *Phytochemical adaptations to stress.* Recenct Advances in Phytochemistry 18. Plenum Press, New York, pp 273–320.

Hanson PJ, Todd DE, Edwards NT, Huston MA (1995) Field performance of the Walker Branch Throughfall Displacement Experiment. In Jenkins A, Ferrier RC, Kirby C (Eds) *Ecosystem manipulation experiments: Scientific approaches, experimental design, and relevant results.* Ecosystems Research Report No. 20. Commission of the European Communities, Brussels, Belgium, pp 307–313.

Hanson PJ, Todd DE Jr., Amthor JS (2001) A six-year study of sapling and large-tree growth and mortality responses to natural and induced variability in precipitation and throughfall. Tree Physiol 21:345–358.

Hargrove WW, Crossley DA Jr., Seastedt TR (1984) Shifts in herbivory in the canopy of black locust, *Robinia pseudoacacia* L., following fertilization. Oikos 43:322–328.

Herms DA, Mattson WJ (1992) The dilemma of plants: To grow or defend. Q Rev Biol 67:283–335.

Horner JD (1990) Nonlinear effects of water deficits on foliar tannin concentration. Biochem Syst Ecol 18:211–213.

Hsiao TC, Fereres E, Acevedo E, Henderson DW (1976) Water stress and dynamics of growth and yield of crop plants. In Lange OL, Kappen L, Schulze ED (Eds) *Water and plant life. Problems and modern approaches.* Springer-Verlag, New York, pp 281–305.

Hunter MD, Forkner RE (1999) Hurricane damage influences foliar polyphenolics and subsequent herbivory on surviving trees. Ecology 80:2676–2682.

Hunter MD, Schultz JC (1995) Fertilization mitigates chemical induction and herbivore responses within damaged oak trees. Ecology 76:1226–1232.

Johnson DW, Van Hook RI (1989) *Analysis of biogeochemical cycling processes in Walker Branch Watershed.* Springer-Verlag, New York.

Joslin JD, Wolfe MH, Hanson PJ (2000) Effects of altered water regimes on forest root systems. New Phytol 147: 117–129.

Kozlowski TT (1982) Water supply and tree growth: Part I. Water deficits. For Abst 43:57–95.

Landsberg J, Ohmart C (1989) Levels of insect defoliation in forests: Patterns and concepts. Trends Ecol Syst 4: 96–100.

Larsson S (1989) Stressful times for the plant stress-insect performance hypothesis. Oikos 56:277–283.

Larsson S, Wiren A, Lundgren L, Ericsson T (1986) Effects of light and nutrient stress on leaf phenolic chemistry in *Salix dasyclados* and susceptibility to *Galerucella lineola* (Coleoptera). Oikos 47:205–210.

Louda SM, Collinge SK (1992) Plant resistance to insect herbivores: A field test of the environmental stress hypothesis. Ecology 73:153–169.

Marquis RJ, Whelan CJ (1994) Insectivorous birds increase growth of white oak through consumption of leaf-chewing insects. Ecology 75:2007–2014.

Mattson WJ (1980) Herbivory in relation to plant nitrogen content. Annu Rev Ecol Syst 11:119–161.

Mattson WJ, Haack RA (1987a) The role of drought stress in provoking outbreaks of phytophagous insects. In Barbosa P, Schultz JC (Eds) *Insect outbreaks*. Academic Press, New York, pp 365–407.

Mattson WJ, Haack RA (1987b) The role of drought in outbreaks of plant eating insects. BioScience 37:110–118.

Mattson WJ, Scriber JM (1987) Nutritional ecology of insect folivores of woody plants: Nitrogen, water, fiber, and mineral considerations. In Slansky F Jr., Rodriquez JG (Eds) *The nutritional ecology of insects, mites, and spiders*. John Wiley and Sons, New York, pp 105–146.

Mole S, Ross JAM, Waterman PG (1988) Light–induced variation in phenolic levels in foliage of rain-forest plants. I. Chemical changes. J Chem Ecol 14:1–21.

Mousseau TA, Fox CW (1998) The adaptive significance of maternal effects. Trends Ecol Syst 13:403–407.

Reichle DE, Goldstein RA, Van Hook RI Jr., Dodson GJ (1973) Analysis of insect consumption in a forest canopy. Ecology 54:1076–1084.

Reynolds BC, Crossley DA Jr. (1997) Spatial variation in herbivory by canopy arthropods along an elevational gradient. Environ Entomol 26:1232–1239.

Risley LS, Crossley DA Jr. (1993) Contribution of herbivore-caused greenfall to litterfall nitrogen flux in several southern Appalachian forested watersheds. Am Midl Nat 129:67–74.

Rossiter M (1994) Maternal effects hypothesis of herbivore outbreak. BioScience 44:752–763.

Schneider SH (1993) Scenarios of global warming. In Kareiva PM, Kingsolver JG, Huey RB (Eds) *Biotic interactions and global change*. Sinauer Associates, Sunderland, Massachusetts, pp 9–23.

Schowalter TD (2000) *Insect ecology: An ecosystem approach*. Academic Press. San Diego, California.

Schowalter TD, Hargrove WW, Crossley DA Jr. (1986) Herbivory in forested ecosystems. Annu Rev Entomol 31: 177–196.

Schowalter TD, Lightfoot DC, Whitford WG (1999) Diversity of arthropod responses to host-plant water stress in a desert ecosystem in southern New Mexico. Am Midl Nat 142:281–290.

Schultz JC (1988) Many factors influence the coevolution of herbivore diets, but plant chemistry is central. Ecology 69:896–897.

Schultz JC, Baldwin IT (1982) Oak leaf quality declines in response to defoliation by gypsy moth larvae. Science 217:149–151.

Scriber JM (1984) Host plant suitability. In Bell WJ, Carde RT (Eds) *Chemical ecology of insects*. Sinauer Associates, Sunderland, Massachusetts, pp 159–202.

Scriber JM, Slansky F Jr. (1981) The nutritional ecology of immature insects. Ann Rev Ent 26: 183–211.

Shure DJ, Mooreside PD, Ogle SM (1998) Rainfall effects on plant-herbivore processes in an upland oak forest. Ecology 79:604–617.

Shure DJ, Wilson LA (1993) Patch-size effects on plant phenolics in successional openings of the Southern Appalachians. Ecology 74:55–67.

Sork VL, Stowe KA, Hochwender C (1993) Evidence for local adaptation in closely adjacent subpopulations of northern red oak (*Quercus rubra* L.) expressed as resistance to herbivores. Amer Nat 142:928–936.

Stowe KA, Sork VL, Farrell AW (1994) Effect of water availability on the phenotypic expression of herbivore resistance in northern red oak seedlings (*Quercus rubra* L.) Oecologia 100:309–315.

Swain T, Hillis WE (1959) The phenolic constituents of *Prunus domestica*. I. The quantitative analysis of phenolic constituents. J Agric Food Sci 10:63–68.

Tuomi J (1992) Toward integration of plant defense theories. Trends Ecol Evol 7:365–367.

Tuomi J, Niemela P, Rousi M, Siren S, Vuorisalo T (1988) Induced accumulation of foliage phenols in mountain birch: Branch response to defoliation? Amer Nat 132:602–608.

Wardlaw IF (1990) Tansley Review No. 27: The control of carbon partitioning in plants. New Phytol 116:341–381.

White TRC (1984) The abundance of invertebrate herbivores in relation to the availability of nitrogen in stressed food plants. Oecologia 63:90–105.

Wold EN, Marquis RJ (1997) Induced defense in white oak: Effects on herbivores and consequences for the plant. Ecology 78:1356–1369.

Zucker WV (1983) Tannins: Does structure determine function? An ecological perspective. Amer Nat 121:335–365.

19. Opportunistically Pathogenic Root Rot Fungi: *Armillaria* Species

Johann N. Bruhn and Jeanne D. Mihail

Introduction

Numerous gap-model studies have attempted to predict responses of tree populations to climate change. Although these models incorporate various disturbance factors, the multifaceted influences of structural-root disease activity on gap dynamics have not been included [e.g., Bachelet and Neilson (2000)]. The fact that apparently similar models yield very different projections for the same forest [e.g., Bugmann (1997)] suggests that most of these models are incomplete and/or may be inappropriately formulated (Bachelet and Neilson 2000). The existence of belowground root gaps corresponding to aboveground canopy gaps and their influence on plant competition and establishment have also received little attention [see Casper and Jackson (1997)]. Yet, *Armillaria* root disease and other structural-root diseases have been associated with gap dynamics [e.g., Worrall and Harrington (1988); Clinton et al. (1993); Lundquist (1993, 1995); Shearer et al. (1997); Rizzo et al. (2000)].

Armillaria (Fr.:Fr.) Staude (Tricholomataceae, Agaricales) is a ubiquitous genus of opportunistic, root-parasitic, white-rot wood decay fungi [see Garraway et al. (1991); Watling et al. (1991); Fox (2000); Pegler (2000)]. *Armillaria* spp. often serve as pivotal contributing factors in stress-mediated forest declines [see Houston (1992); Manion and Lachance (1992); Wargo (1996); Siwecki and Ufnalski (1998)]. Both excess moisture and drought during the growing season have been implicated as stress factors inciting Armillaria root disease [see Wargo and Harrington (1991); Clinton et al. (1993); Lonsdale and Gibbs (1996)]. The subject is more complex than this, however, because moisture extremes are only one among many types of factor that interact with vegetation and fungus populations to affect vulnerability to root disease and determine the distribution of root-disease-related decline [see Houston (1992); Manion and Lachance (1992)]. Other factors reflect geographic and edaphic characteristics [e.g., Worrall and Harrington (1988); Cruikshank et al. (1997); Mallett and Maynard (1998); Bruhn et al. (2000)], entomological considerations [e.g., Worrall and Harrington (1988); Bauce and Allen (1992); Wargo and Harrington (1991); Lundquist (1995)], and anthropogenic disturbances [e.g., Wargo and Harrington (1991); Hood et al. (1991); Bruhn et al. (2002); Guyette and Kabrick (2002)].

Wood decay by *Armillaria* spp. is accomplished by the exploitive activity of intensely branched networks (mycelia) of microscopic filamentous hyphae that penetrate and digest wood cell walls [see Garraway et al. (1991); Fox (2000)]. A portion of the energy derived from wood decay is spent on sexual reproduction in the form of airborne haploid basidiospores produced on the surfaces of mushroom gills [see Guillaumin et al. (1991)]. Each spore that germinates and successfully colonizes a suitable woody substrate (food base) generally mates with another sexually compatible germling to form a genetically unique and stable diploid individual ["genet"; see Harper (1977)], which may establish itself in the landscape as an agent of wood decomposition and possibly also root disease [see Kile et al. (1991); Rayner (1991); Anderson and Kohn (1995); Termorshuizen (2000)].

Another portion of the energy derived from wood decay is spent on genet extension through the forest floor in two ways: (1) by growth across root contacts between colonized and target food bases and (2) by means of exploratory networks of highly differentiated, macroscopic, branching, cordlike organs called rhizomorphs [see Redfern and Filip (1991); Mihail et al. (1995); Fox (2000); Termorshuizen (2000)]. *Armillaria* spp. do not produce asexual spores, but all North American *Armillaria* spp. are capable of generating at least some rhizomorphs from woody food bases under favorable conditions [see Garraway et al. (1991); Redfern and Filip (1991); Fox (2000); Mihail et al. (2002)]. Because most of the resources required to fuel rhizomorph growth are provided by decay of the colonized food base, an *Armillaria* genet's aggressiveness relative to its potential is directly related to the spatial distribution, abundance, and size of woody food bases fueling rhizomorph growth [see Redfern and Filip (1991); Bruhn et al. (1998); Termorshuizen 2000)]. When a rhizomorph encounters a vulnerable food base, it penetrates the bark and dedifferentiates into a mycelium (see Fox 2000). Thus, a single genet has the potential to occupy numerous food bases simultaneously and to maintain physical continuity indefinitely within the volume of forest floor it occupies. Dead or severely stressed roots of living trees may be colonized in this manner, and vigorous live roots may have their resistance tested more or less continuously by rhizomorphs. Established genets appear to have a preemptive advantage over invading conspecific genets in acquiring food-base resources that become available within their boundaries. As a result, it appears that survival of new genets is quite rare within the territories of established conspecific genets [e.g., Smith et al. (1992)]. Therefore, genet maps of a single *Armillaria* sp. often form a mosaic of nonoverlapping territories of various sizes. Perhaps because of differences in host specialization or colonization strategy, genet mosaics of different *Armillaria* spp. often appear to intermingle [e.g., Kile et al. (1991); Smith et al. (1992, 1994); Rizzo and Harrington (1993); Anderson and Kohn (1995); Legrand et al. (1996); Bruhn et al. (1997)].

Armillaria spp. differ in their overall aggressiveness and in their relative aggressiveness toward different kinds of trees [see Gregory et al. (1991); Termorshuizen (2000)], and conspecific *Armillaria* genets can differ in aggressiveness toward individual host tree species (Omdal et al. 1995). A loose relationship has been noted between the rhizomorph growth characteristics of different *Armillaria* spp. and their relative ability to cause root disease [e.g., Morrison (1989); Mihail et al. (1995)]. Thus, the very weakly pathogenic *A. gallica* produces prolific systems of monopodially branching rhizomorphs, whereas *A. mellea* and *A. ostoyae* (which characteristically produce fewer, less intensely, and dichotomously branching rhizomorphs) are much more aggressive toward hardwoods and conifers, respectively [see Gregory et al. (1991); Rishbeth (1991)]. *Armillaria tabescens,* a weakly to moderately pathogenic species appears to produce only very short and sparingly branched rhizomorphs and only under conditions of both high oxygen and moisture content approaching saturation (Mihail et al. 2002). Where species with overlapping host ranges (e.g., *A. gallica, A. mellea,* and *A. tabescens* in the Missouri Ozarks) occur in the same landscape, interspecific or intergenet differences in aggressiveness and colonization strategy may result in niche partitioning (Leibold 1995). Models of interspecific interaction and landscape subdivision for various life-forms have assumed trade-offs among genet or species colonization ability, competitive ability, and longevity [e.g., Tilman (1994); Marshall et al. (2000)]. These trade-offs seem to be essential to niche partitioning and maintenance of biodiversity. In the case of *Armillaria* in the Missouri Ozarks (Bruhn et al. 2000), *A. mellea* and *A. tabescens* are much less effective in exploring the forest floor with rhizomorph production than is *A. gallica,* yet the former two species produce abundant airborne basidiospore inocula and are more aggressive colonists of stressed root systems than is *A. gallica.* As a result, these three *Armillaria* spp. co-occur in Missouri Ozark forest stands.

Studies of *Armillaria* populations on the Throughfall Displacement Experiment (TDE) plots were initiated in 1994 to identify to species all *Armillaria* genets occupying the TDE plots and to determine their spatial distributions with respect to each other and the TDE design.

Methods

The TDE and the environments of the TDE treatment plots during the 8-year study are described by Hanson et al. (Chapter 2, this volume).

Armillaria Population Studies

Armillaria gallica and *A. mellea* fruited between mid-September and early November in 1994 and 1995. The forest-floor surface in each TDE subplot was repeatedly searched during these periods for *Armillaria* basidiomata and for rhizomorphs colonizing woody debris. All collections were mapped. *Armillaria* isolates were identified to genet based on vegetative compatibility tests among the collected field isolates (Adams 1974; Guillaumin et al. 1996). Vegetative compatibility was determined through confrontational pairings of isolates on agar-solidified (2% w/v) malt extract (3% w/v) growth medium. Pairings were duplicated for confirmation of results. A field isolate representing each *Armillaria* genet was then used to identify that genet to species in mating tests with voucher single-basidiospore (haploid) isolates of *Armillaria* spp. (Korhonen 1978; Rizzo and Harrington 1992). The voucher isolates used in these matings were supplied by J. B. Anderson (Department of Botany, Erindale Campus, University of Toronto).

A map showing approximate locations of *Armillaria* genets for the TDE site was produced, in which we represented each genet as a spatially continuous individual. We realize that individual genets may repeatedly fragment and reunite depending on conditions of forest health perceived by the fungi as periods of feast or famine [see Anderson and Kohn (1995); Rayner (1991); Smith et al. (1992, 1994)].

Statistical Analysis

For each *Armillaria* spp., we tested the null hypothesis that the numbers of field isolates collected in the three TDE treatment plots were not significantly different from expected values ($\alpha = 0.05$). A chi-square test of a 3×1 contingency table (Sokal and Rohlf 1995) was used to test this hypothesis for each species. For each species, the expected number of isolates collected from each plot was one-third of the total number of isolates collected from all three plots.

Results

Covariate analyses of soil moisture data indicate that the TDE accomplished both significant drying of the "dry" plot and significant wetting of the "wet" plot during both relatively dry and wet growing seasons (Hanson et al. 2001; Chapter 2, this volume). All *Armillaria* isolate collections were made from basidiomata and rhizomorphs associated with woody debris on the forest floor. Isolates from 36 rhizomorph and 48 basidioma collections of *A. gallica* and from 21 basidioma collections of *A. mellea* represent 5 *A. gallica* genets and 18 *A. mellea* genets. Basidiomata of *A. gallica* were collected between October 13 and November 2, 1994; basidiomata of both species were collected on October 11, 1995. Although rhizomorph collection is not seasonally limited, most rhizomorph collections were also made in the late autumn.

Table 19.1. Numbers of *Armillaria gallica* and *A. mellea* field isolates collected from the three TDE plots (for 1994 and 1995 combined), and the associated Pearson statistic (X^2 for goodness of fit) values

	TDE plot				
Armillaria species	Wet	Ambient	Dry	Total	X^2
A. gallica	16	65	3	84	76.4
A. mellea	9	11	1	21	8.0

Note: The critical value of X^2 is 5.991.

Three points stand out from inspection of the *Armillaria* field isolate map (see Fig. 19.1). First, genets of *A. gallica* on the TDE site have achieved much greater size than have those of *A. mellea,* probably because of *A. gallica*'s greater production of rhizomorphs. In this regard, we note that all rhizomorph collections from the TDE plots belonged to *A. gallica* genets. The two largest *A. gallica* genets were represented by isolates collected from 9 and 46 subplots, respectively, and appear to occupy ~0.33 ha and 0.54 ha, respectively. Fifteen of the 18 *A. mellea* genets were recovered from only 1 subplot (0.0064 ha) each; only 3 *A. mellea* genets were recovered from 2 adjacent TDE subplots (see Fig. 19.1).

Second, genets of *A. gallica* appear to represent continuous and territorial individuals, with little tendency to overlap one another. The large size of *A. gallica* genets "a" and "b" may explain the smaller number of *A. gallica* genets detected compared with *A. mellea.* Although genets of *A. mellea* in the TDE plots are apparently too small to evaluate their territoriality, 11 of the 18 *A. mellea* genets appear to occur within the bounds of the 2 largest *A. gallica* genets.

Third, surprisingly few collections of either *Armillaria* spp. were made in the dry plot, compared with either the wet or ambient plots. The two largest *A. gallica* genets occupy most of the wet and ambient plots. Only 1 genet of each species, represented by 3 of the 84 *A. gallica* collections and 1 of the 21 *A. mellea* collections, was detected in the dry plot. For both *A. mellea* and *A. gallica,* the distributions of isolates among the three TDE plots differed significantly from expectation (Table 19.1). For both species, more collections than expected were made in the ambient plot, and fewer collections than expected were made in the dry plot.

Discussion

It is most appropriate to think of *Armillaria* populations as consisting of typically territorial, long-lived, stable genets of various sizes. This requires a conceptual adjustment for many forest scientists, because (superficially) genets are only apparent as scattered, seasonal, ephemeral, sporadically produced basidiomata (Watling 1995).

The spatial distributions and ecological attributes of *Armillaria* genets in a forest community have important implications for forest-community responses to perturbations ranging from silvicultural operations to climate change. Each *Ar-*

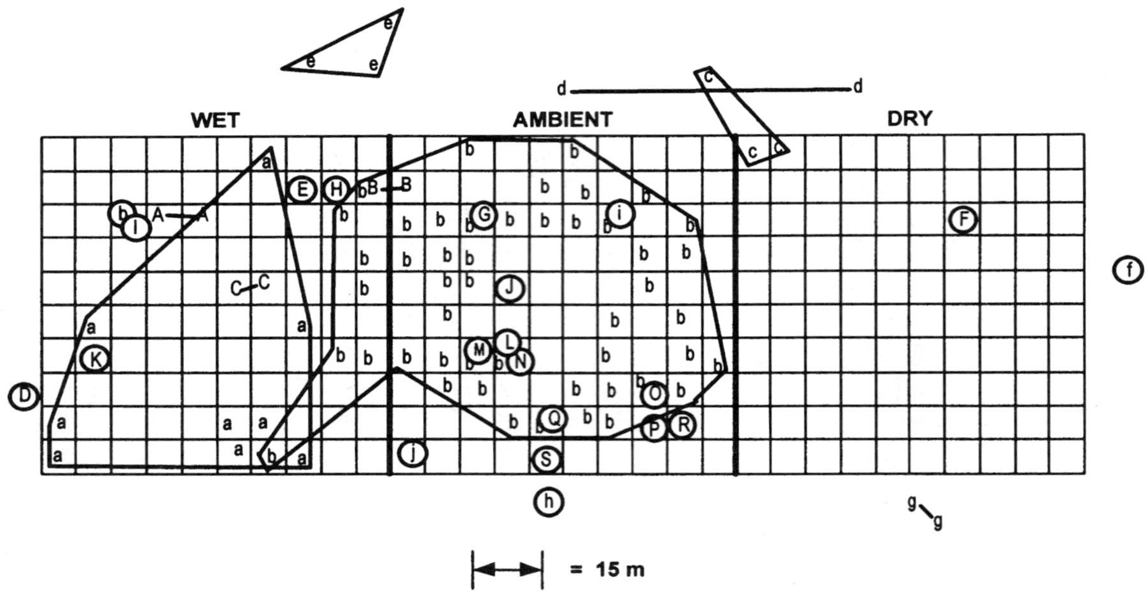

Figure 19.1. Spatial distribution of *Armillaria* genets in the TDE plots. Each collection point is marked with a letter; lowercase letters indicate *A. gallica* isolates, uppercase letters indicate *A. mellea* isolates. Isolates belonging to the same genet share the same uppercase or lowercase letter. Approximate boundaries of each genet are indicated.

millaria genet exercises ecological influence through its unique genetic potential for woody-resource foraging, wood decomposition and nutrient recycling, and root parasitism, as mediated by environmental constraints on the expression of these potentials. Without controlled experimentation, it is difficult to establish whether the differences observed in behavior of neighboring *Armillaria* genets [e.g., Smith et al. (1992, 1994); Rizzo and Harrington (1993); Legrand et al. (1996); Bruhn et al. (1996, 2000)] are caused by differences among genetic potentials [e.g., Omdal et al. 1995] or by limitations of the local environment in which each genet functions (such as edaphic factors, availability of suitable food-base resources, interactions with competitors, etc.). Because *Armillaria* genets are fundamentally territorial, continuously distributed in the forest floor, and long-lived, they exercise their ecological influences in discrete portions of the forest landscape over potentially unlimited time frames [e.g., Anderson et al. (1979); Shaw and Roth (1976); Smith et al. (1992)]. In this way, the spatial pattern of *Armillaria* genets represents one component of a forest landscape's genetic "memory" of how to respond to change, thus helping to determine long-term forest-community structure. As a result, the spatial pattern of *Armillaria* genets even influences wildlife populations through effects on vegetation structure [e.g., Guyette and Kabrick (2002)].

Until more work is done, we can only speculate on the interactions occurring belowground among *Armillaria* genets of the same or different species. Territorial behavior of conspecific *Armillaria* genets coupled with the apparent overlapping boundaries of genets of different species has been noted previously [e.g., Kile et al. (1991); Rizzo and Harrington (1993); Smith et al. (1992, 1994); Legrand et al. (1996); Bruhn et al. (1996, 2000)]. When genets of certain *Armillaria* spp. occupy the same landscape, it may be possible as a result of niche overlap (Leibold 1995) for genets of the different species to influence each other's behavior in areas of genet contact or overlap. For example, because both *A. gallica* and *A. mellea* are hardwood-specializing fungi and *A. gallica* is a much more aggressive producer of rhizomorphs, *A. mellea* may often have less access to the portion of food-base resources suitable for both species within the boundaries of *A. gallica* genets, resulting in a smaller "realized niche" for *A. mellea* than elsewhere in the absence of competition from *A. gallica.* Thus, as a result of niche overlap, we anticipate that *A. gallica* might mitigate the plant-root-parasitic aggressiveness of a co-occurring *A. mellea* genet, perhaps by limiting the opportunities for expansion of the *A. mellea* genet. On the other hand, because of their greater aggressiveness as root parasites, it may be possible for *A. mellea* genets to become established within the bounds of *A. gallica* genets by colonizing predisposed root systems before they become vulnerable to *A. gallica* infection. Mohammed and Guillaumin (1989) found that when *A. gallica* and *A. mellea* occupied the same substrate, *A. mellea* inhibited *A. gallica* without being inhibited in return by *A. gallica.*

The lower activity of *Armillaria* spp. in the TDE dry plot may be caused by (1) the random failure of any *Armillaria* genets to establish over time in that portion of the landscape, (2) a less suitable distribution or lower volume of food-base material to support rhizomorph growth and/or fruiting, or (3) long-term effects of the drier nature of the dry plot. Wood decay and/or rhizomorph growth rate(s) may be reduced in scale and occur deeper in the soil profile on the dry plot than on the ambient and wet plots, thus perhaps also influencing fruiting at the soil surface of the dry plot. *Armillaria* rhizomorph biomass and rates of *Armillaria* wood decay and rhizomorph production and growth at various depths in the TDE soils would be worth studying.

Within the TDE, sapling populations grew faster and experienced less mortality with greater precipitation throughfall, whereas overstory tree diameter growth responded only to annual variation in soil-water status and overstory tree mortality was negligible (Hanson et al. 2001; Chapter 15, this volume). The extent of *Armillaria* involvement in sapling mortality was not determined, but it should not be surprising that little overstory mortality occurred. Hardwood root disease, decline, and mortality associated with *A. mellea* appear to develop very slowly on overstory trees, especially in relatively mesic ecological landtypes (Bruhn et al. 2000). The processes of oak decline and mortality in the Missouri Ozarks are frequently associated with *Armillaria* root disease (Bruhn et al. 2000), occur over periods of decades in response to multiple distinct episodes of severe drought, and can affect scattered trees of all size classes and canopy positions (Dwyer et al. 1995).

Finally, we need to consider the implications of both our sampling time frame (September 1994 through October 1996) and the TDE time frame (1993 through 2000). Because fungal genets do not fruit equally well during all years (Watling 1995), it was important to sample during multiple years. Additional sampling (i.e., more frequent sampling within years and sampling during additional years) would doubtless have provided additional mapped collections. Additional collections would have resulted in more precise genet boundaries and perhaps also in the detection of some additional *Armillaria* genets. Nevertheless, greater mapping intensity would not alter the conclusions drawn here concerning the *Armillaria* species involved on the TDE site and the relative sizes, juxtaposition, and distributions of their genets with respect to each other and the TDE. Our genet map represents an unusual "snapshot" in time, due to the unusually durable nature of *Armillaria* genets in forest landscapes (Anderson and Kohn 1995).

Although it is possible that some new *Armillaria* genets may have become established over the TDE time frame and it is likely that some preexisting genets have experienced adjustments in territory extent and/or biomass and aggressiveness within their territories, it is unlikely that any established *Armillaria* genets have disappeared from the TDE site. Over longer periods of reduced precipitation, territorial contraction and reduced biomass are likely scenarios for *A. gallica* genets, because of this species' relative dependence for territory exploitation on rhizomorph growth through soil. Because of its greater aggressiveness than *A. gallica* in colonizing stressed host trees, *Armillaria mellea* might take advantage of drought stress to establish additional genets and to extend their territories by growth through connected root systems (Wargo and Harrington 1991; Wargo 1996; Lonsdale and Gibbs 1996). Testing of these hypotheses would require controlled field experimentation.

References

Adams DH (1974) Identification of clones of *Armillaria mellea* in young-growth ponderosa pine. Northwest Sci 48: 21–28.

Anderson JB, Kohn LM (1995) Clonality in soilborne, plant-pathogenic fungi. Annu Rev Phytopathol 33:369–391.

Anderson JB, Ullrich RC, Roth LF, Filip GM (1979) Genetic identification of clones of *Armillaria mellea* in coniferous forests in Washington. Phytopathology 69:1109–1111.

Bachelet D, Neilson RP (2000) Biome redistribution under climate change. In Joyce LA, Birdsey R (Eds) *The impact of climate change on America's forests: A technical document supporting the 2000 USDA Forest Service RPA Assessment.* General Technical Report RMRS-59. USDA Forest Service, Rocky Mountain Research Station, Fort Collins, Colorado, pp 18–39.

Bauce E, Allen DC (1992) Role of *Armillaria calvescens* and *Glycobius speciosus* in a sugar maple decline. Can J For Res 22:549–552.

Bruhn JN, Brenneman JA, Wetteroff JJ Jr., Mihail JD, Leininger TD (1997) Spatial patterns of *Armillaria* populations in the Walker Branch Watershed Throughfall Displacement Experiment, Tennessee, USA. In Pallardy SG, Cecich RA, Garrett HG, Johnson PS (Eds), *11th Central Hardwood Forest Conference.* General Technical Report NC-188. USDA Forest Service, North Central Forest Experiment Station, St. Paul, Minnesota, pp 49–57.

Bruhn JN, Mihail JD, Meyer TR (1996) Using spatial and temporal patterns of *Armillaria* root disease to formulate management recommendations for Ontario's black spruce (*Picea mariana*) seed orchards. Can J For Res 26: 298–305.

Bruhn JN, Mihail JD, Meyer TR (1998) *Armillaria* root disease inoculum remaining in young Ontario conifer plantations following root raking or madge rotoclearing. Northern J Appl For 15:191–196.

Bruhn JN, Wetteroff JJ Jr., Mihail JD, Kabrick JM, Pickens JB (2000) Distribution of *Armillaria* species in upland Ozark Mountain forests with respect to site, overstory species composition and oak decline. For Pathol 30: 43–60.

Bruhn JN, Wetteroff JJ Jr., Mihail JD, Jensen RG, Pickens JB (2002) Harvest-associated disturbance in upland Ozark forests of MOFEP. In Shifley SR, Kabrick JM (Eds) *Proceedings of the Second Missouri Ozark Forest EcosystemProject Symposium: Post-treatment results of the landscape experiment.* General Technical Report NC-227. USDA Forest Service, North Central Forest Experiment Station, St. Paul, Minnesota, pp. 130–146.

Bugmann H (1997) Sensitivity of forests in the European Alps to future climatic change. Climate Res 8:35–44.

Casper BB, Jackson RB (1997) Plant competition underground. Annu Rev Ecol Syst 28:545–570.

Clinton BD, Boring LR, Swank WT (1993) Canopy gap characteristics and drought influences in oak forests of the Coweeta Basin. Ecology 74:1551–1558.

Cruickshank MG, Morrison DJ, Punja ZK (1997) Incidence of *Armillaria* species in precommercial thinning stumps and spread of *Armillaria ostoyae* to adjacent Douglas-fir trees. Can J For Res 27:481–490.

Dwyer JP, Cutter BE, Wetteroff JJ Jr. (1995) A dendrochronological study of black and scarlet oak decline in the Missouri Ozarks. For Ecol Manage 75:69–75.

Fox RTV (2000) Biology and life cycle. In Fox RTV (Ed) *Armillaria root rot: Biology and control of honey fungus.* Intercept, Andover, England, pp 3–44.

Garraway MO, Hütterman A, Wargo PM (1991) Ontogeny and physiology. In Shaw CG III, Kile GA (Eds) *Armillaria root disease.* Agricultural Handbook No. 691.USDA Forest Service, Washington, District of Columbia, pp 21–47.

Gregory SC, Rishbeth J, Shaw CG III (1991) Pathogenicity and virulence. In Shaw CG III, Kile GA (Eds) *Armillaria root disease.* Agricultural Handbook No. 691. USDA Forest Service, Washington, District of Columbia, pp 76–87.

Guillaumin J-J, Anderson JB, Korhonen K (1991) Life cycle, interfertility, and biological species. In Shaw CG III, Kile GA (Eds) *Armillaria root disease.* Agricultural Handbook No. 691. USDA Forest Service, Washington, District of Columbia, pp 10–20.

Guillaumin J-J, Anderson JB, Legrand P, Ghahari S, Berthelay S (1996) A comparison of different methods for the identification of genets of *Armillaria* spp. New Phytol 133:333–343.

Guyette RP, Kabrick JN (2002) The legacy and continuity of forest disturbance, succession, and species at the MOFEP sites. In Shifley SR, Kabrick JM (Eds) *Proceedings of the Second Missouri Ozark Forest EcosystemProject Symposium: Post-treatment results of the landscape experiment.* General Technical Report NC-227. USDA Forest Service, North Central Forest Experiment Station, St. Paul, Minnesota, pp 26–44.

Hanson PJ, Todd DE Jr., Amthor JS (2001) A six-year study of sapling and large-tree growth and mortality responses to natural and induced variability in precipitation throughfall. Tree Physiol 21:345–358.

Harper JL (1977) *Population biology of plants.* Academic Press, New York.

Hood IA, Redfern DB, Kile GA (1991) *Armillaria* in planted hosts. In Shaw CG III, Kile GA (Eds) *Armillaria root disease.* Agricultural Handbook No. 691. USDA Forest Service, Washington, District of Columbia, pp 122–149.

Houston DR (1992) A host–stress–saprogen model for forest dieback-decline diseases. In Manion PD, Lachance D (Eds) *Forest decline concepts.* APS Press, St. Paul, Minnesota, pp 3–25.

Kile GA, McDonald GI, Byler JW (1991) Ecology and disease in natural forests. In Shaw CG III, Kile GA (Eds) *Armillaria root disease.* Agricultural Handbook No 691. USDA Forest Service, Washington, District of Columbia, pp 102–121.

Korhonen K (1978) Interfertility and clonal size in the *Armillariella mellea* complex. Karstenia 18:31–42.

Legrand P, Ghahari S, Guillaumin J-J (1996) Occurrence of genets of *Armillaria* spp. in four mountain forests in central France: The colonization strategy of *Armillaria ostoyae.* New Phytol 133:321–332.
Leibold MA (1995) The niche concept revisited: Mechanistic models and community context. Ecology 76:1371–1382.
Lonsdale D, Gibbs JN (1996) Effects of climate change on fungal diseases of trees. In Frankland JC, Magan N, Gadd GM (Eds) *Fungi and environmental change.* Cambridge University Press, Cambridge, England, pp 1–19.
Lundquist JE (1993) Spatial and temporal characteristics of canopy gaps caused by *Armillaria* root disease and their management implications in lowveld forests of South Africa. Eur J For Pathol 23:362–371.
Lundquist JE (1995) Pest interactions and canopy gaps in ponderosa pine stands in the Black Hills, South Dakota, USA. For Ecol Manage 74:37–48.
Mallett KI, Maynard DG (1998) *Armillaria* root disease, stand characteristics, and soil properties in young lodgepole pine. For Ecol Manage 105:37–44.
Manion PD, Lachance D (1992) Forest decline concepts: an overview. In Manion PD, Lacnance D (Eds) *Forest decline concepts.* APS Press, St. Paul, Minnesota, pp 181–190.
Marshall SD, Walker SE, Rypstra AL (2000) A test for a differential colonization and competitive ability in two generalist predators. Ecology 81:3341–3349.
Mihail JD, Bruhn JN, Leininger TD (2002) The effects of moisture and oxygen availability on rhizomorph generation by *Armillaria tabescens* in comparison with *A. gallica* and *A. mellea.* Mycol Res 106:697–704.
Mihail JD, Obert M, Bruhn JN, Taylor SJ (1995) Fractal geometry of diffuse mycelia and rhizomorphs of *Armillaria* species. Mycol Res 99:81–88.
Mohammed C, Guillaumin J-J (1989) Competition phenomena between European species of *Armillaria.* In Morrison DJ (Ed) *Proceedings of the 7th international conference on root and butt rots of forest trees.* IUFRO Working Party S2.06.01. Forestry Canada, Victoria, British Columbia, pp 347–354.
Morrison DJ (1989) Pathogenicity of *Armillaria* species is related to rhizomorph growth habit. In Morrison DJ (Ed) *Proceedings of the 7th international conference on root and butt rots of forest trees.* IUFRO Working Party S2.06.01. Forestry Canada, Victoria, British Columbia, pp 584–589.
Omdal DW, Shaw CG III, Jacobi WR, Wager TC (1995) Variation in pathogenicity and virulence of isolates of *Armillaria ostoyae* on eight tree species. Plant Dis 79:939–944.
Pegler DN (2000) Taxonomy, nomenclature and description of *Armillaria.* In Fox RTV (Ed) *Armillaria root rot: Biology and control of honey fungus.* Intercept, Andover, England, pp 81–93.
Rayner ADM (1991) The challenge of the individualistic mycelium. Mycologia 83:48–71.
Redfern DB, Filip GM (1991) Inoculum and infection. In Shaw CG III, Kile GA (Eds) *Armillaria root disease.* Agricultural Handbook No. 691. USDA Forest Service, Washington, District of Columbia, pp 48–61.
Rishbeth J (1991) *Armillaria* in an ancient broadleaved woodland. Eur J For Pathol 21:239–249.
Rizzo DM, Harrington TC (1992) Nuclear migration in diploid-haploid pairings of *Armillaria ostoyae.* Mycologia 84:863–869.
Rizzo DM, Harrington TC (1993) Delineation and biology of clones of *Armillaria ostoyae, A. gemina* and *A. calvescens.* Mycologia 85:164–174.
Rizzo DM, Slaughter GW, Parmeter JR Jr. (2000) Enlargement of canopy gaps associated with a fungal pathogen in Yosemite Valley, California. Can J For Res 30:1501–1510.
Shaw CG III, Roth LF (1976) Persistence and distribution of a clone of *Armillaria mellea* in a ponderosa pine forest. Phytopathology 66:1210–1213.
Shearer BL, Byrne A, Dillon M, Buehrig, R (1997) Distribution of *Armillaria luteobubalina* and its impact on community diversity and structure in *Eucalyptus wandoo* woodland of southern Western Australia. Aust J Bot 44:151–165.
Siwecki R, Ufnalski K (1998) Review of oak stand decline with special reference to the role of drought in Poland. Eur J For Pathol 28: 99–112.
Smith ML, Bruhn JN, Anderson JB (1992) The fungus *Armillaria bulbosa* is among the largest and oldest living organisms. Nature 356:428–431.
Smith ML, Bruhn JN, Anderson, JB (1994) Relatedness and spatial distribution of *Armillaria* genets infecting red pine seedlings. Phytopathology 84:822–829.
Sokal RR, Rohlf, FJ (1995) *Biometry,* 3rd ed. WH Freeman and Co, New York.
Termorshuizen AJ (2000) Ecology and epidemiology of *Armillaria.* In Fox RTV (Ed) *Armillaria root rot: Biology and control of honey fungus.* Intercept, Andover, England, pp 45–63.
Tilman D (1994) Competition and biodiversity in spatially structured habitats. Ecology 75:2–16.
Wargo PM (1996) Consequences of environmental stress on oak: Predisposition to pathogens. Ann Sci For 53:359–368.
Wargo PM, Harrington TC (1991) Host stress and susceptibility. In Shaw CG III, Kile GA (Eds) *Armillaria root disease.* Agricultural Handbook No. 691. USDA Forest Service, Washington, District of Columbia, pp 88–101.
Watling R (1995) Assessment of fungal diversity: Macromycetes, the problems. Can J Bot 73(Suppl. 1):S15–S24.
Watling R, Kile GA, Burdsall HH Jr. (1991) Nomenclature, Taxonomy, and Identification. In Shaw CG III, Kile GA (Eds) *Armillaria root disease.* Agricultural Handbook No. 691. USDA Forest Service, Washington, District of Columbia, pp 1–9.
Worrall JJ, Harrington TC (1988) Etiology of canopy gaps in spruce-fir forests at Crawford Notch, New Hampshire. Can J For Res 18:1463–1469.

20. The Influence of Precipitation Change on Spiders as Top Predators in the Detrital Community

Kenneth L. Cramer

Introduction

Spiders are significant top predators in the detrital system (Fig. 20.1) of the deciduous forest (Clarke and Grant 1968; Moulder and Reichle 1972; Petersen and Luxton 1982) and may influence decomposition through top-down trophic cascades (Lawrence and Wise 2000). Collembola (springtails) and oribatid mites dominate the detritivore and saprophage guilds of most ecosystems studied (Petersen and Luxton 1982), accounting for up to 95% of total microarthropod numbers (Seasted 1984). Numerous studies have shown that springtails have a significant impact on nutrient cycling in terrestrial systems, both directly by consuming and shredding litter and indirectly by consuming fungal decomposers (Ineson et al. 1982; Leonard and Anderson 1991; Parkinson et al. 1979; Visser et al. 1981). Spiders are significant predators of springtails. Direct collecting of prey found in webs or in the chelicerae of wandering spiders has shown that a variety of spider families prey heavily on Collembola (Nyffeler et al. 1994). Manley et al. (1976) report that up to 55% of springtail mortality in *Acer* forest litter may be attributable to spiders. Moulder and Reichle (1972) found that spiders accounted for the greatest biomass of all microarthropod predators, with small to medium spiders (<10 mg) accounting for 83% of all spider predation. Thus, in their role as top predators in the leaf litter, spiders may affect decomposition through a top-down effect by their influence on Collembola that act as fungivores and detritivores. By manipulating spider densities on small plots, Law-

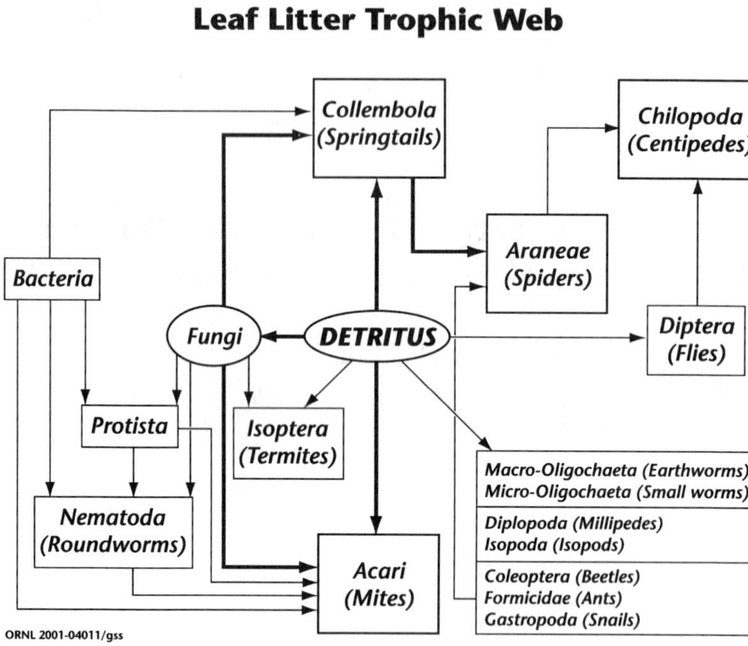

Figure 20.1. A schematic diagram of some of the major routes of energy flow in the detrital food web of a deciduous forest.

rence and Wise (2000) were able to demonstrate that spider removal increased Collembola numbers and also increased rates of decomposition.

The influence of abiotic factors, such as moisture, on leaf-litter spider communities has been surprisingly understudied, but a few studies have investigated the dependence of leaf-litter spiders on adequate moisture. Baehr and Eisenbeis (1985) report that the vertical position of spiders in the leaf litter corresponds to their transpiration rates and resistance to desiccation. Vollmer and MacMahon (1974) studied water loss in a variety of spiders and concluded that larger spiders had higher rates of water loss. Interestingly, they did not consider surface area-to-volume relationships, although they report mean values of surface area to volume for each species studied. Graphing these values against the mean rate of water loss they reported for each species shows a strong negative correlation between spider body sizes and rates of water loss per gram of body weight, as one would predict. Other more descriptive studies [e.g., McIver et al. (1992)] suggest that litter spiders show habitat associations that correlate with variation in moisture. One can certainly conclude that moisture is an important abiotic factor determining spider distributions and, possibly, that smaller spiders will be more sensitive to desiccation.

Another significant abiotic factor likely to affect leaf-litter spiders is the structure of the leaf litter itself. Uetz (1975, 1979) first established that, as litter depth increased, abundance of certain spider families did, as well. Species richness was generally also positively correlated with litter depth. Stevenson and Dindal (1982) increased habitat complexity without affecting litter quality or quantity by adding artificial plastic leaves of various shapes to plots. They found that both species richness and the abundance of one species of theridiid were increased on plots where curled "leaves" had been added. Bultman and Uetz (1984) also used artificial litter to increase structural complexity of litter and found that although the abundance of spiders increased (independently of litter mass), other predators, Collembola, and mites did not. These studies confirm the suspicion that although leaves in general provide habitat for spiders, the structure of fallen leaves (i.e., curled or flat) is important in determining habitat suitability for many spiders.

Two effects of the Throughfall Displacement Experiment (TDE) on leaf litter are well documented in previous studies and supported by the current one: (1) leaf-litter mass has increased on the dry plot (Chapter 17, this volume) and (2) the treatments create significant divergence of soil moisture, especially at certain times of the year. Leaf-litter moisture parallels the findings for soil moisture (Hanson, unpublished data; Cramer, this chapter). Following the findings of previous studies, we can make several predictions about spider community response to these changes. First, increased litter mass on the dry plot could cause an increase in detritivores and saprophages, especially Collembola, causing an increase in spider abundance in a bottom-up response. Increased litter mass could also increase habitat structure and volume and thereby increase species richness and abundance of certain taxa. Wise et al. (1999) suggest other possible reactions resulting from increasing detritus. Second, the decrease in moisture on the dry plots could decrease habitat suitability for some spiders, especially small spiders with high surface area-to-volume ratios that are more susceptible to desiccation. The converse of these predictions would be expected for the wet plot of the TDE.

Methods

Two sampling periods of 4 days each, the first from August 26 to 29 and the second from November 17 to 20, 2000, were designed to obtain a profile of spider communities in summer and winter (after the first hard frost). Samples were collected between 9 and 11 AM each day and no sooner than 24 h after a significant ($>$ 0.5 cm) rainfall. The TDE experimental units of wet, ambient, and dry were further subdivided into upslope and downslope components because of known variation in moisture and tree species composition that correlates with slope (Hanson et al. 1998). The upslope transect ran parallel to the ridge top ~15 m from the edge of highest elevation of the plots, and the downslope ran parallel to the ridge top ~15 m from the edge of lowest elevation of the plots. During each of the two sampling periods, each of the six resulting blocks (3 $\times$ 2) combinations of moisture and slope, respectively) was sampled as follows. A strip transect at each elevation was sampled starting at least 10 m from the plot boundary to avoid possible edge effects. A boardwalk provided a starting point for sampling. The location of each sample was determined using three random numbers. One determined the distance in meters (1–5) from the previous sample along the transect, the second the distance from the boardwalk (at least 1 m), and the third whether the sample would be upslope or downslope from the boardwalk. Areas sampled during the first period were mapped and marked so they would not be resampled during the second sampling period. Twelve samples (0.2-m^2 circular plots) of the entire O horizon (hereafter referred to simply as "leaf litter") were permanently removed from each of the six blocks. Leaf litter samples were bagged in the field and taken to the lab where they were weighed (wet) and then hand-sorted for spiders. After the spiders had been removed and preserved in 95% ethanol, the leaf litter was oven dried for up to 48 h at 100°C. Woody debris (nuts, twigs, bark, etc.) was separated, and both the litter and woody debris were weighed to the nearest 0.1 g.

Spiders were identified to species if adults were found and, in most cases, reliably to genus if only juveniles were present. Kaston's (1978) and Roth's (1993) keys to the genera were used as well as numerous publications on specific groups cited in the latter. Questionable or difficult identifications were referred to experts for verification. Because of the large percentage of juveniles in the summer sample, analysis of diversity and composition is limited to the level of genus.

Densities of all spiders and of individual taxa were analyzed by analysis of variance (ANOVA) and Kruskal–Wallis tests, respectively, with the Minitab statistical package. Species richness was compared across treatments and slope with the rarefaction method provided in the program EcoSim developed by Gotelli and Entsminger (2000). Species-composition data were analyzed with the techniques of multidimensional scaling (MDS), ANOSIM (analysis of similarities), and SIMPER provided in the multivariate nonparametric analysis package PRIMER (Clarke and Warwick 1994). A detailed description and justification of each statistical method is found in the appropriate section of the results.

Results

Changes in Leaf Litter Moisture and Mass

Although Hanson et al. (Chapter 2, this volume) documented a clear change in soil moisture levels from the TDE treatments, measurements were done to confirm that these changes were paralleled by concomitant changes in leaf litter moisture. By comparing the leaf mass before and after drying (dividing the dry weight by wet weight), the percent of the total weight accounted for by leaves versus moisture was determined. Moisture weight varied significantly across the treatments (one-way ANOVA, $F = 3.16, p = 0.045$) with mean moisture weights of 48%, 46%, and 42%, respectively, for the wet, ambient, and dry plots. Fisher's pairwise comparison of wet and dry plots was statistically significant ($p < 0.05$). The high moisture on all plots reflects significant late fall rains before the November collections. However, August results parallel the combined data with 38%, 35%, and 32% moisture content respectively for wet, ambient, and dry plots. Hanson (un-

Table 20.1. Mean values (SE) by treatment and slope for dry leaf mass (g) and spider density by area (number per square meters) and volume (number per kilogram of dry leaf mass)

	Treatment			Slope	
	Wet	Ambient	Dry	Up	Down
Leaf mass	237.6 (7.2)	249.9 (7.9)	273.4 (7.6)	262.4 (6.7)	244.9 (6.0)
Density/area	11.4 (1.2)	9.7 (1.0)	13.0 (1.1)	11.5 (0.8)	11.2 (1.0)
Density/volume	46.9 (4.4)	38.5 (4.0)	47.0 (3.7)	45.9 (3.3)	42.3 (3.4)

published data) confirms that leaf litter moisture varies across treatments, with significant divergence among the plots during periods of moderate rainfall.

More important are the changes in leaf litter mass observed across the treatments. The dry plot had accumulated significantly more litter than the wet plot (ANOVA, $F = 5.72$, $p = 0.004$), about 10% more than the ambient, and 15% greater than the wet samples (Table 20.1).

Seasonal effects on spider communities were the most pronounced of the three factors considered (season, slope, and treatment). Therefore, in subsequent discussions, obvious and expected seasonal changes are considered first, followed by a discussion of consistent treatment or slope effects across seasons.

Spider Population Density

Density was measured in two ways: first by area and second by litter mass (as an estimate of litter volume). Because previous studies had shown an effect of leaf volume on spider numbers and richness (Uetz 1975, 1979), tests for a correlation between the two were conducted to see if calculating density by volume (rather than area) would be justified. Leaf volume was strongly correlated with spider abundance (Pearson's $r = 0.41$, $p < 0.001$), so the results that follow will refer to density by volume (spiders per kilogram of dry leaf mass) except as noted. However, for ease of interpretation and comparison to previous studies, density by area (spiders per square meter) is included in tabular summaries of the data.

Seasonal Differences. As expected, spider densities dropped dramatically after the first frost on all plots, falling by more than 50% from a mean of 60/kg to 28/kg (ANOVA, $F = 90.9$, $p < 0.001$).

Treatment and Slope Effects. A two-way ANOVA revealed no significant change in spider density resulting from treatment ($F = 1.49$, $p = 0.23$) or slope ($F = 0.59$, $p = 0.44$; Table 20.1) and no significant interaction between the two factors ($F = 1.93$, $p = 0.15$). The slight increase in spider density by area on the dry plots is negated when accounting for the larger leaf mass on these plots (spider density by volume, Table 20.1).

Species Richness

The rarefaction method was used to compare species richness at different sites. This method pools all individuals of a given site and randomly draws from them according to their observed proportions in the sample to produce a species-accumulation curve. Multiple iterations (in this case, 500) for each sample size (from one individual proceeding in equal increments to the total number actually collected on the site) produce a mean species richness for each sampling effort up to the observed total. To compare sites, it is only necessary to rarify the total sample size of the larger sample to that of the smaller and see if the observed species richness of the smaller sample falls within the confidence limits of the larger sample rarified to the same total *N*. The method of extrapolation described by Colwell and Coddington (1994) was investigated but not applied to these data because the primary objective was not to extrapolate species richness at the sites but to compare observed species richness. One benefit of the rarefaction technique is that, because it relies completely on the observed species richness and does not extrapolate beyond the number sampled, there are relatively few assumptions. Disadvantages include the necessity of pooling separate samples (although they come from the same treatment or factor), so some information of among sample variability is lost.

Treatment Effects. Fifty-eight species (Fig. 20.2) were observed across the TDE with a species richness of 43 on the ambient plot ($N = 469$), 42 on the dry plot ($N = 626$), and 45 ($N = 548$) on the wet plot. Rarefying the two sites with the larger *N* (dry and wet) to the ambient plot shows no significant difference between the wet and ambient sites (Fig. 20.3; Table 20.2). The dry site, however, shows a lower than expected richness in comparison to both the ambient and wet sites ($p < 0.01$).

Slope Effects. The most dramatic difference in species richness was not among the treatments, but from the upslope and downslope sites (Fig. 20.4). Observed species richness on the lower slope was 51 ($N = 811$) compared to only 42 ($N = 832$) on the upper slope, and because the sample sizes are nearly equal, comparing the two sites by rarefaction hardly surprises intuition by demonstrating a significant difference at $p < 0.001$ (Table 20.2).

Species Composition

Because of the nature of the community data matrix (59 taxa by 144 samples) with many zeros and ones for many species in a given sample, a nonparametric method of ordination and comparison of multivariate datasets (Clarke and Warwick 1994) was used to analyze the data. First, a Bray–Curtis similarity matrix was calculated from the species by sample abundance matrix. In this study, the data were neither transformed nor standardized for reasons given in Clarke and Warwick (1994). For analysis of similarity (ANOSIM), the data were not averaged in order to preserve information on sample variation. ANOSIM is a nonparametric analog of ANOVA that operates on the similarity matrix to determine if significant community differences exist between sites. Finally, from the raw abundance data, the SIMPER program was used to calculate which taxa contributed the most to any observed differences between groups.

Figure 20.2. Some of the more common spiders encountered in the study. (A) *Anahita punctulata* (Ctenidae), (B) *Crustulina* sp. (Theridiidae), (C) *Dictyna* sp. (Dictynidae), (D) *Drassyllus* sp. (Gnaphosidae), (E) *Lepthyphantes nebulosa* (Linyphiidae), (F) *Schizocosa* sp. (Lycosidae), (G) *Metaphidippus protervus* (Salticidae), (H) *Phrurotimpus alarius* (Clubionidae), and (I) *Origanates rostratus* (Linyphiidae). (See also color plate.)

Seasonal Differences. Clear differences in community composition exist between the August and November samples (one-way ANOSIM; $R = 0.218, p = 0.001$). Analysis of the species contributing to dissimilarity (SIMPER) between dates shows that 58% of the difference in seasons is caused by a simple reduction in density of seven common species (of six families) contributing at least 5% to the observed dissimilarity (*Origanates, Schizocosa, Phrurotimpus, Dictyna, Anahita, Lepthyphantes,* and *Drassyllus*) in the November samples. The most abundant species on both dates was the erigonine *Origanates rostratus,* followed by immature lycosids, probably *Schizocosa.* There were a few species unique to the November samples, however, such as a second species of *Dictyna,* two salticids (*Metaphidippus protervus* and *Sassacus* sp.), and the anyphaenid *Aysha gracilis.*

Slope and Treatment Effects. Because of the significant difference in composition among season, further analyses for slope and treatment effects for each date were conducted separately. Differences in community structure across slope and treatments are demonstrated by a two-way ANOSIM (August, $R = 0.303, p = 0.001$; November, $R = 0.150, p = 0.001$), and a look at SIMPER shows that the same

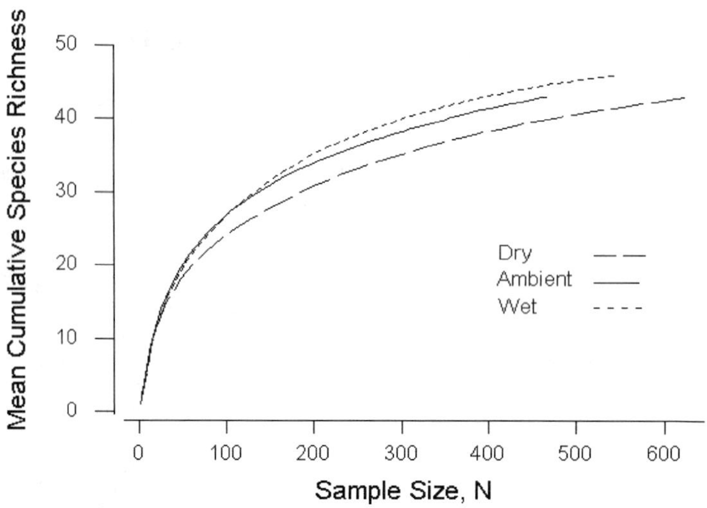

Figure 20.3. Mean species-accumulation curves (500 iterations) by treatment.

Table 20.2. Comparison of treatment and slope species richness with rarefaction

Treatment comparison	Observed richness	Mean (95% CI) of rarefied sample	p
Wet versus ambient	45 versus **43**	43.7 (41.6–45.7)	NS
Dry versus ambient	42 versus **43**	38.9 (36.0–41.8)	0.01
Dry versus wet	42 versus **45**	40.6 (38.5–42.7)	0.001
Up versus down	42 versus **51**	41.8 (40.9–42.7)	< 0.001

Note: Observed richness of the sample with the smaller *N* (in **bold**) is compared to the 95% confidence interval (CI) of the rarefied sample mean estimate.

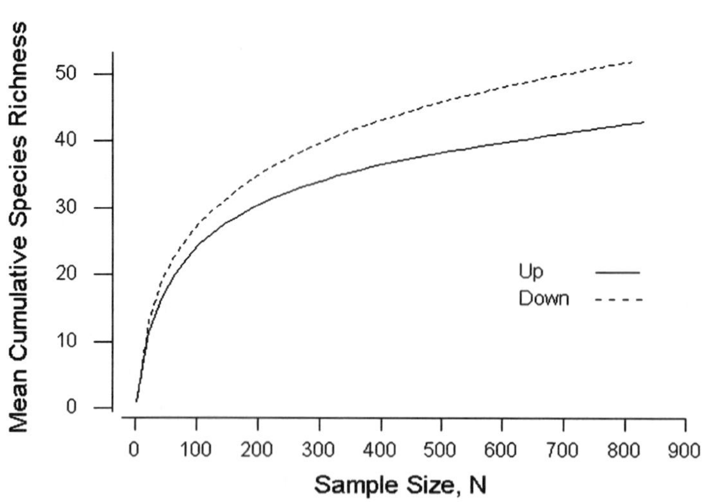

Figure 20.4. Mean species-accumulation curves (500 iterations) by slope.

Table 20.3. Kruskal–Wallis tests (*H*; adjusted for ties) for slope preference among common species

Species	*H*	*p*	Slope preference
Schizocosa	20.4	< 0.001	Up
Dictyna	4.6	0.03	Up
Anahita	12.8	< 0.001	Down
Scotinella	7.0	0.008	Down
Crustulina	20.6	< 0.001	Down

Note: Wandering spiders include *Schizocosa* and *Anahita*; the others spin webs.

species are responsible for this difference at both times. Of the 10 species accounting for 67% of the dissimilarity, 5 species show clear preferences for either upslope or downslope, as determined by a Kruskal–Wallis rank test (Table 20.3). Together, these five account for 31% of the dissimilarity of sites, the rest of the difference accumulating among the five other species, two of which are very abundant across the entire site (Fig. 20.5). The second most abundant species, *Schizocosa,* prefers the upslope site as well as *Anahita* and (marginally) *Dictyna.* The association of *Dictyna* is more suspect because of its relatively strong correlation with leaf mass ($R = 0.197, p = 0.018$) that is also greater in the upslope samples. Notably, two less abundant species, *Scotinella* and *Crustulina,* contribute significantly to the slope differences because they are disproportionately represented in the downslope samples. There is no clear pattern among web-spinning versus wandering spiders in these habitat preferences.

The treatments are also distinct in their community structure at both sampling dates. ANOSIM shows that, in August, not only is there a significant treatment

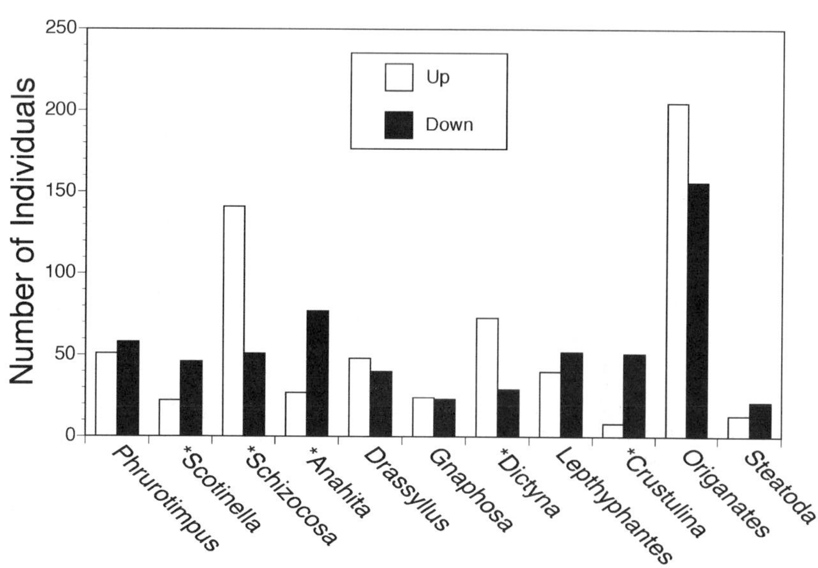

Figure 20.5. Histogram of absolute abundances of 11 common species by slope.

Table 20.4. ANOSIM paired comparisons for treatment effects on community composition

	August		November	
Treatments compared	R	p	R	p
Ambient–dry	0.076	0.032	0.211	0.001
Ambient–wet	0.168	0.001	0.085	0.054 NS
Dry–wet	0.219	0.001	0.090	0.029

effect ($R = 0.157$, $p = 0.001$) but also each paired comparison of moisture treatments is significant (Table 20.4). For November, the difference holds, although it is slightly less significant, probably because of the much smaller sample size ($R = 0.132$, $p = 0.002$), and the ambient-wet paired comparison is not significant ($R = 0.085$, $p = 0.054$, Table 20.4).

SIMPER points to 10 species that account for 66–68% of the dissimilarity among pairs of treatment sites. Six of these species show trends in abundance that roughly track the gradient of moisture produced by the TDE treatments, although not in the same order or preference (Fig. 20.6). Kruskal–Wallis one-way rank tests across treatment are significant at $p < 0.05$ for five of these (Table 20.5). *Dictyna, Anahita,* and *Drassyllus* all show a preference for the dry site. Although both *Dictyna* and *Drassyllus* show a correlation to leaf mass that is higher on the dry plots, the strength of the association with the dry plot is unlikely to be caused solely by the greater leaf mass. Interestingly, *Scotinella* appears to prefer the undisturbed plot, and its numbers decline markedly in both the wet and dry plots. Finally, *Lepthyphantes* also is marginally significant for a preference

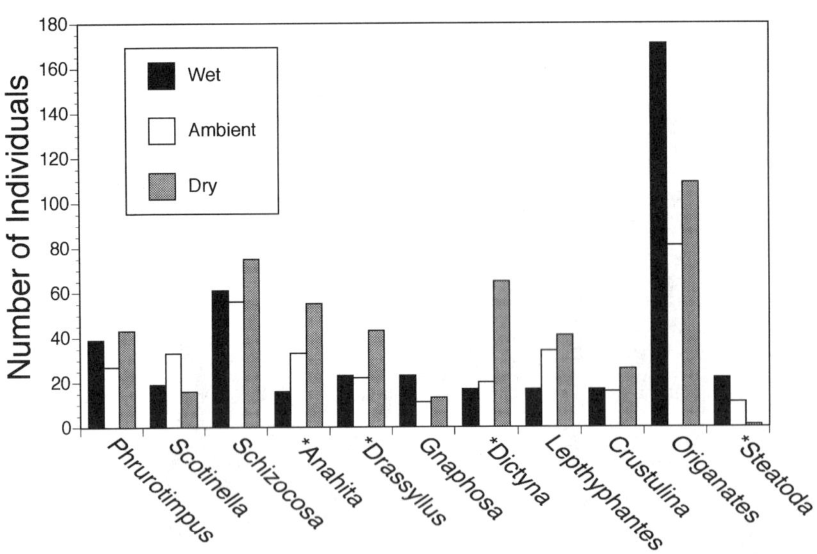

Figure 20.6. Histogram of absolute abundances of 12 common species by treatment.

Table 20.5. Kruskal–Wallis tests (adjusted for ties) for treatment preferences among common species

Species	H	p	Treatment preference
Anahita	10.8	0.005	Dry
Drassyllus	9.2	0.01	Dry
Scotinella	6.0	0.05	Ambient
Dictyna	23.3	< 0.001	Dry
Lepthyphantes	5.6	0.06	Dry
Steatoda	13.0	0.001	Wet

Note: Anahita, Drassyllus, and *Scotinella* are wandering spiders; the others spin webs.

of the dry area. Collectively, these six species contribute 27–33% of the dissimilarity among treatments depending on the pair of treatments compared.

Discussion

The fact that there appears to be no effect of the treatments on spider density and a slight effect on species richness is significant. With increased litter mass, an increased density of spiders was postulated, and although this occurs, it was not statistically significant. One might expect that with greater biomass from litter fall, a greater biomass of all components of the leaf-litter community could be supported, including spiders, but this was not the case. It could be that the increase in litter mass is subtle enough (about 10% above ambient) that a detectable response in spiders is not to be expected [unlike the manipulative experiments of Chen and Wise (1999)].

Species richness might also be expected to decline on the disturbed (wet and dry) plots if there are species especially sensitive to changes in moisture. Rare species might drop out, but rare species by definition are often undersampled. In fact, extrapolation techniques (Colwell and Coddington 1994) applied to my data showed that continued sampling would probably encounter significantly more rare species; the asymptote of the collector's curve had not been reached. The size of the experimental plots and their proximity to undisturbed areas for recruitment of rare species also mitigates the likelihood of showing a decline in rare species because of the treatments. In spite of these hurdles, the dry plot was significantly less species rich despite an increased abundance of spiders there. So, in fact, richness was compromised on at least one of the disturbed plots.

The most interesting aspects of the data are the changes observed in species composition. First, it is notable that all of the multivariate nonparametric techniques employed to elucidate patterns in the data showed that the clearest patterns were seasonal, followed by slope (habitat) differences, and, finally, by treatment effects. In other words, the treatments apparently have had less impact on species composition than do seasonal fluctuations or even slight habitat changes with slope. The seasonal patterns are not surprising and have been reported by many arachnologists [e.g., Weeks and Holtzer (2000)].

Several species showed strong affinities for upslope or downslope locations. The downslope area of the TDE is, of course, an increased moisture microenvironment, but also has some biotic differences in habitat. The downslope area is more riparian in nature and poplars become more abundant relative to the more common *Quercus–Acer* assemblage of the upslope site. The herbaceous layer also appears more diverse and with greater coverage. So little is known of most of the spiders studied that it is hard to predict if individual species are actually demonstrating habitat selection, but certainly there are confirmed differences in where the majority of some species are found.

Although there was no overall increase in spider density on the dry plots as predicted, there were several individual species that showed a significant preference for the dry plot. *Dictyna,* a cribellate spider that is said to favor building its web in curled leaves (Chamberlain and Gertsch 1958), had the strongest response. *Lepthyphantes,* another web-building spider, also showed increased numbers on the dry plot, so perhaps increased surface area or structure for building webs favored these species. However, the wandering spiders *Anahita* and *Drassyllus* also preferred the dry site. Possibly, increased numbers of prey caused a numerical response in these hunting spiders. A final interesting trend was an increased abundance of the numerically dominant erigonine, *Origanates rostratus,* on the wet plots. Although not a statistically significant trend, it is tempting to speculate that avoidance of desiccation plays a role in habitat choice by these minute spiders.

As a caveat, it should be noted that these data are mere snapshots in time (two 5-d periods) of the leaf-litter spider community. As such, they represent a narrow temporal sampling of an experiment that has been ongoing for 7 years and may not be sensitive to what could be large annual variations in response to the treatments. Nonetheless, it is clear that there was a significant effect of the treatments on spider-community composition, although the effects are dwarfed by differences caused by seasonal and habitat (slope) variation.

References

Baehr B and Eisenbeis G (1985) Comparative investigations on the resistance to desiccation in *Lycosidae, Hahniidae, Linyphiidae* and *Micryphantidae* (Arachnida: Araneae). Zool Jahrb (Syst Oekol Geogr Tiere) 112:225–34.

Bultman TL and Uetz GW (1984) Effect of structure and nutritional quality of litter on abundances of litter-dwelling arthropods. Am Midl Nat 111(1):165–172.

Chamberlain RV and Gertsch WJ (1958) The spider family *Dictynidae* in America north of Mexico. Bull Am Mus Nat Hist 116(1):1–152.

Chen B, Wise DH (1999) Bottom-up limitation of predaceous arthropods in a detritus-based terrestrial food web. Ecology 80(3):761–772.

Clarke RD, Grant PR (1968) An experimental study of the role of spiders as predators in a forest litter community. Part 1. Ecology 49(6):1152–1154.

Clarke KR, Warwick RM (1994) *Change in marine communities: An approach to statistical analysis and interpretation.* Natural Environment Research Council, Bournemouth, England.

Colwell RK, Coddington JA (1994) Estimating terrestrial biodiversity through extrapolation. Philos Trans R Soc London, Ser B 345:101–118.

Gotelli NJ, Entsminger GL (2000) *EcoSim: Null models software for ecology. Version 5.0.* Acquired Intelligence Inc. and Kesey-Bear, Burlington, Vermont. http://homepages.together.net/~gentsmin/ecosim.htm.

Hanson PJ, Todd DE, Huston MA, Joslin JD, Croker J, Augé RM (1998) *Description and field performance of the Walker Branch Throughfall Displacement Experiment: 1993–1996.* ORNL/TM-13586. Oak Ridge National Laboratory, Oak Ridge, Tennessee.

Ineson P, Leonard MA, Anderson JM (1982) Effect of collembolan grazing upon nitrogen and cation leaching from decomposing leaf litter. Soil Biol Biochem 14:601–605.

Kaston BJ (1978) *How to know the spiders.* Wm C Brown/McGraw-Hill, Boston, Massachusetts.

Lawrence KL, Wise DH (2000) Spider predation on forest-floor Collembola and evidence for indirect effects on decomposition. Pedobiologia 44:33–39.

Leonard MJ, Anderson JM (1991) Grazing interactions between a collembolan and fungi in a leaf litter matrix. Pedobiologia 35:239–246.

Manley GV, Butcher JW, Zabik, M (1976) DDT transfer and metabolism in a forest litter macro-arthropod food chain. Pedobiologia 16(8):81–98.

McIver JD, Parsons GL, Moldenke AR (1992) Litter spider succession after clear-cutting in a western coniferous forest. Can J For Res 22:984–992.

Moulder BC, Reichle DE (1972) Significance of spider predation in the energy dynamics of forest-floor arthropod communities. Ecol Monogr 42(4):473–498.

Nyffeler M, Sterling WL, Dean DA (1994) How spiders make a living. Environ Entomol 23(6):1357–1367.

Parkinson D, Visser S, Whittaker JB (1979) Effects of collembolan grazing on fungal colonization of leaf litter. Soil Biol Biochem 11:529–535.

Petersen H, Luxton M (1982) A comparative analysis of soil fauna populations and their role in decomposition processes. Oikos 39(3):287–388.

Roth VD (1993) *Spider genera of North America.* 3rd edition. American Arachnological Society, Gainsville, Florida.

Scheu S, Shaefer M (1998) Bottom-up control of the soil macrofauna community in a beechwood on limestone: Manipulation of food resources. Ecology 79(5):1573–1585.

Seastedt TR (1984) The role of microarthropods in decomposition and mineralization processes. Annu Rev Entomol 29:25–46.

Stevenson BG, Dindal DL (1982) Effect of leaf shape on forest litter spiders: Community organization and micro-habitat selection of immature *Enoplognatha ovata* (Clerck) (Theridiidae). J Arachnol 10:165–178.

Uetz GW (1975) Temporal and spatial variation in species diversity of wandering spiders (Araneae) in deciduous forest litter. Environ Entomol 4(5):719–724.

Uetz GW (1979) The influence of variation in litter habitats on spider communities. Oecologia 40:29–42.

Visser S, Whittaker JB, Parkinson D (1981) Effects of collembolan grazing on nutrient release and respiration of a leaf litter inhabiting fungus. Soil Biol Biochem 13:215–218.

Vollmer AT, MacMahon JA (1974) Comparative water relations of five species of spiders from different habitats. Comp Biochem Physiol 47A:753–765.

Weeks RD, Holtzer TO (2000) Habitat and season in structuring ground-dwelling spider (Araneae) communities in a shortgrass steppe ecosystem. Environ Entomol 29(6):1164–1172.

Wise DH, Snyder WE, Tuntibunpakul P (1999) Spiders in decomposition food webs of agroecosystems: Theory and evidence. J Arachnol 27:363–370.

7. Forest Stand-Level Syntheses

21. Forest Water Use and the Influence of Precipitation Change

Stan D. Wullschleger, Paul J. Hanson, and Donald E. Todd

Introduction

Altered patterns of precipitation associated with regional or global climatic change can be expected to affect ecosystem water use in a manner that reflects, in the case of drought, the response of individual components of that system to soil-water deficits. For temperate deciduous forests, these components include overstory and understory vegetation, the distribution of size classes among seedlings, saplings, and mature trees, and the differences among the many species that occupy these different canopy strata. In addition, consideration must also be given to leaf area dynamics and potential differences among species in their response to environmental variables, including radiation, vapor pressure deficit, and soil-water content.

Considerable time and energy has been invested in (1) trying to understand the mechanisms that underlie drought tolerance and species-specific responses to the environment and (2) quantifying how positional differences among trees within the forest canopy influence carbon and water exchange (Calder 1998; Falge et al. 2000; Kostner et al. 1992). The challenge has been not only to measure these processes in heterogeneously complex forest stands but also to integrate this information in order to understand how stand water use may change in a future climate. Few studies, however, have attempted to integrate the response of multiple species within a forest to altered precipitation regimes, despite the fact that interesting ecological insights might emerge from such an analysis (Bréda et al.

1993; Cienciala et al. 1994; Pataki et al. 2000). Recently, Wullschleger et al. (2001) used sap-flow techniques to measure water use for six hardwood species growing in an upland oak forest of eastern Tennessee and integrated those data over a season to derive a species-weighted estimate of total stand transpiration. However, this analysis was applied only to data collected during a relatively wet year; thus, the response of stand water use to changes in soil water content could not be addressed.

Because forest water use in a drier climate will reflect the response of individual species to drought, studies such as the Throughfall Displacement Experiment (TDE) can provide much needed information on seedling, sapling, and mature tree responses to imposed water limitations. Here, we extend the analysis of Wullschleger et al. (2001), adding complexities associated with the response of multiple species to altered precipitation. Our analysis encompasses water use by both overstory and understory species, and we quantify how soil-water availability impacts stand water use for the three treatments of the TDE. Moreover, a simple model driven by daily vapor-pressure deficit, fractional leaf area index, and soil-water potential is used to assess TDE-imposed and interannual variation in stand water use for the 8 years (1993–2000) since the experiment was initiated. Modeled data are then compared with catchment-scale estimates of net gain for two deciduous forests in the eastern United States, and regional implications of drought for seasonal water use are explored.

Methods

Forest water use is calculated as the product of sap velocity, cross-sectional sapwood area for the stand, and the fraction of sapwood functional in water transport, according to

$$\text{Stand water use} = f\,(V_s,\ \text{SA},\ f_s), \tag{1}$$

where V_s is sap velocity (m s^{-1}), SA is sapwood area for individual species comprising the stand (m^2 ha^{-1}), and f_s is the fraction of sapwood functional in water transport. Stand water use is expressed in units of mm d^{-1} and, as used in this study, refers to water use on a ground-area basis.

Techniques for Measuring Sap Flow and Radial Profiles of Sap Velocity

Thermal dissipation probes were used to estimate rates of sap velocity. These probes consist of two probes that are inserted into the sapwood. The upper probe was installed at a height of 1.3 m and contained a heating element, whereas the lower probe served as an unheated reference (Granier 1987). Each probe contained a thermocouple, and the temperature difference between the probes was used to calculate rates of sap velocity.

Heat-pulse probes (Greenspan Technology Pty. Ltd., Warwick, Queensland, Australia) were used to estimate the fraction of sapwood functional in water transport for overstory species on the TDE. Each probe consisted of two thermistors, one 10 mm upstream and the other 5 mm downstream from a central heater, and probes were implanted to each of four predetermined depths in the sapwood (Hatton et al. 1990). Insertion depths were calculated so that thermistors were optimally located within concentric annuli of equal sapwood area. Sap velocity rates were estimated as outlined by Barrett et al. (1995), and all estimates were corrected for probe-implantation effects (Swanson and Whitfield 1981). An increment bore was used to extract a sample of sapwood near where the probes were implanted for determination of wood density and water content. The fraction of sapwood functional in water transport (a value between 0 and 1) was calculated for each measurement tree (Wullschleger and King 2000).

Species-Specific Estimates of Sapwood Area

Species-specific allometric equations were developed for estimating sapwood area from tree diameters. Diameter and sapwood thickness were measured on trees growing in closed-canopy stands on Walker Branch Watershed near Oak Ridge, Tennessee, Morgan Monroe State Forest near Bloomington, Indiana, and the Fernow Experimental Forest near Parsons, West Virginia. Trees from a lumber mill in Oak Ridge, Tennessee were also sampled. Stem diameter was measured with a diameter tape, whereas sapwood and bark thickness were measured with a digital caliper. Calculated sapwood areas were fitted to the equation (Vertessy et al. 1995)

$$\mathrm{SA} = B_0 \times \mathrm{dbh}^{B_1}, \tag{2}$$

where SA is the sapwood area (cm^2), dbh is measured tree diameter at breast height (cm), and B_0 and B_1 are species-specific coefficients, as determined with regression techniques.

Comparison of Results to Long-Term Measurements of Net Gain and to Model Simulations

Estimates of annual water use or, more specifically, transpiration, which were derived from measurements of sap flow and cross-sectional sapwood area, were compared to annual estimates of net gain for two forested catchments in the eastern United States. Net gain is the difference between precipitation and stream flow, which, on an annual basis, is related to the loss of water from the landscape via evapotranspiration (ET) (Luxmoore and Huff 1989). Two long-term datasets were used to compare catchment-scale estimates of ET with sap-flow-derived estimates of transpiration. One dataset was a 28-year record of ET for Walker Branch Watershed in eastern Tennessee (Wilson et al. 2001) and the other was a 40-year record for the Fernow Experimental Forest in northern West Virginia (Adams et al. 1994). Because transpiration is only one component of ET (the others being evaporation and interception), we also compared stand transpiration for the TDE with simulated rates of annual transpiration obtained using the LINKAGES v2.2 model of forest succession (Chapter 25, this volume). This model was derived from the original version of LINKAGES (Pastor and Post 1985) and, as discussed in Chapter 25, was modified to study the effects of climate change (i.e., temperature and precipitation) and interannual and intra-annual variations in climate on long-term forest dynamics.

Results

Allometric Equations and Estimates of Stand Sapwood Area

Sapwood areas for species on the TDE were reliably described as a function of measured stem diameter (Table 21.1) (Wullschleger et al. 2001). Sapwood area in the diffuse-porous species *Acer rubrum* varied from 50 to 2280 cm^2 across a range of stem diameters (8–66 cm), whereas sapwood area in the ring-porous species *Quercus prinus* were, for the diameters measured (up to 81 cm), typically 960 cm^2 or less. Sapwood areas for similar-sized individuals were threefold to fourfold greater in diffuse-porous species like *A. rubrum, A. saccharum, Liriodendron tulipifera,* and *Nyssa sylvatica* than in ring-porous trees, which included *Quercus* species. Sapwood thickness increased with increasing stem diameter in the diffuse-porous trees (up to 15 cm for a 78-cm dbh *L. tulipifera* tree), but varied little across a range of stem diameters for the ring-porous *Quercus* (average 2.4 cm). The relationship between sapwood area and stem diameter for many of the species on the TDE could be mathematically described using Eq. (1). Coefficients of determination (R^2) for these relationships varied from 0.98 in *A. saccharum* to 0.77 in *Q. rubra* and were generally higher for the diffuse-porous trees than they were for ring-porous species (Table 21.1).

Not all sapwood area predicted from the allometric relationships was equally functional in water transport (i.e., f_s). Measured rates of sap velocity were greatest

Table 21.1. Allometric equations describing the relationship between cross-sectional sapwood area (cm^2) and stem diameter at breast height (cm) for 11 temperate tree species present on the TDE plots

Species	Sample size	Eq. (1) coefficients			f_s
		B_0	B_1	R^2	
A. rubrum	33	0.7579	1.9171	0.97	0.74
A. saccharum	52	0.6981	1.9677	0.98	0.65
Carya spp.	51	3.1640	1.3754	0.89	0.71
L. tulipifera	75	0.3117	2.0561	0.93	0.66
N. sylvatica	14	1.0372	1.7958	0.95	0.74
Prunus serotina	52	0.4758	1.7207	0.86	0.74
Pinus taeda	51	0.2277	2.2472	0.97	0.69
Q. alba	45	0.8175	1.5483	0.89	na
Q. coccinea	35	0.0823	2.0883	0.75	na
Q. prinus	49	0.4249	1.7154	0.92	na
Q. rubra	61	0.8454	1.5075	0.77	na

Note: The fraction of sapwood area functional in water transport (f_s) is also shown.

for probes located in the two outer sapwood annuli and lowest for probes in closest proximity to the heartwood. Radial profiles of sap velocity indicated that f_s varied from 0.66 for *L. tulipifera* to 0.74 for *Prunus serotina, A. rubrum,* and *N. sylvatica* (Table 21.1). Gradients in sap velocity were also observed for ring-porous *Quercus* (data not shown), but because the thermal dissipation probes used to measure sap velocity spanned the entire width of the sapwood in these species, no adjustment factor was used in scaling sap-flow data to stand-level estimates of water use.

Application of the allometric equations to stand inventory data indicated that the total overstory sapwood area was 8.9 m^2 ha^{-1} (Fig. 21.1). An estimated 6.4 m^2 ha^{-1} of this area was contributed by diffuse-porous trees and 2.5 m^2 ha^{-1} by ring-porous species. *A. rubrum* (2.3 m^2 ha^{-1}), *N. sylvatica* (2.1 m^2 ha^{-1}), *Q. prinus* (1.1 m^2 ha^{-1}), and *Q. alba* (0.9 m^2 ha^{-1}) accounted for 70% of overstory sapwood area. Saplings in the understory, mostly dominated by *A. rubrum,* contributed an additional 3.0 m^2 ha^{-1} sapwood area. Total sapwood area, therefore, for the stand was 11.9 m^2 ha^{-1} or 53% of the stand basal area (22.4 m^2 ha^{-1}).

Estimates of Forest Water Use

Forest water use scaled from measurements of sap flow on individual trees approached and/or slightly exceeded 3 mm d^{-1} for all plots on the TDE (Fig. 21.2).

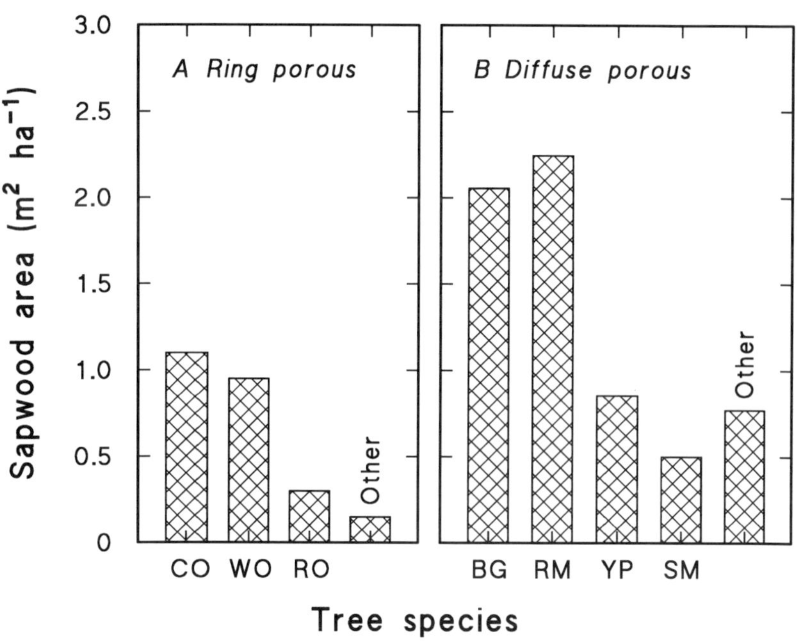

Figure 21.1. Estimated sapwood areas for (A) ring-porous and (B) diffuse-porous trees. Species are abbreviated: CO, *Q. prinus;* WO, *Q. alba;* RO, *Q. rubra;* BG, *N. sylvatica;* RM, *A. rubrum;* YP, *L. tulipifera;* and SM, *A. saccharum.* (From Wullschleger et al. 2001, with permission.)

Such rates were typical for sunny days from May to July, and there were strong linear relationships between stand water use and both mean daily vapor-pressure deficit and radiation during this period (Figs. 21.3A and 21.3B). Linear regressions fitted to these data sets yielded two equations; one describing the dependency of stand water use on mean daily vapor-pressure deficit (VPD),

$$\text{Daily water use (mm d}^{-1}) = -0.022 + 1.99\text{VPD}, \qquad R^2 = 0.91, \qquad (3)$$

and another describing the dependency of stand water use on mean daily radiation (Rn),

$$\text{Daily water use (mm d}^{-1}) = -1.08 + 0.011\text{Rn}, \qquad R^2 = 0.87, \qquad (4)$$

where mean daily vapor-pressure deficit and radiation refer to values averaged only over daylight hours. Although water use rates were similar for the three TDE plots during much of the year, beginning in mid-July and continuing into September, stand water use on the dry plot declined relative to the two other plots (Fig. 21.2). As a result, considerable scatter was introduced into the relationships between water use and VPD, and between water use and Rn (Figs. 21.3 C and 21.3D).

Treatment-specific differences in stand water use increased throughout August and September, when rates of water use on the dry plot declined to ~25% of those estimated for the wet and ambient plots (Fig. 21.4). Integrated over the growing season, water use for overstory trees on the dry plot was 25 to 27% lower than those estimated for the ambient and wet TDE treatments (Table 21.2). When the estimated water use of understory saplings was included, annual forest water use was 243 mm for the ambient plot, 257 mm for the wet plot, and 180 mm for the dry plot. These differences translated to 25–30% lower rates of seasonal water use for dry plots compared to wet and ambient treatment plots. Analysis of the data restricted to after the start of the drought (July 15) and continuing through the remainder of the growing season revealed that treatment differences were much more dramatic. Summed over ~105 days, stand water use was 130 mm

Table 21.2. Stand water use (mm season^{-1}) for each of the three TDE plots during the 2000 field season

Component	Wet	Ambient	Dry
Overstory			
Ring-porous species	30	29	22
Diffuse-porous species	178	172	128
Total overstory	208	201	150
Understory	49	42	30
Stand water use	257	243	180

Note: Sap-flow rates for individual trees were measured with thermal dissipation probes and then scaled to the stand on the basis of estimates of sapwood area.

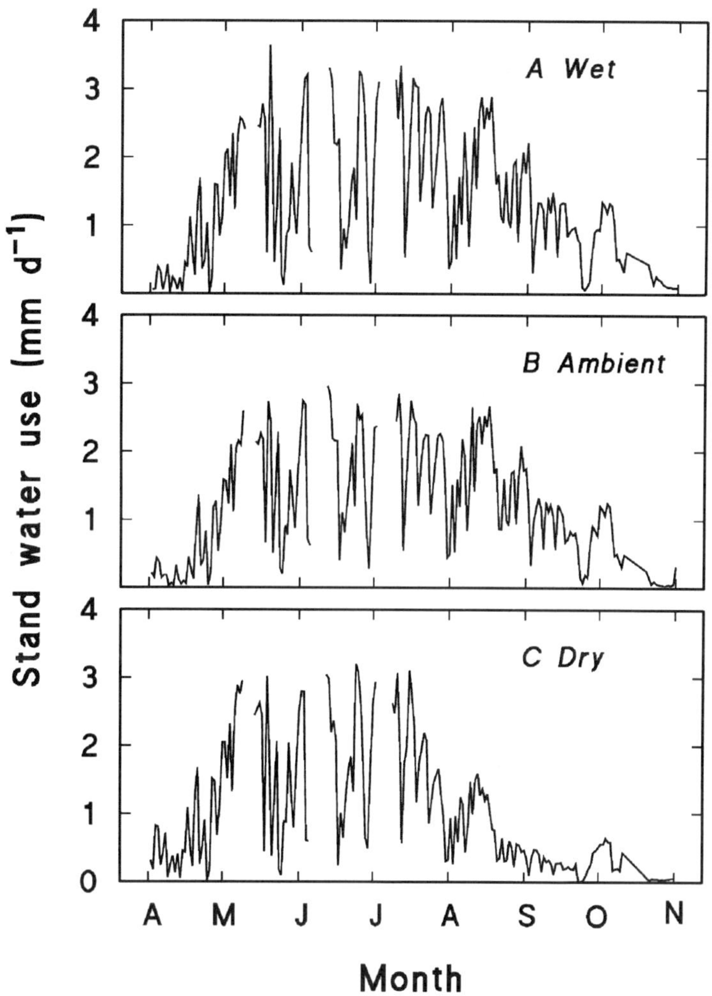

Figure 21.2. Seasonal pattern of forest water use for the (A) wet, (B) ambient, and (C) dry TDE plots during the 2000 growing season.

and 118 mm for trees on the ambient and wet plots, respectively, but only 65 mm for trees on the dry plot.

Forest water use expressed on a daily basis during the 2000 season declined strongly as a function of soil-water potential (Fig. 21.5). Rates of stand water use were maintained at maximum levels ($T/T_{max} = 1.0$) only as long as soil-water potential was greater than ~ -0.3 MPa. However, reductions in soil-water potential below this level brought about rapid and marked reductions in water use, such that, at a soil water potential of -1.0 MPa, the relative water use was 0.30 or 30% of unstressed levels. Further declines in soil-water potential to -1.5 MPa

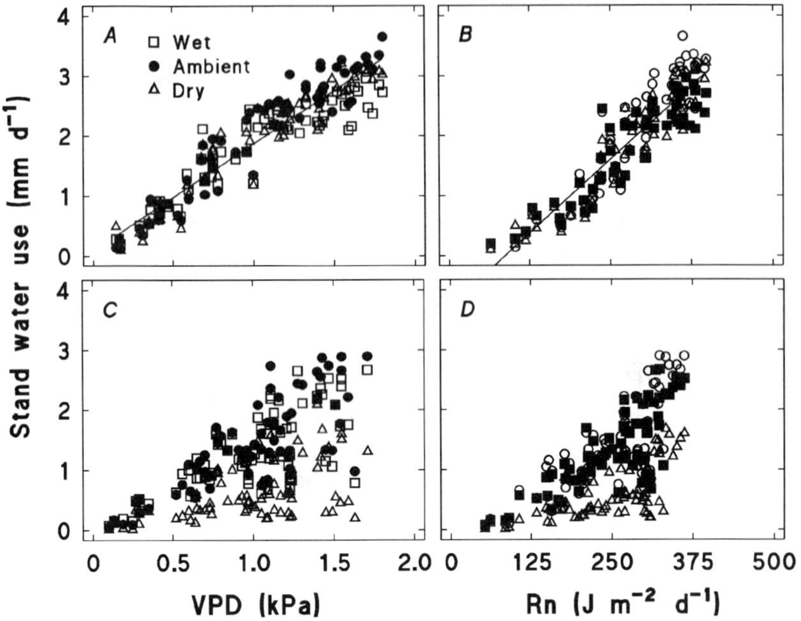

Figure 21.3. Relationship of stand water use to vapor-pressure deficit (VPD) and radiation (Rn) either before (A and B) or during (C and D) a late-season drought. Linear regression was used to determine best-fit equations for the relationship of stand transpiration to VPD or Rn.

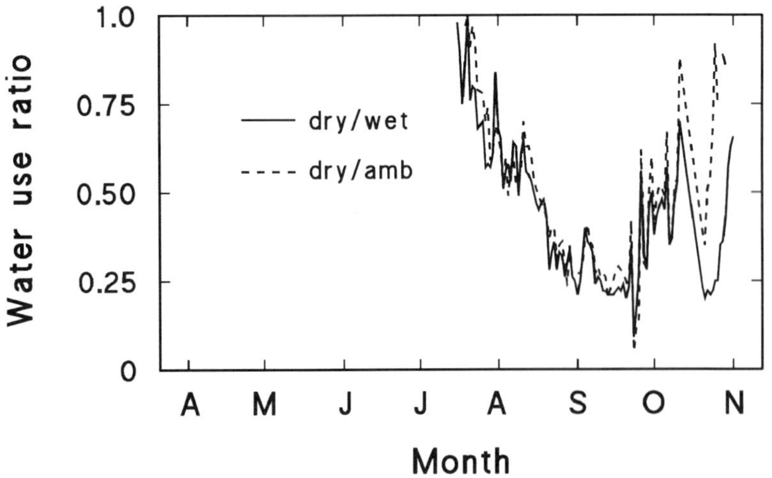

Figure 21.4. Seasonal differences in stand transpiration for the TDE plots. Relative differences are shown for the wet and dry treatments and for the ambient and dry treatments.

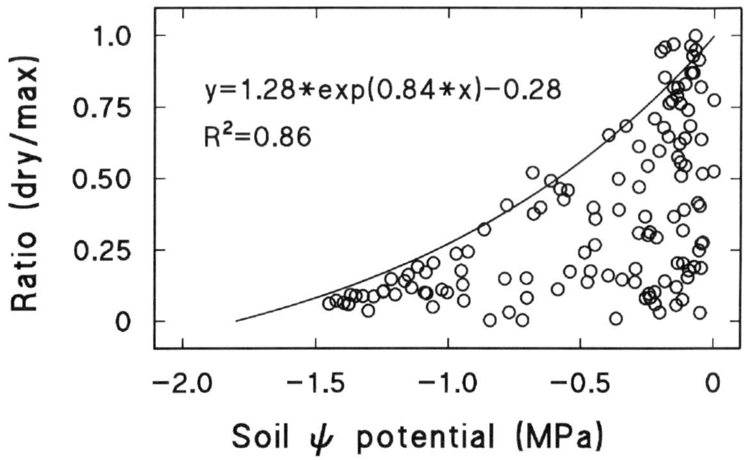

Figure 21.5. Stand water use (expressed as a relative estimate) as a function of soil-water potential for the TDE.

and below (typical of dry years) reduced observed rates of forest water use to less than 15% of unstressed rates.

Discussion

Whole-tree measurements of sap flow with thermal dissipation probes showed that stand water use was highly responsive to differences in soil and climatic conditions during the years of the TDE study. During wet growing seasons like 1996 and 1997, significant treatment responses were not observed for trees on the TDE (Wullschleger et al. 2000; Wullschleger et al. 2001), whereas treatment differences were documented in years with mid- to late-season droughts. In 2000, treatment differences strongly coincided with below-average precipitation during the later half of the season (after July 15) and the subsequent development of differences in soil-water potential among the three TDE treatments. Minimum soil-water potentials, which in 2000 occurred during late August and September, were −1.2 MPa for the wet plot, −1.1 for the ambient plot, and −1.6 for the dry plot.

In an effort to characterize the seasonal severity of interannual and intra-annual drought on the TDE, Hanson et al. (2001; Chapter 2, this volume) used both measured and modeled soil-water potentials (0–35 cm) to derive what they termed a water stress integral. The water stress integral (expressed in units of MPa d) differed between wet (−24 in 1994) and dry years (−133 in 1995) and, not surprisingly, differed among treatments on the TDE. For the year 2000, we calculate that the annual water stress integral for the wet, ambient, and dry plots was −40, −49, and −122, respectively. The water stress integral for the dry plot was ~149% lower than that observed for the ambient plot, whereas for the wet plot, it was only 18% higher compared to the ambient plot. This greater separation in treatment effects between the ambient and dry plots compared to the ambient and wet plots is consistent with our stand water use data where differences were comparatively greater between the ambient and dry plots (−26%) than they were for the ambient and wet plots (6%). Furthermore, Hanson et al. (2001) noted that in almost all of the dry years, late-season droughts contributed disproportionately to the annual water stress integral for plots on the TDE. In support of this observation, we estimate, that for the year 2000, almost 85% of the annual water stress integral accumulated after July 15. This statistic is also well supported by the late-season reductions in stand water use that we observed for the wet, ambient, and especially the dry TDE plots.

Although studies on forest water use have been conducted and comparisons made between wet and dry years (Bréda et al. 1993; Cienciala et al. 1997; David et al. 1997), few have actually manipulated precipitation to quantify impacts on stand transpiration. Cienciala et al. (1997) evaluated the effects of water availability on transpiration in a Norway spruce stand subjected to irrigated, control, and drought conditions (two-thirds removal of precipitation) and observed that seasonal water use was 380 mm, 325 mm, and 209 mm for the three treatments, respectively. Stand water use in the drought-treated plot was reduced by slightly more than 35% compared to the control treatment. In another manipulative study, Irvine et al. (1998) compared stand water use in a Scots pine plantation for drought-treated plots and those that received average precipitation and observed that, between mid-May and late August, 140 mm of water was transpired in the control plot, whereas only 93 mm was transpired in the drought treatment (a 33% difference). These data are consistent with the results of our studies, where, integrated over the growing season, water use for the dry plot was 25–30% lower than that estimated for the ambient and wet TDE treatments.

Many studies, including ours, have documented the sensitivity of forest water use to soil-water availability (David et al. 1997; Cienciala et al. 1997; Irvin et al. 1998). Although there is a high degree of consensus regarding the fact that forest water use declines with increasing soil-water deficits, the manner by which this dependency should be characterized is not clear. Some express a relative change in water use on the basis of available soil water (Teskey and Sheriff 1996) or soil-water content (Granier et al. 2000; Irvine et al. 1998), whereas others describe the response based on measurements of mid-day or predawn leaf water potential (David et al. 1997; Lu et al. 1995). Irvine et al. (1998) reported that the transpiration rate declined once volumetric water content for the top 20 cm of soil reached a threshold of 12%. In our study, we expressed the decline in relative water use in terms of soil-water potential (see Fig. 21.5). However, if we reexamine our data based on original measurements of soil-water content, we observe an abrupt reduction in relative stand water use as the volumetric water content in the top 35 cm of soil declines below about 14%. Therefore, even modest decreases in soil-water content or potential can, for the stands examined in the TDE, result in marked reductions in forest water use.

Although we did not monitor stand water use on the TDE for the entire 8 years of the project, relationships developed on the 2000 dataset can be used to evaluate the impact of natural and imposed drought on forest water use on the TDE. With the equation that describes the relationship between stand water use and mean daily VPD [Eq. (3)], data on soil-water potential (Fig. 21.5), and fractional leaf area development (Chapter 17, this volume), an estimate of stand water use for each TDE plot and each year could be derived (Table 21.3). In general, between 1993 and 2000, stand water use for the ambient plot varied from 219 mm in 1995 (a dry year) to 277 mm in 1997 (a wet year). Rates of stand water use also increased as a result of applying supplemental precipitation to the wet plot, with rates increasing 3–14% compared to the ambient plot. In three of the wettest years (1994, 1996, and 1997) differences between the wet and ambient plots were 3–4%. Simulations of water use for the dry plot, however, indicated that throughfall removal resulted in a dramatic decrease in water use compared to the ambient plot, on the order of −8% to −28% (Table 21.3).

The estimates of transpiration that we derived in this study using a simple model driven by daily VPD, fractional leaf area index, and soil-water potential represent only a fraction of the total amount of water used by forest stands. As such, transpiration should be less than ET, with the difference between ET and transpiration being accounted for by water lost through evaporation and interception (Chapter 25, this volume). Comparing estimates of annual transpiration for the ambient plot of the TDE to long-term datasets from Walker Branch Watershed and the Fernow Experimental Forest shows that transpiration was indeed less than estimates of ET derived at the catchment scale (Fig. 21.6). For the period 1993–2000, simulated rates of transpiration under ambient precipitation on the TDE equaled 249 mm and ranged from 219 in 1995 to 277 mm in 1997. During this same period, annual ET for Walker Branch Watershed averaged 611 mm and, thus, transpiration was ~41% of ET.

Luxmoore and Huff (1989) summarize many years of hydrologic research on the Walker Branch Watershed and used long-term records of net gain, along with model simulations and measurements of soil-water content, to derive estimates of forest water balance. These authors conclude that annual ET for this mixed-

Table 21.3. Simulated water use (mm season^{-1}) for each of the three TDE plots since 1993

				Percent change	
Year	Ambient	Wet	Dry	Ambient to dry	Ambient to wet
1993	261	288	213	−23	+10
1994	259	267	237	−8	+3
1995	219	255	158	−28	+16
1996	244	254	219	−10	+4
1997	277 (267)[a]	287	250	−10	+4
1998	236 (230)[b]	268	176	−25	+14
1999	263 (269)[b]	289	210	−20	+10
2000	232 (243)[c]	258 (257)[c]	204 (180)[c]	−21	+11

Note: A series of equations were established that related climate, soil-water potential, leaf area development, and canopy transpiration, and then growing season estimates of water use were determined.

[a] Estimate of stand water use from Wullschleger et al. (2000).

[b] Estimate of stand water use from Wilson et al. (2001).

[c] This study.

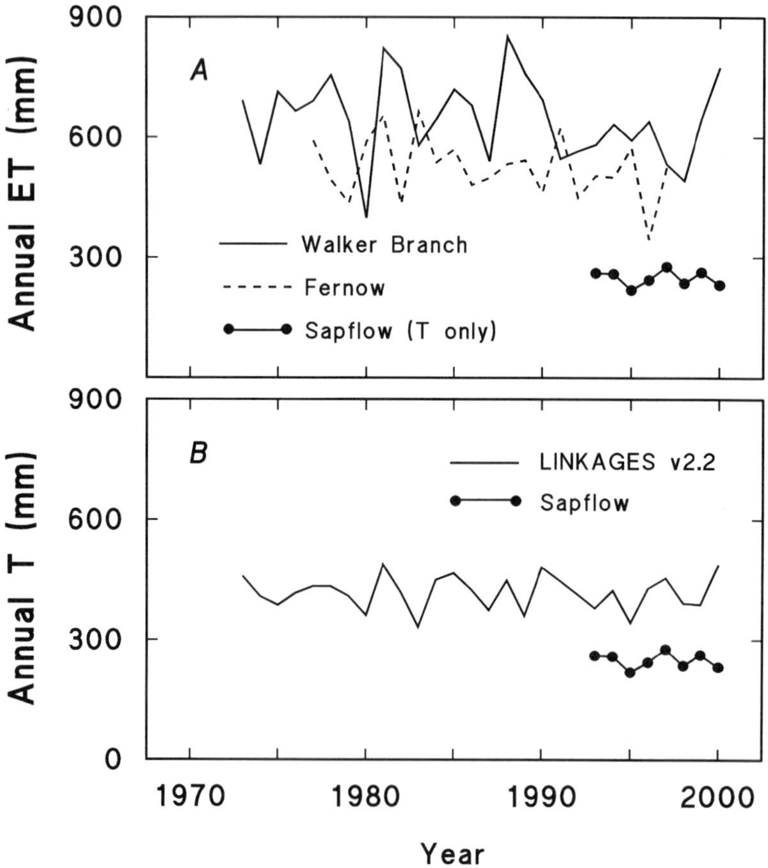

Figure 21.6. Sap-flow-derived estimates of transpiration compared to estimates of (A) annual stand ET, as derived from long-term measurements of precipitation and streamflow (i.e., net gain) at two forest sites, and (B) annual transpiration, as simulated with the LINKAGES v2.2 model of forest succession.

deciduous forest equaled 730 mm, with individual components being ~145 mm for canopy interception, 50 mm for soil evaporation, and the remaining 535 mm for transpiration. From these data, we calculate that transpiration was 73% of forest ET, a much higher fraction than the 41% we derive from sap-flow measurements. Wilson et al. (2001) used eddy-covariance techniques and computer simulations to estimate individual components of ET for Walker Branch Watershed and showed, for the years 1998 and 1999, that transpiration varied between 357 and 409 mm or 65% and 68% of ET in the two years, respectively. These estimates of transpiration, obtained using independent techniques, are higher than those derived from sap-flow measurements. Similarly, estimates of transpiration as modeled with the forest succession model LINKAGES v2.2 are also higher than those estimated from sap-flow measurements (Fig. 21.6). We are, unfortunately, currently unable to reconcile these discrepancies.

Although the TDE was designed to address mechanisms whereby trees and stands adjust to altered precipitation regimes, it was always an objective to address how site-specific studies, such as those conducted here, could contribute to understanding the response of forest water use to drought over broad regions of the United States. Hanson and Weltzin (2000) classified much of the eastern United States as a region where random, occasional droughts occur, yet they are limited in duration and spatial extent and tend to occur late in the growing season. These areas, of which the TDE site is typical, are embedded in a broader region where late-summer droughts are common and can be expressed across quite large spatial scales. Evidence for such regional definitions are well supported by the drought-induced growth responses derived from tree-ring chronologies (Abrams et al. 1998; Orwig and Abrams 1997). Evidence for interannual variation in stand water use is also seen in 2 years for multiple pine and hardwood stands in the eastern United States (Fig. 21.7). Results from those measurement campaigns in 1998 and 1999 showed that year-to-year differences in water use ranged from −5% to

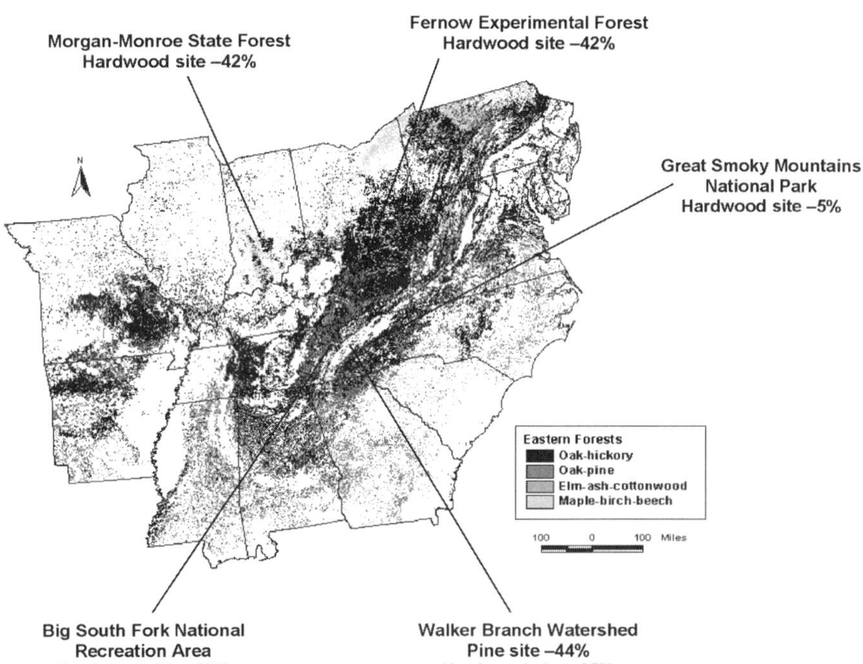

Figure 21.7. Regional perspective of stand water use as obtained in a series of distributed sap-flow measurement campaigns in 1998 and 1999. The data from each of six sites were obtained as part of a NASA-sponsored project (Hanson et al., unpublished data), and data show percent differences in forest water use between the 2 years. Forest type and distribution are as described in Fig. 2.1 (Chapter 2, this volume).

−44% between wet and dry years, and that marked variation was observed across broad regions of the eastern United States.

Summary

Differences in stand water use between trees on the ambient and dry plot averaged 20–25% on a seasonal basis and 45% if only the latter portion of the season was considered. Changes of this magnitude could have serious implications for catchment water yield, hydrologic budgets across broad spatial scales, impacts on terrestrial vegetation, and consequences for ecosystem goods and services upon which society depends. Furthermore, such differences clearly point out the significance that late-season drought can have on the water resources of temperate forests in the eastern United States. Unfortunately, end-of-season droughts are problematic to study given that changes in canopy water use and leaf area index (LAI), as a result of natural or premature senescence, take place simultaneously. Therefore it is important to track changes in forest water use caused by soil water and LAI independently. This task can prove difficult. One way to separate these processes would be to manipulate soil water such that changes in forest water use occur much earlier in the growing season, prior to leaf senescence. This would have the additional feature that drought and growth increment would be better coordinated in time and, therefore, their impacts would be easier to diagnose.

References

Abrams MD, Ruffner CM, Morgan TA (1998) Tree-ring responses to drought across species and contrasting sites in the ridge and valley of central Pennsylvania. For Sci 44:550–558.

Adams MB, Kochenderfer JN, Wood F, Wood TR, Edwards P (1994) *Forty years of hydrometeorological data from the Fernow Experimental Forest, West Virginia.* General Technical Report NE-184. USDA, Forest Service, Northeastern Forest Experimental Station, Radnor, Pennsylvania.

Barrett DJ, Hatton TJ, Ash JE, Ball MC (1995) Evaluation of the heat pulse velocity technique for measurement of sap flow in rainforest and eucalypt forest species of south-eastern Australia. Plant Cell Environ 18: 463–469.

Bréda N, Cochard H, Dreyer E, Granier A (1993) Water transfer in a mature oak stand (*Quercus petraea*): Seasonal evolution and effects of a severe drought. Can J For Res 23:1136–1143.

Calder IR (1998) Water use by forests, limits and controls. Tree Physiol 18:625–631.

Cienciala E, Lindroth A, Cermák J, Hallgren JE, Kucera J (1994) The effects of water availability on transpiration, water potential and growth of *Picea abies* during a growing season. J Hydrol 155:57–71

Cienciala E, Kucera J, Lindroth A, Cermák J, Grelle A, Halldin S (1997) Canopy transpiration from a boreal forest in Sweden during a dry year. Agric For Meteorol 86:157–167.

David TS, Ferreira MI, David JS, Pereira JS (1997) Transpiration from a mature *Eucalyptus globulus* plantation in Portugal during a spring-summer period of progressively higher water deficit. Oecologia 110:153–159.

Falge E, Tenhunen JD, Ryel R, Alsheimer M, Köstner B (2000) Modeling age-and density-related gas exchange of *Picea abies* canopies in the Fichtelgebirge, Germany. Ann For Sci 57:229–243.

Granier A (1987) Evaluation of transpiration in a Douglas-fir stand by means of sap flow measurements. Tree Physiol 3:309–320.

Granier A, Loustau D, Bréda N (2000) A generic model of forest canopy conductance dependent on climate, soil water availability and leaf area index. Ann For Sci 57:755–765.

Hanson PJ, Weltzin JF (2000) Drought disturbances from climate change: Response of United States forests. Sci Total Environ 262:205–220.

Hanson PJ, Todd DE, Amthor JS (2001) A six-year study of sapling and large-tree growth and mortality responses to natural and induced variability in precipitation and throughfall. Tree Physiol 21:345–358.

Hatton TJ, Catchpole EA, Vertessy RA (1990) Integration of sapflow velocity to estimate plant water use. Tree Physiol 6:201–209.

Irvine J, Perks MP, Magnani F, Grace J (1998) The response of *Pinus sylvestris* to drought: Stomatal control of transpiration and hydraulic conductance. Tree Physiol 18:393–402.

Kostner BM, Schulze ED, Kelliher FM, Hollinger DY, Byers JN, Hunt JE, McSeveny TM, Meserth R, Weir PL (1992) Transpiration and canopy conductance in a pristine broad-leaved forest of *Nothofagus*—An analysis of xylem sap flow and eddy-correlation measurements. Oecologia 91:350–359.

Lu P, Biron P, Bréda N, Granier A (1995) Water relations of adult Norway spruce [*Picea abies* (L) Karst] under soil drought in the Vosges mountains: Water potential, stomatal conductance and transpiration. Ann Sci For 52: 117–129.

Luxmoore RJ, Huff DD (1989) Water. In Johnson DW, van Hook RI (Eds) *Analysis of biogeochemical cycling processes in Walker Branch Watershed.* Springer-Verlag, New York, pp 165–196.

Orwig DA, Abrams MD (1997) Variation in radial growth responses to drought among species, site, and canopy strata. Trees 11:474–484.

Pastor J, Post WM (1985) *Development of a linked forest productivity-soil process model.* ORNL/TM-9519. Oak Ridge National Laboratory, Oak Ridge, Tennessee.

Pataki DE, Oren R, Smith WK (2000) Sap flux of co-occurring species in a western subalpine forest during seasonal soil drought. Ecology 81:2557–2566.

Swanson RH, Whitfield DWA (1981) A numerical analysis of heat pulse velocity theory and practice. J Exp Bot 32:221–239.

Teskey RO, Sheriff DW (1996) Water use by *Pinus radiata* trees in a plantation. Tree Physiol 16:273–279.

Vertessy RA, Benyon RG, O'Sullivan SK, Gribben PR (1995) Relationship between stem diameter, sapwood area, leaf-area and transpiration in a young mountain ash forest. Tree Physiol 15:559–567.

Wilson KB, Hanson PJ, Mulholland PJ, Baldocchi DD, Wullschleger SD (2001) A comparison of methods for determining forest evapotranspiration and its components: sap-flow, soil water budget, eddy covariance and catchment water balance. Agric For Meteorol 106:153–168.

Wullschleger SD, King AW (2000) Radial variation in sap velocity as a function of stem diameter and sapwood thickness in yellow-poplar trees. Tree Physiol 20:511–518.

Wullschleger SD, Wilson KB, Hanson PJ (2000) Environmental control of whole-plant transpiration, canopy conductance and estimates of the decoupling coefficient for large red maple trees. Agric For Meteorol 104:157–168.

Wullschleger SD, Hanson PJ, Todd DE (2001) Transpiration from a multi-species deciduous forest as estimated by xylem sap flow techniques. For Ecol Management 143:205–213.

22. Estimating the Net Primary and Net Ecosystem Production of a Southeastern Upland *Quercus* Forest from an 8-Year Biometric Record

Paul J. Hanson, Nelson T. Edwards, Timothy J. Tschaplinski, Stan D. Wullschleger, and J. Devereux Joslin

Introduction

Although 8 years of manipulated precipitation levels have not produced significant tree-growth responses in an upland *Quercus* forest in eastern Tennessee (Hanson et al. 2001), interannual differences in physiological processes (Chapters 3, 4, 6–8, and 10, this volume), storage carbohydrates (Chapter 5, this volume), tree and sapling growth rates (Chapter 15, this volume), root growth (Chapter 16, this volume), and foliar-litter production (Chapter 17, this volume) suggest that interannual differences in net primary production (NPP) and net ecosystem production (NEP) are likely. By accepted definition, NPP is the difference between carbon gain from autotrophic organisms (i.e., gross primary production, GPP) minus autotrophic respiration (R_{auto}). NEP is the annual net change in ecosystem carbon storage defined as NPP minus heterotrophic respiration (R_{hetero}). At the local scale, NEP may also be reduced from non-CO_2 losses of volatile organic compounds (VOCs) from vegetation (Guenther et al. 1996; Isebrands et al. 1999) and soils (Hanson and Hoffman 1994) and under ambient or flooded conditions from the gain or loss of CH_4 from soils, respectively (LeMer and Roger 2000). The following equations capture these processes:

$$\text{NPP} = \text{GPP} - R_{auto}, \tag{1}$$

$$\text{NEP} = \text{NPP} - R_{hetero} - \text{VOC} \pm \text{CH}_4. \tag{2}$$

Heterotrophic respiration includes all nonautotrophic production of CO_2, but vertebrate organisms are often ignored because their impact on carbon flux is small with respect to the magnitude of ecosystem carbon fluxes. Faunal-induced movement of carbon sources (e.g., seed transport by birds and squirrels) is assumed to be random, with losses to a particular area balancing gains from other locations. Loss of dissolved organic carbon into deep soils below the rooted soil profile is also commonly ignored, but specific quantitative estimates of this process are lacking.

Although the definitions behind Eqs. (1) and (2) are simple to conceptualize, direct measurements of NPP or NEP are not straightforward. Clark et al. (2001a) concluded that, although it is impossible to measure the annual NPP of an ecosystem directly, close approximations of NPP (i.e., NPP*) could be attained from detailed measurements of the components of stand production, including the following: stem, branch, and stump growth; leaf, flower, and seed production; fine and coarse root production, estimates of leaf production lost to leaf herbivory; and other, minor components (e.g., pollen production). Curtis et al. (2002) proposed that NPP could be derived from incremental changes of live plant mass, detritus, and estimates of herbivory losses and termed such estimates of NPP* as biometric measurements of NPP. They concluded that a biometric estimate of NEP (NEP*) could also be derived as follows:

$$\mathrm{NEP}^* = \mathrm{NPP}^* - R_{\mathrm{hetero}}, \tag{3}$$

where R_{hetero} is obtained from integrated annual measurements of soil respiration minus the autotrophic root respiration (Hanson et al. 2000).

In this chapter, we apply biometric approaches (Curtis et al. 2002) to multiyear growth measurements of the Walker Branch Throughfall Displacement Experiment (TDE) (Hanson et al. 2001; Chapters 2, 15, and 17, this volume) for the determination of standing carbon stocks of an upland *Quercus* forest and for the interannual estimates of net primary production (Clark et al. 2001a) for 1993–2000. Additional estimates of NEP* are calculated from the difference between NPP* and interpolated data for ecosystem heterotrophic respiration. The magnitude and multiyear patterns of NPP* for the upland *Quercus* forest of the TDE are compared to data published for other temperate deciduous forests. Similarly, NEP* data for 1995–1999 are contrasted to the magnitude and the interannual pattern of independent eddy-covariance-based net ecosystem exchange data for Walker Branch Watershed (Wilson and Baldocchi 2001).

Methods

Site Description

The upland *Quercus* forest considered in this chapter is located on the Walker Branch watershed (35°58′ N and 84°17′ W) in the Ridge and Valley Province of East Tennessee (Johnson and Van Hook 1989). The local topography is represented by a series of repeating parallel ridges and valleys oriented from southwest to northeast and elevation ranges from 270 to 340 m. The soils are fairly acidic (pH 3.5–4.6), typic paleudults with a depth to bedrock of ~30 m. The long-term mean annual precipitation is 1358 mm and the mean temperature is 14.2°C. The site is dominated by *Quercus alba* L., *Q. prinus* L., and *Acer rubrum* L., and understory saplings are predominantly *A. rubrum* and *Cornus florida* L., with scattered individuals of less common species. Total stand basal-area averages 20–25 m^2 ha^{-1}. Further details on the stand composition can be found in Hanson et al. (2001) and in Chapter 2 (this volume).

Forest Carbon Stocks and Biometric NPP*

Direct biometric estimates of standing carbon pools and annual NPP* were determined from an 8-year (1993–2000) record of plant growth and litter production as a part of a multiyear TDE (Hanson et al. 2001; Joslin et al. 2000). The components of annual NPP* that were considered for use in this analysis include the following items consistent with the recommendations of Clark et al. (2001a):

1. Allometric estimates of overstory branch, stem, and stump growth
2. Allometric estimates of sapling growth
3. Leaf production corrected for herbivory losses
4. Terminal branch litter production
5. Acorn production
6. Gross coarse and fine root production
7. The production of volatile organic compounds (VOCs)
8. Changes in nonstructural carbohydrate storage
9. Limited production of pollen and trichomes

Of these, only VOC, pollen, and trichome production are not included because they represent insignificant contributions to NEP* in this upland *Quercus* forest. On the basis of the experimental data of Harley et al. (1997), VOC production in the form of isoprene emissions was calculated to contribute less than 15 g C m^{-2} y^{-1} to annual NPP; because it is volatile, it leaves the local forest stand, resulting in no net change in ecosystem carbon stocks. Pier and McDuffie (1997) and Sharkey et al. (1996) independently estimated that VOC emissions would be unlikely to exceed 20 g C m^{-2} y^{-1}. VOC losses in the form of monoterpene evaporation from soils represent less than 1 g C m^{-2} y^{-1} (Hanson and Hoffman 1994). Methane gains or losses from the well-drained upland soils at this site are inconsequential with respect to the ecosystem carbon budget (Bradford et al. 2001).

Aboveground carbon stocks in wood were estimated at all sites with allometric equations relating diameter at breast height (dbh at 1.3 m) to bole, branch, and stump biomass. Allometric relationships for canopy trees (Harris et al. 1973) and saplings (Sollins and Anderson 1971) were developed from local tree harvests. Additional whole-tree harvests, conducted in 2001, confirmed that the allometric relationships of Harris et al. (1973) were not substantially different after 30 years

Table 22.1. Summary of regression coefficients for the relationship between mean tree or sapling component mass on tree diameter at breast height

	Constants				
Tree/sapling component (*Y*)	*a*	*b*	R^2	*N*	*k*
Tree coefficients					
Bole	−2.437	2.418	0.97	298	1.08
Branches	−3.188	2.226	0.91	298	1.26
Stump[a]	−2.392	1.845	0.77	18	1.06
Sapling coefficients					
Stem and branches	−3.166	2.866	0.94	122	na[b]

Note: The data for trees are reproduced with permission from Harris et al. (1973), and the coefficients for saplings were derived from sapling-dry-mass data published in Sollins and Anderson (1971). The regression takes the following form:

$$Y = e^{(a + \ln(X)\, b)}\, k,$$

where *Y* is the component mass (in kg), *X* is the dbh (in cm), and *a, b,* and *k* are constants.

[a] Harris et al. (1973) defined stump as the central root plus large support laterals to a radius of 60 cm.

[b] na = not applicable.

(data not shown). Coefficients for the allometric relationships for trees and saplings are provided in Table 22.1. Stem growth of trees > 0.1 m at dbh from 1993 to 2000 was evaluated with annual diameter measurements of all trees in the 1.92-ha experimental area (729 trees); however, stem-mass production for individual years was apportioned among years according to dendrometer-band observations collected on a subset of the trees (230 trees) (Hanson et al. 2001; Chapter 15, this volume). The annual increment of sapling-size vegetation (dbh < 0.1 m) was measured on randomly selected saplings along eight transects across the three plots. Herbaceous-layer plants (ferns and spring ephemerals) are not common on the southeast-facing slope that comprises the TDE study area and were not summarized as a separate carbon stock.

Fallen leaves and other materials (twigs, seeds, etc.) were collected from 147 baskets at fixed locations distributed across the TDE experimental area from 1992 through 2000, as described by Hanson et al. (Chapter 17, this volume). Once collected, the litter samples were dried and sorted to separate the foliage, terminal branch, and acorn components of annual NPP*. Loss of leaf area to herbivores was estimated by collecting leaves prior to abscission and comparing their remaining area to that of undamaged leaves of the same size. Herbivore losses for the TDE forest were reported as the mean of those measured in 1992, 1993, and 1994 (Shure et al. 1998; Chapter 18, this volume). They showed that annual reductions in foliar leaf area ranged from 6% to as much as 13% for *A. rubrum* and *Quercus* species, respectively.

Changes in the branch and bole sapwood total nonstructural carbohydrate pool (i.e., sugars and starch; TNC) were evaluated by Tschaplinski and Hanson (Chapter 5, this volume) during February of 1993, 1996, 1998, 1999, and 2000. These measures of carbon content were extrapolated to the stand level with the dbh-based relationships described by Wullschleger et al. (Chapter 21, this volume).

The amount of carbon stored belowground as wood (i.e., stumps and very large lateral roots) was derived from allometric equations relating dbh to woody-root mass (Harris et al. 1973). The annual increment in woody-root carbon was estimated from the annual change in dbh. The fine (< 2 mm) and coarse (2–5 mm) root carbon pools were measured from soil cores collected on an 8 × 8-m grid across the TDE area (Chapter 16, this volume). Annual fine root production was estimated from direct observation of root turnover with minirhizotron cameras (Joslin et al. 2000), and the annual increment of woody roots was assumed to be 20% of the aboveground wood increment (Curtis et al. 2002).

Initial mineral–soil carbon pools were derived from soil-profile data (organic matter and bulk density) for the Fullerton cherty-silt loam soil on Walker Branch watershed (Peters et al. 1970), and the dynamics of change with time were inferred from a 21-year study on other Walker Branch permanent plots (Trettin et al. 1999). The forest-floor organic layers were sampled in 1992 and 1999 with replicate (n = 15–30) 0.25-m^2 circular plots distributed across each treatment area of the TDE area (Johnson et al. 2002). Data from TDE site surveys of dead, coarse, woody litter in April 1992 (i.e., dead boles and branches) and organic soil horizons in January 1992 were used to establish the initial values for the respective forest carbon pools. Coarse-woody-debris carbon pools were quantified across the TDE area with the methods of Harmon and Sexton (1996).

Although the carbon content of the leaf, bark, wood, fine roots, and soil organic matter vary slightly from species to species and tissue to tissue (data not shown), we assume that all biomass is 47% carbon and calculate stand-level pools and annual fluxes of carbon as Mg C ha^{-1} and g C m^{-2} per ground area per year, respectively.

Interannual Estimates of NEP*

Annual estimates of NEP* require a measurement of ecosystem heterotrophic respiration for subtraction from NPP*. In practice, stand heterotrophic respiration is calculated as the sum of decomposition losses from soil and coarse, woody debris. Aboveground heterotrophic respiration is ignored in the current analysis because it has been estimated to be less than 1 g C m^{-2} y^{-1} (Reichle 1971; Edwards et al. 1980). Annual carbon losses from decomposition of the standing woody-debris pool were assumed to be ~10% per year based on observations that mass losses from bole sections of *Carya* and *Q. prinus* were 38 and 44%, respectively, over a 4-year period (Chapter 12, this volume). In a recent review of forest literature on soil respiration, Hanson et al. (2000) concluded that root respiration is often a dominant component of soil carbon cycles, typically accounting for 50% or more of the CO_2 leaving the forest floor. Direct point-in-time measurements of root contribution to soil respiration for the TDE site show substantial intra-annual variation, ranging from a high of 85% in the spring to a low of less than 20% in the fall (Trumbore et al., 2002). For the calculations in this chapter, we assumed that annual heterotrophic losses from soils were 50% of the annual soil-respiration totals from Hanson et al. (Chapter 10, this volume), consistent with the mean forest response reported by Hanson et al. (2000).

Results and Discussion

The Upland *Quercus* Forest and Its Carbon Pools

Prior to 1940, the experimental site was occupied by forest vegetation, but gaps were present because of past individual-tree logging associated with subsistence farming and grazing. Since 1940, natural successional processes have governed stand development, and the stand currently has a closed canopy with a mean annual leaf area index of $\sim$6 m^2 m^{-2}. Individual tree ages in this stand range from 50 to 100+ years, but the majority of the overstory trees range in age from 70 to 100 years. Carbon pools for the TDE forest are summarized in Table 22.2. To a soil depth of 1 m, the forest stand contains 171 Mg C ha^{-1}. Of this total, 58% is found in the vegetation biomass (nearly 50% in branch, bole, and stump wood), with the balance being stored in organic wood and forest-floor litter ($\sim$5%) and the mineral soil (37%). Because the potentially rooted soils at the TDE site are deep (up to 30 m), we also estimated the size of the soil carbon pool for a deep layer from 1 to 9 m, assuming a mean soil carbon concentration of 0.3%, a bulk density of 1.2 g cm^{-3}, and a coarse fraction content of 8%. Such a pool contains an additional 88 Mg C ha^{-1} and when included in the total stand carbon pool, it shifts the majority of stand carbon to the soils consistent with typical global patterns (Post et al. 1982, 1990; Dixon et al. 1994). Notwithstanding the function of deep soils as a large carbon reservoir, the turnover times of such carbon is very slow (Chapter 11, this volume) and changes in the deep-soil carbon pool are not considered to play a role in year-to-year estimates of NEP*. The turnover times of the 0–1-m mineral–soil carbon pool is also long (> 80 years; Chapter 11, this volume) and changes in the mineral soil carbon pool are not considered in the biometric estimates calculated below. In support of this assumption, Trettin et al. (1999) found little evidence for significant changes in the mineral–soil carbon pools of adjacent sites on Walker Branch from 1973 to 1993.

The total standing pool of carbon in the TDE forest is similar to that for six temperate forest stands of similar age found in the eastern United States (Table 22.2). Total stand carbon pools range from 90 to 130 Mg C ha^{-1}, and the aboveground totals range from 72 to 106 Mg C ha^{-1}. The *Quercus*-dominated and mixed-hardwood stands have 50% or more of their carbon stored in the living biomass, but the *Liriodendron* stands have the majority in the soil pool. The soil carbon of a *Populus* spp.-dominated northern hardwood stand (Curtis et al. 2002), not shown in Table 22.2, was also dominated by the soil carbon pool (i.e., only 39% of the carbon in living biomass). The carbon pools of these example stands are similar to recent estimates for the eastern United States derived from remote-sensing methods (Myneni et al. 2001).

Estimates of NPP*

Biometric estimates of NPP based on traditional methods ranged from 604 to 840 g C m^{-2} y^{-1} from 1993 to 2000 for a mean annual NPP value of 729 g C m^{-2} y^{-1} (Table 22.3). Percent contributions of leaf, wood, root, and acorn production

Table 22.2. Carbon pools in Mg ha^{-1} for the *Quercus* dominated TDE site in 1999 and similar temperate deciduous forests from other studies and time periods

Individual stands / Components	*Quercus* (this study)	*Quercus/Carya* (Edwards et al. 1989)	*Q. prinus* (Edwards et al. 1989)	*Quercus/Tsuga* (Curtis et al. 2002)	*Liriodendron* (Harris et al. 1975)	*Liriodendron* (Edwards et al. 1989)	Mixed species (Curtis et al. 2002)
Location	Tennessee	Tennessee	Tennessee	Massachusetts	Tennessee	Tennessee	Indiana
Vegetation (Mg C ha^{-1})							
Leaves	2.3	1.6	2.0	1.3	2.0	1.7	3.7
Branches	16.3	14.8	17.7	—[a]	—	16.8	—
Stems	59.9	56.0	67.6	105.0[b]	78.1[b]	63.7	98.2
Stumps	7.7	4.9	5.7	—	—	5.5	—
Coarse roots	9.4	na[c]	na	20.0	—	na	9.5
Fine roots	3.6	na	na	3.6	7.6	na	6.8
Total vegetation C	99.3	77.2[d]	93.0[d]	129.9	87.7	87.7[d]	118.2
Soils (Mg C ha^{-1})							
Coarse wood litter	4.1	na	na	na	na	na	na
Organic layers	4.3	15.0	15.5	—	2.4	8.5	—
Soil C (0–1 m)	~63.9	54.5	54.5	88.0	122.5	88.8	108.0
Total soil C	73.0	69.6	70.0	88.0	124.9	97.3	108.0
Total (Mg C ha^{-1})	171.6	146.8	163.0	217.9	212.6	184.9	226.2
Stand age or range (y)	58–100	41–83	41–83	60	48	41–83	80
Stand C (Mg C ha^{-1})	107.7	92.3	108.5	129.9	90.1	96.1	118.2
Rate (g C m^{-2} y^{-1})	108–186	111–225	131–265	216	188	116–234	148

Note: Stand-age range and the rate of carbon accumulation for the biomass plus soil organic layer pools over that time period (i.e., recent carbon accumulation in g C m^{-2} y^{-1}) are also calculated.

[a] Data are included in another category.

[b] Includes branches, boles, and sometimes stumps.

[c] na = not available

[d] Does not include carbon in the coarse or fine root pools.

Table 22.3. Components of forest NPP and NEP for the upland *Quercus* forest of the TDE from 1993 to 2000

Components	1993	1994	1995	1996	1997	1998	1999	2000	Annual mean ± CI ($g\ C\ m^{-2}\ y^{-1}$)
NPP									
Leaf production	233.1	236.4	238.3	245.8	243.5	229.4	233.6	257.6	240 ± 7
Herbivory[a]	18.6	25.1	21.4	~22.1	21.9	20.6	21.0	23.2	22 ± 2
Wood production[b]	168.1	266.3	191.3	293.9	215.5	211	240.2	209.5	224 ± 32
Saplings	7.1	7.1	7.1	7.1	14.6	7.5	7.1	5.3	8 ± 2
Terminal branches	40.4	55.5	112.8	43.7	55.9	40.4	95.9	32.4	60 ± 23
Acorn production	0.5	0.5	28.3	4.4	15.1	0.5	40.7	1.0	11 ± 11
Coarse roots (> 5 mm)	8.1	11.1	22.6	8.7	11.2	8.1	19.2	6.5	12 ± 5
Small roots (< 5 mm)[c]	128.3	160.7	119.9	207.7	135.8	120.3	182.4	167.3	153 ± 51
Total NPP	604	763	742	833	713	637.8	840.0	702.8	729 ± 69
Heterotrophic respiration[d]									
Soils	416	476	422	459	457	470	469	476	456 ± 18
Dead wood	99	93	86	87	91	89	79	69	87 ± 7
Total HR	515	569	508	546	548	559	548	545	542 ± 16
NEP (total NPP − HR)	89	194	234	287	165	79	292	158	187 ± 67

Note: The 95% confidence interval (CI) representing interannual variation does not include the error associated with measurement approaches or spatial heterogeneity. Soil and litter-layer carbon pools are assumed to be constant for this analysis (Trettin et al. 1999).

[a] Loss of leaf area and associated mass was measured in 1993, 1994, and 1995 by Shure et al. (1998). Values for 1996–2000 are estimates based on the mean of the first 3 years.

[b] Annual wood increment includes growth of branches, boles, and coarse structural roots.

[c] Fine root production was estimated from the direct observation of root turnover using minirhizotron cameras (Joslin et al. 2000).

[d] Heterotrophic respiration was assumed to be 50% of total soil respiration (Hanson et al. 2000) on the basis of the interpolated annual estimates from Hanson et al. (Chapter 10, this volume).

to mean annual NPP* were 36%, 40%, 23%, and 1%, respectively. In peak mast years (e.g., 1999), acorn production can contribute as much as 5% of NPP*. The two lowest values of NPP were associated with drought years in 1993 and 1998 having water stress integrals (WSIs) of −81 and −175 MPa, respectively (Chapter 2, this volume). Whereas 1995 was also a significant drought year (WSI = −133 MPa), it was also a year with high late-season mast production (28.3 g C m^{-2} y^{-1}), which may have offset losses to wood production limiting the correlation to midseason drought. Early onset of spring leafout in 1996 (Chapter 17, this volume) may explain the high wood production for that year (Table 22.3).

The calculated estimates of NPP* for this study are in the middle range of values expected for temperate deciduous forests (Leith 1975a; DeAngelis et al. 1981, 1997; Edwards et al. 1989; Vogt 1991; Mellilo et al. 1993; Esser et al. 1997; Esser 1998; Amthor et al. 1998) and reach only the low end of the range for NPP* of old-gowth tropical forests (i.e, 200–2000 g C m^{-2} y^{-1}) (Clark et al. (2001b). Edwards et al. (1980) reported an average temperate forest NPP value of 561 g C m^{-2} y^{-1}, but showed specific values for a *Quercus–Pinus* forest in New York (Woodwell and Botkin 1970) of 280 g C m^{-2} y^{-1}, and a *Quercus* forest in Britain of 354 g C m^{-2} y^{-1}. Figure 22.1 shows the range of published NPP* data for temperate deciduous forests, with the TDE estimates falling in the upper half. Within the range of climate conditions that support the growth of temperate deciduous forests (e.g., 4–16°C mean annual temperature and 400–2000 mm in annual precipitation), the correlation of NPP* with temperature or precipitation is not strong, but it is consistent with the interbiome relationships of the Miami model proposed by Lieth (1975b). Knapp and Smith (2001) found a positive correlation between aboveground NPP and annual precipitation for a range of ecosystems ranging from NPP values around 200 g C m^{-2} y^{-1} at 300–400 mm of precipitation to NPP values greater than 600 g C m^{-2} y^{-1} for annual precipitation levels exceeding 1100 mm. The mean annual aboveground NPP* for the TDE forest of 729 g C m^{-2} y^{-1} consistent with the Miami model (Fig. 22.1) and the relationships reported by Knapp and Smith (2001).

Estimates of NEP*

Estimates of NEP* were derived from interpolated data for R_{hetero}, which ranged from 515 to 569 g C m^{-2} y^{-1} and showed little year-to-year variation even though temperatures gradually increased from 1993 to 2000. NEP* ranged from a low of 79 g C m^{-2} y^{-1} in 1998 to a high of 292 g C m^{-2} y^{-1} in 1999 for an overall 8-year mean NEP* of 187 g C m^{-2} y^{-1}. The range of NEP* estimates from 1993 to 2000 demonstrates the variable interannual nature of NEP. Birdsey (1992) used forest-inventory data to estimate the range of net accumulation of carbon by forests of the eastern United States and found a range from 100 to 240 g C m^{-2} y^{-1}, which is in good agreement with the NEP* values reported here.

Johnson et al. (1987) concluded that no significant change in soil carbon accumulation (including litter layers) had occurred in *Quercus–Carya* and *Q. prinus* stands on Walker Branch watershed from 1972 through 1982, and more recent

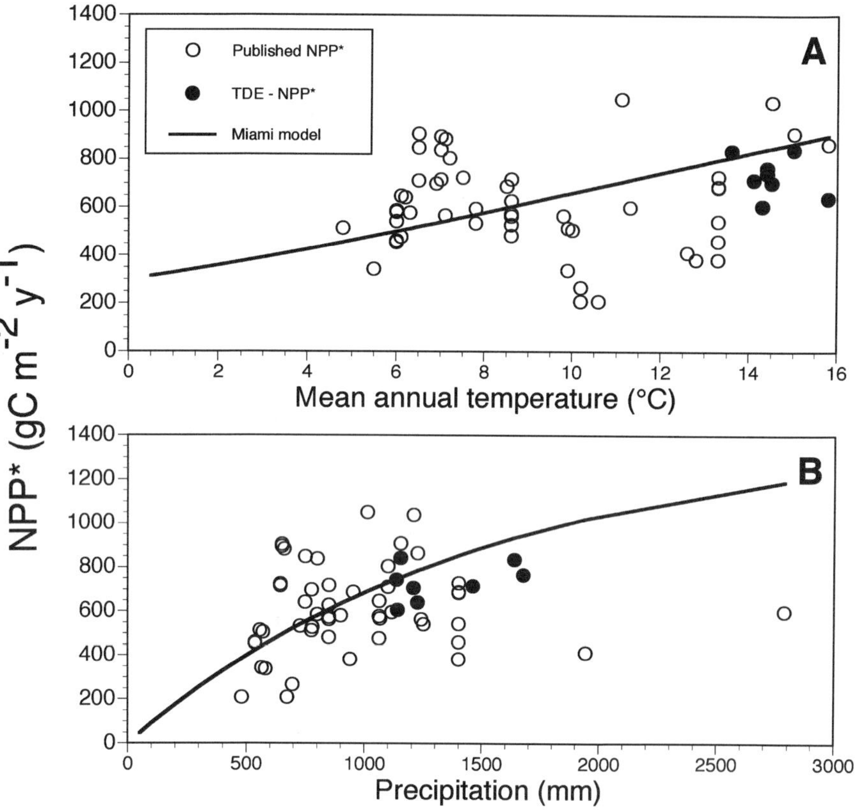

Figure 22.1. Relationship between NPP* and mean annual air temperature (A) or precipitation (B) for data from this study (filled symbols) and data from the published literature (open symbols) Published estimates of NPP* are from DeAngelis et al. (1981, 1997), Edwards et al. (1989), Valentini et al. (1996), Esser et al, (1997), Esser (1998), and Norby et al. (2002). The lines in each graph are the relationships between NPP and mean annual temperature or precipitation, as proposed by Lieth (1975b).

analyses of 1992 soil samples lead to the same conclusion (Trettin et al. 1999). Therefore, we conclude that any net annual carbon accumulation in these upland *Quercus* stands represents an addition to the live biomass pool made up mostly of large woody branches, roots, and tree boles.

Potential Confounding of NPP* and NEP* from Variable Nonstructural Carbohydrate Storage

Measured changes in dormant-season TNC stocks were substantial from 1993 to 2000 (Fig. 22.2A). The dormant-season TNC levels ranged from a low of ~7% in February of 1993 to a high of ~23% in February of 1998. Following the severe summer drought of 1998, TNC values fell to around 12%. Such large interannual changes of the size of the TNC pool reflect a significant change in tissue mass

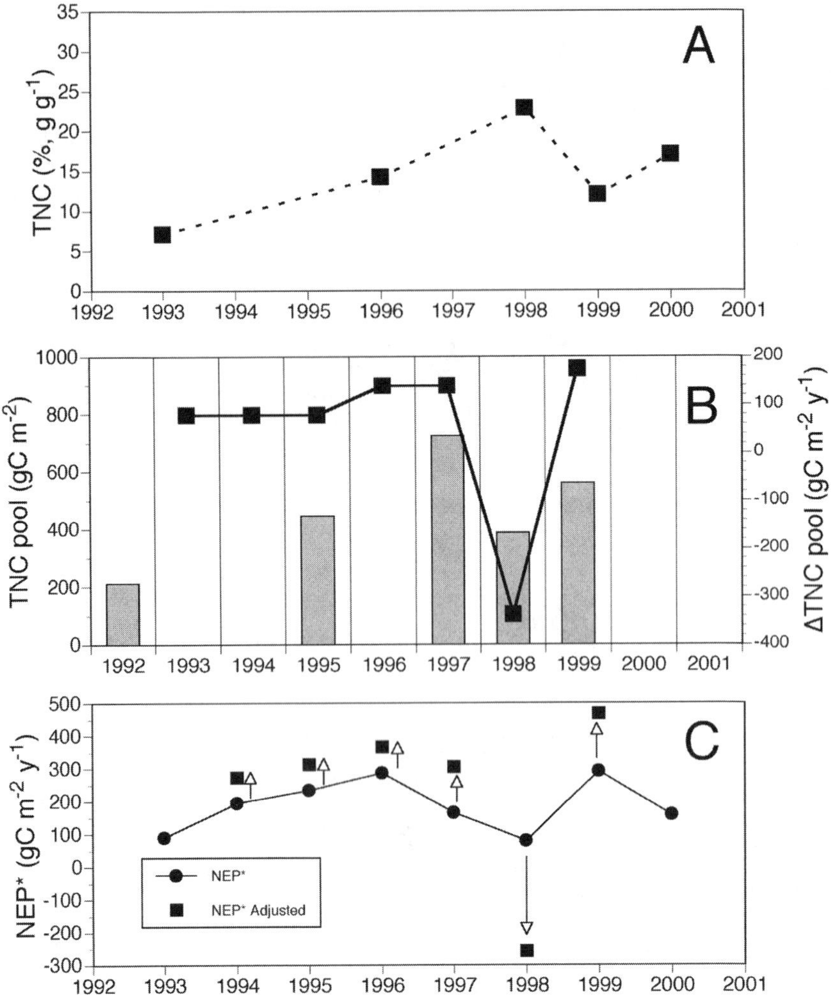

Figure 22.2. (A) Multiyear patterns of dormant-season TNC measured in February of the respective years; (B) estimated stocks of TNC and estimated annual rates of change in the TNC stocks; (C) influence of changing stocks of TNC estimated in graph B on the biometric estimates of net ecosystem production (NEP*).

that is not reflected in the application of fixed allometric relationships to the calculation of NPP* and NEP*. Furthermore, because TNC changes are bidirectional, the bias introduced might be positive in one annual cycle and negative in the next. To determine the potential magnitude of changing TNC concentrations (Fig. 22.2A) on stand carbon pools, we multiplied the known TNC concentrations by sapwood mass per unit ground area (derived from estimates of sapwood area [Chapter 21, this volume] converted to mass) and divided the change in pool size between measurement intervals by the appropriate increment of time (i.e., 1–3 years) (Fig. 22.2B) to yield the annual change in TNC pools (ΔTNC). The net change in the TNC carbon pool ranged from -338 to 174 g C m^{-2} y^{-1}, which is similar in magnitude to annual increments of foliar or wood production (Fig. 22.2B). Changes of this magnitude are substantial and would produce dramatic changes in NEP* (Fig. 22.2C). However, when year-to-year changes in TNC stocks are averaged over time, the net change in TNC stocks tends to cancel out, and the long-term impacts of changes on the TNC pool would likely be small. Temporally resolved data for TNC levels of multiple tissue types will be required to fully evaluate the impact of changing TNC pools on biometric estimates of NEP*.

Estimates of Error Associated with NPP* and NEP*

Complete estimates of error for the individual components of NPP* and NEP* would include stochastic variation associated with measurement approaches, spatial heterogeneity, and temporal heterogeneity. The application of consistent measurement approaches to identical plots from 1993 to 2000 allowed for the calculation of confidence intervals for year-to-year differences in NPP* and NEP* (Table 22.3). However, developing an accurate estimate of cumulative error for the biometric estimates of NPP* and NEP*and their components is not possible because the estimated and/or empirical error terms for all components are not equivalent.

Calculated relative-error estimates of measured variation (i.e., a 95% confidence interval) are as follows. The spatial and measurement error associated with leaf production or leaf standing stocks is ±4%. The error associated with the cumulative estimate of aboveground biomass is ±20%. The error associated with the estimate of standing coarse-and fine-root carbon stocks are ±54% and ±16%, respectively. The error associated with the measurement of organic-layer carbon is ±13%. Estimates of error associated with the spatial variation of carbon stocks and annual carbon increments for upland *Quercus* forests could, and should, be done by the independent evaluation of multiple forest stands.

Interannual variability in NEP* for Walker Branch is in the range of ±36%. Although we can have confidence in the biometric estimates of NPP* because they are largely based on directly measured carbon pools, the ability to estimate NEP* is limited by our inability to differentiate autotrophic versus heterotrophic sources of soil respiration (Hanson et al. 2000); therefore, the true variability around NEP* is probably higher than the ±36%.

NEP* Versus Eddy-Covariance Estimates

While the TDE study has been operating, continuous eddy-covariance observations of landscape-scale CO_2 and water-vapor flux (Baldocchi et al. 1996) were collected from 1995 to 1999 over a nearly identical vegetation footprint 1 km to the southwest of the TDE study area (Wilson and Baldocchi 2001). Curtis et al. (2002) suggest that biometric NEP* data should be directly comparable to eddy-covariance-based net-ecosystem-exchange data when integrated over an annual time step (NEE_a). Furthermore, the two estimates of NEP should be highly correlated over time and should produce the same mean NEP estimates when accounting for the errors of both approaches. Unfortunately, neither condition appears to hold for the NEP* and NEE_a comparison at Walker Branch (Fig. 22.3). Although the measurement footprint for the eddy-covariance tower and the TDE site do not overlap, the species composition, leaf area index, soils, and weather for the respective stands are very similar, and site-to-site differences are not expected to confound the comparison of each approaches' estimate of NEP.

The Walker Branch NEE_a estimates from Wilson and Baldocchi (2001) are 230–450 g C m^{-2} y^{-1} higher than the NEP* estimates developed in this chapter (Table 22.3). Law et al. (2002) conclude that NEE_a data have a total error of $\sim\pm 12\%$ and the NEP* data have a minimum error of $\pm 37\%$. Combining these two assumptions, we should anticipate NEP* to agree within at least 100 g C m^{-2} y^{-1}. Because this was not the case for the NEP* and NEE_a comparison for Walker Branch, we cannot conclude that the NEE_a estimates have been validated by the NEP* values. Furthermore, the interannual pattern of accumulation between the NEP* and NEE_a was not significantly correlated ($R = 0.059$; $P = 0.91$), suggesting that the assumptions driving the interannual differences in NEP

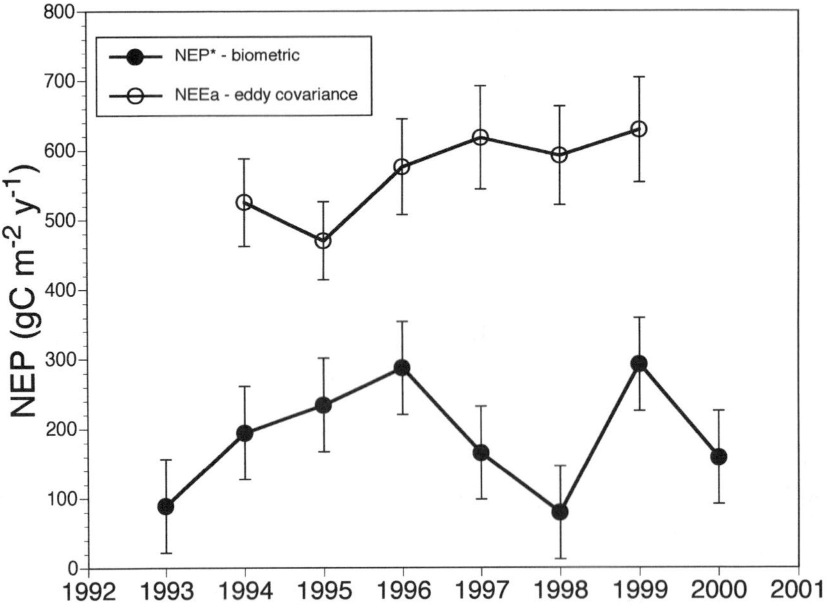

Figure 22.3. Net ecosystem production (NEP*) estimates from biometric approaches (Table 22.3) for 1993 to 2000 and analogous estimates based on eddy-covariance data (NEE_a) [Greco and Baldocchi (1996); Valentini et al. (1996); Wilson and Baldocchi (2001)] for the period from 1994 to 1999. Error bars for NEE_a are only approximate and follow the range suggested by Law et al. (2002). Error bars for NEP* represent a confidence interval associated with interannual variation (Table 22.3) and are only a minimum estimate of the total variation around the calculated value.

as derived by both approaches do not agree. Lack of correlation between interannual patterns of NEP* and NEE_a might be partly explained by hidden changes in nonstructural storage carbohydrates and the transport of dissolved organic carbon to deep soil storage (as discussed previously).

Concern over the lack of agreement between the NEP* and NEE_a data for the Walker Branch site must be put in perspective. First, NEP is likely to be a highly variable quantity, simply because it is a small difference between two large and diametrically opposed fluxes. The reader should not conclude from this single-site example that better comparisons between NEP* and NEE are not possible. When five temperate deciduous hardwood sites, including Walker Branch, were combined in a similar comparison (Curtis et al. 2002), reasonable agreement between biometric NEP* and eddy-covariance NEE_a was found for several sites. Barford et al. (2001) found mean annual NEP* data for a *Quercus/Tsuga* site in Massachusetts to be only 40 g C m^{-2} y^{-1} lower than the mean NEE_a data for a 9-year period. Ehman et al. (2002) found NEP* to overestimate NEE_a by 34 and 90 g C m^{-2} y^{-1} for 1998 and 1999, respectively. It would appear that the difficult site and environmental conditions (i.e., sloping terrain and low night turbulence) on Walker Branch represent a severe limitation to the direct integration of short-term NEE data to annual totals (Baldocchi et al. 2000; Curtis et al. 2002), and it is therefore not easy to reconcile NEE_a derived from the integrating approaches of Wilson and Baldocchi (2001) with the NEP* data presented in this chapter.

Although both the NEP* and NEE_a approaches are subject to error, two lines of reasoning lead us to believe that the NEP* data represent the more accurate estimate of NEP for the upland *Quercus* forests at Walker Branch, Tennessee. First, a simple evaluation of the standing pool of carbon at Walker Branch divided by the age of the stand (Table 22.2) suggests that mean annual carbon accumulation would be unlikely to exceed 200 g C m^{-2} y^{-1}. In fact, similar checks for other temperate deciduous forests throughout the eastern United States also support a mean annual carbon accumulation over the life of the forest of no more than 300 g C m^{-2} y^{-1} (Table 22.2). Second, it is not clear where the additional carbon (i.e., 230–450 g C m^{-2} y^{-1}) would be accumulating in the Walker Branch forest. Gaudinski and Trumbore (Chapter 11, this volume) showed that changing soil carbon was not a likely storage pool, and Chapter 16 (this volume) concluded that no change in belowground root mass was observed during the 8 years of their observations on the TDE. Dissolved-organic-carbon losses from the leaching of water below the rooting zone (Jardine et al. 1990) represents an unquantified pathway for carbon transport that might help resolve differences in NEP and NEE within sites. New research is underway to evaluate the magnitude of deep-soil carbon storage via the translocation of dissolved organic carbon using background-level ^{14}C tracers (Trumbore et al. 2002), but carbon-storage levels of sufficient size to account for the differences between NEP* and NEE_a for the Walker Branch site are not expected. Finally, errors associated with the allometric estimation of the annual aboveground biomass increment might contribute an additional 50–70 g C m^{-2} y^{-1}, as discussed previously, but would not bridge the gap between the NEP* and NEE_a estimates.

A comparison of published NEP* and NEE_a data for temperate deciduous hardwood forests (Valentini et al. 2000; Law et al. 2002; Norby et al. 2002) and data from this study (Fig. 22.4) also demonstrates that biometrically based NEP* data tend to be lower than some but not all NEE_a data, especially at high mean annual temperatures and precipitation levels. However, neither method yields a strong relationship between NEP values for temperate deciduous forests and temperature or precipitation. Janssens et al. (2001) used EUROFLUX network NEE_a data collected across a range of both deciduous and coniferous forest stands to conclude that productivity, not temperature, was the primary driver for explaining site-to-site differences in NEE_a. Although it is convenient and satisfying to identify simple relationships between environmental variables and NEP, long-term goals should focus on the development of mechanistic explanations for observed NEP. Models designed to explain the magnitude and interannual variation of NEP will undoubtedly require mechanisms that respond to a full range of environmental conditions, including radiation inputs, temperature, soil-water and nutrient availability, growing-season duration, and herbivory.

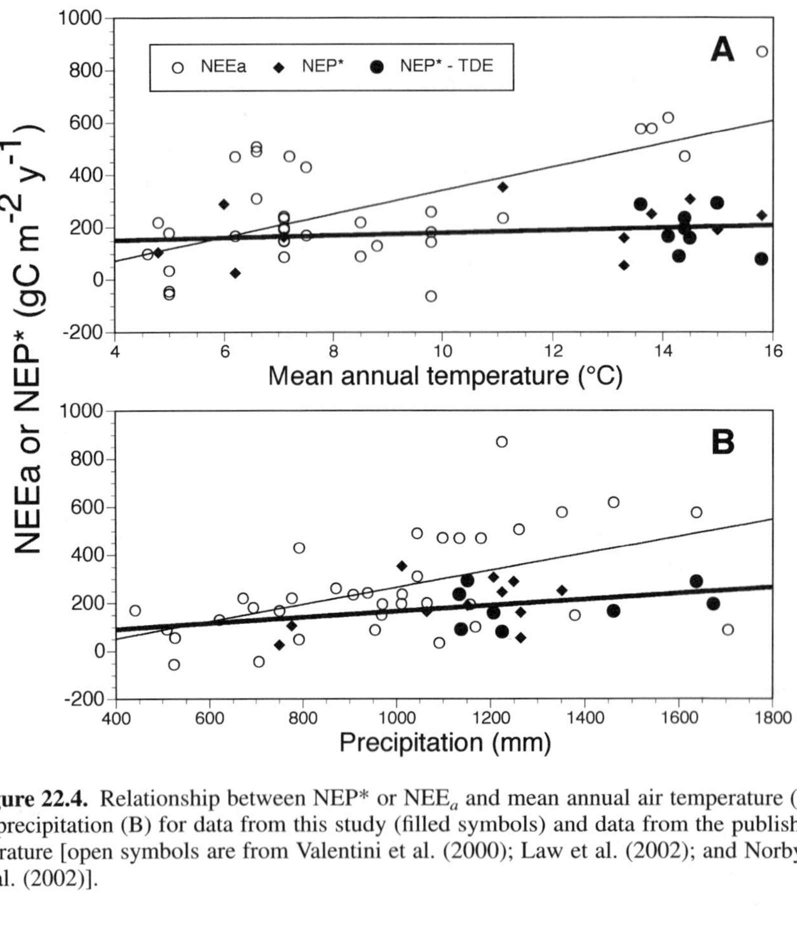

Figure 22.4. Relationship between NEP* or NEE_a and mean annual air temperature (A) or precipitation (B) for data from this study (filled symbols) and data from the published literature [open symbols are from Valentini et al. (2000); Law et al. (2002); and Norby et al. (2002)].

Conclusions

The standing pools of carbon in the 58- to 100-year upland *Quercus* forest of Walker Branch Watershed show the majority to be tied up in the living biomass. Because little change in belowground root density and soil-carbon content has been observed in recent years (Trettin et al. 1999; Chapter 16, this volume), we anticipate that annual NEP for this stand represents accumulation of carbon in coarse woody material. Estimates of NPP* and NEP* from biometric methods yielded mean annual values of 729 and 187 g C m^{-2} y^{-1}, respectively. Interannual differences in NPP* and NEP* were ± 10 or 36%, and the extremely low estimates were attributed to changes in forest water status. Interannual changes in the TNC carbon stocks were shown to represent a potentially large hidden change in carbon storage that is often overlooked in the calculation of NPP* and NEP*. Finally, for the TDE and Walker Branch study areas, interpolated eddy-covariance measurements of NEE_a appear to systematically overestimate NEP and must be interpreted with caution.

References

Amthor JS and members of the Ecosystems Working Group (1998) *Terrestrial ecosystem responses to global change: A research strategy.* ORNL/TM-1998/27. Oak Ridge National Laboratory, Oak Ridge, Tennessee.

Baldocchi DD, Valentini R, Running S, Oechel W, Dahlman R (1996) Strategies for measuring and modeling carbon dioxide and water vapour fluxes over terrestrial ecosystems. Global Change Biol 2:159–168.

Baldocchi DD, Finnigan J, Wilson KB, Paw UK, Falge E, (2000) On measuring net ecosystem carbon exchange over tall vegetation on complex terrain. Boundary-layer Meteorol 96:257–291.

Barford CC, Wofsy SC, Goulden ML, Munger JW, Pyle EH, Urbanski SP, Hutyra L, Saleska SR, Fitzjarrald D, Moore K (2001) Factors controlling long-and short-term sequestration of atmospheric CO_2 in a mid-latitude forest. Science 294:1688–1691.

Birdsey RA (1992) *Carbon storage and accumulation in United States forest ecosystems.* General Technical Report WO-59. USDA Forest Service, Washington, District of Columbia.

Bradford MA, Ineson P, Wookey PA, Lappin-Scott HM (2001) Role of CH_4 oxidation, production and transport in forest soil CH_4 flux. Soil Biol Biochem 33:1625–1631.

Clark DA, Brown S, Kicklighter DW, Chambers JQ, Thomlinson JR, Ni J (2001a) Measuring net primary production in forests: Concepts and field methods. Ecol Appl 11:356–370.

Clark DA, Brown S, Kicklighter DW, Chambers JQ, Thomlinson JR, Ni J, Holland EA (2001b) Net primary production in tropical forests: An evaluation and synthesis of existing field data. Ecol. Appl 11:371–384.

Curtis PS, Hanson PJ, Bolstad P, Barford C, Randolph JC, Schmid HP, Wilson KB (2002) Biometric and eddy-covariance estimates of annual carbon storage in five eastern North American deciduous forests. Agric For Meteorol 113:3–15.

DeAngelis DL, Gardner RH, Shugart HH (1981) Productivity of forest ecosystems studied during the IBP: The woodlands data set. In Reichle DE (Ed) *Dynamics of forest ecosystems.* IBP 23. Cambridge University Press, Cambridge, England, pp 567–672.

DeAngelis DL, Gardner RH, Shugart HH (1997) *NPP multi-biome: Global IBP woodlands data, 1955–1975.* Available online [http://www.daac.ornl.gov/] from Distributed Active Archive Center, Oak Ridge National Laboratory, Oak Ridge, Tennessee.

Dixon RK, Brown S, Houghton RA, Solomon AM, Trexler MC, Wisniewski J (1994) Carbon pools and flux of global forest ecosystems. Nature 263:185 –190.

Edwards NT, Shugart HH Jr., McLaughlin SB, Harris WF, Reichle DE (1980) Carbon metabolism in terrestrial ecosystems. In Reichle DE (Ed) *Dynamic properties of forest ecosystems.* Cambridge University Press, Cambridge, England, pp 499–536.

Edwards NT, Johnson DW, McLaughlin SB, Harris WF (1989) Carbon dynamics and productivity, In Johnson DW, VanHook RI (Eds) *Analysis of biogeochemical cycling processes in Walker Branch Watershed.* Springer-Verlag, New York, pp 197–232.

Ehman JL, Schmid HP, Grimmond CSB, Randolph JC, Hanson PJ, Wayson CA, Cropley FD (2002) An initial intercomparison of micrometeorological and ecological inventory estimates of carbon sequestration in a mid-latitude deciduous forest. Global Change Biol 8:575–589.

Esser, G (1998) *NPP multi-biome: Global Osnabruck data, 1937–1981.* Available online [http://www.daac.ornl.gov/] from Distributed Active Archive Center, Oak Ridge National Laboratory, Oak Ridge, Tennessee.

Esser G, Lieth HFH, Scurlock JMO, Olson RJ (1997) *Worldwide estimates and bibliography of net primary productivity derived from pre-1982 publications.* ORNL/TM-13485. Oak Ridge National Laboratory, Oak Ridge, Tennessee.

Greco S, Baldocchi DD (1996) Seasonal variations of CO_2 and water vapour exchange rates over a temperate deciduous forest. Global Change Biol 2:183–197.

Guenther A, Greenberg J, Harley P, Helmig D, Klinger L, Vierling L, Zimmerman P, Geron C (1996) Leaf, branch, stand and landscape scale measurements of volatile organic compound fluxes from U.S. woodlands. Tree Physiol 16:17–24.

Hanson PJ, Hoffman WA (1994) Emissions of non-methane organic compounds and carbon dioxide from forest floor cores. Soil Sci Soc Am J 58:552–555.

Hanson PJ, Edwards NT, Garten CT Jr., Andrews JA (2000) Separating root and soil microbial contributions to soil respiration: A review of methods and observations. Biogeochemistry 48:115–146.

Hanson PJ, Todd DE Jr, Amthor JS (2001) A six year study of sapling and large-tree growth and mortality responses to natural and induced variability in precipitation and throughfall. Tree Physiol 21:345–358.

Harley P, Guenther A, Zimmerman P (1997) Environmental controls over isoprene emission in deciduous oak canopies. Tree Physiol 17: 705–714.

Harmon ME, Sexton J (1996) *Guidelines for measurements of woody detritus in forest ecosystems.* Publication No. 20, US LTER Network Office, University of Washington, Seattle, Washington.

Harris WF, Goldstein RA, Henderson GS (1973) Analysis of forest biomass pools, annual primary production and turnover of biomass for a mixed deciduous forest watershed. In Young HE (Ed) *Proceedings IUFRO symposium working party on forest biomass.* University of Maine Press, Orono, Maine, pp 41–64.

Harris WF, Sollins P, Edwards NT, Dinger BE, Shugart HH (1975) Analysis of carbon flow and productivity in a temperate deciduous forest ecosystem. In Reichle DE, Franklin JF, Goodall DW (Eds) *Productivity of world ecosystems.* National Academy of Sciences, Washington, District of Columbia, pp 116–122.

Isebrands J, Guenther AB, Harley P, Helmig D, Klinger L, Vierling L, Zimmerman P, Geron C (1999) Volatile organic compound emission rates from mixed deciduous and coniferous forests in Northern Wisconsin, USA. Atmos Environ 33:2527–2536.

Janssens IA, Lankreijer H, Matteucci G, Kowalski AS, Buchmann N, Epron D, Pilegaard K, Kutsch W, Longdoz B, Grunwald T, Montagnani L, Dore S, Rebmann C, Moors EJ, Grelle A, Rannik U, Morgenstern K, Oltchev S, Clement R, Gudmundsson J, Minerbi S, Berbigier P, Ibrom A, Moncrieff J, Aubinet M, Bernhofer C, Jensen NO, Vesala T, Granier A, Schulze ED, Lindroth A, Dolman AJ, Jarvis PG, Ceulemans R, Valentini R (2001) Productivity overshadows temperature in determining soil and ecosystem respiration across European forests. Global Change Biol 7:269–278.

Jardine PM, Wilson GV, McCarthy JF, Luxmoore RJ, Taylor DL (1990) Hydrogeochemical processes controlling the transport of dissolved organic carbon through a forested hillslope. J Contaminant Hydrol 6:3–19.

Johnson DW, VanHook RI (Eds) (1989) *Analysis of biogeochemical cycling processes in Walker Branch Watershed.* Springer-Verlag, New York.

Johnson DW, Hanson PJ, Todd DE Jr. (2002) The effects of throughfall manipulation on soil leaching in a deciduous forest. J Environ Qual 31:204–216.

Johnson DW, Henderson GS, Harris WF (1987) Changes in aboveground biomass and nutrient content on Walker Branch Watershed from 1967 to 1983. In Hay RL, Woods FW, DeSelm H (Eds) *Proceedings of the central hardwood forest conference VI,* February 24–26, 1987, Knoxville, Tennessee, pp 487–495.

Joslin JD, Wolfe MH, Hanson PJ (2000) Effects of shifting water regimes on forest root systems. New Phytol 147: 117–129.

Knapp AK, Smith MD (2001) Variation among biomes in temporal dynamics of aboveground primary production. Science 291:481–484.

Law BE, Falge E, Gu L, Baldocchi DD, Bakwin P, Berbigier P, Davis K, Dolman AJ, Falk M, Fuentes JD, Goldstein A, Granier A, Grelle A, Hollinger D, Janssens IA, Jarvis P, Jensen NO, Katul G, Mahli Y, Matteucci G, Meyers T, Monson R, Munger W, Oechel W, Olson R, Pilegaard K, Paw KT, Thorgeirsson H, Valtentini R, Verma S, Vesala T, Wilson K, Wofsy S (2002) Environmental controls over carbon dioxide and water vapor exchange of terrestrial vegetation. Agric For Meteorol 113:97–120.

LeMer J, Roger P (2000) Production, oxidation, emission and consumption of methane by soils: A review. Eur J Soil Biol 37:25–50.

Lieth HFH (1975a) Primary production of the major vegetation units of the world. In Lieth H, Whittaker RH (Eds) *Primary productivity of the biosphere.* Ecological Studies 14. Springer-Verlag, New York, pp 203–215.

Lieth HFH (1975b) Modeling the primary productivity of the world. In Lieth H, Whittaker RH (Eds) *Primary productivity of the biosphere.* Ecological Studies 14. Springer-Verlag, New York, pp 237–263.

Mellilo JM, McGuire AD, Kicklighter DW, Moore B III, Vorosmarty CJ, Schloss AL (1993) Global climate change and terrestrial net primary production. Nature 363:234–239.

Myneni RB, Dong J, Tucker CJ, Kaufmann RK, Kauppi PE, Liski J, Zhou L, Alexeyev V, Hughes MK (2001) A large carbon sink in the woody biomass of northern forests. Proc Natl Acad Sci 98:14784–14789.

Norby RJ, Hanson PJ, O'Neill EG, Tschaplinski TJ, Weltzin JF, Hansen RT, Cheng W, Wullschleger SD, Gunderson CA, Edwards NT, Johnson DW (2002) Net primary productivity of a CO_2-enriched deciduous forest and the implications for carbon storage. Ecol Appl 12:1261–1266.

Peters LN, Grigal DF, Curlin JW, Selvidge WJ (1970) *Walker Branch Watershed project: Chemical, physical, and morphological properties of the soils of Walker Branch Watershed.* ORNL/TM-2968. Oak Ridge National Laboratory, Oak Ridge, Tennessee.

Pier PA, McDuffie C Jr. (1997) Seasonal isoprene emission rates and model comparisons using whole-tree emissions from white oak. J Geophys Res 102:23963–23971.

Post WM, Emanuel WR, Zinke PJ, Stangenberger AG (1982) Soil carbon pools and world life zones. Nature 298: 156–159.

Post WM, Peng T-H, Emanuel WR, King AW, Dale VH, DeAngelis DL (1990) The global carbon cycle. Am Sci 78:310–326.

Reichle DE (1971) Energy and nutrient metabolism of soil and litter invertebrates. In Duvigneaud P (Ed) *Productivity of forest ecosystems.* UNESCO, Paris, pp 465–477.

Sharkey TD, Singsaas EL, Vanderveer PJ, Geron C (1996) Field measurements of isoprene emission from trees in response to temperature and light. Tree Physiol 16:649–654.

Shure DJ, Mooreside PD, Ogle SM (1998) Rainfall effects on plant-herbivore processes in an upland oak forest. Ecology 79:604–617.

Sollins P, Anderson RM (1971) *Dry-weight and other data for trees and woody shrubs of the southeastern United States.* ORNL-IBP-71–6. International Biological Program, Oak Ridge National Laboratory, Oak Ridge, Tennessee.

Trettin CC, Johnson DW, Todd DE Jr. (1999) Forest nutrient and carbon pools at Walker Branch Watershed: Changes during a 21 year period. Soil Sci Soc Amer J 63:1436–1448.

Trumbore S, Gaudinski JB, Hanson PJ, Southon JR (2002) A whole-ecosystem carbon-14 label in a temperate forest. EOS Trans Am Geophys Union 83:265,267–268.

Sharkey TD, Singsaas EL, Vanderveer PJ, Geron C (1996) Field measurements of isoprene emission from trees in response to temperature and light. Tree Physiol 16:649–654.

Valentini R, DeAngelis P, Matteucci G, Monaco R, Dore S, Scarascia-Mugnozza GE (1996) Seasonal net carbon dioxide exchange of a beech forest with the atmosphere. Global Change Biol 2:199–207.

Valentini R, Matteucci G, Dolman AJ, Schulze ED, Rebmann C, Moors EJ, Granier A, Gross P, Jensen NO, Pilegaard K, Lindroth A, Grelle A, Bernhofer C, Grunwald, T, Aubinet M, Ceulemans R, Kowalski AS, Vesala T, Rannik U, Berbigier P, Loustau D, Guomundsson J, Thorgeirsson H, Ibrom A, Morgenstern K, Clement R, Moncrieff J, Montagnani L, Minerbi S, Jarvis PG (2000) Respiration as the main determinant of carbon balance in European forests. Nature 404:861–865.

Vogt K (1991) Carbon budgets of temperate forest ecosystems. Tree Physiol 9:69–86.

Wilson KB, Baldocchi DD (2001) Comparing independent estimates of carbon dioxide exchange over 5 years at a deciduous forest in the southeastern United States. J Geophys Res 106(D24):34167–34178.

Woodwell GM, Botkin DB (1970) Metabolism of terrestrial ecosystems by gas exchange techniques: The Brookhaven approach. In Reichle DE (Ed) *Analysis of temperate forest ecosystems.* Ecological studies 1. Springer-Verlag, New York, pp 73–85.

23. Nutrient Availability and Cycling

Dale W. Johnson, Paul J. Hanson, and Donald E. Todd

Introduction

Precipitation changes can affect forest nutrient cycles in a number of complex ways. Increases or decreases in precipitation may cause changes in productivity and nutrient uptake rates and will surely cause changes in soil-water flux and the hydrologic fluxes of nutrients. To date, most studies have emphasized the effects of temperature or elevated carbon dioxide on these processes, but changes in precipitation may have equal or greater effects (Kirshbaum et al. 1992). Specifically, changes in precipitation could cause changes in (1) the productivity of water-limited ecosystems, (2) water yield, and (3) water quality and soil leaching rates.

As part of the Throughfall Displacement Experiment (TDE), we have been monitoring soil solution concentrations, organic (O) horizon nitrogen and phosphorous fluxes, and simulating both these fluxes and long-term changes in soils, litter, and vegetation using the Nutrient Cycling Model (NuCM). Results of these studies are described in detail in other articles (Johnson et al. 1998, 2001, 2002). In this chapter, we summarize the results of the field studies to date and compare them with the results of the simulations, recently updated for the purposes of this chapter. Many of the responses noted were caused by the disproportionate effects of throughfall changes on soil-water flux. Luxmoore and Huff (1989) noted that, over a period of years, streamflow was much more closely correlated to precipitation than was apparent evapotranspiration (ET; the difference between precipitation inputs and streamflow outputs). This observation suggests that ET was relatively constant and that changes in precipitation caused disproportionately large changes in soil-water flux and streamflow. Thus, in essence, the effects of throughfall water fluxes were magnified as they entered the soil system.

Literature Review on Effects of Precipitation on Soils and Nutrients

In the Solling forest ecosystem in Germany, Ulrich (1983) noted pulses of NO_3^- and aluminum (Al) during warm, dry years. He hypothesized that drought intensifies nitrogen (N) mineralization and nitrification during summer, resulting in NO_3^- pulses during rewetting periods. In acidic soils, the NO_3^- pulse is accompanied by a pulse in soil-solution Al concentrations. Lamersdorf et al. (1995) describe a field research project designed to test Ulrich's hypothesis on a large scale. The EXperimental MANipulation of Forest Ecosystems (EXMAN) project involves the imposition of drought through roofs in the forest subcanopy to five forest ecosystems in Europe. Initial results showed that this artificially imposed drought had no effect upon nitrification, but some suggest that natural periods of drought could produce such pulses, as observed by Ulrich (1983).

In addition to affecting N mineralization and nitrification, drought can be expected to cause increased ionic concentrations in soil solutions, especially for conservative ions such as Cl^- and Na^+ (i.e., ions which are neither produced nor consumed in soils to any large degree). Changes in the concentrations of other ions may be either buffered by soil chemical processes or controlled by mechanisms that are not sensitive to drought. In the cases of $H_2PO_4^-$ and SO_4^{2-}, for example, adsorption to Fe and Al hydrous oxides may mitigate concentration increases caused by drought. In the case of HCO_3^-, concentrations are controlled by the partial pressure of CO_2 (pCO_2) in soil and soil solution pH and could be affected either positively or negatively by drought. If drought causes reduced root and microbial respiration, pCO_2 could decrease. In addition, reduced effective soil CO_2 diffusivity with lower soil-water concentrations would tend to result in lower pCO_2 (Johnson et al. 1994). On the other hand, drought may cause increased HCO_3^- concentrations by causing increased pH in less acidic soils, especially over the longer term if base saturation increases. The effects of drought on cations in soil solution should, in theory, be a function of changes in total ionic concentration. As noted by Reuss (1983), increased total ionic concentrations cause trivalent cations to increase disproportionately to divalent and monovalent cations (to the 3/2rd and 3rd power, respectively) and divalent cations to increase as the square of monovalent cations.

Hypotheses Tested in This Study

Based on the above-reviewed literature, we hypothesized that the dry treatments would cause (1) disproportionate reductions in soil-water flux, (2) increased total ionic concentrations in soil solution, (3) decreased ionic fluxes in soil, and (4) increased soil exchangeable nutrients over the long term. We also hypothesized that (5) the reverse will be true for the wet treatments

Methods

Field: Tension Lysimeters (After Johnson et al. 2002)

Soil solutions were collected with ceramic cup tension lysimeters installed at depths of 25 cm (E horizon) and 70 cm (Bt horizon). The lysimeters were installed prior to the initiation of treatment in a three-by-three array in each plot at spacing of 7.9 m (nine replicates per treatment and depth). Some pretreatment collections were made in 1992 and reported earlier (Johnson et al. 1998). Posttreatment sampling was initiated in 1996 when funding became available. In Spring 1996, soil solution collections were initiated from the Bt horizon lysimeters, and in Autumn 1997 collections were initiated from the E horizon.

Prior to collection of samples, each lysimeter was set to a tension of −40 kPa. Soil solutions were analyzed for electrical conductivity (EC), pH and HCO_3^- (by titration to pH 5.0), Ca^{2+}, Mg^{2+}, K^+, and Na^+ by atomic absorption spectroscopy, for NH_4^+, NO_3^-, and orthophosphate by automated colorimetric analysis, and for Cl^- and SO_4^{2-} by Dionex ion chromatography. In most cases (64% of all samples), pH values were 5.0 or less; thus, HCO_3^- concentrations were 0.

Ion fluxes were calculated from average ion concentrations for each water year multiplied by simulated soil-water fluxes. Daily estimates of soil-water content, water potential, and rates of percolation were estimated from a revised version of the TEHM/PROSPER model (Huff et al. 1977; Luxmoore 1983) coded with "Ithink" modeling software (High Performance Systems, Hanover, NH) for the Macintosh computer. A comparison of measured and modeled soil-water potentials for 1993 through 2000 showed good agreement for a range of wet and dry years (Johnson et al. 2002).

Soil-water content (%, v/v) and soil-water potential (MPa) were measured and calculated, repectively, for the 0–0.35-m and 0–0.7-m depths as described by Hanson et al. (1998; Chapter 2, this volume).

Field: Resin Lysimeters

In addition to the ceramic-cup lysimeters, we installed, processed and analyzed resin lysimeters beneath the O horizons of all treatments to collect cumulative soil N and P leaching. Resin lysimeters were used because they do not suffer from the problems in maintaining vacuum during frequent wetting and drying that tension lysimeters do and also because resin lysimeters have very low cost and maintenance. The resin lysimeters consisted of a 5.5-cm-long, 4-cm-inside-diameter PVC pipe within which a resin bag containing 10 g of oven-dried Rexyn® I-300 (H-OH) resin was sandwiched between layers of washed silica sand (Johnson et al. 2002).

The lysimeters were installed in June 1998 by excavating a small hole and tunneling beneath the O horizons. The resin lysimeters were collocated with the ceramic cup lysimeters ($n = 9$ per treatment and strata). For the dry and wet treatments, the resin lysimeters were stratified according to position. In the dry treatment, at each location, one set of resin lysimeters was placed directly beneath a trough and one set was placed between troughs ($n = 9$ for each, for a total of 18 replicates). For the wet treatment, one set of resin lysimeters was placed beneath an area which was receiving water from a driphole and one which was not ($n = 9$ for each, for a total of 18 replicates).

After collection, resins were removed from the lysimeters, extracted, and analyzed for NH_4^+, NO_3^-, and ortho-phosphate by automated colorimetric analysis. Fluxes were calculated from the amount of NH_4^+, NO_3^-, and ortho-P extracted from the resins (minus blanks) divided by the surface area of the lysimeters (12.6 cm^2). Further details of the processing of the resin can be found in Johnson et al. (2002).

The Nutrient Cycling Model and Its Use for This Study

The NuCM model has been described in detail elsewhere (Liu et al. 1991; Johnson et al. 2001), and only a few relevant details are repeated. The model depicts nutrient cycling at a stand level, where the ecosystem is represented as a series of vegetation and soil components. Provision is made for one generic conifer and one generic deciduous species of specified biomass and nutrient concentration (foliage, branch, bole, and roots). The model also provides for an overstory, which can be divided into canopy, bole, and roots. Tree growth is a function of user-defined stand developmental stage as well as the availability of nutrients and moisture. The model provides for up to 10 soil layers, and each layer can have different physical and chemical characteristics. The model routes precipitation through the canopy and soil layers, simulating evapotranspiration, deep seepage, and lateral flow. The movement of water through the system is simulated with the continuity equation, Darcy's equation for permeable media flow, and Manning's equation for free surface flow. Percolation occurs between layers as a function of layer permeabilities and differences in moisture content. Nutrient pools associated with soil solution, the ion-exchange complex, minerals, and soil organic matter are all tracked explicitly by NuCM. The processes that govern interactions among these pools include user-specified rates for decay, nitrification, anion adsorption, cation exchange, and mineral weathering.

The model simulates the noncompetitive adsorption of sulfate, phosphate, and organic acid. Sulfate adsorption can be simulated by NuCM with either linear or Langmuir adsorption isotherms. The Langmuir isotherm was used in these simulations. Phosphate adsorption in the model is represented by a linear isotherm. Cation exchange is represented by the Gapon equation. Mineral weathering reactions in the model use rate expressions that depend on the mass of mineral present and the solution-phase H^+ concentration raised to a fractional power.

Temperature-dependent processes in NuCM include evapotranspiration, occurrence of precipitation as rainfall versus snowfall, snowpack melting, litter decay, and nitrification. Potential evapotranspiration (ETp) is calculated as

$$\text{Etp} = \left(\frac{\text{Fet}}{n}\right)(T_m C_e H_c),$$

where Fet is the evapotranspiration factor, which is a function of latitude (r) (Hargreaves 1974): Fet $= 2.322 - 0.0115r$, $r \leq 35°$; Fet $= 3.434 - 0.0434r$, $r \geq 35°$; T_m is the mean ambient daily temperature (°F); H_c is the humidity correction factor (no units); and C_e is the calibration factor (no units). See Munsen et al. (1992) for further details.

Litter decay is represented by NuCM as a four-stage process in which (1) litter converts to fine litter, (2) fine litter converts to humus and cations, (3) humus converts to organic acid, NH_4^+, SO_4^{2-}, H^+, and CO_2, and (4) organic acid converts to NH_4^+, SO_4^{2-}, H^+, and CO_2. Each stage is a first-order equation of the form

$$-\frac{dC}{dt} = -k\theta_t^{(T-20)}C,$$

where C is the component in question, t is time; k is the reaction rate coefficient (user-defined); θ_t is the temperature correction factor; and T is the soil temperature (°C).

Nitrification is represented in the form of a Michaelis–Menton rate expression:

$$u = \alpha v\left(\frac{[NH_4^+]}{K_s} + [NH_4^+]\right)\theta_N^{(T-20)},$$

where u is the nitrification rate, α is the pH-dependency factor, v is the maximum nitrification rate at 20°C, K_s is the half saturation constant, and $\theta_N^{(T-20)}$ is the temperature-dependency factor, where T is the soil temperature (in °C).

The NuCM model was calibrated for the Walker Branch site with data from Johnson and Todd (1990) and Johnson and Van Hook (1989). Meteorological data from the period September 1993 to August 2000 was used in the simulations and repeated to obtain long-term simulation results.

The scenarios used to test the hypotheses included no change (ambient), −33% precipitation (dry), and +33% precipitation (wet). Changes in precipitation and air temperature were made directly to the meteorological files (by multiplying daily values by 0.67 or 1.33 for precipitation).

Results

Field Results

Hydrologic Fluxes

As noted in previous articles (Luxmoore and Huff 1989; Johnson et al. 1998, 2001), both watershed-scale measurements and simulations indicated that soil-water flux (SWF) varies more on an annual basis than does ET, and, therefore, SWF is affected disproportionately by treatments. Reducing precipitation by 33% in the dry treatment caused decreases in simulated ET of 2.5–14.9 cm or 5–23%, whereas simulated SWF decreased by 27.2–50.6 cm or 46–59% (Fig. 23.1). Increasing precipitation by 33% in the wet treatment caused simulated ET to increase by 5.1–11.2 cm or 6.7–17.0%, whereas simulated SWF increased 23.2–48.7 cm or by 47–56% (Fig. 23.1).

Soil Solution Concentrations

As noted by Johnson et al. (2002), there were no statistically significant effects of treatments on soil solution concentrations in the E horizons. Treatment effects were statistically significant ($p < 0.01$) for all measured ions except for HCO_3^-,

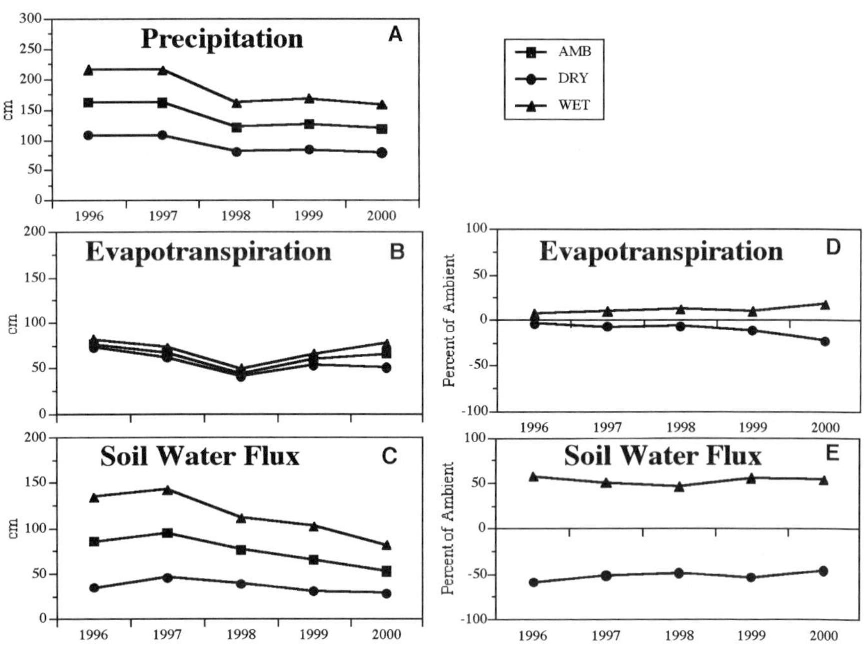

Figure 23.1. Precipitation (A) and simulated evapotranspiration (ET) (B), soil-water flux (SWF) (C) in the TDE experiment during the sampling period. Panels D and E show ET and SWF as percentages of ambient.

ortho-P, NH_4^+, and NO_3^- in the Bt horizons, however (Figs. 23.2 and 23.3 and Table 23.1). The observed treatment effects in the Bt horizon were caused by differences between the dry treatment and the other treatments; differences between the wet and ambient treatments were not significant (Table 23.1). Soil solutions from the Bt horizons in the dry treatment had consistently higher concentrations of Ca^{2+}, Mg^{2+}, K^+, Cl^-, and SO_4^{2+} during the sampling period. In the case of Na^+, treatment differences were less pronounced, but there was a sharp increase in concentrations over time between the 1997–1998 and 1999–2000 water years in both the E and Bt horizons (Figs. 23.2 and 23.3). Concen-

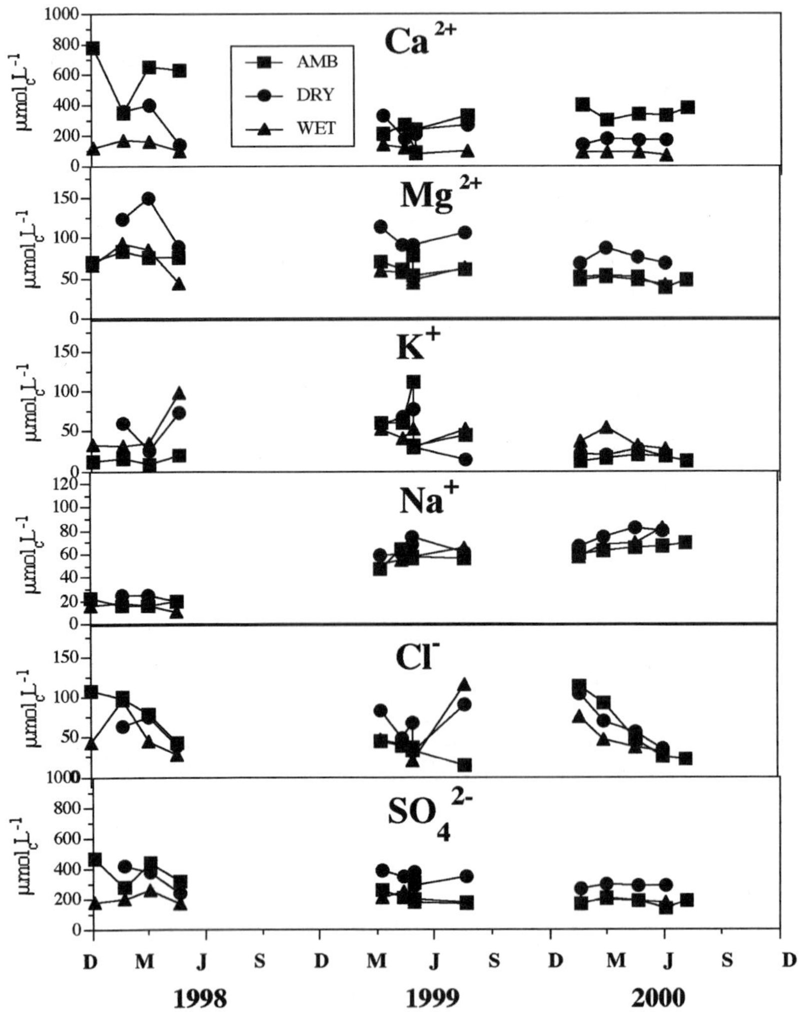

Figure 23.2. Soil solution concentrations in the E horizon. (After Johnson et al. 2002, repeated by permission.)

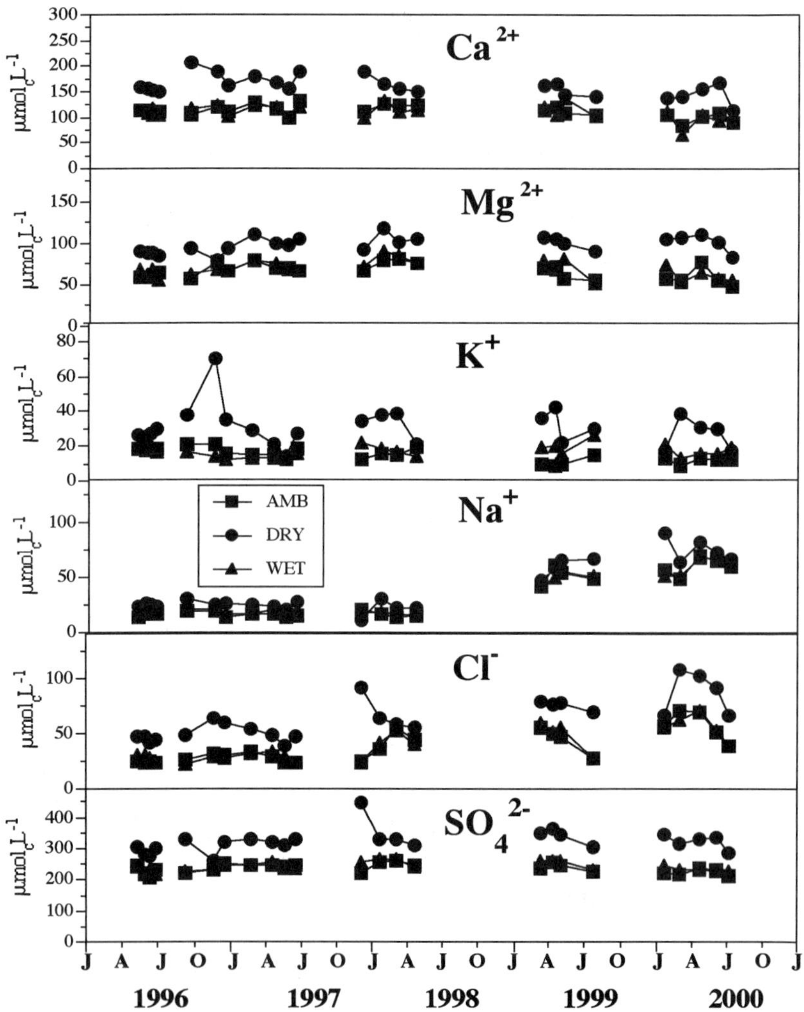

Figure 23.3. Soil solution concentrations in the Bt horizon. (After Johnson et al. 2002, repeated by permission.)

trations of Cl^- also increased significantly ($r^2 = 0.28$, $p < 0.0001$) in the Bt horizons between the 1997–1998 and 1999–2000 water years, but to a lesser degree than for Na^+. The increases in Na^+ concentrations (and, to a lesser extent, Cl^-) in the spring of 1999 probably reflect a response to the strong 1998 drought. The concentrations of conservative ions, such as Na^+ and Cl^-, would be expected to reflect this drought more than the concentrations of other ions, which are buffered by exchange, adsorption, and biological uptake. There were no significant trends in the concentrations of other ions except for a very slight decrease in Ca^{2+} ($r^2 = 0.01$, $p < 0.0128$) in the Bt horizon. The collection date was a significant

Table 23.1. Probability values for repeated measures ANOVA and Bonferonni post hoc tests on the effects of treatment on soil solution concentrations from the Bt horizons (70 cm)

				Bonferonni post hoc tests		
Ion	Treat	Date	Treat * Date	Ambient-Dry	Ambient-Wet	Wet-Dry
pH	***	***	NS	**	NS	*
EC	***	***	NS	***	NS	***
Ca^{2+}	***	***	NS	***	NS	***
K^+	***	***	***	***	NS	***
Mg^{2+}	***	***	NS	***	NS	***
Na^+	***	***	NS	***	NS	***
$SO_4{}^{2-}$	***	***	NS	***	NS	***
Cl^-	***	***	NS	***	NS	***
$HCO_3{}^-$	NS	***	NS	NS	NS	NS
$NO_3{}^-$	NS	***	NS	NS	NS	NS
$NH_4{}^+$	NS	***	NS	NS	NS	NS

NS = not significant.
* = $p < 0.10$.
** = $p < 0.05$.
*** = $p < 0.01$.

factor for all measured soil solution properties, but the interaction term (Treatment*Date) was significant ($p < 0.05$) only for K^+, suggesting that the effects of treatment changed over time only in the case of K^+.

Calculated Ion Fluxes in Mineral Soil

As noted by Johnson et al. (2002), the drought of 1998–2000 caused reductions in calculated Ca^{2+}, K^+, Mg^{2+}, and $SO_4{}^{2-}$ fluxes over the sampling period because of the reduction in soil-water flux; Bt horizon concentrations of these ions showed no trend over time (Figs. 23.3 and 23.4). In contrast, the calculated flux of Na^+ increased sharply in 1999 and 2000, and in this case, it was caused by the marked increase in soil solution Na^+ concentration noted above (Fig. 23.4). Calculated Cl^- flux increased only slightly over the sampling period.

O horizon Fluxes Measured by Resin Lysimeters

As noted by Johnson et al. (2002), results for the resin lysimeters were sensitive to their placement with regard to the troughs in the dry treatment and with regard to the drip locations on the wet treatment. Between troughs of the dry treatment (dry-O) and between the drip holes of the wet treatment (wet-O) resin-derived fluxes from the O layer were not significantly different from the fluxes in the ambient treatment. However, fluxes of $NH_4{}^+$ and $NO_3{}^-$ beneath the troughs in the dry treatment (dry-T) were significantly ($p < 0.05$) lower than those in the other treatments or locations. Fluxes of $NH_4{}^+$ and $NO_3{}^-$ near dripholes in the wet treatment (wet-T) were significantly ($p < 0.05$) greater than in the other treatments. The area-weighted average fluxes of $NH_4{}^+$ and $NO_3{}^-$ in the dry treatment (i.e., 0.67 times the between-trough flux plus 0.33 times the beneath-trough flux) were 45–52% lower than those in the ambient treatment in 1998–

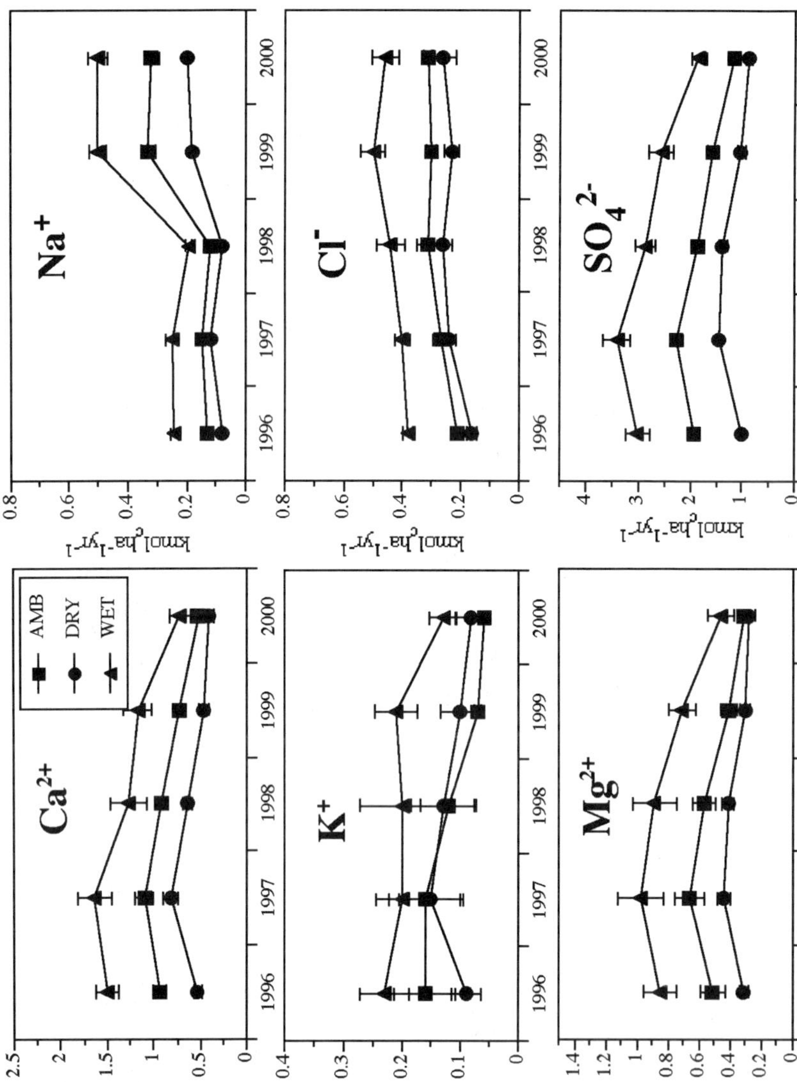

Figure 23.4. Calculated soil ionic fluxes in the Bt horizon. (After Johnson et al. 2002, repeated by permission.)

1999, but only 10–20% lower than the ambient treatment in 1999–2000 (Fig. 23.5). It was not possible to accurately calculate aerially weighted average fluxes for the wet treatment because the areas affected by dripholes were both uncertain and variable with time, depending on rainfall intensity and duration.

Changes in Litter N Content

As noted by Johnson et al. (2002), mass and N content in the Oe + Oa, nonleaf component, and in total O horizon were significantly greater in the dry treatment than in the other treatments. The dry and wet treatments also had lower Oi horizon N concentrations and greater Oe + Oa horizon N concentrations than the ambient treatment. Differences in standing litter layer mass were not attributable to dif-

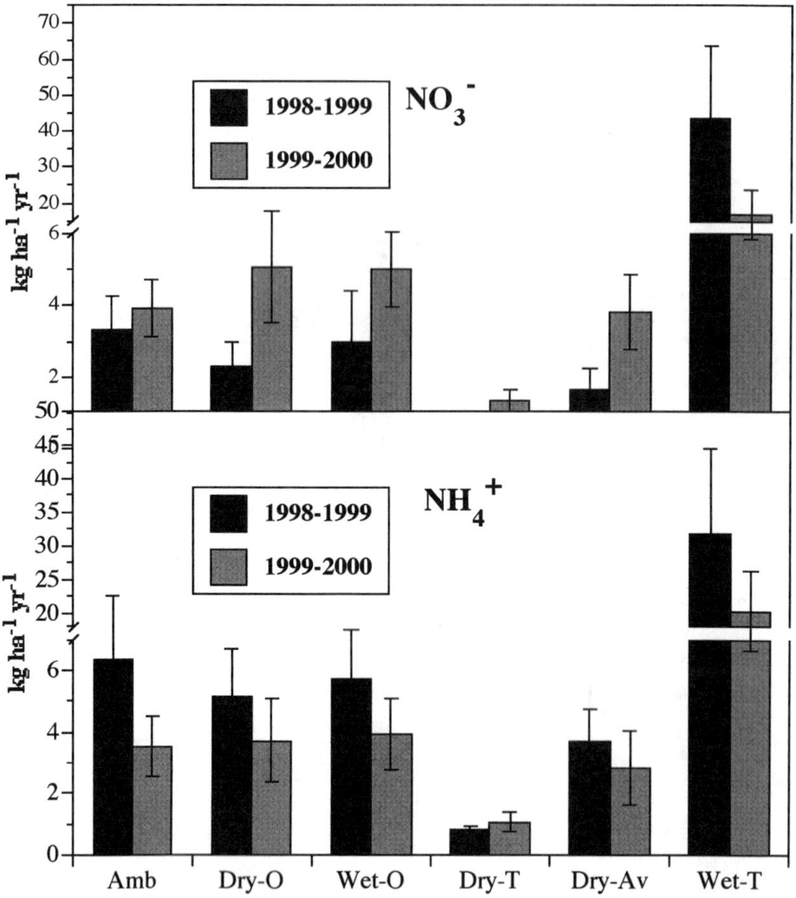

Figure 23.5. N and P fluxes measured with resin lysimeters. (After Johnson et al. 2002, repeated by permission.)

ferences in litter inputs, which did not differ among treatment plots from 1994 through 1998 (data not shown).

NuCM Simulations

Comparison to Field Data

Simulated soil solutions for the E and Bt horizons are shown in Figs. 23.6 and 23.7. The simulations show higher concentrations in the dry than in the ambient

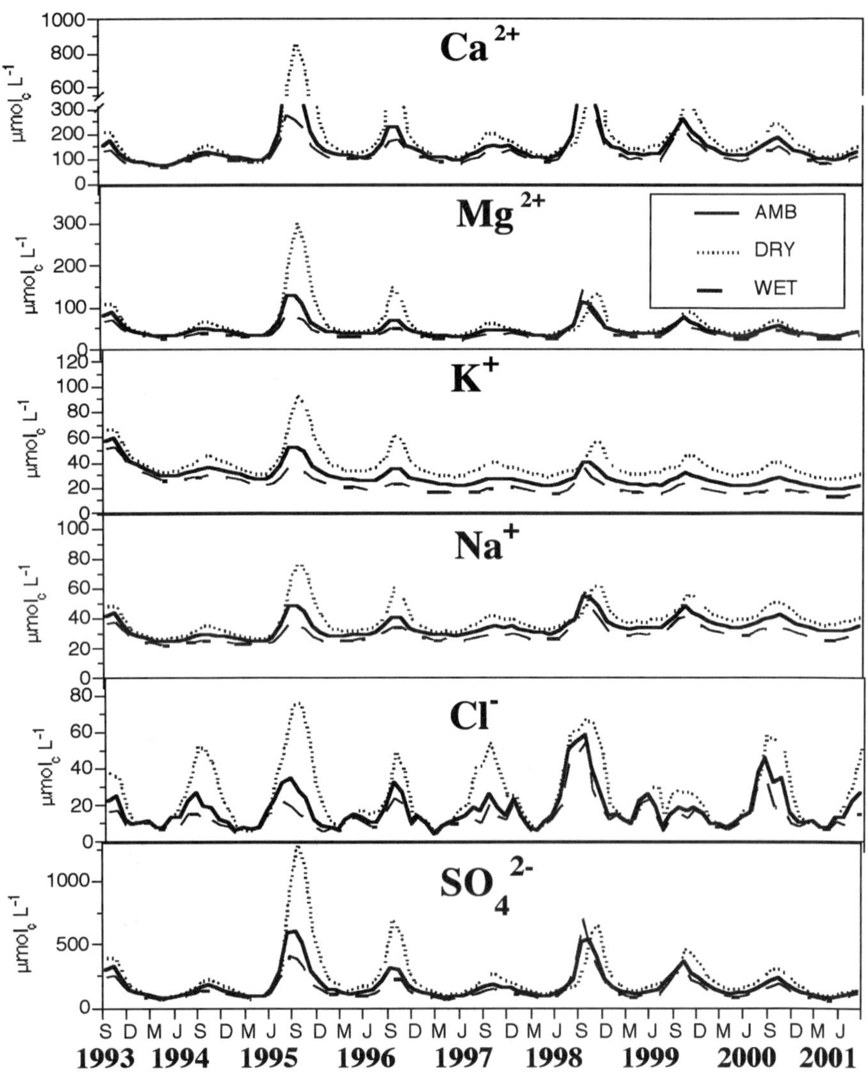

Figure 23.6. Simulated solution concentrations in the E horizon.

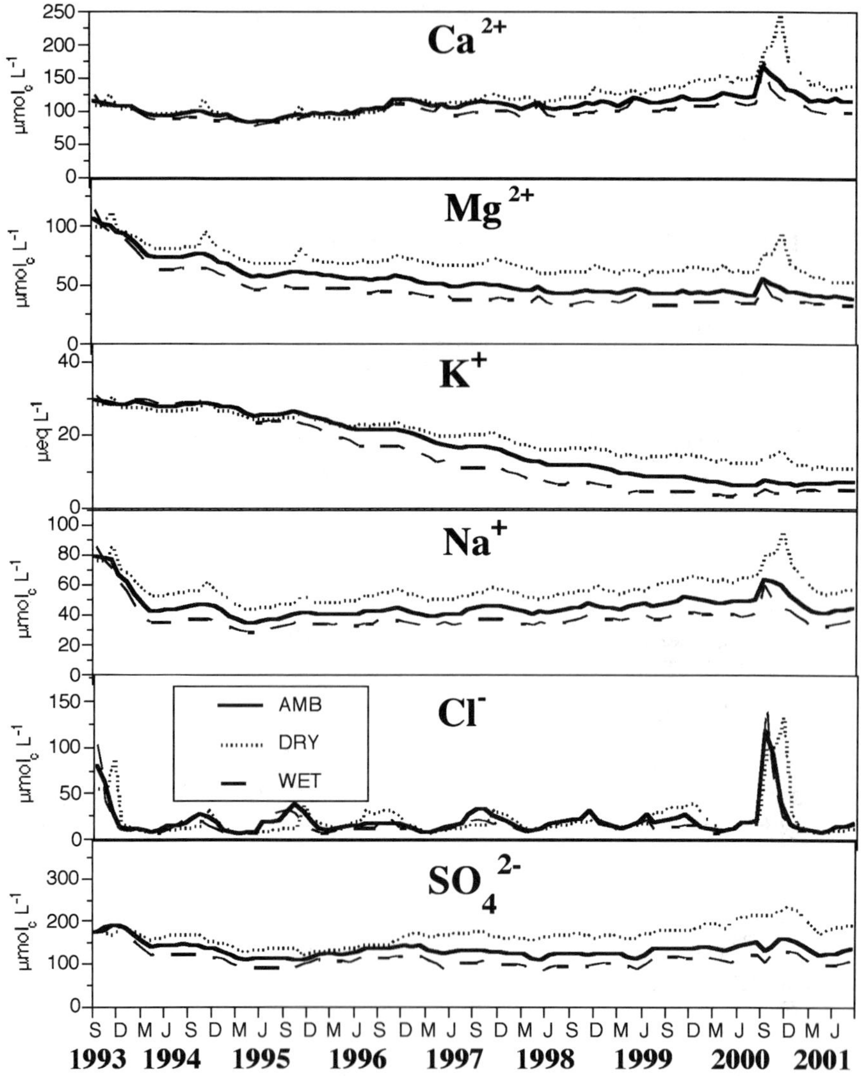

Figure 23.7. Simulated solution concentrations in the Bt horizon.

and lower concentrations in the wet than in the ambient treatments. These results contrast to the field data, which showed no statistically significant effects in the E horizons and no differences between the wet and ambient treatments in either horizon. With the simulations, statistical significance is not relevant; however, many of the differences attributed to treatment in the simulations would not have been statistically significant in the field.

The simulations also show (1) much greater treatment effects during the summer seasons than in other times of the year and (2) increasing treatment effects with time. Neither of these effects was observed in the field but may have occurred. Soil solutions are not collected from the field during summer (the conditions are too dry), and those represented by simulations would be in water held at tensions much greater than field moisture capacity; such water would not be collected in the field with normal tension lysimeters. Thus, heightened treatment effects could well occur in the field during summer, but special extractions of strongly bound soil solutions would be needed to test this hypothesis.

The pattern of increasing treatment effects with time is most pronounced in the Bt horizon simulations and is caused by changes in soil-exchangeable cation pools (see Discussion). Thus far, the only statistically significant increase in treatment effects in the field was for Bt horizon K^+ (Fig. 23.3).

The simulations did not show the increase in E horizon soil solution Na^+ observed in the field, but they did show a spike in Bt horizon soil solution Na^+ and Cl^- in 2000. The simulations also suggested that soil solution Na^+ was initially high before treatment and decreased in all treatments within the first year. Limited pretreatment data from 1992 did not indicate elevated soil solution Na^+ concentrations (values were 25–40 $mol_c\ L^{-1}$) (Johnson et al. 1998). The simulations also showed reductions in Bt horizon soil solution Mg^{2+} and K^+ over time that were not observed in the field. The reductions in simulated soil solution Mg^{2+} and K^+ reflected reductions in the exchangeable pools of these cations (see Discussion); in the field, deep rooting and recycling may have prevented such changes from occurring.

Simulated soil solution fluxes are shown in Fig. 23.8. The values of the fluxes are within the ranges of those calculated for the field data, and the temporal patterns in simulated flux largely reflect the temporal patterns in Bt horizon concentration.

Results from TDE Extrapolated over Long Time Periods

Table 23.2 shows the simulated biomass and nutrient contents at 35 years. The effects of changing precipitation on simulated biomass and vegetation nutrient contents are slight ($< 2\%$). Similarly, the effects on forest-floor nutrient contents are generally small (5–10%). The largest effects are those on soil extractable P and exchangeable Ca^{2+}, K^+, and Mg^{2+}, where differences range from 7% to 30%. As expected, the dry treatment had the greatest simulated extractable P and exchangeable Ca^{2+}, K^+, and Mg^{2+} contents, and the wet treatment had the lowest (Table 23.2). Treatment effects on simulated extractable $SO_4{}^{2-}$ contents were very slight ($<1\%$) because soils in all cases came to near equilibrium with atmospheric S inputs.

Discussion

Collective Implications of Field Results

Although soil-water fluxes are derived from simulation modeling and are therefore subject to considerably uncertainty, it seems clear that they must have

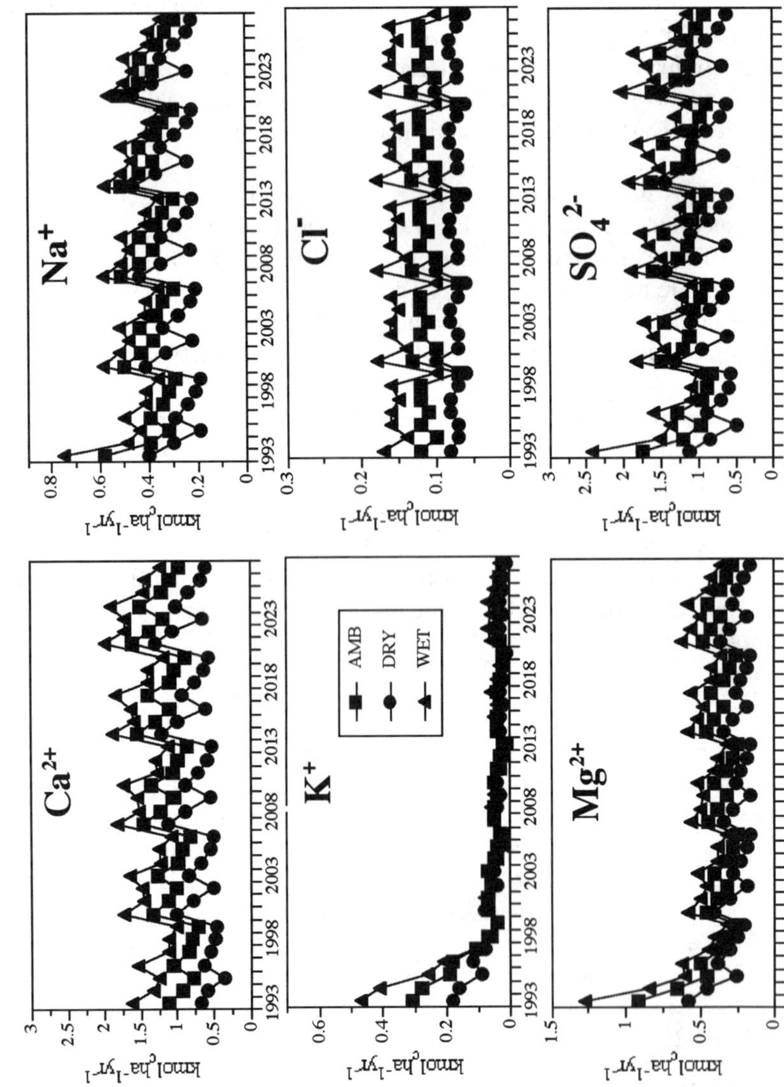

Figure 23.8. Simulated soil ionic fluxes.

Table 23.2. Simulated biomass and nutrient contents

		After 35 years (kg ha^{-1})		
	Initial	Ambient	Dry	Wet
Biomass	212,200	255,100	251,700	256,400
Nitrogen				
Vegetation	650	767	753	775
Litter	234	263	249	272
Soil, exchangeable	0.00	0.14	0.14	0.14
Phosphorus				
Vegetation	75	88	86	88
Litter	13	15	14	16
Soil, exchangeable	9.6	1.9	2.5	2.5
Calcium				
Vegetation	1235	1450	1428	1456
Litter	184	405	380	421
Soil, exchangeable	181	257	332	208
Potassium				
Vegetation	502	553	547	555
Litter	113	154	144	161
Soil, exchangeable	84	40	65	29
Magnesium				
Vegetation	152	180	177	181
Litter	26	60	56	63
Soil, exchangeable	48	49	62	42
Sulfur				
Vegetation	98	132	131	132
Litter	199	269	261	273
Soil, exchangeable	384	392	391	395

changed disproportionately in response to changes in throughfall inputs. Thus, soil solution concentrations should show major responses to changes in precipitation. It is not entirely clear why soil solution concentrations in the wet treatment were not more substantially affected than they were. The lack of response in soil solution concentrations in the wet treatment caused soil leaching flux estimates to increase substantially above those in the ambient treatment. Some of this increase was certainly caused by increased throughfall fluxes. Regardless of the contributions of throughfall fluxes to soil leaching fluxes, however, a change in soil leaching flux represents a net loss from the system, whereas throughfall fluxes include a proportion of ions that have been recycled (especially in the case of K^+) as well as atmospheric dry deposition. Thus, changes in soil leaching fluxes can be expected to have long-term consequences for soil change.

Nitrogen fluxes were significantly greater in the wet treatment and lower in the dry treatment than in the ambient treatment. This effect was exclusively produced by those lysimeters placed either under troughs (dry) or near drip points (wet); the lack of difference between those lysimeters placed in the open (ambient, WO, DO) is reassuring that the plots were comparable in other ways.

The dry treatment continued to cause increases in ion concentrations in Bt horizon soil solutions, as noted in Johnson et al. (1998). However, there were no statistically significant differences in E horizon solutions. The lack of treatment effect on the E horizon solutions may have been caused in part by the smaller dataset (i.e., the shorter data collection period). Treatment effects were evident in Bt horizon soil solutions at a stage when that dataset was of a comparable size to that of the E horizon in this phase of the study, however (Johnson et al. 1998). Treatment effects on soil-water flux likely increased with depth, and this may have contributed to the greater response in the Bt than in the E horizon soil solutions. Although roots are concentrated in the upper 30 cm of soil in this forest ecosystem (Joslin and Wolfe 1998), the relative difference in soil water flux may become greater with depth as ET takes an increasing toll.

The Bt horizon soil solution data continued to show that the dry treatment effects were far more pronounced than those of the wet treatment. This result was probably caused by the disproportionate impact of the dry treatment on soil-water flux (Johnson et al. 1998, 2002). The sharp increase in Na^+ concentrations (and, to a lesser extent, Cl^-) in 1999 may also have been caused by drought.

With the resin lysimeters, we were able to detect large and statistically significant treatment effects on N fluxes from the forest floor (O horizons). The magnitudes of the N fluxes measured were within values that would be expected in this ecosystem. Nitrogen fluxes via litterfall and crownwash in this forest type are 34 and 3 kg N ha^{-1} y^{-1}, respectively (Johnson and Van Hook 1989); thus, the resin lysimeter values suggested that there was a net uptake of N (by soil organisms, tree roots, or both) from the forest floor in the ambient and dry treatments, as would be expected (with average fluxes of ~10 and 5 kg N ha^{-1} y^{-1}, respectively).

The differences in forest-floor N content between treatments provides corroborating evidence that decomposition and N mineralization from litter are being reduced by the dry treatment. Nitrogen content of litter in the dry treatment was about 70 kg N ha^{-1} greater than the ambient treatment suggesting a mean annual rate of N immobilization in the dry plot organic horizons of 12 kg N ha^{-1} y^{-1}. This rate of immobilization is equal to about three times the difference in aerially weighted mineral N flux between the ambient and dry plots in 1998–1999 (4.3 kg ha^{-1} y^{-1}; Fig. 23.3). In the 1999–2000 resin lysimeter collection, the aerially-weighted fluxes in the dry and ambient treatments were virtually identical (7.4 and 6.7 kg ha^{-1} y^{-1}, respectively) even though the fluxes beneath the troughs remained substantially and significantly lower in the dry treatment. These differences could reflect error in estimation or they could reflect a leveling off of forest-floor mass and N in the dry treatment; only future samplings will reveal which. In any event, both the forest-floor sampling in 1999 and the resin lysimeter data suggest greater sequestration of N in the forest floor in the dry treatment.

NuCM Simulations

The NuCM simulations conducted for this chapter indicated the same general patterns as those observed previously (and using a previous calibration): Soils in the dry treatment should become enriched in base cations, whereas those in the wet treatment should become depleted over time compared to the ambient treatment. The magnitudes of these changes did not suggest any measurable effect on plant productivity, however, and simulated effects on biomass were very small. These predictions remain to be tested. The NuCM model also predicted that because of slightly higher litter production in the wet treatment, forest-floor mass and nutrient contents would increase relative to the ambient treatment, and the reverse would occur in the dry treatment (lower forest-floor mass and nutrient contents). This prediction was clearly not supported by the results of this study, where a significant increase in mass and N content in the dry treatment was found and no effects were found in the wet treatment.

Conclusions

Throughfall manipulations caused disproportionate changes in soil-water flux, which, in turn, caused changes in estimated ion leaching rates. Soil solution concentrations in the dry treatment were more concentrated than in the ambient or wet treatments, but this did not outweigh the reduction in estimated soil-water flux and thus estimated ion leaching rates were, therefore, reduced. Soil solutions in the wet treatment were unaffected by treatment, and thus estimated ion leaching rates were considerably higher than in the ambient or wet treatments. Because soil-water fluxes cannot be directly measured, ion leaching rates cannot be known with certainty, however.

The NuCM simulations mimicked the patterns observed in the field in some cases (soil solution concentration increases in the dry treatment), but failed to do so in other cases (the observed increase in forest-floor mass and N content in the field was not indicated by NuCM). The NuCM predictions of long-term changes in base saturation in soils seem reasonable but remain to be tested.

Perhaps the most significant effects relative to plant productivity noted in this study were the changes in O horizon leaching rates and the corresponding increases in forest-floor N content in the dry treatment. Although these effects are statistically significant and excessive immobilization (or reduced mineralization) of N in the forest-floor might reduce N availability for growth if sustained over long time periods, the rates of N immobilization observed here were relatively small, and we hypothesize that they will eventually decline as a new steady-state in the forest-floor mass is reached.

References

Drungil CEC, Gish TJ and Abt K (1987) Soil moisture determination in gravelly soils with time domain reflectometry. Trans ASAE 32:177–180.

Hanson PJ, Todd DE, Huston MA, Joslin JD, Croker J, and Augé RM (1998) *Description and field performance of the Walker Branch Throughfall Displacement Experiment: 1993–1996.* ORNL/TM-13586. Oak Ridge National Laboratory, Oak Ridge, Tennessee.

Hargreaves GH (1974) Estimation of potential and crop evapotranspiration. Trans Am Soc Agr Eng 17:701–704.

Huff DD, Luxmoore RJ, Mankin JB, and Begovich CL (1977) *TEHM: A terrestrial ecosystem hydrology model.* ORNL/NSF/EATC-27. Oak Ridge National Laboratory, Oak Ridge, Tennessee.

Johnson DW, and Todd DE (1990) Nutrient cycling in forests of Walker Branch Watershed: Roles of uptake and leaching in causing soil change. J Environ Qual 19:97–104.

Johnson DW, and Van Hook RI (Eds) (1989) *Analysis of biogeochemical cycling processes in Walker Branch Watershed.* Springer-Verlag, New York.

Johnson DW, Geisinger DR, Walker RF, Newma J, Vose JM, Elliot KJ, and Ball JT (1994) Soil pCO_2, soil respiration, and root activity in CO_2-fumigated and nitrogen-fertilized ponderosa pine. Plant Soil 165:111–121.

Johnson DW, Hanson PJ, Todd DE Jr., Susfalk RB, and Trettin CF (1998) Precipitation change and soil leaching: Field results and simulations from Walker Branch Watershed, Tennessee. Water Air Soil Pollut 105:251–262.

Johnson DW, Susfalk RB, Gholz HL, and Hanson PJ (2001) Simulated effects of temperature and precipitation change in several forest ecosystems. J Hydrology 235:183–204.

Johnson DW, Hanson PJ, and Todd DE Jr. (2002) The effects of throughfall manipulation on soil leaching in a deciduous forest. J Environ Qual 31:204–216.

Joslin JD, and Wolfe M H. (1998) Impacts of long-term water input manipulations on fine root production and mortality in mature hardwood forests. Plant Soil 204:165–174.

Kirschbaum MU (1995) The temperature dependence of soil organic matter decomposition, and the effect of global warming on soil organic C storage. Soil Biol Biochem 27:753–760.

Lamersdorf NP, Bier C, Blanck K, Bredemeier M, Cummins T, Farrel EP, Rasmusson L, and Ryan M (1995) Reactions of soil solution chemistry to drought: Results of the EXMAN project. In Jenkins A, Ferrier RD, Kirby C(Eds) *Ecosystem manipulation experiments: Scientific approaches, experimental design, and relevant results.* Ecosystem Research Report #20. Commission of the European Communities, Brussels, Belgium, pp 86–95.

Liu S., Munson R, Johnson D, Gherini S, Summers K, Hudson R, Wilkinson K, and Pitelka L (1991) Application of a nutrient cycling model (NuCM) to northern mixed hardwood and southern coniferous forest. Tree Physiol 9:173–182.

Luxmoore RJ (1983) Water budget of an eastern deciduous forest stand. Soil Sci Soc Am J 47:785–791.

Luxmoore RJ, and Huff DD (1989) Water. In Johnson DW, VanHook RI (Eds) *Analysis of biogeochemical cycling processes in Walker Branch Watershed.* Springer-Verlag, New York, pp 197–232.

Munsen RK, Liu S, Gherini SA, Johnson DW, Wilkinson KJ, Hudson RJM, White KS, Summers KV (1992) *NuCM Code Version 2.0: An IBM PC code for simulating nutrient cycling in forest ecosystems.* Tetra-Tech, Hadley, Massachusetts.

Reuss JO (1983) Implications of the Ca-Al exchange system for the effect of acid precipitation on soils. J Environ Qual 12:591–595.

Ulrich B (1983) Soil acidity and its relation to acid deposition. In Ulrich B, Pankrath J (Eds) *Effects of accumulation of air pollutants in ecosystems.* D. Reidel Co, Dordrecht, The Netherlands, pp 127–146.

8. Extrapolation

24. Long-Term Forest Dynamics and Tree Growth at the TDE Site on Walker Branch Watershed

Michael A. Huston, Donald E. Todd, and D. Gregory Barlar

Introduction

The environmental history of the Throughfall Displacement Experiment (TDE) site is typical of much of the land on the Oak Ridge Reservation. The land was used for a mix of traditional and subsistence agriculture, including cleared pastures, woodland pastures, cultivated crops, and home sites from the mid-1800s until 1942. Following the 1942 acquisition of the area for the Manhattan Project, all agriculture was abandoned, and most of the landscape reverted to forest.

Most of the TDE site was open woodland pasture in 1942 with a cultivated field to the south across the small valley and a dirt road running along the adjacent ridgetop. Today, scattered "wolf" trees and old farm roads indicate the relatively intensive use that these forests experienced. As a consequence of this use and subsequent abandonment, the forest on the TDE site is still increasing in biomass and changing in species composition.

The life span of most or all of the tree stems on the site started during the twentieth century. For East Tennessee (NCDC Climate Division 1), the climate of that period, indicated by the Palmer Drought Severity Index (PDSI) (Fig. 24.1) Palmer 1965; Karl and Koscielny 1982; Alley 1984, 1985) was quite different in the first and second halves of the century. In the first half, the "normal" periods were punctuated by short (2–3 year) droughts, one each decade. The longest and most severe drought of that period lasted from 1938 to 1943.

The second half of the century (Figs. 24.1 and 24.2) was characterized by wider swings in climate. It began with a 3-year wet period followed by a 3.5-year

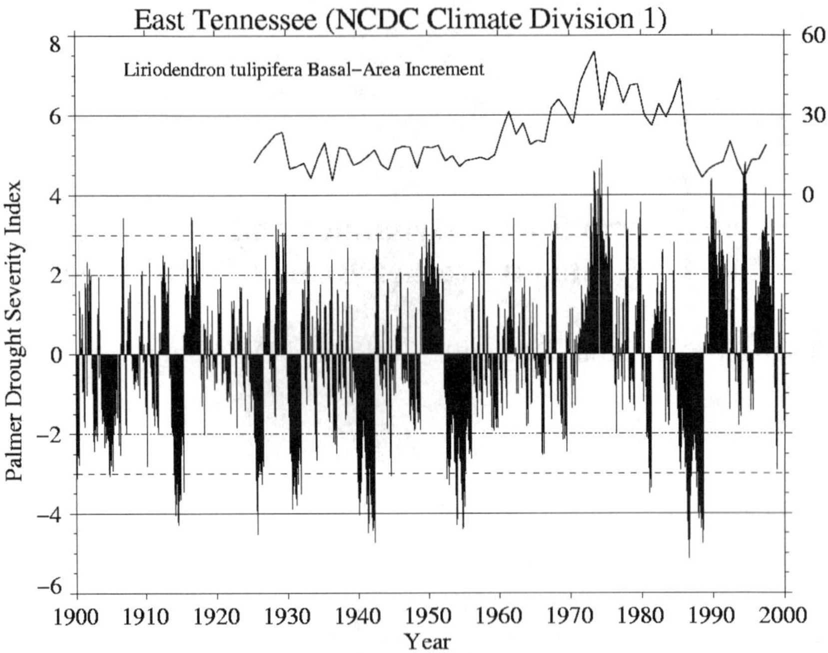

Figure 24.1. Seventy-year record of the growth of *Liriodendron tulipifera* on Walker Branch Watershed in relation to soil-water conditions estimated with the Palmer Drought Severity Index (PDSI) from 1900 to 2000. Upper jagged line is the mean annual basal-area increment of 106 *L. tulipifera* trees, calulated from increment cores collected in 1998 (scale to right). Vertical black bars indicate monthly values of the PDSI for Eastern Tennessee (Climate Division 1). Negative values represent drier conditions, and positive values represent wetter conditions. Values between -1.49 and 1.49 are "near-normal" conditions. Magnitudes of 1.50–2.99 indicate "mild to moderate" drought or wetness; 3.00–3.99 indicates "severe" drought or wetness; values greater than 4.00 indicate "extreme" wetness or drought. Note the positive response of *L. tulipifera* growth to the normal-to-wet soil-water conditions from 1960 to 1985 and the large decline in the late 1980s.

drought, the most severe of the century to that time. Following that drought, the region experienced normal to significantly above-normal soil-water conditions from 1957 to 1984. This remarkably consistent period of favorable conditions ended dramatically with the most severe drought of the century, which included 2 years (1986 and 1988) during which most of the growing season was classified as "severe" (PDSI < -3.0) to "extreme" (PDSI < -4.0). The last decade of the century, when the TDE was conducted, was characterized by soil-water conditions ranging from "near normal" (-1.49 to 1.49) to "extreme wetness" (> 4.0), with a "mild to moderate" drought beginning in 1998 (Fig. 24.2).

The effect of soil-water conditions on tree growth is dramatically shown by the dendrochronology of *Liriodendron tulipifera* sampled across Walker Branch Watershed (Fig. 24.1) (Barlar 2000). Tree growth, presented as annual basal-area increment [the increase in cross-sectional area (in cm^2)] was determined from the

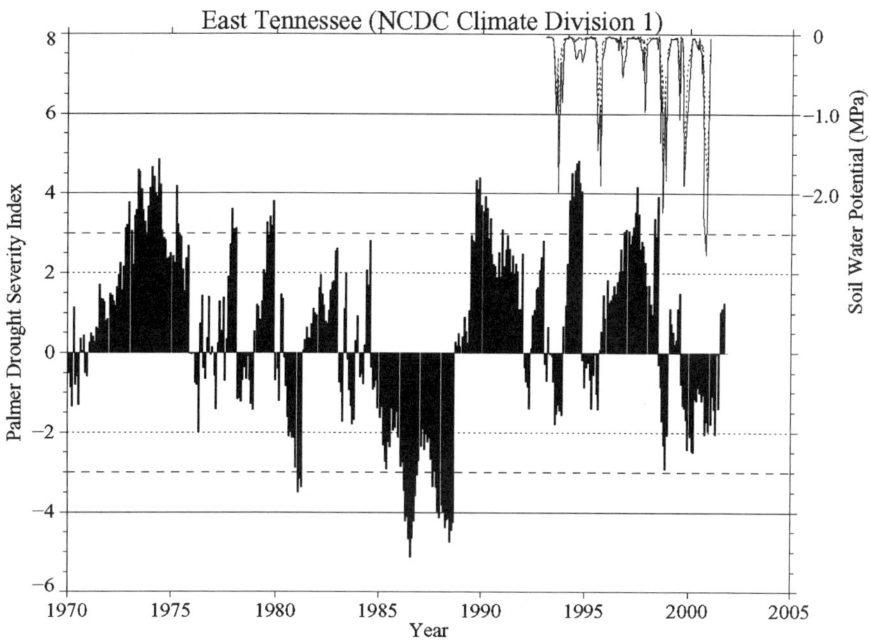

Figure 24.2. Soil-water potentials on the TDE site (Chapter 2, this volume) in relation to the PDSI from 1970 to 2001. Upper line of downward "spikes" is mean soil-water potential for the TDE ambient treatment (solid line) and dry treatment (dotted line), with the scale on the right axis. Vertical black lines indicate monthly PDSI values (see legend of Fig. 24.1). Vertical white lines show breaks between calendar years. Note the strong relationship between climatic conditions summarized in the PDSI and the soil-water potential of the TDE treatments (more-negative values indicate drier soil conditions).

dendrochronology data. The growth pattern is one of relatively constant, slow growth during the middle third of the century, increasing during the 1960s, and peaking during the early 1970s at the beginning of the century's longest period of excessively wet soil-water conditions. Growth remained high until it dropped precipitously in 1986, the first of 2 years of extreme drought. In spite of a return to wet conditions following 1988, the growth rates of the sampled trees had not increased above pre-1960 levels by 1997 (the year before the trees were cored).

Although *L. tulipifera* trees on Walker Branch were most strongly affected by the severe drought of the mid to late 1980s, the greatest impact on the forest as a whole resulted from a shorter drought in 1981 to 1982. This drought appears relatively minor in comparison to other droughts of the century, although the local drought around Oak Ridge may have been more severe than that indicated by the Division 1 PDSI, which represents the average soil-water conditions for roughly the eastern third of Tennessee. Nonetheless, this drought represents the first significant drought in 25 years, which is more than half the life span of the trees that established after the 1942 abandonment of agriculture.

It is inevitable that the species composition of the forest, as well as the biomass-allocation pattern of many of the trees (e.g., root-to-shoot ratios), was influenced by the long period of normal-to-wet soil-water conditions. For this reason, and probably others, the 1981–1982 drought resulted in a major episode of tree mortality on Walker Branch. It also produced a sharp reduction in the growth rates of most tree species on the watershed. Most of the mature *Carya* trees died by the mid-1980s as the result of a *Carya* bark-beetle outbreak that followed the drought. Following this drought, pine mortality also increased, accelerated by periodic outbreaks of southern pine-bark beetles. Most of the pure pine stands that had regenerated following the 1942 abandonment of agriculture had disintegrated by the end of the twentieth century, and they are being replaced by the hardwood understory that grew beneath them.

Forest Dynamics on the TDE Site

Ecological studies and data collection on Walker Branch Watershed began in 1967 in the middle of the most favorable climatic period of the century, which was immediately followed by the most severe drought of the century. The TDE was carried out during a period of "moderate-to-extreme" wetness that followed the severe drought (Fig. 24.2). Although both the PDSI and measured soil-water potentials indicated drier conditions on the TDE from 1998 to 2001, these conditions, even in the dry treatment plots, are not comparable in either duration or severity to the conditions experienced from 1986 to 1988 or during other severe droughts during the century.

Twelve forest-inventory plots in and near the TDE site (Fig. 24.3) have been sampled periodically since 1967, when 298 plots were established across Walker Branch Watershed (Curlin and Nelson 1968). Each plot consists of four concentric circles of radius 16.05 (0.2 acre), 11.35 (0.1 acre), 3.59 (0.01 acre), and 1.14 m (0.001 acre), within which the diameter at breast height (dbh) of all woody plants greater than 24.4 (30 in. in circumference), 9.14, 1.5, and 0 cm, respectively, are measured. All stems greater than 1.5 cm dbh are marked with permanent metal tags. Inventories have been conducted in 1967, 1970, 1973, 1979, 1983, 1987, 1991, and 1997. The 12 plots near the TDE site were remeasured in March 2001. None of these plots was sampled in the 1979 inventory, and only three were sampled in 1983 and 1987. All were sampled in 1991, 1997, and 2001.

The effects of the 1981–1982 drought are evident in the basal-area dynamics of most of the forest-inventory plots near the TDE site. Seven of the 12 plots declined in basal-area between 1973 and 1991 (Fig. 24.4). It is likely that several more of the plots would have shown a decline if they had been inventoried in 1979, just before the drought. The high rainfall period prior to the drought was one of rapid basal-area increase in most plots across the watershed. Unfortunately, none of the TDE plots was inventoried in 1979, so none shows this increase. Of the three plots that were measured during the 1980s, two (291 and 292) increased in basal area from 1987 to 1991, the period spanning the prolonged, severe drought. In most plots, the negative effects of the short drought of the early 1980s were much more severe than those of the later, longer drought, possibly because most of the drought-sensitive individuals were eliminated by the first drought.

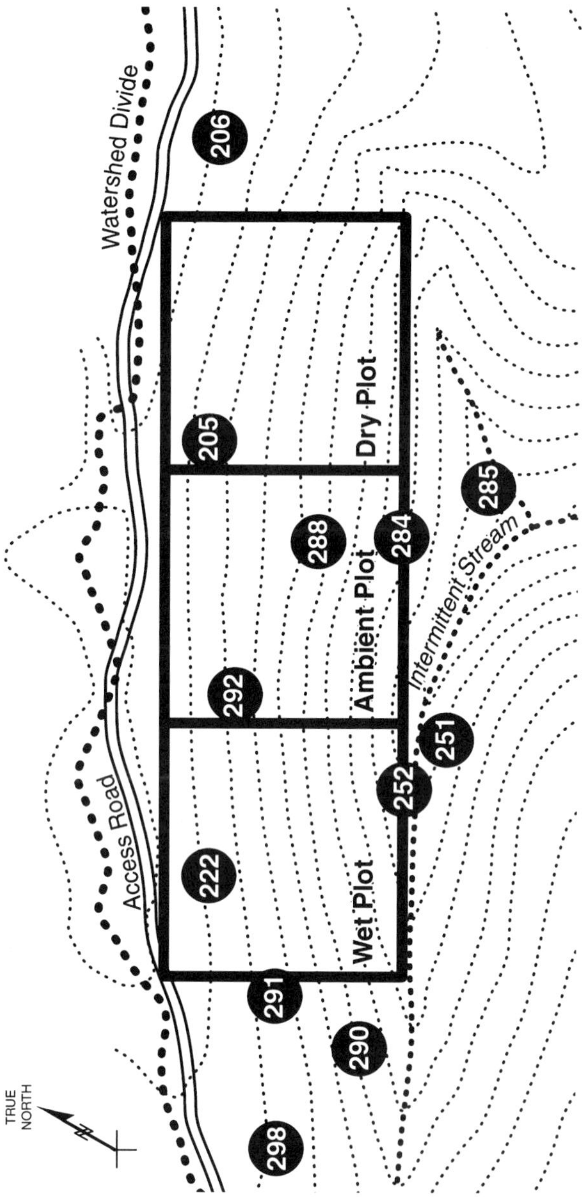

Figure 24.3. Topographic map of the TDE site, showing location of the 12 forest-inventory plots discussed in this chapter. Contour interval 3.05 m (10 ft).

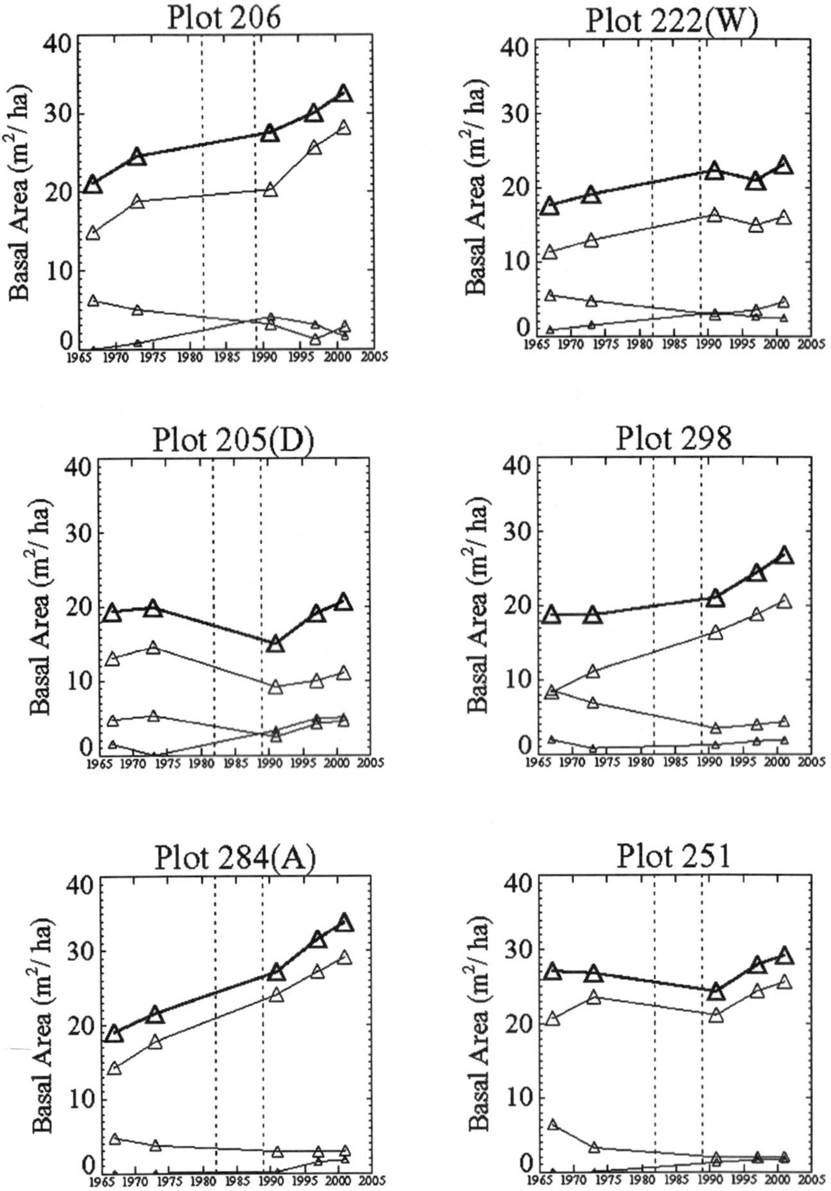

Figure 24.4. Changes over time in the total basal-area of forest-inventory plots in and around the TDE site on Walker Branch Watershed. Triangles indicate years in which a particular plot was inventoried. Horizontal dotted line is total basal area of all stems between 1.5 and 9.14 cm dbh; lower thin line is total for all stems between 9.14 and 24.4 cm dbh; middle medium line is total for all stems greater then 24.4 cm dbh. Upper bold line is sum of basal-area of all three size classes. Vertical dotted lines indicate the

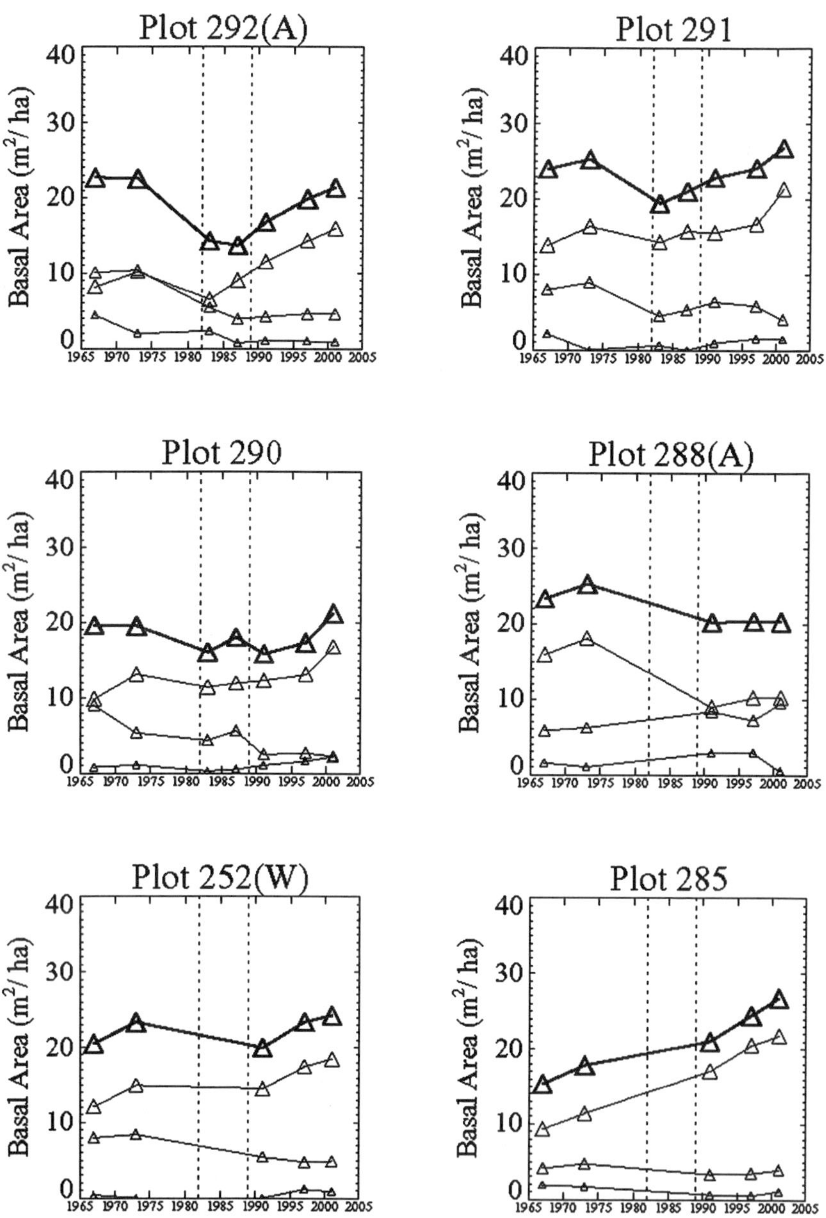

end of severe droughts in 1982 and 1989. Plots located within the TDE are indicated with a letter referring to the treatment in which they are located (wet, ambient, and dry). Note that most plots were not inventoried in 1979, 1983, or 1987. Plots are arranged from the highest topographic positions (presumably drier upper slope) in the upper left corner to the lowest topographic positions (presumably wetter valley bottom) at the lower right. Plots in the middle row are from mid-slope positions (see Fig. 24.3).

The high rainfall and wet soil conditions of the 1990s were associated with a steadily increasing basal area in all of the TDE inventory plots except two (222 and 288), both of which lost large trees prior to the 1996 inventory (Fig. 24.4). Only two of the plots exceeded a basal area of 30 m^2 ha^{-1} (extrapolated to 1 ha from the 0.08-ha plot), although all are likely to continue increasing for some years, unless trees die prior to achieving their potential maximum size.

In contrast to the relatively gradual changes in basal area, species composition in many of the plots changed significantly (Fig. 24.5). Several of the plots (292, 291, 205, 290, 288, and 252) experienced a major loss in *Carya* basal area following the 1981–1982 drought. *Carya* remained abundant only in plot 285, which was in the wettest topographic position of all the plots.

In 7 of the 12 plots, *Quercus prinus* had the highest basal area in 2001, whereas in 2 of the 12 *Q. alba* had the highest. All 12 of these plots had been dominated in 1967 by either the 2001 dominant *Quercus* species or by *Carya* (plots 290 and 291), and in most of these plots, the dominance of *Quercus* increased consistently over the 30+ years of measurement. Two inventory plots in the ambient TDE treatment area (plots 288 and 292) experienced a rapid linear increase in the basal area of *Acer rubrum,* beginning with a low basal area in 1973 and increasing to a level at least 50% greater than the next most abundant species. A third plot (252) had about the same basal area of *Acer rubrum* in 2001 as in 1973, whereas several other species increased during that period, including *Q. prinus,* which more than doubled in basal area.

The topographic moisture index of the hydrologic model TOPMODEL (Beven and Kirby 1979; Beven and Wood 1983; Beven et al. 1984; Beven 1986; Beven and Moore 1993; Huston and Fontaine 1984; Garten et al. 1994) was used to estimate the long-term soil-water environment of the 12 plots in the vicinity of the TDE. Forest structure and dynamics of the plots could then be evaluated in terms of their relative soil-water environment, which was altered by TDE treatments in only 3 of the 12 plots [205(D), 222(W), and 252(W)].

On upper-slope plots, dominance by *Quercus* species tended to increase over time, particularly if the plots were not strongly dominated by *Carya* species prior to 1980. In several of these plots, *Q. prinus* and *Q. alba* increased steadily in basal area; most other species changed very little (e.g., plots 206, 222, 298).

Initial *Carya* dominance and subsequent *Carya* mortality were highest in the upper- and mid-slope plots (292, 291, 205, 290, and 188). Because of the high *Carya* mortality, the temporal dynamics of these plots tended to be more complex than in plots with less *Carya.* Relative abundances changed over time, and the distribution of basal area among species was more even in 2001 than it was in 1967. These results are what would be expected for a relatively early successional stand prior to competitive exclusion (Huston 1979, 1994, 1999; Huston et al. 1999).

Except for the loss of *Carya* in some plots, relatively few major changes in relative abundance occurred between 1967 and 2001. In nearly every plot, the second most abundant species at the time of *Carya* mortality became the most abundant species in 2001. The only exception was plot 292, where *Nyssa sylvatica* Marsh. was surpassed by *Acer rubrum,* presumably because of *Nyssa*'s slow growth rate. Indeed, black gum basal area was relatively constant over time in most plots.

The two lower-slope plots without significant *Carya* mortality (284 and 285) show a strikingly different temporal pattern from the other plots. In both plots, all species increased in basal area at similar rates (somewhat higher for the species with highest basal area), resulting in a similar ranking of species in 2001 as in 1967. The other two lower-slope plots (251 and 252) experienced some *Carya* mortality, but they also showed a pattern of most species increasing in basal area. However, these two plots had fewer species (five and six, respectively) than did plots 284 (nine) and 285 (eight), possibly because of disturbances from an old farm road near these plots.

In summary, the forest-inventory plots in and around the TDE site show several patterns of forest dynamics. Plots that were originally dominated by *Carya,* most of which died following the 1981–1982 droughts, have had complex temporal dynamics with major changes in relative abundance and currently have higher tree diversity (particularly in terms of the evenness of basal-area distribution) than would be expected without the disturbance. Undisturbed plots show contrasting patterns, depending on slope position. Upper-slope plots generally experienced an increasing dominance by *Quercus* species between 1967 and 2001, with little increase in species other than *Quercus.* Lower-slope plots generally had little change in species dominance or relative abundance, with the basal area of most species increasing consistently through time.

Growth Rates of Individual Tree Species

Because all trees in the forest-inventory plots are individually labeled, it is possible to calculate temporal variation in tree growth in relation to topographic position and species identity. Unfortunately, none of the plots around the TDE was inventoried in 1979 (the last inventory prior to the severe drought of 1981–1982), and only three plots were inventoried during the 1980s (in 1983, immediately after the first drought, and in 1989, immediately after the second drought). Consequently, none of the data for the plots captured the high growth rates of the 1970s (see Fig. 24.1) or the decline in growth caused by the 1981–1982 drought. However, three plots (290, 291, and 292) were measured in all inventories from 1983 to the present and reveal the responses in tree growth to variation in soil-water conditions from 1983 onward. Growth rate here is presented as annual basal-area increment (cm^2 per tree per year), calculated by dividing the basal-area change between inventories by the number of years between the inventories.

The two dominant *Quercus* species, *Q. prinus* and *Q. alba,* show similar responses to soil moisture variation in the 1990s, with consistent variation between plots. However, clear differences between individuals in same plot are also observed (data not shown). Both species had increased rates of basal-area growth between the 1991 and 1997 inventories, followed by a decline in growth rate between 1997 and 2001, reflecting the shift from the wet conditions between

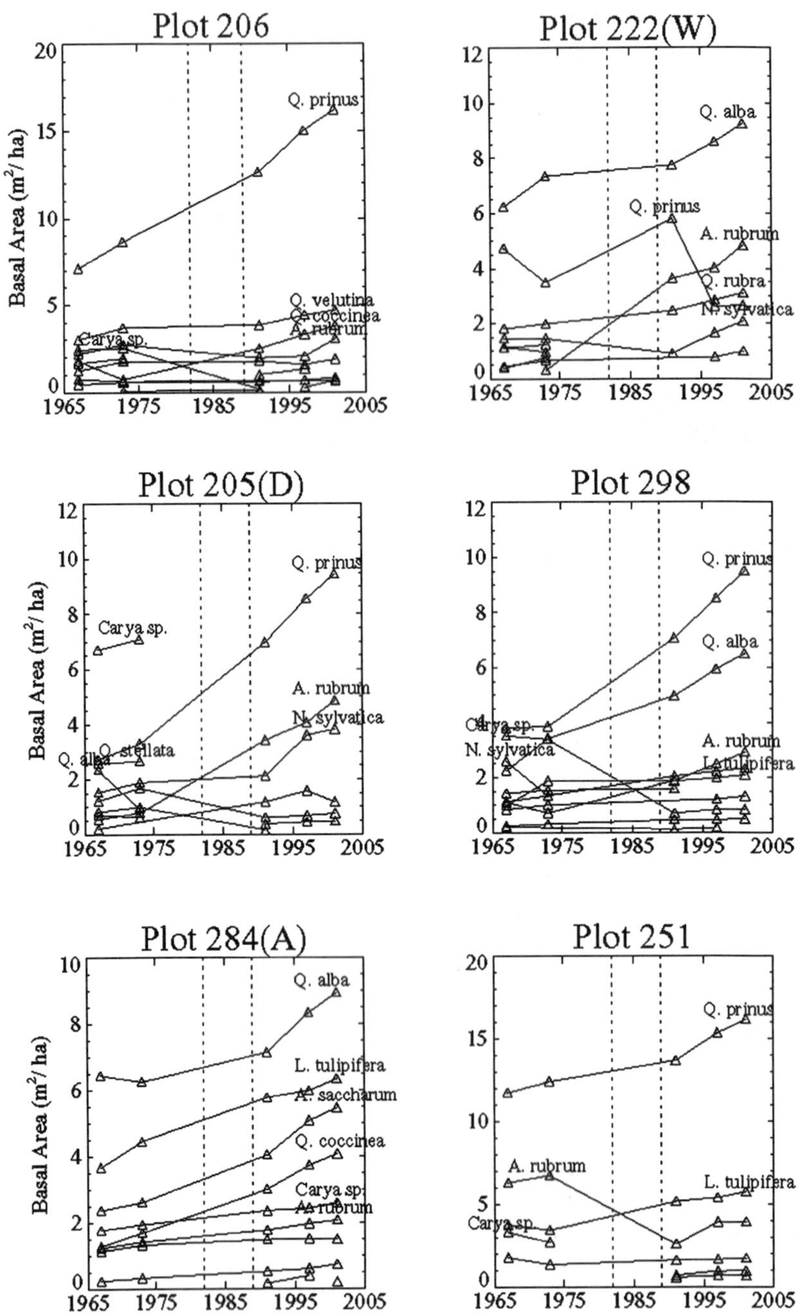

Figure 24.5. Changes over time in the total basal area by species in forest inventory plots in and around the TDE site on Walker Branch Watershed. Triangles indicate years in which a particular plot was inventoried. Lines for the major species in each plot are

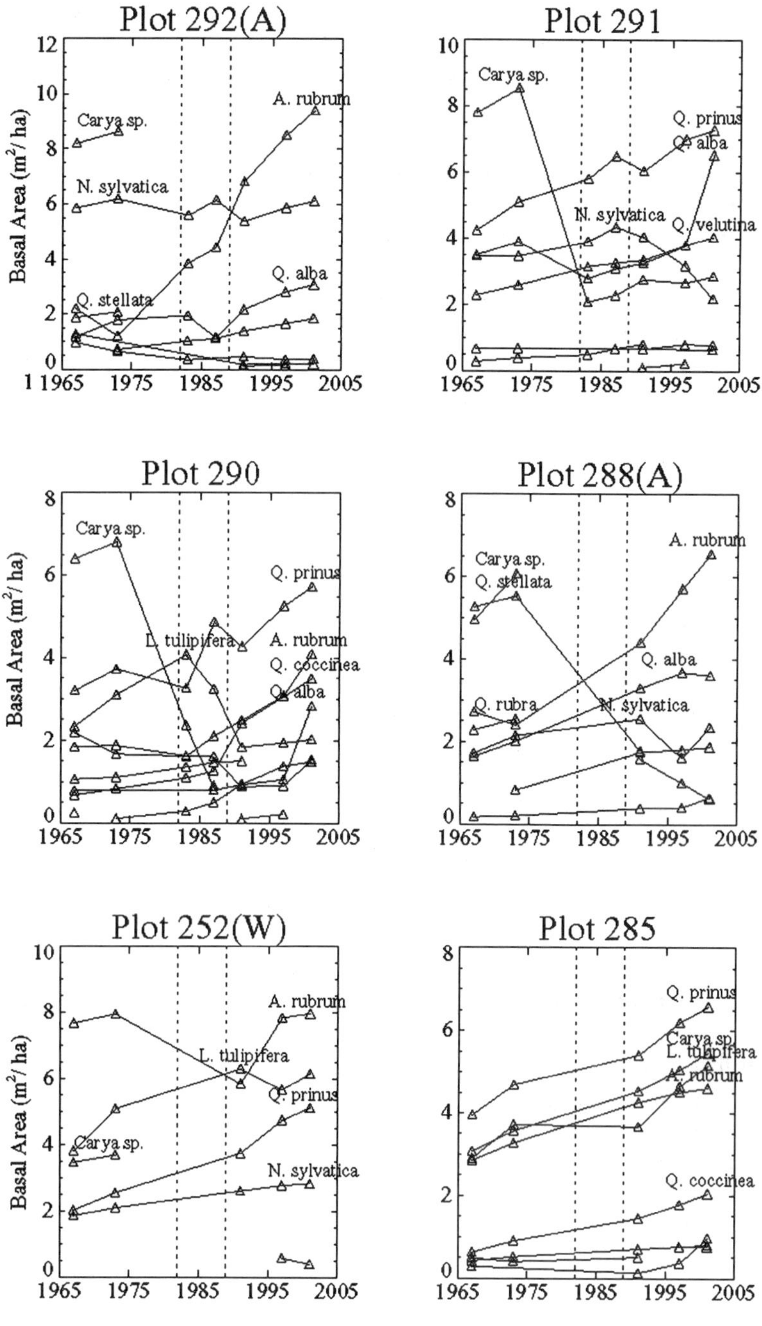

labeled at the point where the species reaches its maximum basal area. Vertical dotted lines indicate the end of severe droughts in 1982 and 1989. Plots are arranged from upslope to downslope, as in Fig. 24.4.

1990 and 1998 to the drier conditions of 1999 and 2000. For *Q. prinus,* trees in the upper-slope plots showed a stronger positive response to the wet conditions in 1994 and a stronger negative response to the drier conditions in 1995 than did trees in the lower-slope positions. For some reason, both *Quercus* species in plot 292 showed an anomalous increase during the severe late-1980s drought.

Annual growth derived from dendrochronological data for the two *Quercus* species plus *A. rubrum* on the TDE site shows a strong positive response to wet conditions, but relatively little negative response to extreme drought (Fig. 24.6). Average basal-area increment increased with increasingly wet conditions in both the early 1970s and again in the early 1990s. *A. rubrum* (Figs. 24.5 and 24.6) growth had a similar pattern to that of the *Quercus* species but with a less pronounced increase in growth during the early to mid-1990s. Black gum grew very slowly for the entire period and varied little in response to temporal or topographic variation in soil water, with the exception of a slight increase in basal-area increment in the interdrought period of the 1980s (see data for plots 291 and 292 in Fig. 25.5).

During the 1990s, *L. tulipifera* had a growth pattern that was dramatically different from that any of the other tree species (Fig. 24.6). In most plots, *L. tulipifera*'s annual basal-area increments decreased between 1991 and 1997, when soil conditions were wetter than normal, and actually increased from 1997 to 2001, when soil conditions were drier. This response is surprising, given the well-known drought intolerance of this species. However, growth rates did tend to be higher in the plots at lower topographic positions, which is consistent with the high soil moisture and nutrient requirements of this species.

The unusual 1990s growth pattern of *L. tulipifera* can perhaps be best interpreted in the context of the climatic patterns of the past 30 years or longer. As is clearly illustrated in Fig. 24.6, the period from the early 1970s to the late 1980s represents a transition from extreme wetness to extreme drought. *L. tulipifera* growth on Walker Branch reached the highest rates of the century in the early 1970s and began to decline over the late 1970s and eary 1980s until dropping precipitously in the late 1980s (Barlar 2000). This dramatic decline in growth rates was associated with high mortality of the smaller size classes, deterioration of the crowns of canopy individuals, and severe outbreaks of a host-specific weevel that heavily damaged the leaves (Barlar and Huston, unpublished data). Thus, the decline of poplar growth in the 1990s, despite highly favorable soil moisture conditions, may represent the lingering effects of the late-1980s drought and associated damage to the trees. Similarly, the increase in growth in the last few years of the century, despite the mild droughts of 1999 and 2000, may reflect a delayed recovery from the severe drought more than 10 years earlier.

Liriodendron tulipifera and *Q. prinus* show strong, and opposite, patterns of basal-area growth in relation to topographic position. *Q. prinus* had its highest growth rates and greatest temporal variation in growth in upper-slope positions. In contrast, *L. tulipifera* had its highest growth rates at lower-slope positions, where a delayed recovery from the late 1980s droughts was apparent in the late 1990s.

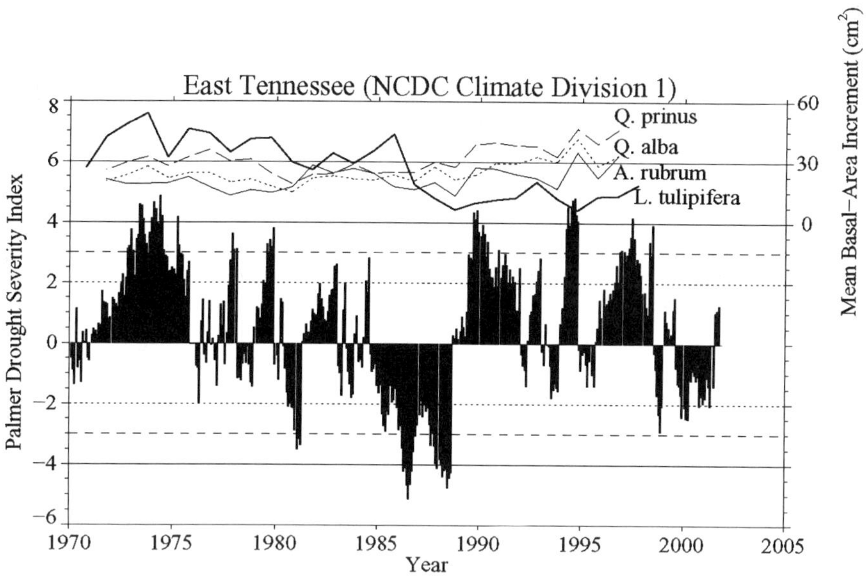

Figure 24.6. Mean annual basal-area growth of four major tree species on the TDE site in relation to the PDSI from 1970 to 2001. Four upper lines show the mean growth of *Q. prinus* (dashed line, $N = 15$), *Q. alba* (dotted line, $N = 14$) and *A. rubrum* (thin line, $N = 10$) from the vicinity of the TDE, and *L. tulipifera* (bold line, $N = 106$) from across Walker Branch Watershed, with the scale on the right axis. Vertical black bars indicate monthly PDSI values (see legend of Fig. 24.1). Vertical white lines show breaks between calendar years. Note the strong relationship between climatic conditions summarized in the PDSI and the basal-area increment, particularly for drier years following wet years.

Discussion

The effects of drought or other stresses on a forest will obviously vary with the severity of the stress. The relatively minor effects of the TDE drought treatment on the forest are consistent with the unusually wet conditions that characterized most of the experimental period and the difficulty of experimentally producing conditions as extreme as those created naturally by climate. Topographic effects on individual tree growth were generally as strong as treatment effects.

Nonetheless, severe droughts do occur fairly frequently and can have major and long-lasting effects on the forest. The relatively short drought of the early 1980s had a large impact on the forest of Walker Branch Watershed. It resulted in an average 17% reduction of stand basal area in the ~150 long-term inventory plots across the watershed that were measured in both 1979 and 1983. This rapid reduction in live basal area resulted primarily from mortality of *Carya* and pines. However, elevated levels of mortality have continued to afflict a number of tree species, with the consequence that the average stand basal-area increments had reached only 30% of the 1979 rate by the time of the 1997 inventory. Basal-area increment was still low 15 years after the 1981–1982 drought, despite the fact that soil-water conditions were very favorable from 1990 through 1997.

The long-term negative effects on Walker Branch Watershed from the drought of the 1980s were manifested primarily through pests and pathogens, although physical damage from branch and root mortality may have been a factor, as well. The sudden mortality of *Carya* in the early 1990s was associated with an outbreak of the native *Carya* bark beetles. Dogwood anthracnose, an introduced fungus, nearly eliminated dogwood as an understory species. Pines, particularly the Virginia and shortleaf pines established in the 1940s, continued to suffer mortality during periodic outbreaks of southern pine bark beetles, and today most of the pine stands on the watershed have been eliminated. *Quercus* dieback, possibly caused by *Armillaria* fungus, continues to kill scattered mature *Quercus* trees across the watershed. *L. tulipifera,* which continued to suffer epidemic outbreaks of the *L. tulipifera* weevil (*Odontopus calceatus*; cf. Burns and Gibson, 1968; Hyche 1990, 1992) through the mid-1990s, continues to decline as individuals with greatly reduced canopies slowly die. Such effects of drought are not unusual and have been implicated in various episodes of "forest decline" that have disappeared with the return of normal precipitation (e.g., Kolb and McCormick 1993; Tainter et al. 1990). Droughts have both immediate (e.g., Abrams et al. 1998; Liu and Muller 1993; Tainter et al. 1990) and delayed (e.g., Pedersen 1998) effects on tree growth and mortality.

Some of the drought impacts on Walker Branch Watershed, particularly the high mortality of the early successional pines and *L. tulipifera,* were intensified by the land-use history of the forest that increased the abundance of these drought-sensitive species. However, catastrophic drought effects have also been observed in undisturbed old-growth forests (McGee 1984). Four dry years that fell between 1976 and 1981 resulted in a 15% decline in stand basal area in Dick Cove, a mature mixed-mesophytic old-growth forest on the western slope of the Cumberland Plateau near Sewanee, Tennessee, 120 miles southwest of Oak Ridge. The highest mortality was of *Carya* sp. (26.5% of all stems greater than 43 cm dbh), *Q. rubra* (17.4%), and *Q. alba* (18.8%). For all stems greater than 1 m in diameter (including trees between 200 and 375 years of age), mortality was nearly 40% by 1982.

Such catastrophic mortality of dominant and codominant trees can result in rapid changes in forest structure, including the loss of biomass (and the release of carbon through subsequent decompositon) as well as changes in species composition and successional dynamics. It seems plausible that long periods of relatively wet conditions may predispose trees to higher mortality in subsequent droughts. This effect would be expected to be stronger in relatively young forests, such as Walker Branch, where most of the trees had matured under favorable climatic conditions. However, in the case of the old-growth forest at Dick Cove, the 4 years preceeding the start of the drought period averaged 26% higher in precipitation than the 25-year average of 61 in.

Thus, in both secondary and old-growth forests, drought impacts on forests seem to be strongest when droughts follow long wet periods. To the degree that future climate resembles the prolonged wet to normal conditions with infrequent severe droughts of the late twentieth century, drought impacts on forests are likely to be greater than they would be with shorter-term fluctuations between wet and dry periods.

Conclusion

The temporal dynamics of the total basal area, the relative basal area of individual tree species, and the growth rates of individual trees and tree species in long-term forest-inventory plots near the TDE site show the strong effects of long-term patterns of soil-water availability, but no detectable effects of the TDE treatments themselves. The lack of any clear effect of the TDE is not surprising, given the small number of long-term plots that, by chance, fell within the TDE site. However, the long-term changes in forest structure and composition in response to the climatic patterns of the twentieth century provide the context in which the experimental results must be interpreted.

The climatic context of the TDE is that it was conducted during the longest drought-free period of normal to "severely wet" soil conditions of the entire century, which immediately followed the longest and most severe drought of the century. The TDE drought treatments were relatively weak in comparison to the natural conditions of "severe drought" that preceded the experiment. The most drought-sensitive individuals and species were probably eliminated by the droughts of the 1980s. Furthermore, the passive design of the experiment allowed differences among the treatments to disappear during prolonged wet or dry periods.

Consequently, the observed responses to the TDE treatments should be interpreted as indications of processes at the individual tree level, the community level, and the ecosystem level that respond to moderate variation in soil-water conditions. However, the observed responses are inevitably much smaller than those that would occur during prolonged natural droughts and may not include many of the major processes that influence forest structure and dynamics during severe droughts.

References

Abrams MD, Ruffner CM, Morgan TA (1998) Tree-ring response to drought across species and contrasting sites in the ridge and valley of central Pennsylvania. For Sci 44:550–558.

Alley WM (1984) Palmer Drought Severity Index: Limitations and assumptions. J Clim Meteorol 23:1100–1109.

Alley WM (1985) Palmer Drought Severity Index as a measure of hydrologic drought. Water Res Bull 21:105–114.

Barlar DG (2000) *Temporal and spatial variation in annual growth of yellow-poplar (Liriodendron tulipifera L.) at Walker Branch Watershed, Tennessee.* M.S. Thesis, The University of Tennessee, Knoxville, Tennessee.

Beven KJ (1986) Hillslope runoff processes and flood frequency characteristics. In Abrahams AD (Ed) *Hillslope processes.* Allen and Unwin, Boston, pp 187–202.

Beven KJ, Kirby MJ (1979) A physically-based, variable contributing area model of basin hydrology. Hydrol Sci Bull 24:43–69.

Beven KJ, Moore ID (Eds) (1993) *Terrain analysis and distributed modeling in hydrology.* Wiley, New York.

Beven KJ, Wood EF (1983) Catchment geomorphology and the dynamics of runoff contributing areas. J Hydrol 65: 139–58.

Beven KJ, Kirby MJ, Schoffield N, Tagg A (1984) Testing a physically-based flood forecasting model (TOPMODEL) for three UK Catchments. J Hydrol 69:119–143.

Burns DP, Gibson LP (1968) The leaf-mining weevil of yellow-poplar. Canadian Entomologist 100:421–429.

Curlin JW, Nelson DJ (1968) *Walker Branch Watershed Project: Objectives, facilities, and ecological characteristics.* ORNL-TM-2271. Oak Ridge National Laboratory, Oak Ridge, Tennessee.

Garten CT, Huston MA, Thoms C (1994) Topographic variation of soil nitrogen dynamics at Walker Branch Watershed, Tennessee. For Sci 40:497–513.

Huston MA (1979) A general hypothesis of species diversity. Am Nat 113:81–101.

Huston MA (1994) *Biological diversity: The coexistence of species on changing landscapes.* Cambridge University Press, Cambridge, England.

Huston MA (1999) Forest productivity and diversity: Using ecological theory and landscape models to guide sustainable forest management. In Aguirre-Bravo C, Franco CR (Eds) *North American Science Symposium: Toward a unified framework for inventorying and monitoring forest ecosystem resources.* USDA Forest Service Proceedings RMRS-P-12. Rocky Mountain Research Station, Fort Collins, Colorado. pp 329–341.

Huston MA, Fontaine T (1994) Predicting temporal and spatial flood dynamics using a pre-calibrated model. Water Res Bull 30:651–661.

Huston MA, McVicker G, Nielsen J (1999) A functional approach to ecosystem management: Implications for Species Diversity. In Szaro RC, Johnson NC, Sexton WT, Malk AJ (Eds) *Ecological stewardship: A common reference for ecosystem management, Vol. II.* Elsevier, New York, pp 45–85.

Hyche LL (1990) Ovipositional habits of *Odontopus calceatus* (Coleoptera: Curculionidae) on yellow-poplar. J Ent Sci 25:493–495.

Hyche LL (1992) *The yellow-poplar leaf-mining weevil: A guide to recognition of habits in Alabama.* Bulletin 622. Alabama Agricultural Experiment Station, Auburn, Alabama.

Karl TR, Koscielny AJ (1982) Drought in the United States: 1895–1981. J Clim 2:313–329.

Kolb TE, McCormick LH (1993) Etiology of sugar maple decline in four Pennsylvania stands. Can J For Res 23: 2395–2402.

Liu Y, Muller RN (1993) Effect of drought and frost on radial growth in red oaks is associated with short-term changes in climate. Eur J For Path 20:95–105.

McGee CE (1984) *Heavy mortality and succession in a virgin mixed mesophytic forest.* Research Paper SO-209. USDA Forest Service, Southern Research Station, New Orleans, Louisiana.

Palmer WC (1965) *Meteorologic drought.* Res. Pap. No. 45. US Department of Commerce, Weather Bureau, Washington, District of Columbia.

Pedersen BS (1998) The role of stress in the mortality of Midwestern oaks as indicated by growth prior to death. Ecology 79:79–93.

Tainter FH, Retzlaff WA, Starkey DA, Oak SW (1990) Decline of radial growth in red oaks is associated with short-term changes in climate. Eur J For Path 20:95–105.

25. Simulated Patterns of Forest Succession and Productivity as a Consequence of Altered Precipitation

Stan D. Wullschleger, Carla A. Gunderson, M. Lynn Tharp, Darrell C. West, and Wilfred M. Post

Introduction

Individual-based models of forest succession have long been used to address how productivity and species composition might respond to long-term changes in climate (Pastor and Post 1988; Solomon 1986; Urban et al. 1993; Bugmann et al. 2001). Because population dynamics are so strongly coupled with the forest water, carbon, and nitrogen cycles, gap models are well suited for the study of forest responses to environmental change. These models simulate a range of plant and ecosystem dynamics by considering interactions between physiological processes and individual tree growth, demographic processes and tree-population dynamics, microbial processes and nitrogen availability, and the roles played by climate and soils as they combine to influence site water balance (Shugart et al. 1992).

Most gap models simulate a dramatic decline in steady-state and transient forest productivity with climatic change, specifically as a result of rising temperatures and temperature-induced impacts on soil-water availability (Pastor and Post 1985; Solomon 1986; Badeck et al. 2001; Bugmann et al. 2001; Shao et al. 2001). The empirical nature of gap models has, however, been criticized, and questions have been raised about the ultimate reliability of such predictions. Loehle and LeBlanc (1996) highlight several limitations of gap models and argue that changes must be made in the representation of various processes in these models in order to achieve greater confidence in model predictions. Fortunately, many of the perceived limitations of gap models can be overcome by the inclusion of physiological detail and improved representations of process-level constraints on ecosystem function. These improvements offer to move gap models from empirical-based tools to more process-based models that mechanistically describe the cycling of carbon, water, and nutrients within terrestrial ecosystems (Friend et al. 1997; Norby et al. 2001; Wullschleger et al. 2001b).

In this chapter we examine the structure of existing gap models, and based on our experience from the Throughfall Displacement Experiment (TDE), we make several modifications to the forest succession model LINKAGES (Pastor and Post 1985). Once these changes are made, computer simulations are run for an upland oak forest typical of East Tennessee with both the original LINKAGES and the new LINKAGES v2.2. The overall ability of LINKAGES v2.2 to predict forest water use, productivity, and species composition under the current climate is examined, and the long-term consequences of increasing (+33%) and decreasing (−33%) precipitation are evaluated.

Methods

Models Used for Simulating Patterns of Forest Succession and Productivity

The gap model LINKAGES, as originally developed by Pastor and Post (1985), was used to provide a baseline for subsequent comparison to LINKAGES v2.2. The LINKAGES model is based on the individual-tree model FORET (Shugart and West 1977), but it incorporates additional ecosystem functions of soil-water balance, litter return and decomposition along with nitrogen dynamics, and the effect of these factors on tree establishment and growth. In LINKAGES, diameter growth is defined by the size of an individual relative to a species-specific maximum, and trees respond to light, soil moisture, soil fertility, and temperature according to scaling functions that reduce tree growth as influenced by less than optimum soil, climate, and stand conditions. In this model, trees respond to soil-water deficit by reducing their diameter increment as a species-specific maximum drought tolerance is approached. Drought days are an important determinant of this response and are defined as the number of days during which soil-water potential falls below −1.5 MPa (Pastor and Post 1985). In LINKAGES, drought days can occur at any time during the year and, regardless of when they occur, have equal weighting in terms of their impact on diameter growth. Evapotranspiration is computed monthly on the basis of inputs of temperature and precipitation (Thornthwaite and Mather 1957). The decomposition and nitrogen dynamics of LINKAGES are, in many respects, similar to those in the FORTNITE model (Aber and Melillo 1982).

LINKAGES v2.2 incorporated several changes that reflect improved process knowledge:

1. The monthly description of evapotranspiration (ET) was replaced with a daily scheme in which evaporation from the soil surface and canopy transpiration were treated separately (Shuttleworth and Wallace 1985; Federer et al. 1996). A maximum stomatal conductance was specified for the stand, and transpiration was modeled on the basis of the response of stomatal conductance to radiation, temperature, vapor-pressure deficit, and extractable soil water. Interception losses were determined for the canopy from leaf area and stem area indexes (Federer 1995).
2. The bucket model of soil-water extraction in LINKAGES, which was predicted on the basis of a single soil layer, was replaced with an approach that used multiple soil layers and a scheme that extracted water for transpiration from each soil layer as modified by the relative distribution of roots within the soil profile. Eighty-one percent of the total root biomass was allocated to the upper 50 cm of soil, another 11% to the layers from 50 to 70 cm, and the remaining 8% to deeper-soil profiles (70–100 cm).

In addition to these enhancements, several lessons learned from research specifically on the TDE were used to modify the LINKAGES model:

3. The calculation of drought days was weighted according to the time of year that soil water deficits occurred (Fig. 25.1). This weighting factor reflected the fact that although late-season droughts are frequently observed on the TDE, they are of little consequence to the growth increment of the species studied on the TDE in comparison to early-season droughts. As characterized by Han-

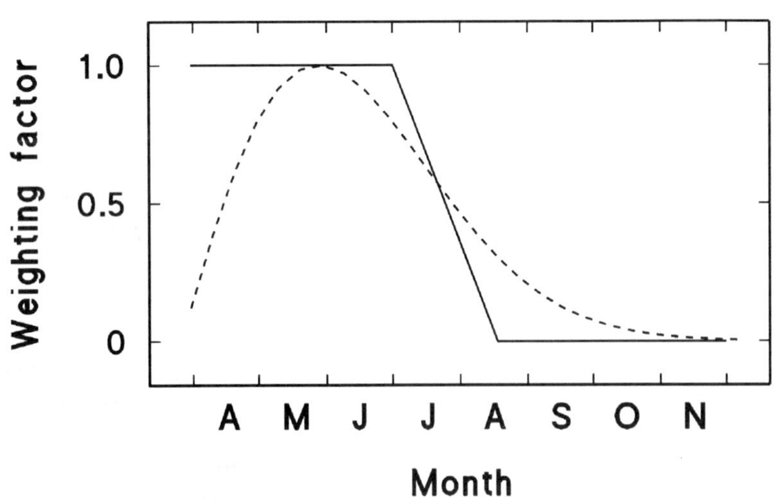

Figure 25.1. The scheme by which drought days in LINKAGES v2.2 were weighted according to when they occurred during the growing season. The dashed line shows the seasonal pattern of growth increment (relative to maximum) as determined by Hanson et al. (2001).

son and Weltzin (2000) and Hanson et al. (2001), there is a temporal mismatch between late-season drought and diameter growth. The weighting factor was an attempt to account for this temporal mismatch.

4. Rather than allow seedlings, saplings, and mature trees to have equal access to all soil layers and thus to all extractable soil water, roots were restricted to specific soil layers based on whether plants were designated seedlings, saplings, or mature trees. Seedlings occupied the soil layers from 0 to 50 cm, saplings occupied soil layers from 0 to 70 cm, and mature trees occupied the entire soil profile. The number of drought days for each size class was estimated separately, and as a result, the LINKAGES v2.2 model mimicked the differential susceptibility of seedlings, saplings, and mature trees to drought.

Other modifications to LINKAGES were also made. These changes or improvements included replacing the parabolic dependency of diameter growth on growing degree days with an asymptotic function for each species that mimicked the lower portion of the temperature–response function but did not impose a growth reduction under higher temperatures (Talkkari et al. 1999; Bugmann and Solomon 2000). Furthermore, canopy leaf area development was made a function of air temperature, thus allowing for year-to-year variation in stand phenology. Other than these changes, LINKAGES v2.2 retains all other components of the original LINKAGES model. Neither model accounts for major ecosystem disturbances, such as pest, fire, or disease.

Simulations and Model Evaluation

An environmental dataset was compiled for use in these studies from 28 years (1973–2000) of weather data collected on or near Walker Branch Watershed. Hourly data were available for radiation, air temperature, precipitation, relative humidity, and wind speed. From these data, monthly averages of air temperature and precipitation were calculated for use in LINKAGES, and daily averages were derived for use in the higher-resolution LINKAGES v2.2. Long-term simulations with LINKAGES and LINKAGES v2.2 were accomplished by running the models using repeated cycles of the 28-year climate dataset.

Simulations for LINKAGES and LINKAGES v.2.2 were run from bare ground to steady state (600 years). Increases (+33%) and decreases (−33%) in precipitation were then gradually changed over a 100-year period, and simulations were allowed to run for an additional 500 years. All data reported from these simulations are for 200 model plots, and averages are presented. Species composition, basal area, and stems ha^{-1} calculated by an earlier version of this model (LINKAGES v2.1) have been compared against inventory data from the Walker Branch Watershed (Bugmann et al. 2001). Simulated rates of annual ET from LINKAGES v2.2 were evaluated against catchment-scale estimates of net gain for the Walker Branch Watershed (Wilson et al. 2001). Net gain is the difference between precipitation and stream flow, which on an annual basis is closely related to the loss of water from the landscape via ET (Luxmoore and Huff 1989).

Results

Model Validation: Estimates of Evapotranspiration

Annual rates of ET as simulated by LINKAGES v2.2 for the years between 1973 and 2000 agreed well with catchment-scale estimates of net gain from v-notched weirs (Fig. 25.2). Modeled ET varied between 528 and 727 mm y^{-1} and averaged 638 mm y^{-1} for the 28-year period. This annual average compared favorably with the average 646 mm y^{-1} as obtained from the v-notched-weir data. Although rates were comparable in magnitude, estimates of net gain from catchment-scale measurements were much more variable than those simulated by LINKAGES v2.2 (Fig. 25.2).

Considerable interannual variation was also observed for transpiration, evaporation, and canopy interception as simulated by LINKAGES v2.2 (Table 25.1). Annual estimates of transpiration varied between 332 mm y^{-1} in 1983 and 488 mm y^{-1} in 1981 and 2000, whereas evaporation and interception losses across all years averaged 104 and 116 mm y^{-1}, respectively. Simulated rates of transpiration averaged 65% of total ET, whereas ET as a fraction of precipitation varied between 34% and 65% and averaged 49%. Annual runoff, as the difference between precipitation and ET, ranged from 345 and 1204 mm y^{-1}, and averaged 700 mm y^{-1} (Table 25.1).

Model Simulations: Original LINKAGES

Species composition and biomass production for the Walker Branch Watershed was simulated for 600 years with the original LINKAGES model of forest succession. Fifty years after the simulation started from bare ground, standing bio-

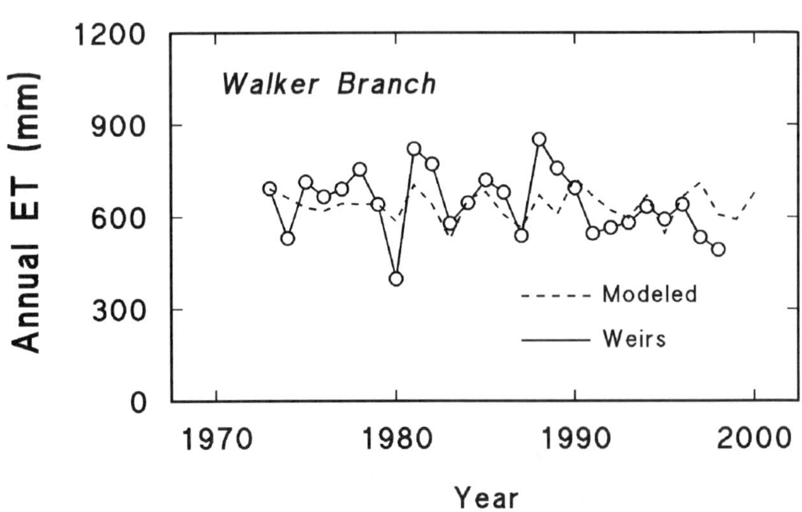

Figure 25.2. Annual estimates of evapotranspiration as simulated by LINKAGES v2.2 for Walker Branch Watershed and a 28-year dataset for measured catchment water yield.

Table 25.1. Individual components of evapotranspiration, precipitation, and runoff as simulated by LINKAGES v2.2 for the 28-year climate record used to model the response of forest succession and productivity to altered precipitation regimes

Year	Transpiration (mm y^{-1})	Evaporation (mm y^{-1})	Interception (mm y^{-1})	Precipitation (mm y^{-1})	Runoff (mm y^{-1})
1973	458	115	119	1895	1204
1974	408	125	130	1495	797
1975	386	131	113	1536	907
1976	416	93	111	1202	583
1977	433	94	117	1705	1063
1978	433	102	106	1241	600
1979	409	102	132	1661	1019
1980	361	138	87	1019	437
1981	488	101	115	1082	378
1982	419	95	126	1525	887
1983	332	99	96	1121	594
1984	449	127	84	1261	602
1985	466	103	112	1080	400
1986	425	85	97	992	388
1987	374	92	98	908	345
1988	448	101	122	1242	572
1989	360	99	148	1799	1192
1990	481	100	146	1546	820
1991	448	93	126	1471	806
1992	414	96	112	1018	399
1993	379	112	105	1149	557
1994	423	96	149	1656	988
1995	343	92	113	1151	604
1996	428	111	126	1651	988
1997	454	100	157	1461	751
1998	391	121	95	1217	613
1999	388	108	95	1130	581
2000	488	81	109	1207	532
Average	418	104	116	1335	700

Note: Simulations are for a canopy leaf area index of 6.2.

mass was estimated to be 145 Mg ha^{-1}. Species composition at this time was dominated by *A. rubrum* (28%), *Q. alba* (15%), *Q. velutina* (11%), *Oxydendron arboreum* (11%), and *Fagus grandifolia* (9%) with lesser amounts of *Fraxinus americana* (6%), *Liriodendron tulipifera* (5%), and *Carya glabra* (5%). Maximum biomass was simulated to be approximately 245 Mg ha^{-1} after 160 years, with *Q. alba, Q. velutina,* and *A. rubrum* dominating the site. Biomass at the end of the 1200-year simulation period, assuming no change in ambient precipitation, was 220 Mg ha^{-1} (Fig. 25.3A). Species composition at this time was dominated by *Q. velutina* (28%) and *Q. alba* (23%), with lesser contributions from *F. americana* and *C. glabra* (Table 25.2).

Species composition and biomass, as simulated by LINKAGES, were both sensitive to changes in precipitation (Fig. 25.3A). A 33% increase in precipitation increased stand biomass at 1200 years by ~5% (to 230 Mg ha^{-1}) compared to the ambient precipitation scenario. Species composition changed only slightly (Table 25.2). By contrast, a 33% decrease in precipitation resulted in a precipitous decline in simulated biomass. Standing biomass at 1200 years was reduced by

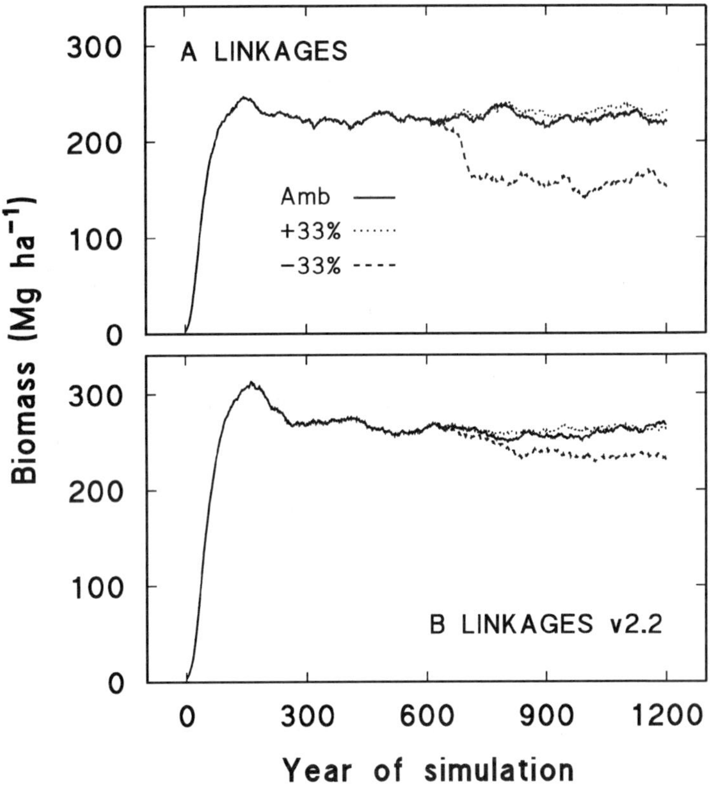

Figure 25.3. Total forest biomass as simulated by (A) the monthly version of LINKAGES and (B) LINKAGES v2.2 for Walker Branch Watershed. Results are shown for ambient precipitation, +33% average precipitation, and –33% average precipitation.

Table 25.2. Species composition as simulated by the LINKAGES and LINKAGES v2.2 models for each precipitation scenario

	LINKAGES			LINKAGES v2.2		
Species	Amb	+33%	−33%	Amb	+33%	−33%
Acer rubrum	6	6	8	na	na	na
Carya glabra	13	15	2	3	4	2
Fagus grandifolia	5	6	1	12	13	3
Fraxinus americana	15	15	7	6	7	2
Quercus alba	23	22	44	50	48	86
Quercus prinus	5	4	3	6	6	1
Quercus rubra	na	na	na	11	12	1
Quercus velutina	28	28	33	7	7	3
Oxydendron arboreum	4	4	na	2	1	na
Others	1	0	2	3	2	2

Note: The percent contribution of species to total stand biomass at 1200 years is shown for ambient, −33% precipitation, and +33% precipitation scenario.

30% (to 155 Mg ha^{-1}) compared to the ambient scenario, and species composition changed considerably. In particular, there was a general decrease in the abundance of *F. americana;* a gradual disappearance of *F. grandifolia, C. glabra,* and *O. arboreum;* and an overall increase in the abundance of *Quercus* species, especially the more drought-tolerant species *Q. alba* and *Q. velutina* (Table 25.2).

Model Simulations: LINKAGES v2.2

Biomass at stand age of 50 years was estimated by the LINKAGES v2.2 model to be 158 Mg ha^{-1} (Fig. 25.3B). Species composition for a stand at this age was dominated by *L. tulipifera* (23%), *Q. prinus* (12%), *O. arboreum* (11%), *A. rubrum* (10%), and *Q. alba* (8%) with lesser amounts of *F. grandifolia, C. glabra,* and *Q. rubra.* Maximum biomass was simulated to be 310 Mg ha^{-1} after 160 years, which represented a higher estimate of biomass than that simulated by LINKAGES. *Liriodendron tulipifera, A. rubrum, Q. alba,* and *Q. prinus* dominated the site at this time. The pioneer species *L. tulipifera* dominated early (0–100 years) in succession, but disappeared later in the simulations as it was replaced by *Q. alba.* Biomass at the end of the 1200-year simulation period, assuming no change in ambient precipitation, was 268 Mg ha^{-1} (Fig. 25.3B). Species composition at this time was dominated by *Q. alba* (50%), *F. grandifolia* (12%), and *A. rubrum* (11%), with lesser amounts of *Q. velutina, Q. prinus,* and *F. americana* (Table 25.2).

In contrast to long-term simulations with the LINKAGES model, species composition and the production of biomass were less sensitive to changes in precipitation as simulated with LINKAGES v2.2 (Fig. 25.3B). Increasing precipitation in the model by 33% had little effect on biomass or species composition at the

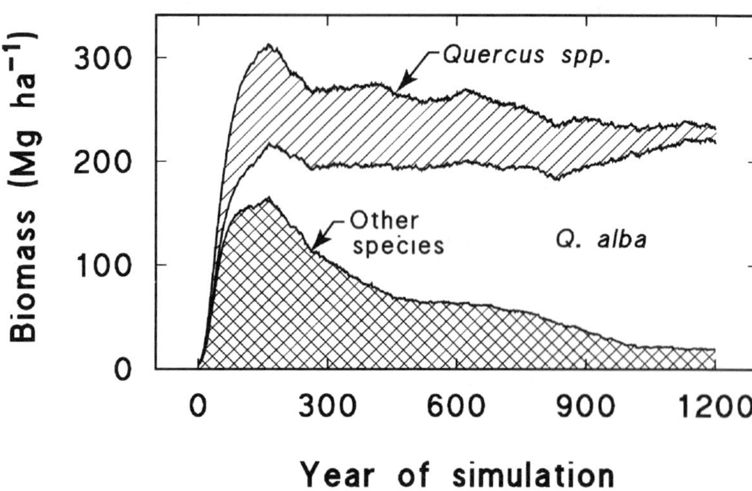

Figure 25.4. Species composition over a 1200-year period as simulated by LINKAGES v2.2 for ambient precipitation, +33% average precipitation, and –33% average precipitation.

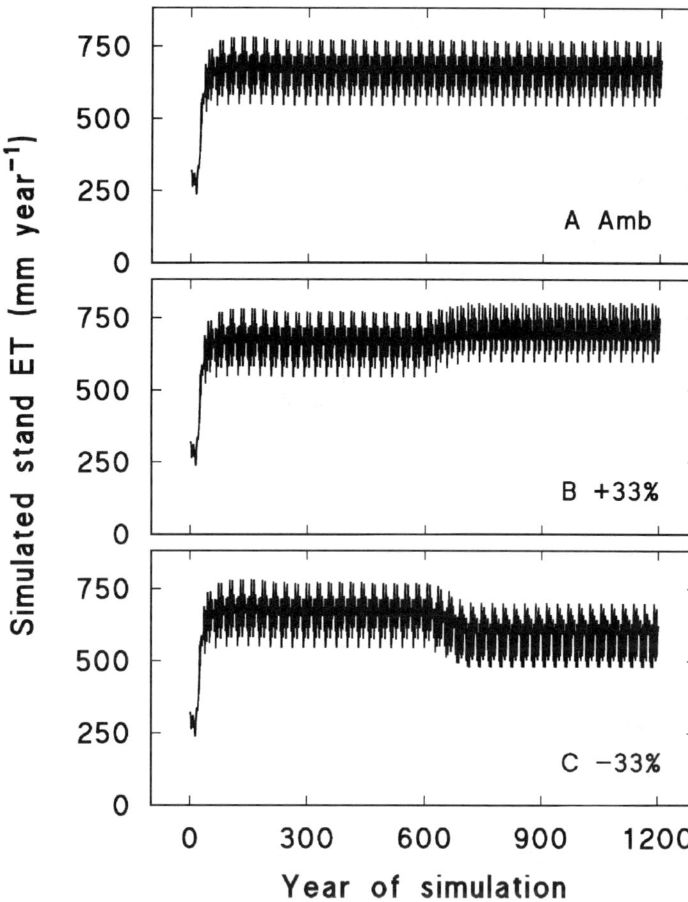

Figure 25.5. Simulated rates of annual evapotranspiration for Walker Branch Watershed as obtained using the LINKAGES v2.2 model. Simulations are shown for (A) average precipitation, (B) a 33% increase, and (C) a 33% decrease in precipitation.

end of the 1200-year simulation, and a 33% decrease in precipitation resulted in only a modest 10–15% reduction in standing biomass. However, although a 33% reduction in precipitation resulted in little change in simulated biomass, species composition at 1200 years shifted from being a mixture of *Q. alba, F. grandifolia,* and *Q. rubra* under ambient precipitation, to a stand dominated almost exclusively by *Q. alba* (Table 25.2). This increase in the abundance of *Q. alba* occurred slowly during the final 500 years of the simulation, as other *Quercus* species, including *Q. rubra, Q. prinus,* and *Q. velutina,* gradually disappeared, and the abundance of *F. grandifolia, F. americana,* and *C. glabra* generally declined (Fig. 25.4).

Modeled estimates of ET for the first few years of the simulations were 250–300 mm y^{-1} (Fig. 25.5). Because the leaf area index (LAI) during this period of stand development was low, the main contributor to ET was evaporation. At a stand age of 50 years, however, estimates of ET had increased to 620 mm y^{-1}, with the fractional contribution of transpiration increasing as the canopy leaf area developed. Interannual variation in ET was high, varying between 550 and 775 mm y^{-1} because of year-to-year differences in radiation, temperature, vapor-pressure deficit, soil-water content, and, to a lesser extent, simulated effects of climate on LAI duration (i.e., phenology).

Increases and decreases in precipitation had only a modest effect on ET as modeled by LINKAGES v2.2 (Fig. 25.5). Stand-level estimates of ET averaged over the last 28 years of the simulations were 668 mm y^{-1} for the ambient scenario, 707 mm y^{-1} for the +33% scenario, and 586 mm y^{-1} for the −33% precipitation scenario. Estimates of ET for the +33% scenario were ~6% higher than those for the ambient scenario, whereas estimates of ET for the −33% scenario were ~12% lower. Again, interannual variation in ET was considerable. During the last 28 years of the simulations, differences in ET between the ambient and +33% scenario varied from 2% to 12%, and differences between the ambient and −33% scenario varied from 6% to 19%.

Discussion

Estimates of ET as simulated by LINKAGES v2.2 agreed reasonably well with catchment-scale measurements of net gain, although interannual variation in simulated ET was not as great as that seen in measurements of net gain. Annual average ET for Walker Branch Watershed based on the eddy covariance method was 582 ± 28 mm for the 5 years between 1995 and 1999 (Wilson et al. 2001), whereas Luxmoore and Huff (1989) estimated that ET on the Walker Branch Watershed was 730 mm. Annual rates of ET, as simulated by LINKAGES v2.2, generally fell within this range, with rates ranging from a low of 528 mm in 1983 to a high of 727 mm in 1991 and averaging 638 mm for the 28 years. In a forested watershed in West Virginia similar in many respects to Walker Branch Watershed, annual average ET as estimated from precipitation and stream flow data (i.e., net gain) was 640 mm (Adams et al. 1994). Annual estimates of ET for the Hubbard Brook Experimental Forest in New Hampshire, admittedly a more northerly site with cooler temperatures and a shorter growing season, ranged from 459 to 623 mm from 1964 to 1986 and averaged 531 mm (Federer et al. 1990).

Estimates of evaporation and interception derived from LINKAGES v2.2 also agreed well with published data for Walker Branch Watershed. Forest-floor eddy-covariance measurements suggested that evaporation rates were 86 mm in 1998 and 91 mm in 1999 (Wilson et al. 2001), and these values compare reasonably well to estimates of 81–138 mm, as simulated by the LINKAGES v2.2 model. Interception losses for the upland oak stand in the study of Wilson et al. (2001) were estimated to be 114 in 1998 and 115 mm in 1999 and averaged ~20% of annual ET. These estimates are again quite comparable to modeled estimates of interception, which, during a 28-year period, ranged from 84 to 157 mm, averaged 116 mm, and accounted for roughly 18% of annual ET.

In contrast to evaporation and interception, annual estimates of transpiration obtained with LINKAGES v2.2 were less easily validated. Simulated rates of transpiration that ranged from 332 to 488 mm are high compared to rates reported from sap-flow studies at this same site (see Chapter 21, this volume). Wullschleger et al. (2001a) measured sap velocity for canopy-dominant species on the Walker Branch Watershed and derived a value of 267 mm for annual transpiration. Estimates of a similar magnitude (230 and 269 mm) were reported for 2 years of sap-flow data (Wilson et al. 2001), and in Chapter 21, we report an estimate of 243 mm as derived from measured sap flow. Unfortunately, although the relative agreement among daily estimates of transpiration from sap-flow data, model simulations with LINKAGES v2.2, and eddy covariance is good (Wilson et al. 2001), modeled and sap-flow-derived estimates of transpiration in this study differed by 100–250 mm. As a result, questions have been raised concerning what appears to be a systematic underestimation of transpiration with the sap-flow technique (Wilson et al. 2001). To date, the exact cause of this perceived underestimation remains unresolved.

LINKAGES v2.2 simulated realistic species composition during early succession, with 50-year-old stands dominated by *L. tulipifera, Q. prinus, O. arboreum, A. rubrum,* and *Q. alba,* totaling 64% of the total stand biomass. These patterns are similar to those reported for earlier simulations conducted with data from this site (Bugmann et al. 2001) and are broadly representative of the species composition at Walker Branch Watershed (Lafon et al. 2000). Furthermore, early successional species, including *L. tulipifera, Pinus echinata,* and *A. rubrum,* that occupied the site during the first 100–200 years of stand development subsequently disappeared from the simulated stands during later stages of succession. These patterns are also consistent with general expectations of stand development on Walker Branch Watershed (Küchler 1975).

The equilibrium species composition as simulated by LINKAGES v2.2 at 1200 years was dominated by *Quercus* species (*Q. alba* > *Q. rubra* > *Q. velutina* > *Q. prinus* > *Q. coccinea*), with lesser amounts of *F. grandifolia* and *F. americana.* The various *Quercus* species accounted for ~77% of the total equilibrium biomass. This agrees closely to results from Bugmann et al. (2001), who reported that *Quercus* species accounted for 75% and 65% of the total equilibrium biomass based on simulations for this same site using the gap models FORCLIM and LINKAGES v2.0, respectively. Our simulated results are also in good agreement with the general expectations that the steady-state composition of mixed mesophytic forests would favor *Q. alba, Q. rubra,* and *F. grandifolia* as the dominant species (Küchler 1975).

Many gap models predict that the changes in climate (temperature and/or precipitation) associated with a doubling of atmospheric CO_2 concentration will be of a sufficient magnitude to bring about abrupt declines in forest productivity and species composition (Solomon 1986; Pastor and Post 1988; Badeck et al. 2001; Bugmann et al. 2001). In Urban and Shugart (1989; see Shugart et al. 1992), simulated results from the gap model ZELIG indicated that 18 tree species could no longer grow in the southeastern United States under conditions of CO_2-

induced climatic change and that much of the southern half of the southeast, especially areas characterized by shallow soils and poor water-holding capacity, would not support the growth of trees. For our site-specific application, the original version of LINKAGES predicted a moderate decline in forest productivity with reductions in precipitation. Following a −33% change in precipitation, standing biomass declined by ~30%, and drought-induced mortality of large trees favored an increase in the presence of *Q. alba* and *Q. velutina* and a general decline in *F. grandifolia, C. glabra,* and *O. arboreum.* Simulations with LINKAGES v2.2, however, revealed that reductions in biomass production from a 33% reduction in precipitation were much less than those observed with LINKAGES. Such a difference was explained by the fact that in contrast to LINKAGES, where reductions in soil-water potential below a −1.5-MPa threshold act to reduce growth regardless of when drought occurs during the growing season, the seasonal weighting factor that was incorporated into LINKAGES v2.2 dampened the magnitude of this response by more realistically matching the temporal impacts of drought on growth increment.

Although inclusion of a seasonal weighting factor in LINKAGES v2.2 lessened the overall impact of drought on standing biomass compared to the predictions of the original LINKAGES, other changes to the model also introduced surprising or unexpected findings. Most obvious was the observation that following a 33% reduction in precipitation, despite little change in biomass production, species composition of the simulated stand shifted from a mixed deciduous forest to a stand dominated by the moderately drought-tolerant *Q. alba.* Such an increase in the abundance of *Q. alba* was not the result of large-tree mortality because at no time during the simulation did water potential for the top 100 cm of soil fall below the −1.5-MPa threshold (i.e., accumulated drought days equaled zero). However, the addition of soil layers to LINKAGES v2.2 and the placement of seedlings and saplings in soils of shallower depths created conditions whereby seedlings and, to a lesser extent, saplings were exposed to more intense and more frequent drought compared to the larger trees. Thus, the disappearance of species like *F. grandifolia* and *O. arboreum* and the increase in others, like *Q. alba,* resulted not from large-tree mortality, but from drought-induced effects on seedling establishment, regeneration, and growth of smaller sized trees in the forest understory. This simulated observation is consistent with measurements of seedling and sapling mortality on the TDE, as summarized by Hanson et al. (2001).

Summary

Modifications made herein to the gap model LINKAGES fall into two broad categories. The first category is simply a general replacement of empirical functions found in the original version of LINKAGES with more mechanistic representations of canopy transpiration, incorporation of spatial complexity in soil hydrology, and processes associated with soil-water extraction. Such changes make it easier to parameterize models for site-specific simulations and easier to compare modeled results directly to experimental datasets. The second category of changes to the model are conceptually more important and arise from lessons learned throughout 8 years of experimentation on the TDE. These include the need to better match seasonal patterns of growth with the occurrence of drought and the placement of seedlings, saplings, and mature trees in distinct soil layers so that different size classes of trees are able to respond to unique soil-water environments. Having made these changes, we conclude that, in comparison to earlier predictions, the productivity of forest stands in areas similar in soils and climate to east Tennessee show resilience to what otherwise would be perceived as large reductions in average annual precipitation. However, although the effects of uniformly altering precipitation throughout the growing season on the standing biomass of forests in a future, drier world are likely to be minimal, species composition, as influenced through processes of seedling regeneration, stand establishment, and seedling and sapling mortality would be more responsive to changes in precipitation. These processes, at least in the simulations conducted here with LINKAGES v2.2, largely shaped species composition under conditions of changing precipitation. As such, there is a clear need to improve these aspects of gap models (Keane et al. 2001; Price et al. 2001).

References

Aber JD, Melillo JM (1982) *FORTNITE: A computer model of organic matter and nitrogen dynamics in forest ecosystems.* University of Wisconsin Research Bulletin R3130. University of Wisconsin, Madison, Wisconsin.

Adams MB, Kochenderfer JN, Wood F, Wood TR, Edwards P (1994) *Forty years of hydrometerological data from the Fernow Experimental Forest, West Virginia.* General Technical Report NE-184. USDA Forest Service, Northeastern Forest Experimental Station, Radnor, Pennsylvania.

Badeck FW, Lischke H, Bugmann H, Hickler T, Honninger K, Lasch P, Lexer MJ, Mouillot F, Schaber J, Smith B (2001) Tree species composition in European pristine forests: Comparison of stand data to model predictions. Clim Change 51:307–347.

Bugmann HKM, Solomon AM (2000) Explaining forest composition and biomass across multiple biogeographical regions. Ecol Appl 10:95–114.

Bugmann HKM, Wullschleger SD, Price DT, Ogle K, Clark DF, Solomon AM (2001) Comparing the performance of forest gap models in North America. Clim Change 51:349–388.

Federer CA (1995) *BROOK90: A simulation model for evaporation, soil water, and streamflow, Version 3.1.* USDA Forest Service, Durham, New Hampshire.

Federer CA, Flynn LD, Martin CW, Hornbeck JW, Pierce RS (1990) *Thirty years of hydrologic data at the Hubbard Brook Experimental Forest, New Hampshire.* General Technical Report NE–141. USDA Forest Service, Northeastern Forest Experiment Station, Durham, New Hampshire.

Federer CA, C Vörösmarty, Fekete B (1996) Intercomparison of methods for calculating potential evapotranspiration in regional and global water balance models. Water Resour Res 32:2315–2321.

Friend AD, Stevens AK, Knox RG, Cannell MGR (1997) A process-based, terrestrial biosphere model of ecosystem dynamics (Hybrid v3.0). Ecol Model 95:249–287.

Hanson PJ, Weltzin JF (2000) Drought disturbances from climate change: Response of United States forests. Sci Total Environ 262:205–220.

Hanson PJ, Todd DE, Amthor JS (2001) A six-year study of sapling and large-tree growth and mortality responses to natural and induced variability in precipitation and throughfall. Tree Physiol 21:345–358.

Keane RE, Austin M, Field C, Huth A, Lexer MJ, Peters D, Solomon A, Wyckoff P (2001) Tree mortality in gap models: Application to climate change. Clim Change 51:509–540.

Küchler AW (1975) *Potential natural vegetation of the conterminous United States.* Special Publication No. 36. American Geophysical Society, New York.

Lafon CW, Huston MA, Horn SP (2000) Effects of agricultural soil loss on forest succession rates and tree diversity in East Tennessee. Oikos 90:431–444.

Loehle C, LeBlanc D (1996) Model-based assessment of climate change effects on forests: A critical review. Ecol Model 90:1–31

Luxmoore RJ, Huff DD (1989) Water. In Johnson DW, van Hook RI (Eds) *Analysis of biogeochemical cycling processes in Walker Branch Watershed.* Springer-Verlag, New York, pp 165–196.

Norby RJ, Ogle K, Curtis PS, Badeck FW, Huth A, Hurtt GC, Kohyama T, Penuelas J (2001) Aboveground growth and competition in forest gap models: An analysis for studies of climatic change. Clim Change 51:415–447.

Pastor J, Post WM (1985) *Development of a linked forest productivity-soil process model,* ORNL/TM-9519. Oak Ridge National Laboratory, Oak Ridge, Tennessee.

Pastor J, Post WM (1988) Response of northern forests to CO_2-induced climate change. Nature 334:55–58.

Price DT, Zimmermann NE, van der Meer PJ, Lexer MJ, Leadley P, Jorritsma ITM, Schaber J, Clark DF, Lasch P, McNulty S, Wu J, Smith B (2001) Regeneration in gap models: Priority issues for studying forest responses to climate change. Clim Change 51:475–508

Shao GF, Bugmann H, Yan XD (2001) A comparative analysis of the structure and behavior of three gap models at sites in northeastern China. Clim Change 51:389–413.

Shugart HH, West DC (1977) Development of an Appalachian deciduous forest simulation model and its application to assessment of the impact of the chestnut blight. J Environ Manage 5:161–179.

Shugart HH, Smith TM, Post WM (1992) The potential for application of individual-based simulation models for assessing the effects of global change. Ann Rev Ecol Syst 23:15–38.

Shuttleworth WJ, Wallace JS (1985) Evaporation from sparse crops—An energy combination theory. Q J Roy Meteorol Soc 111:839–855.

Solomon AM (1986) Transient response of forests to CO_2-induced climate change—Simulation modeling experiments in eastern North America. Oecologia 68:567–579.

Talkkari A, Kellomäki S, Peltola H (1999) Bridging the gap between a gap model and a physiological model for calculating the effect of temperature on forest growth under boreal conditions. For Ecol Manage 119:137–150.

Thornthwaite CW, Mather JR (1957) Instructions and tables for computing potential evapotranspiration and the water balance. Publications in Climatology 10: 183–311.

Urban DL, Shugart HH (1989) Forest response to climatic change: A simulation study for southeastern forests. In Smith JB, Tirpak DA (Eds) *The potential effects of global climate change on the United States: Appendix D—Forests.* Office of Policy, Planning, and Evaluation, U.S. Environmental Protection Agency, Washington, District of Columbia.

Urban DL, Harmon ME, Halpern CB (1993) Potential response of Pacific Northwest forests to climatic change, effects of stand age and initial composition. Clim Change 23:247–266.

Wilson KB, Hanson PJ, Mulholland PJ, Baldocchi DD, Wullschleger SD (2001) A comparison of methods for determining forest evapotranspiration and its components: Sap-flow, soil water budget, eddy covariance and catchment water balance. Agric For Meteorol 106:153–168.

Wullschleger SD, Hanson PJ, Todd DE (2001a) Transpiration from a multi-species deciduous forest as estimated by xylem sap flow techniques. For Ecol Manage 143:205–213.

Wullschleger SD, Jackson RB, Currie WS, Friend AD, Luo Y, Mouillot F, Pan Y, Shao GF (2001b) Below-ground processes in gap models for simulating forest response to global change. Clim Change 51:449–473.

26. Regional Implications of the Throughfall Displacement Experiment on Forest Productivity

E. Raymond Hunt, Jr.

Introduction

Global climate models generally indicate that the southeastern region of the United States should become wetter on average during the next century (Cooter 1998). Model analyses on the responses of southern forest ecosystems to climatic change are based on the expected average conditions (Weinstein et al. 1998); however, these analyses do not consider the changes in climatic extremes. These changes of climatic extremes may accentuate or attenuate deciduous-forest responses to changes in mean climate. The interactions with changing climatic extremes for precipitation need to be assessed regionally because of the economic importance of forestry in the southeastern United States.

The Walker Branch Throughfall Displacement Experiment (TDE) is well designed to study changes in the climatic extremes, as well as changes in average annual precipitation, because the subtraction of precipitation throughfall from the dry plot and its addition of to the wet plot was imposed on the natural variability of precipitation (Fig. 26.1). Long-term average precipitation at Oak Ridge, Tennessee, is about 137 cm, and there was no single year during the TDE that had between 125 and 145 cm of precipitation. The detailed experimental studies in TDE are separating some of the interactions (Chapter 2, this volume). The essential problem is how to scale the results from a single ambient, dry, or wet plot to the regional scale encompassing the southeastern United States.

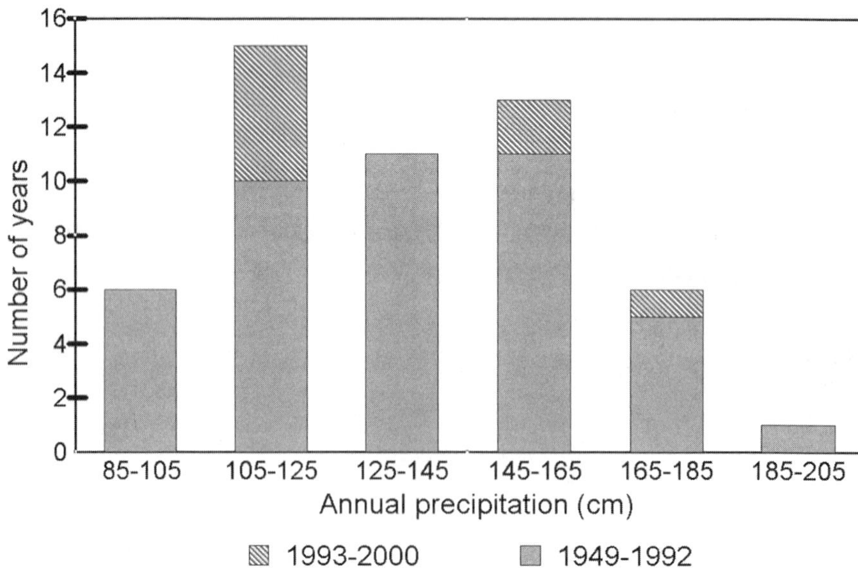

Figure 26.1. Frequency of annual precipitation for Oak Ridge, Tennessee. Number of years for each 20-cm range is given for two time intervals, 1949 to 1992 (NOAA Atmospheric Turbulence and Diffusion Laboratory) and 1993 to 2000 (Walker Branch Throughfall Displacement Experiment Site).

Two of the major techniques for scaling involve the use of geospatial technologies, particularly remote sensing and geographic information systems (GIS). Remote sensing, especially with satellite sensors, is used to determine the distribution of vegetation types and estimate the amount of biomass from vegetation indices. The TDE cannot address these typical issues of remote sensing for regional scales because it deals with one vegetation type (deciduous oak forest) and does not have large differences in leaf area index. Extrapolating the results of the TDE with available GIS data layers is again limited because the results are restricted to a single location (single soil type, single slope class, etc.), so there is no basis for geospatial distribution. Then, why discuss regional extrapolations?

The goal of this chapter is to show two examples on how data from the TDE can be used regionally. The first uses satellite observations made with the Advanced Very High Resolution Radiometer (AVHRR) of canopy leafout, data that are important signals of global climate change. The second uses GIS data to obtain the ecoclimatic zones that can be used with ecosystem simulation models to obtain regional extrapolations of forest productivity.

Deciduous Forest Phenology with Remote Sensing

Two of the major events in the phenological development of forests are leafout in the springtime and senescence during autumn. The timing for leafout and senescence involves a compromise between the risk of frost damage and the benefit of a longer growing season. Hänninen (1991) suggests that the risk of frost damage may increase with the warming that might occur as part of global climatic change, because of increased frequency of extreme frosts when trees are leafing out.

Satellite Data

Meteorological satellites have been routinely used to estimate the time of leafout and the time of senescence, by using vegetation indices (Schwartz 1993, 1994; Reed 1994). One commonly used index is the Normalized Difference Vegetation Index (NDVI):

$$NDVI = \frac{NIR - Red}{NIR + Red}, \tag{1}$$

where NIR is the reflectance at near-infrared wavelengths from about 0.7 to 1.1 μm and Red is the reflectance at red wavelengths from about 0.6 to 0.7 μm. The actual wavelengths are different for each sensor, AVHRR uses different wavelengths than, say, the Landsat Thematic Mapper. This index was originally used to enhance the contrast between vegetation and the soil background reflectance and to reduce differences from slope, aspect, zenith, and azimuth (Rouse et al. 1974). Subsequently, NDVI has been shown to have a strong relationship with the fraction of photosynthetically active radiation (PAR) that is absorbed by the canopy (Asrar et al. 1984). NDVI is not strongly related to leaf area index (LAI), particularly for deciduous forests because NDVI saturates at LAI above 4.0 $m^2\ m^{-2}$ (Fassnacht et al. 1997; Franklin et al. 1997), which is well below the LAI for the TDE site.

Growing-season length, the time between leafout and senescence, is an important determinant of primary production and other ecosystem processes (Schwartz 1999; White et al. 1999). Satellites easily pick up differences in the dates of leafout and senescence when compared across climatic gradients at continental scales. The problem in developing models from these data is that the local vegetation has adapted to the conditions at each site; two species growing at the same location are more similar to each other with regard to timing than a single species growing at two locations far apart (Sharik and Barnes 1976). Thus, the real physiological and climatic controls over phenological timing may be difficult to determine, and the major challenge is to understand the year-to-year variation at a single site (Schwartz 1998, 1999).

The AVHRR data from 1993 to 1999 were from Conterminous U.S. AVHRR Biweekly and Weekly Composites produced by the U.S. Geological Survey (Eidenshink 1992). The AVHRR data for 1993 to 1996 were biweekly composites and the data for 1997 to 1999 were weekly composites. Unlike in other analyses that have used these data, the seasonal trajectory of NDVI for the Walker Branch Watershed did not have to be smoothed to remove the period-to-period variation from changes in the sun-target-sensor viewing geometry.

Comparison of Leaf Phenology to NDVI

Each year, the timing of leafout and senescence were determined as the fraction of maximum LAI on a given day of the year (Hanson et al. 2001). These data were determined as the ratio of PAR from 1.5 m above the ground to PAR above the canopy.

The leafout data were compared to the fraction of maximum NDVI on a given day of the year as determined for the single AVHRR pixel (1000 m × 1000 m) of the Walker Branch Watershed surrounding the TDE. Maximum NDVI for this pixel was 0.65 (the June average for all years), and the minimum NDVI was taken to be 0.22 (the average snow-free value during December and January). The fraction of maximum NDVI was determined as (current − minimum)/(maximum − minimum).

In Fig. 26.2, data for two contrasting years, 1996 and 1997, show the variation in onset of LAI and NDVI by day of the year (DOY). First, the fraction of maximum NDVI increases from 0.0 to about 0.2 early in March (actual NDVI increased from 0.22 to 0.31 between DOY 60 and 80), sooner than leafout of the tree canopy for all years. It is hypothesized that this increase in NDVI may be produced by the onset of leaves in the understory. Because of the warmer weather

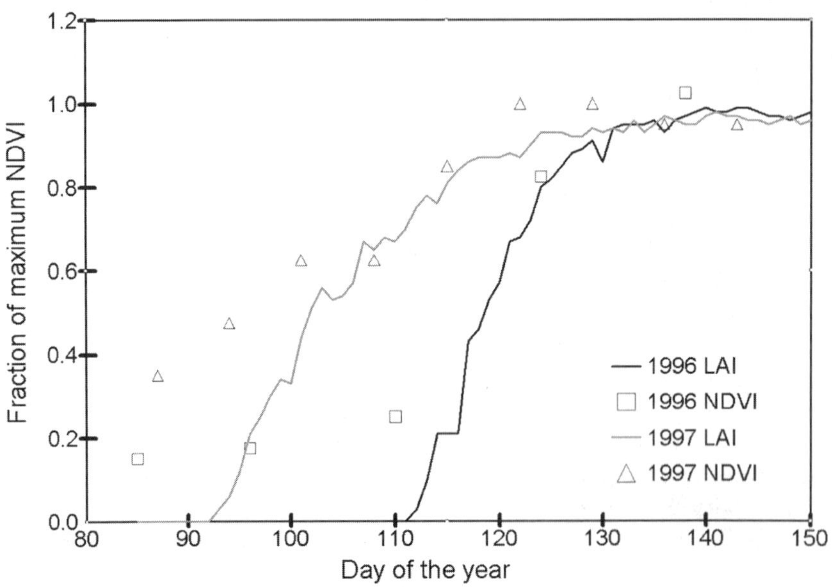

Figure 26.2. Fraction of maximum NDVI and fraction of maximum LAI for two contrasting years, 1996 and 1997. Fraction of maximum LAI was from Hanson et al. (2001). Maximum NDVI was 0.65 and minimum NDVI was 0.22 using data from the U.S. Conterminous AVHRR Biweekly and Weekly Composites produced by the USGS EROS Data Center (Sioux Falls, SD).

in 1997, both LAI and NDVI increased 2 weeks before the increase of LAI and NDVI for 1996 (Fig. 26.2). Thus, high-temporal-resolution AVHRR NDVI is able to accurately determine year-to-year variability in leafout.

There is a linear relationship between the LAI and NDVI during the onset of the growing season (Fig. 26.3). Possibly, the variation about the *y* intercept shows the variability of understory growth, from low (two points about 0.0) to high (two points about 0.3). If a single date is required for leafout by the canopy, then the date for 50% of maximum LAI can be determined easily from the date of 50% maximum NDVI. Extrapolation of NDVI to the date that it starts to increase above the minimum value is much more difficult because of possible confusion with the understory.

These two datasets should not be used to construct a relationship between NDVI and LAI for deciduous forests because the data for LAI are actually an estimate of absorbed PAR. Absorbed PAR/LAI relationships are variable because of differences in species composition, tree density, foliage distribution within a canopy, and soil background. For a single site like the Walker Branch Watershed, the relationship between absorbed PAR and LAI would not help predict LAI for other sites.

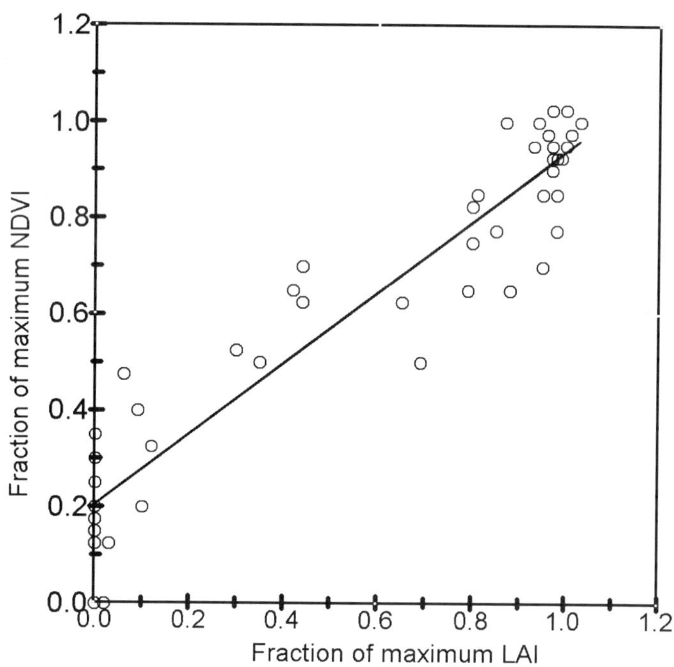

Figure 26.3. Fraction of maximum NDVI versus fraction of maximum LAI from 1993 to 1999. Data were from the springtime greenup covering from DOY 80 to DOY 150. The line is $y = 0.20 + 0.73x$, with an R^2 of 0.89.

Definition of Climatic Zones for Regional Extrapolations

The primary factors that control biogeochemical cycles are climate, topography, soils, and land cover type. Global climate models or general circulation models use grid cells of a given number of degrees latitude and longitude; each grid cell may contain a large amount of variation in the above four factors. Ecosystem-simulation models, such as Biome-BGC (Running and Hunt 1993; Hunt et al. 1996), are often used with grid cells of 1° latitude by 1° longitude (110 km by 110 km at the equator), which also contain considerable variation. On the other hand, large regions of the world are so uniform in these four factors that simulations using 1° grid cells have a large amount of redundancy. The problem is the arbitrary nature of the gridding process (Coughlan and Running 1996).

Walter (1981) surveyed the work from earlier biogeographers and showed how vegetation types are related to climate, soils, and topography. Jenny (1941) emphasized that soils are related to climate, vegetation, topography, parent material, and time. Thus, much of the variation among climate, topography, soils, and vegetation are correlated, and these correlations are used to define uniform regions for ecosystem simulation models. The size of a region is defined by the variability in these four landscape factors.

Ecoregions in the Southeastern United States

Baily (1995, 1996) defined ecoregions in a hierarchy, starting with domains based on precipitation. The southeastern United States is in the humid–temperate domain, except for the Everglades, which is in the humid–tropical domain. The next hierarchical level is the division, which is based on winter temperature. There are three divisions in the humid-temperate domain: the hot continental division, the hot continental regime mountains, and the subtropical division. The third level is termed the province, which is based on the dominant vegetation type. Ecoregion boundaries are available as GIS data layers (Baily 1995) so these regions could be combined with other map-based data; Fig. 26.4A and Table 26.1 show the provinces that encompass the southeastern United States.

Below the hierarchical level of province is the section, which is defined on the basis of geomorphology (McNab and Avers 1994). The important question is: "Which hierarchical level is the appropriate basis for extrapolating the results of the Walker Branch TDE?" Analysis of long-term meteorological data for various stations within each province showed there is variation in precipitation within a province, but this variation is not related to the section boundaries. Furthermore, many of the analyses of climatic-change effects on forest productivity seek to express those effects in economic terms, and economic data are usually aggregated at the state level. Therefore, a single weather station was used in each state to represent a province that covered a sufficiently large area in that state. The weather stations that were selected had long-term complete records and were more or less centered in the state by ecoregion area (Fig. 26.4). These stations are identified in Table 26.1.

BGC++ Simulations

The model used for the simulations is BGC++ (Hunt et al. 1999), which is a derivative of Biome-BGC (Hunt and Running 1992; Running and Hunt 1993; Hunt et al. 1996). The primary difference between the current and older version of Biome-BGC is improved dynamic allocation of carbon and nitrogen among leaves, stems, coarse roots, and fine roots so the wood production over a forest rotation can be simulated under a variety of fertilization and irrigation treatments (Ryan et al. 1996a, 1996b).

Daily climatic data are the primary inputs into BGC++, because the daily time step is the longest interval for which physiological and biophysical algorithms can be used for the simulation of different ecosystem processes, particularly the hydrologic cycle. This time step has disadvantages at large spatial scales because precipitation averages over those large areas are hard to obtain. For example, convective thunderstorms cover a small area for any specific event, and these storms are patchy for a given day. Over a monthly time interval, the entire area gets close to the average rainfall for each station. However, at a daily time interval, spatial averages of patchy daily rainfall result in a constant daily drizzle, which is not even close to actual weather records.

Because air temperatures are somewhat cooler on days with rain, simply taking an average year and adjusting precipitation will lead to some bias in representing the climatology. Therefore, each year of record was examined for suitability to represent an average year, a dry year, or a wet year. First, annual average and monthly average precipitation and temperatures were calculated. Then, average annual and monthly precipitation were increased 33% and reduced 33% to calculate annual and monthly averages for wet and dry years, respectively. Out of a 45-year period, there were usually 3 years with annual rainfall that was similar to average, wet, and dry conditions. From those years, the year that was closest to the monthly average precipitation and monthly mean temperatures was selected as the best representative. For the wet and dry years, the selected year was the one that avoided extremely high or low totals for a single month (i.e., hurricanes or extreme drought). The selected years and the annual precipitation are shown in Table 26.1.

For the following BGC++ simulations, the initial LAI was 0.5 $m^2\ m^{-2}$ and the initial wood mass was 1.0 Mg C ha^{-1}. Soil types available from regional GIS data layers were not useful in obtaining the initial conditions for soil carbon and nitrogen; hence, the data from prior work on the Walker Branch Watershed (Johnson and Van Hook 1989) were used for all other initial conditions. Finally, a nitrogen input was assumed to be 10 kg N ha^{-1} annually.

Regional Patterns of Forest Productivity from Changing Precipitation Regimes

The states influenced by the Gulf of Mexico show the highest amount of annual rainfall for the year representing average conditions; the areas include four of the

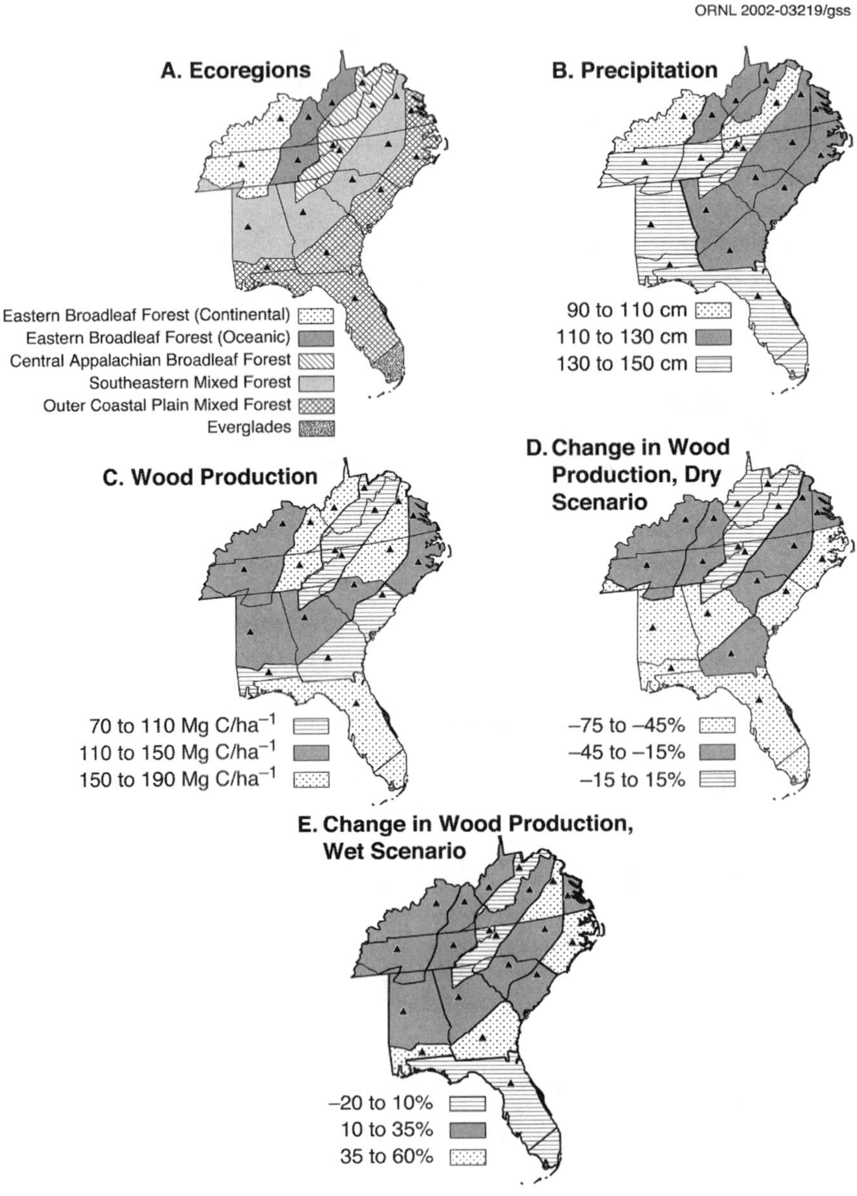

Figure 26.4. Ecoregions of the southeastern United States and their observed or modeled characteristics. Triangles indicate the cities from Table 26.1 that were selected to represent each state in a given ecoregion. Graph A shows the ecoregions defined by Baily (1996). Graph B is mean annual precipitation for the southeastern United States. Graph C is ambient wood production over a 30-year rotation, with the designated ambient year being used continuously. Graph D is the change in wood production over a 30-year rotation, with the designated dry year being used continuously. Graph E is change in wood production over a 30-year rotation with the designated wet year being used continuously.

Table 26.1. Representative meteorological stations for the ecoregions in each state

Ecoregion City, State	No. of years	Average precipitation	Ambient year	Ambient precipitation	Dry year	Dry precipitation	Wet year	Wet precipitation
Outer Coastal Plain Mixed Forest Province								
Andalusia, AL	47	148	1993	146	1968	95	1983	195
Fitzgerald, GA	47	115	1958	115	1954	61	1975	154
Ocala, FL	47	134	1966	135	1990	86	1959	179
Sumter, SC	47	123	1985	122	1952	83	1965	172
Kinston, NC	47	128	1982	128	1954	85	1971	170
Richmond, VA	46	110	1983	110	1970	72	1979	145
Southeastern Mixed Forest Province								
Tuscaloosa, AL	47	134	1953	136	1986	92	1979	173
Atlanta, GA	47	124	1971	125	1954	81	1975	168
Chester, SC	47	119	1952	120	1951	83	1964	164
Chapel Hill, NC	47	116	1963	116	1986	84	1989	146
Charlottesville, VA	47	117	1970	114	1963	77	1979	156
Central Appalachian Broadleaf Forest Province								
Boone, NC	47	134	1960	140	1988	91	1957	177
Bristol, TN	47	105	1955	105	1988	78	1950	138
Lexington, VA	47	97	1951	97	1963	70	1972	134
Elkins, WV	69	116	1978	116	1930	72	1972	149
Eastern Broadleaf Forest (Oceanic) Province								
Oak Ridge, TN	47	137	1971	136	1958	95	1967	175
West Liberty, KY	47	112	1957	115	1963	72	1990	151
Charleston, WV	47	108	1983	108	1988	80	1989	147
Eastern Broadleaf Forest (Continental) Province								
Dickson, TN	47	134	1959	134	1960	93	1989	181
Frankfort, KY	47	107	1967	108	1963	78	1950	147

Note: For each station, the number of years and the average annual precipitation (cm) are given, followed by the year selected to represent ambient conditions and the rainfall (cm) for that year. The years selected to represent dry conditions (33% below average) and wet conditions (33% above average) are presented with the rainfall (cm) for those years.

five forest provinces (Fig. 26.4B), showing the variability within each ecoregion. The Central Appalachian Broadleaf Forest has the largest variability within a province, ranging from the lowest to among the highest amounts of annual precipitation (Fig. 26.4B). Simulations for wood production over a managed-forest rotation of 30 years were conducted; the representative year was used continuously in these simulations.

The highest amounts of forest production were in the regions surrounding the Appalachian Mountains and in Florida (Fig. 26.4C). Florida had the highest production of wood because the growing season is the longest and precipitation is plentiful the entire year. The areas around the Appalachian Mountains had high wood production with a shorter growing season because cooler temperatures reduced the amount of autotrophic respiration. The Central Appalachian Broadleaf Forest has the lowest wood production (Fig. 26.4C) because this province had the shortest growing season. The Outer Coastal Plain Mixed Forest also had lower wood production than did the other forests because of reduced growing-season rainfall and high air temperatures.

When BGC++ is run with the representative dry year (Table 26.1) continuously for 30 years, the largest reductions of wood production occur for the Outer Coastal Plain Mixed Forest and Southeastern Mixed Forest in Alabama and Georgia (Fig. 26.4D). The combination of high temperatures and reduced precipitation during the growing season increased water stress. On the other hand, the Central Appalachian Broadleaf Forest showed no change, and some locations showed a slight increase in wood production (Fig. 26.4D). In those cases, the increase in PAR increased the amount of photosynthesis, whereas the cooler temperatures ameliorated the effects of reduced precipitation.

Wood production in the Central Appalachian Broadleaf Forest goes down when BGC++ is run for 30 years with the representative wet year, whereas wood production goes up in the Outer Coastal Plain Mixed Forest, except for Florida (Fig. 26.4E). Florida and the Central Appalachian Broadleaf Forest both have lower production of wood because lower amounts of PAR reduce the amount of photosynthesis. Other areas on the Outer Coastal Plain Mixed Forest show increased production because the cooler weather associated with the increased rainfall reduces autotrophic respiration.

In summary, the Southeastern United States has considerable rainfall, with averages over 100 cm annually, but the hot summertime temperatures increase respiration and aggravate water stress during drought. With cooler, wetter weather, there is a strong increase in clouds, and PAR becomes limiting for photosynthesis. Hence, simple spatial extrapolation of the Walker Branch TDE is frustrated by multiple, interacting climatic variables.

Effects of Drought Frequency Predicted from an Ecosystem Simulation Model

The above simulations used a representative ambient, dry, or wet year continuously over a 30-year rotation. Variation in climatic extremes also will affect wood production through multiple interactions involving nutrient and carbon allocation, decomposition, and nutrient availability. BGC++ is sensitive to year-to-year variability, so examination of the effects of climatic extremes is easier.

The problem with examining year-to-year variability of real climate is that each year is in some way unique. After all, usually only 3 years of station records were sufficiently similar to the average precipitation even to be considered as representative. Selection of 1 year on the basis of both temperature and precipitation was difficult (Table 26.1).

To examine the effects of drought frequency, the representative ambient year was combined with the representative dry year in different sequences. Hence, for a drought frequency of 0.1, 1 dry year was added for every 9 ambient years; and for a drought frequency of 0.5, a dry year followed every ambient year (Fig. 26.5). The simulations were run for 30 years, and wood production was normalized to the 30-year total with the ambient year.

With increased drought frequency, some climatic stations had increased wood production, and others had lower wood production (Fig. 26.5). The legend to Fig. 26.5 lists from the coolest location (top) to the warmest location (bottom) in terms of mean annual temperature, so it is the cooler locations that have increased wood production with increased drought frequency, just as the previous simulations had indicated.

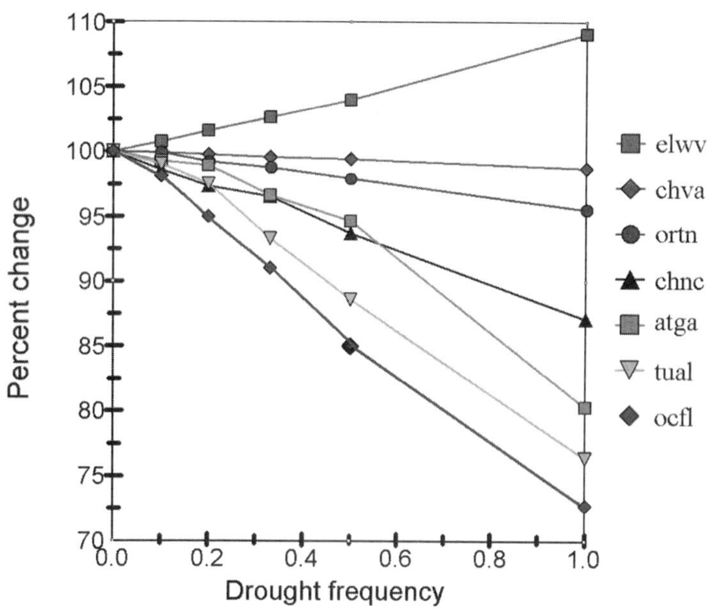

Figure 26.5. Change in wood production over a 30-year rotation with increasing drought frequency. Legend ranks cities from coolest (top) to warmest (bottom) mean annual temperatures: Elkins, WV (elwv), Charlottesville, VA (chva), Oak Ridge, TN (ortn), Chapel Hill, NC (chnc), Atlanta, GA (atga), Tuscaloosa, AL (tual), and Ocala, FL (ocfl). Other cities in Table 26.1 follow the same trend with mean annual temperature.

However, for the stations that had decreased production with increased drought frequency, the final wood production was almost always higher than expected from simply averaging the annual production for the number of dry years and the number of ambient years (Fig. 26.5). The interaction term was determined to be the result of decomposition being less drought sensitive than productivity. So, although nitrogen availability decreased, there was relatively more nitrogen than carbon in the new foliage the next year, which resulted in increased photosynthesis rates over normal for the ambient year. The extra carbon produced by this increased photosynthesis was then allocated into new wood.

The interaction of drought with nitrogen availability suggested that long-term production is influenced by the exact sequence of dry and ambient years for a given drought frequency. Using a drought frequency of 0.4, the 2 dry years and 3 ambient years were combined in different sequences and with different starting points for the same sequence (Table 26.2). The highest simulated wood production was from the 2 drought years being adjacent, and the lowest production was with the longest interval between dry years (Table 26.2).

Conclusions

There was not a year from 1993 to 2000 at Walker Branch Watershed that had average rainfall within 5 cm. Thus, even in the southeastern United States, year-to-year variation may be driving ecosystem processes. The long-term ecosystem manipulation of the TDE offers the opportunity to tease apart experimentally some of interacting processes. The BGC++ model suggests that the important interaction occurring is the differential effects of drought on decomposition and primary production, leading to an increase in vegetation nitrogen concentration by mass balance. The question is: "How similar are these simulations to what

Table 26.2. Wood accumulation after 30 years with a drought frequency of 0.4 at Oak Ridge, Tennessee

Year order	Yield (Mg C ha^{-1})	Comment
DDAAA	51.5	
DADAA	49.3	
DAADA	46.5	
ADDAA	51.4	
ADADA	48.3	
ADAAD	48.1	
AADDA	54.1	
AADAD	49.5	
AAADD	54.5	
Mean	50.6	
AAAAA	51.2	Continuous average year
DDDDD	42.0	Continuous dry year

Note: A specific order of dry and average years is repeated six times. The order of drought and average years strongly influences the total wood accumulation. The year labeled D is 1958 and the year labeled A is 1971 (Table 26.1).

actually occurs in an upland oak forest?" There is storage of nitrogen internally, a physiological process that BGC++ does not simulate, so that the year-to-year variations in vegetation nitrogen could be substantially dampened. Furthermore, there is less consensus in the scientific community relating soil nitrogen availability with leaf nitrogen content. Model errors can be fruitful when they lead to increased understanding.

There was also year-to-year variation in temperature at Walker Branch, as shown by the differences in leafout phenology. Unlike decomposition, year-to-year differences in when the leaves are visible can be determined regionally by meteorological satellites.

Regional extrapolation of the TDE data with the climatic delineations from Baily's (1996) ecoregions produced two surprises. First, some areas of the southeastern United States may actually have increased wood production with less precipitation. These areas are cloudy and may be radiation limited, so the increase in production was from the increase in radiation. Other areas showed decreased wood production with increasing drought frequency, an intuitive result. However, the second surprise was that the productivity was somewhat higher than expected from interactions with the nitrogen cycle and decomposition. Future long-term manipulation experiments should be designed to examine more interactions among ecosystem biogeochemical cycles.

References

Asrar G, Fuchs M, Kanemasu ET, Hatfield JL (1984) Estimating absorbed photosynthetically active radiation and leaf area index from spectral reflectance in wheat. Agron J 76:300–306.

Baily RG (1995) *Description of the ecoregions of the United States.* 2nd edition. Miscellaneous Publication Number 1391. UDSA Forest Service, Washington, District of Columbia.

Baily RG (1996) *Ecosystem geography.* Springer, New York.

Cooter EJ (1998) General circulation model scenarios for the southern United States. In RA Mickler, S Fox (Eds) *The productivity and sustainability of southern forest ecosystems in a changing environment.* Ecological Studies 128. Springer, New York. pp 15–54.

Coughlan JC, Running SW (1996) Biophysical aggregations of a forested landscape using an ecological diagnostic system. Trans GIS 1:25–39.

Eidenshink JC (1992) The 1990 conterminous U.S. AVHRR data set. Photogramm Eng Remote Sensing 58:809–813.

Fassnacht KS, Gower ST, MacKenzie MD, Nordheim EV, Lillesand TM (1997) Estimating the leaf area index of north central Wisconsin forests using the Landsat Thematic Mapper. Remote Sensing Environ 61:229–245.

Franklin SE, Lavigne MB, Deuling MJ, Wulder MA, Hunt ER Jr. (1997) Estimation of forest leaf area index using remote sensing and GIS data for modelling net primary production. Int J Remote Sensing 18:3459–3471.

Hänninen H (1991) Does climatic warming increase the risk of frost damage in northern trees? Plant Cell Environ 14:449–454.

Hanson PJ, Todd DE, Riggs JS, Wolfe ME, O'Neill EG (2001) Walker Branch Throughfall Displacement Experiment data report: Site characterization, system performance, weather, species composition, and growth. ORNL/CDIAC-134, NDP-078A. Oak Ridge National Laboratory, Oak Ridge, Tennessee.

Hunt ER Jr., Running SW (1992) Simulated dry matter yields for aspen and spruce stands in the North American boreal forest. Can J Remote Sens 18:126–133.

Hunt ER Jr., Piper SC, Nemani R, Keeling CD, Otto RD, Running SW (1996) Global net carbon exchange and intra-annual atmospheric CO_2 concentrations predicted by an ecosystem process model and three-dimensional atmospheric transport model. Global Biogeochem Cycles 10:431–456.

Hunt ER Jr., Lavigne MB, Franklin SE (1999) Factors controlling decline of growth efficiency and net primary production for balsam fir in Newfoundland. Ecol Model 122:151–164.

Jenny H (1941) *Factors of Soil Formation.* Dover Publications, Inc., Mineola, New York.

Johnson DW, Van Hook RI (Eds) (1989) Analysis of biogeochemical cycling processes in Walker Branch Watershed. Springer-Verlag, New York.

McNab WH, Avers PE (Eds) (1994) Ecological subregions of the United States: Section descriptions. Publication WO-WSA-5. USDA Forest Service, Washington, District of Columbia.

Reed BC, Brown JF, VanderZee D, Loveland TR, Merchant JW, Ohlen DO (1994) Measuring phenological variability from satellite imagery. J Veg Sci 5:703–714.

Rouse JW, Haas RH, Schell JA, Deering DW (1974) Monitoring vegetation systems in the Great Plains with ERTS. In Freden SC, Mercanti EP, Becker M (Eds) *Third earth resources technology satellite–1 symposium. Volume I: Technical presentations.* NASA SP-351. National Aeronautics and Space Administration, Washington, District of Columbia, pp 309–317.

Running SW, Hunt ER Jr. (1993) Generalization of a forest ecosystem process model for other biomes, BIOME-BGC, and an application for global-scale models. In Ehleringer JR, Field C (Eds) *Scaling physiological processes: Leaf to globe.* Academic Press, Orlando, Florida, pp 141–158.

Ryan MG, Hunt ER Jr., McMurtrie RE, Ågren GI, Aber JD, Friend AD, Rastetter EB, Parton WJ, Raison RJ, Linder S (1996a) Comparing models of ecosystem function for temperate coniferous forests. I. Model descriptions and validation. In Breymeyer A, Hall DO, Melillo JR, Ågren GI (Eds) *Global change: Effects on coniferous forests and grasslands.* Scientific Committee on Problems of the Environment (SCOPE) 56. John Wiley & Sons, Chichester, England, pp 313–362.

Ryan MG, McMurtrie RE, Ågren GI, Hunt ER Jr., Aber JD, Friend AD, Rastetter EB, Parton WJ (1996b) Comparing models of ecosystem function for temperate coniferous forests. II. Simulations of the effect of climate change. In Breymeyer A, Hall DO, Melillo JR, Ågren GI (Eds) *Global change: Effects on coniferous forests and grasslands.* Scientific Committee on Problems of the Environment (SCOPE) 56, John Wiley & Sons, Chichester, England, pp 363–387.

Schwartz MD (1993) Assessing the onset of spring: A climatological perspective. Phys Geogr 14:536–550.

Schwartz MD (1994) Monitoring global change phenology: The case of the spring green wave. Int J Biometeorol 38:18–22.

Schwartz MD (1998) Greenwave phenology. Nature 394:839–840.

Schwartz MD (1999) Advancing to full bloom: Planning phenological research for the 21st century. Int J Biometeorol 42:113–118.

Sharik TL, Barnes BV (1976) Phenology of shoot growth among diverse populations of yellow birch (*Betula alleghaniensis*) and sweet birch (*B. lenta*). Can J Bot 54:2122–2129.

Walter H (1979) *Vegetation of the Earth and Ecological Systems of the Geo-biosphere.* 2nd Ed. Springer-Verlag, New York.

Weinstein DA, Cropper WP Jr., McNulty SG (1998) Summary of simulated forest responses to climate change in the southeastern United States. In Mickler RA, Fox S (Eds) *The productivity and sustainability of southern forest ecosystems in a changing environment.* Ecological Studies 128. Springer, New York, pp 479–500.

White MA, Running SW, Thornton PE (1999) The impact of growing-season length variability on carbon assimilation and evapotranspiration over 88 years in the eastern US deciduous forest. Int J Biometeorol 42:139–145.

Appendix
Scientific and Common Names of Higher Plants

Abies amabilis (Dougl.) Forb.	Pacific silver fir
Abies balsamea (L.) Mill.	Balsam fir
Acacia confuse	Thoughtful tree
Acer pseudoplatanus L.	Sycamore maple
Acer rubrum L.	Red maple
Acer saccharinum L.	Silver maple
Acer saccharum Marsh.	Sugar maple
Albizia julibrissin Durazz.	Mimosa, silk-tree
Amelanchier arborea (Michx.f.) Fern.	Service berry
Betula alleghaniensis Britt.	Yellow birch
Betula lenta L.	Sweet birch, black birch
Betula pendula L.	Silver birch
Carpinus carolinina Walt.	Blue beech
Carya glabra (Mill.) Sweet	Pignut hickory
Carya ovata (Mill.) K.Koch	Shagbark hickory
Carya tomentosa Nutt.	Mockernut hickory
Castanea dentate (Marsh.) Borkh.	Chestnut
Cedrus atlantica Man.	Atlas cedar
Centrosema spp.	Butterfly pea
Cercis canadensis L.	Eastern redbud
Chionanthus virginicus L.	Fringe tree

Scientific name	Common name
Coccoloba spp.	Sea grape
Coffea Arabica	Arabian coffee
Commelina communis L.	Asiatic dayflower, blue dayflower
Cornus florida L.	Flowering dogwood
Cornus sericea	Red osier dogwood
Diospyros virginiana L.	Common persimmon
Drypetes glauca	Drypetes
Encelia farinose	Encelia farinose, gray brittle bush, incienso
Eucalyptus globulus	Bluegum eucalyptus, Tasmanian blue gum
Eucalyptus pauciflora Sieber	Cabbage gum, snow gum, white sale, ghost gum
Eucalyptus regnans	Swamp gum, mountain ash
Eucalyptus viminalis	Ribbon gum, manna gum
Eucalyptus wandoo	White gum, wandoo
Fagus grandifolia Ehrh.	American beech
Fagus sylvatica L.	European beech
Fraxinus americana L.	White ash
Fraxinus excelsior L.	European ash
Halesia carolina L.	Carolina silverbell
Helianthus annus L.	Garden sunflower
Hevea brasiliensis Muell-Arg.	Brazilian rubber tree
Ilex Montana (T.&G.) Gray.	Holly
Jacquinia pungens	Cudjoe wood
Juglans nigra L.	Black walnut
Juniperus virginiana L.	Eastern redcedar
Larix decidua × *leptolepis*	European–Japanese larch hybrid
Liquidambar styraciflua L.	Sweetgum, red gum
Liriodendron tulipifera L.	Yellow poplar
Litsea glutinosa	Indian laurel, pond spice
Malus × *domestica*	Apple
Morus rubra L.	Red mulberry
Morus alba L.	White mulberry
Nothofagus spp.	Nothofagus
Nyssa sylvatica Marsh.	Blackgum, black tupelo
Oxydendrum arboreum [L.] DC.	Sourwood
Picea abies (L.) Karst.	Norway spruce
Picea engelmannii Parry	Engelmann spruce
Picea glauca (Moench) Voss	White spruce
Picea mariana (Mill.) B.S.P.	Black spruce
Picea rubens Sarg.	Red spruce
Pinus contorta Dougl.	Lodgepole pine
Pinus densiflora Sieb. & Zucc.	Japanese red pine
Pinus echinata Mill.	Shortleaf pine

Pinus elliottii Engelm.	Slash pine
Pinus jeffreyi Grev. Balf.	Jeffrey pine
Pinus nigra Arnold	European black pine
Pinus ponderosa Laws.	Ponderosa pine
Pinus radiata D.Don	Monterey pine
Pinus resinosa Ait.	Red pine
Pinus rigida Mill.	Pitch pine
Pinus strobus L.	Eastern white pine
Pinus sylvestris L.	Scotch pine
Pinus taeda L.	Loblolly pine
Plantago spp.	Common plantain
Populus spp.	Aspen, cottonwood, poplar
Populus deltoides Bartr.	Eastern cottonwood
Populus tremuloides Michx.	Quaking aspen
Prunus spp.	Cherry
Prunus avium × *pseudocerasus*	Sweet cherry × Japanese cherry tree
Prunus domestica L.	Plum
Prunus dulcis	Almond
Prunus serotina Ehrh.	Wild black cherry
Pseudotsuga menziesii Franco	Douglas fir
Psychotria L. spp.	Wild coffee
Quercus acutissima Carruthers	Sawtooth oak
Quercus alba L.	White oak
Quercus coccinea Muenchh.	Scarlet oak
Quercus macrocarpa Michx.	Bur oak
Quercus muehlenbergii Willd.	Chinquapin oak
Quercus palustris Muenchh.	Pin oak
Quercus petraea (Matt.) Lielbl.	Sessile oak, durmast oak
Quercus prinus L.	Chestnut oak
Quercus robur L.	English oak
Quercus rubra L.	Northern red oak
Quercus stellata Wang.	Post oak
Quercus velutina Lam.	Black oak
Rhamnus caroliniana L.	Carolina buckthorn
Robinia pseudoacacia L.	Black locust
Rosa spp.	Rose
Salix dasyclados [Wimm.]	Holme willow
Salix nigra Marsh.	Black willow
Sassafras albidum (Nutt.) Nees	Sassafras
Tamarix jordanis L.	Salt cedar
Thouinia paucidentata [Radlk.]	Soapberry plant
Tilia spp.	Basswood
Tsuga canadensis (L.) Carr.	Eastern hemlock
Ulmus rubra Mühl.	Slippery elm
Vitis vinifera L.	Wine grape

Index

C

D

Ecological Studies

Volumes published since 1996

Volume 119
Freshwaters of Alaska: Ecological Synthesis (1997)
A.M. Milner and M.W. Oswood (Eds.)

Volume 120
Landscape Function and Disturbance in Arctic Tundra (1996)
J.F. Reynolds and J.D. Tenhunen (Eds.)

Volume 121
Biodiversity and Savanna Ecosystem Processes: A Global Perspective (1996)
O.T. Solbrig, E. Medina, and J.F. Silva (Eds.)

Volume 122
Biodiversity and Ecosystem Processes in Tropical Forests (1996)
G.H. Orians, R. Dirzo, and J.H. Cushman (Eds.)

Volume 123
Marine Benthic Vegetation: Recent Changes and the Effects of Eutrophication (1996)
W. Schramm and P.H. Nienhuis (Eds.)

Volume 124
Global Change and Arctic Terrestrial Ecosystems (1996)
W.C. Oechel (Ed.)

Volume 125
Ecology and Conservation of Great Plains Vertebrates (1997)
F.L. Knopf and F.B. Samson (Eds.)

Volume 126
The Central Amazon Floodplain: Ecology of a Pulsing System (1997)
W.J. Junk (Ed.)

Volume 127
Forest Design and Ozone: A Comparison of Controlled Chamber and Field Experiments (1997)
H. Sanderman, A.R. Wellburn, and R.L. Heath (Eds.)

Volume 128
The Productivity and Sustainability of Southern Forest Ecosystems in a Changing Environment (1998)
R.A. Mickler and S. Fox (Eds.)

Volume 129
Pelagic Nutrient Cycles (1997)
T. Andersen

Volume 130
Vertical Food Web Interactions (1997)
K. Dettner, G. Bauer, and W. Völkl (Eds.)

Volume 131
The Structuring Role of Submerged Macrophytes in Lakes (1998)
E. Jeppesen, M. Søndergaard, M. Søndergaard, and K. Christoffersen (Eds.)

Volume 132
Vegetation of the Tropical Pacific Islands (1998)
D. Mueller-Dombois and F.R. Fosberg

Volume 133
Aquatic Humic Substances (1998)
D.O. Hessen and L.J. Tranvik

Volume 134
Oxidant Air Pollution Impacts in the Montane Forests of Southern California: A Case Study of the San Bernardino Mountains (1999)
P.R. Miller and J.R. McBride (Eds.)

Volume 135
Predation in Vertebrate Communities: The Bialowieza Primeval Forest as a Case Study (1998)
B. Jedrzejewska and W. Jedrzejewski

Volume 136
Landscape Disturbance and Biodiversity in Mediterranean-Type Ecosystems (1998)
P.W. Rundel, G. Montenegro, and F.M. Jaksic (Eds.)

Volume 137
Ecology of Mediterranean Evergreen Oak Forests (1999)
F. Roda, J. Retana, C.A. Gracia, and J. Bellot (Eds.)

Volume 138
Fire, Climate Change, and Carbon Cycling in the Boreal Forest (2000)
E.S. Kasischke and B.J. Stocks (Eds.)

Volume 139
Responses of Northern U.S. Forests to Environmental Change (2000)
R.A. Mickler, R.A. Birdsey, and J. Hom (Eds.)

Volume 140
Rainforest Ecosystems of East Kalimantan: El Niño, Drought, Fire and Human Impacts (2000)
E. Guhardja et al. (Eds.)

Volume 141
Activity Patterns in Small Mammals: An Ecological Approach (2000)
S. Halle and N.-C. Stenseth (Eds.)

Volume 142
Carbon and Nitrogen Cycling in European Forest Ecosystems (2000)
E.-D. Schulze (Ed.)

Volume 143
Global Climate Change and Human Impacts on Forest Ecosystems: Postglacial Development, Present Situation and Future Trends in Central Europe (2001)
P. Joachim and U. Bernhard

Volume 144
Coastal Marine Ecosystems of Latin America (2001)
U. Seeliger and B. Kjerfve (Eds.)

Volume 145
Ecology and Evolution of the Freshwater Mussels Unionoida (2001)
B. Gerhard and W. Klaus (Eds.)

Volume 146
Inselbergs: Biotic Diversity of Isolated Rock Outcrops in Tropical and Temperate Regions (2001)
S. Porembski and W. Barthlott (Eds.)

Volume 147
Ecosystem Approaches to Landscape Management in Central Europe (2001)
J.D. Tenhunen, R. Lenz, and R. Hantschel (Eds.)

Volume 148
A Systems Analysis of the Baltic Sea (2001)
F.V. Wulff, L.A. Rahm, and P. Larsson (Eds.)

Volume 149
Banded Vegetation Patterning in Arid and Semiarid Environments: Ecological Processes and Consequences for Management (2001)
D.J. Tongway, C. Valentin, and J. Seghieri (Eds.)

Volume 150
Biological Soil Crusts: Structure, Function and Management (2001)
J. Belnap and O.L. Lange (Eds.)

Volume 151
Ecological Comparisons of Sedimentary Shores (2001)
K. Reise (Ed.)

Volume 152
Global Biodiversity in a Changing Environment: Scenarios for the 21st Century (2001)
F.S. Chapin III, O.E. Sala, and E. Huber-Sannwald (Eds.)

Volume 153
UV Radiation and Arctic Ecosystems (2002)
D.O. Hessen (Ed.)

Volume 154
Geoecology of Antarctic Ice-Free Coastal Landscapes (2002)
L. Beyer and M. Bölter (Eds.)

Volume 155
Conserving Biodiversity in East African Forests: A Study of the Eastern Arc Mountains (2002)
W.D. Newmark (Ed.)

Volume 156
Urban Air Pollution and Forests: Resources at Risk in the Mexico City Air Basin (2002)
M.E. Fenn, L.I. de Bauer, and T. Hernández-Tejeda (Eds.)

Volume 157
Mycorrhizal Ecology (2002)
M.G.A. van der Heijden and I.R. Sanders

Volume 158
Diversity and Interaction in a Temperate Forest Community: Ogawa Forest Reserve of Japan (2002)
T. Nakashizuka and Y. Matsumoto

Volume 159
Big-Leaf Mahogany: Genetics, Ecology, and Management (2003)
A.E. Lugo, J.C. Figueroa Colón, and M. Alayón (Eds.)

Volume 160
Fire and Climatic Change in Temperate Ecosystems of the Western Americas (2003)
T.T. Veblen, W.L. Baker, G. Montenegro, and T.W. Swetnam (Eds.)

Volume 161
Competition and Coexistence (2003)
S. Ulrich and B. Worm (Eds.)

Volume 162
How Landscapes Change (2003)
G.A. Bradshaw and P.A. Marquet (Eds.)

Volume 163
Fluxes of Carbon, Water and Energy of European Forests (2003)
R. Valentini (Ed.)

Volume 164
Herbivory of Leaf-Cutting Ants (2003)
R. Wirth, H. Herz, R.J. Ryel, W. Beyschlag, and B. Hölldobler (Eds.)

Volume 165
Population Viability in Plants (2003)
C.A. Brigham and M.W. Schwartz (Eds.)

Volume 166
North American Temperate Deciduous Forest Responses to Changing Precipitation Regimes (2003)
P. Hanson and S. D. Wullschleger (Eds.)

Volume 167
Alpine Biodiversity in Europe (2003)
L. Nagy, G. Grabherr, C. Krner, and D.B.A. Thompson (Eds.)

Volume 168
Root Ecology (2003)
H. de Kroon and E.J.W. Visser (Eds.)

Volume 169
Fire in Tropical Savannas: The Kapalga Experiment (2003)
A.N. Anderson, G.D. Cook, and R.J. Williams (Eds.)

COLOR PLATES

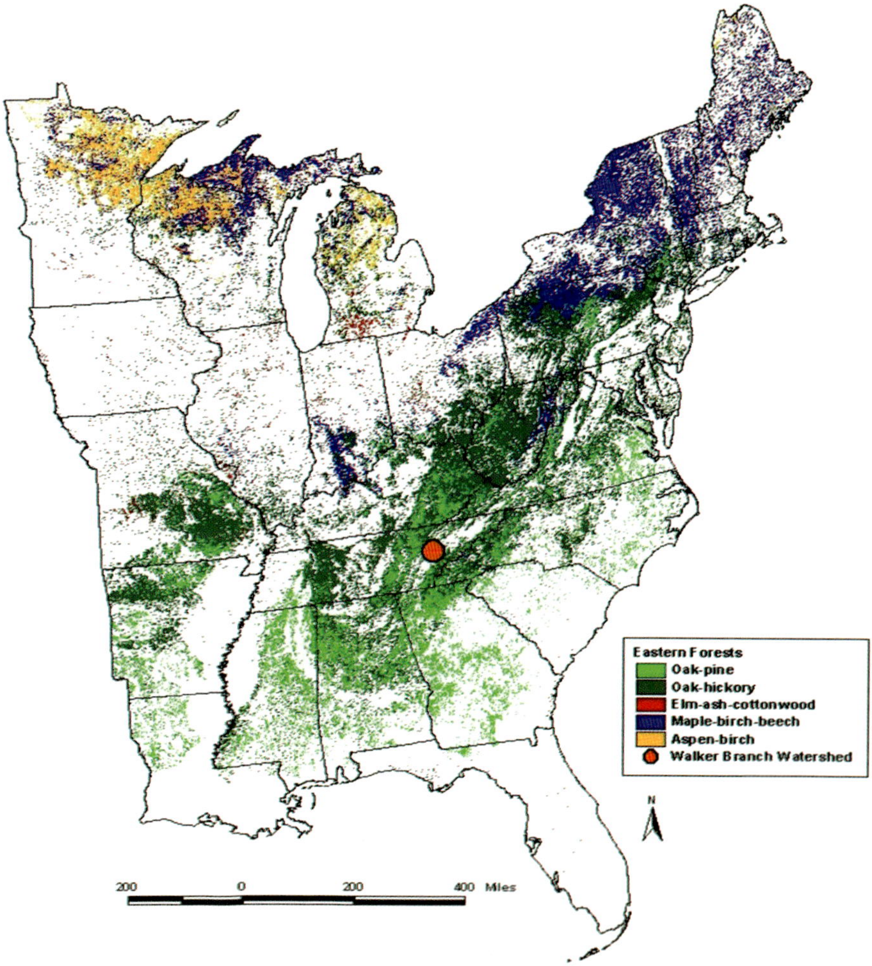

Figure 2.1. Current distribution of deciduous hardwood forests throughout the eastern United States and the location of the Walker Branch Throughfall Displacement Experiment (circle). Individual pixels represent 1 km^2. [Data are from the United States Department of Agriculture, Forest Service (http://www.srsfia.usfs.msstate.edu/rpa/rpa93.htm) and are based on Advanced Very High Resolution Radiometer data from the National Oceanic and Atmospheric Administration satellites. Map prepared by Holly K. Gibbs, Environmental Sciences Division, Oak Ridge National Laboratory.]

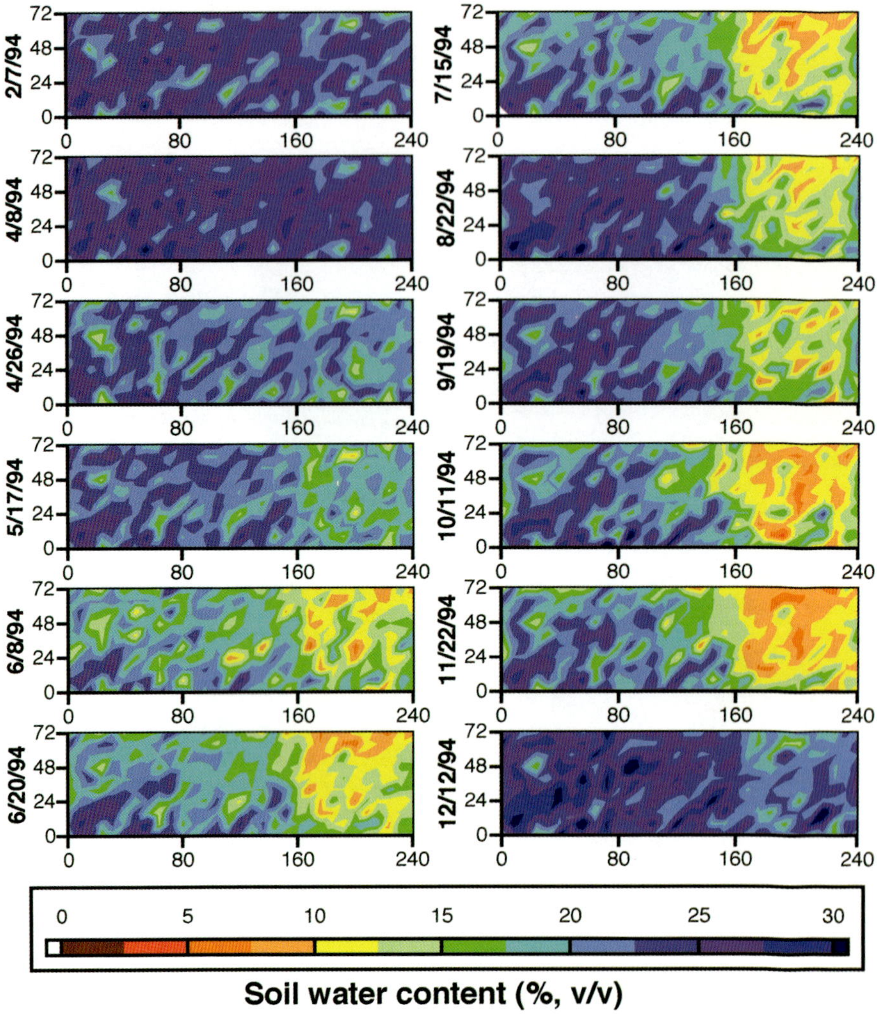

Figure 2.7. Contour plots of the mean soil water content from 0 to 0.35 m across the TDE area throughout 1994. March 4 and August 8 data were left off the figure. Numbers along the x and y axes of each plot are meters, as in Fig. 2.5.

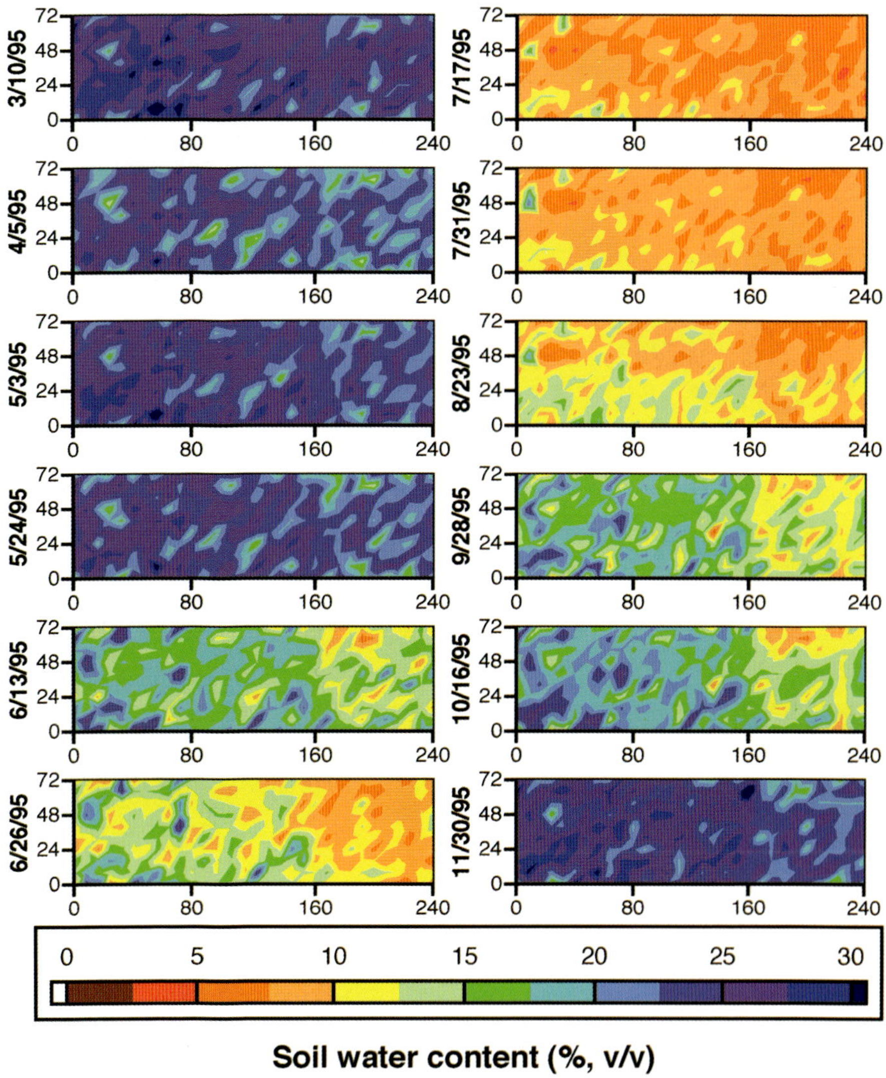

Figure 2.8. Contour plots of the mean soil-water content from 0 to 0.35 m across the TDE area throughout 1995. The January 17 data were left off the figure. Numbers along the *x* and *y* axes of each plot are meters, as in Fig. 2.5.

ORNL-2001-03610/gss

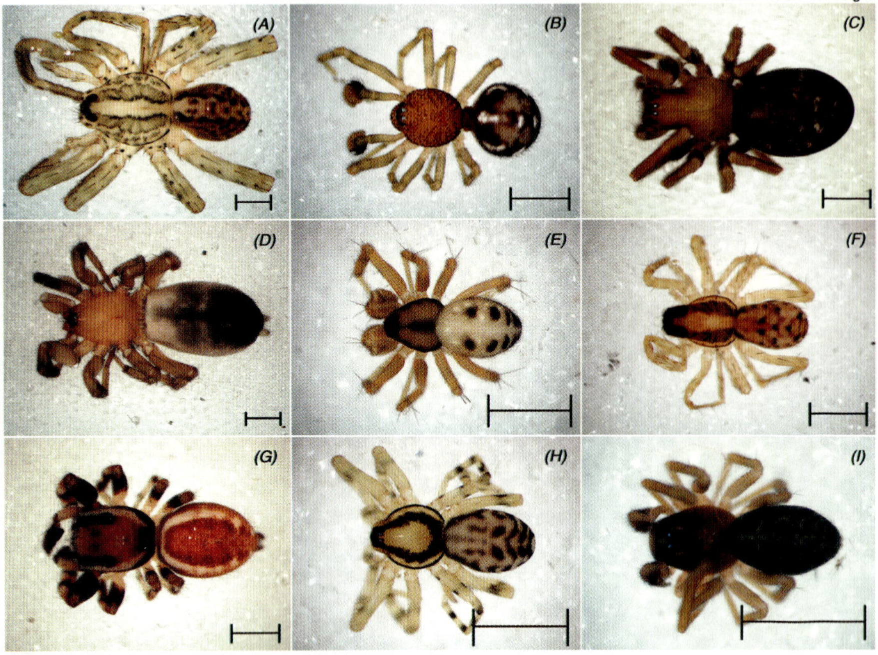

Figure 20.2. Some of the more common spiders encountered in the study. (A) *Anahita punctulata* (Ctenidae), (B) *Crustulina* sp. (Theridiidae), (C) *Dictyna* sp. (Dictynidae), (D) *Drassyllus* sp. (Gnaphosidae), (E) *Lepthyphantes nebulosa* (Linyphiidae), (F) *Schizocosa* sp. (Lycosidae), (G) *Metaphidippus protervus* (Salticidae), (H) *Phrurotimpus alarius* (Clubionidae), and (I) *Origanates rostratus* (Linyphiidae).